TABLE 3 *(Continued)*

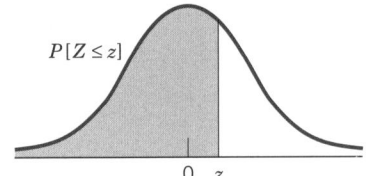

$P[Z \le z]$

0 z

z	.00	.01	.02	.03	.04	.05	.06	.07	.08	.09
.0	.5000	.5040	.5080	.5120	.5160	.5199	.5239	.5279	.5319	.5359
.1	.5398	.5438	.5478	.5517	.5557	.5596	.5636	.5675	.5714	.5753
.2	.5793	.5832	.5871	.5910	.5948	.5987	.6026	.6064	.6103	.6141
.3	.6179	.6217	.6255	.6293	.6331	.6368	.6406	.6443	.6480	.6517
.4	.6554	.6591	.6628	.6664	.6700	.6736	.6772	.6808	.6844	.6879
.5	.6915	.6950	.6985	.7019	.7054	.7088	.7123	.7157	.7190	.7224
.6	.7257	.7291	.7324	.7357	.7389	.7422	.7454	.7486	.7517	.7549
.7	.7580	.7611	.7642	.7673	.7703	.7734	.7764	.7794	.7823	.7852
.8	.7881	.7910	.7939	.7967	.7995	.8023	.8051	.8078	.8106	.8133
.9	.8159	.8186	.8212	.8238	.8264	.8289	.8315	.8340	.8365	.8389
1.0	.8413	.8438	.8461	.8485	.8508	.8531	.8554	.8577	.8599	.8621
1.1	.8643	.8665	.8686	.8708	.8729	.8749	.8770	.8790	.8810	.8830
1.2	.8849	.8869	.8888	.8907	.8925	.8944	.8962	.8980	.8997	.9015
1.3	.9032	.9049	.9066	.9082	.9099	.9115	.9131	.9147	.9162	.9177
1.4	.9192	.9207	.9222	.9236	.9251	.9265	.9279	.9292	.9306	.9319
1.5	.9332	.9345	.9357	.9370	.9382	.9394	.9406	.9418	.9429	.9441
1.6	.9452	.9463	.9474	.9484	.9495	.9505	.9515	.9525	.9535	.9545
1.7	.9554	.9564	.9573	.9582	.9591	.9599	.9608	.9616	.9625	.9633
1.8	.9641	.9649	.9656	.9664	.9671	.9678	.9686	.9693	.9699	.9706
1.9	.9713	.9719	.9726	.9732	.9738	.9744	.9750	.9756	.9761	.9767
2.0	.9772	.9778	.9783	.9788	.9793	.9798	.9803	.9808	.9812	.9817
2.1	.9821	.9826	.9830	.9834	.9838	.9842	.9846	.9850	.9854	.9857
2.2	.9861	.9864	.9868	.9871	.9875	.9878	.9881	.9884	.9887	.9890
2.3	.9893	.9896	.9898	.9901	.9904	.9906	.9909	.9911	.9913	.9916
2.4	.9918	.9920	.9922	.9925	.9927	.9929	.9931	.9932	.9934	.9936
2.5	.9938	.9940	.9941	.9943	.9945	.9946	.9948	.9949	.9951	.9952
2.6	.9953	.9955	.9956	.9957	.9959	.9960	.9961	.9962	.9963	.9964
2.7	.9965	.9966	.9967	.9968	.9969	.9970	.9971	.9972	.9973	.9974
2.8	.9974	.9975	.9976	.9977	.9977	.9978	.9979	.9979	.9980	.9981
2.9	.9981	.9982	.9982	.9983	.9984	.9984	.9985	.9985	.9986	.9986
3.0	.9987	.9987	.9987	.9988	.9988	.9989	.9989	.9989	.9990	.9990
3.1	.9990	.9991	.9991	.9991	.9992	.9992	.9992	.9992	.9993	.9993
3.2	.9993	.9993	.9994	.9994	.9994	.9994	.9994	.9995	.9995	.9995
3.3	.9995	.9995	.9995	.9996	.9996	.9996	.9996	.9996	.9996	.9997
3.4	.9997	.9997	.9997	.9997	.9997	.9997	.9997	.9997	.9997	.9998
3.5	.9998	.9998	.9998	.9998	.9998	.9998	.9998	.9998	.9998	.9998

Preface

THE NATURE OF THE BOOK

Statistics—the subject of data analysis and data-based reasoning—is playing an increasingly vital role in virtually all professions. Some familiarity with this subject is now an essential component of any college education. Yet, pressures to accommodate a growing list of academic requirements often necessitate that this exposure be brief. Keeping these conditions in mind, we have written this book to provide students with a first exposure to the powerful ideas of modern statistics. It presents the key statistical concepts and the most commonly applied methods of statistical analysis. Moreover, to keep it accessible to freshmen and sophomores from a wide range of disciplines, we have avoided mathematical derivations. They usually pose a stumbling block to learning the essentials in a short period of time.

This book is intended for students who do not have a strong background in mathematics but seek to learn the basic ideas of statistics and their application in a variety of practical settings. The core material of this book is common to almost all first courses in statistics and is designed to be covered well within a one-semester course in introductory statistics for freshmen–seniors. It is supplemented with some additional special-topics chapters.

ORIENTATION

The topics treated in this text are, by and large, the ones typically covered in an introductory statistics course. They span three major areas: (i) descriptive statistics, which deals with summarization and description of data; (ii) ideas of probability and an understanding of the manner in which sample-to-sample variation influences our conclusions; and (iii) a collection of statistical methods for analyzing the types of data that are of common occurrence. However, it is the treatment of these topics that makes the text distinctive. By means of good motivation, sound explanations, and an abundance of illustrations given in a real-world context, it emphasizes more than just a superficial understanding.

Each concept or technique is motivated by first setting out its goal and indicating its scope by an illustration of its application. The subsequent discussion is not only limited to showing how a method works but includes an explanation of the why. Even without recourse to mathematics, we are able to make the reader aware of possible pitfalls in the statistical analysis. Students can gain a proper appreciation of statistics only when they are provided with a careful explanation of the underlying logic. Without this understanding, a learning of elementary statistics is bound to be rote and transient.

When describing the various methods of statistical analysis, the reader is continually reminded that the validity of a statistical inference is contingent upon certain model assumptions. Misleading conclusions may result when these assumptions are violated. We feel that the teaching of statistics, even at an introductory level, should not be limited to the prescription of methods. Students should be encouraged to develop a critical attitude in applying the methods and to be cautious when interpreting the results. This attitude is especially important in the study of relationship among variables, which is perhaps the most widely used (and also abused) area of statistics. In addition to discussing inference procedures in this context, we have particularly stressed critical examination of the model assumptions and careful interpretation of the conclusions.

SPECIAL FEATURES

1. Crucial elements are boxed to highlight important concepts and methods. These boxes provide an ongoing summary of the important items essential for learning statistics. At the end of each chapter, all of its **key ideas and formulas** are summarized.

2. A rich collection of examples and exercises is included. These are drawn from a large variety of **real-life settings.** In fact, many data sets stem from genuine experiments, surveys, or reports.

3. Exercises are provided at the end of **each major section.** These provide the reader with the opportunity to practice the ideas just learned. Occasionally, they supplement some points raised in the text. A larger collection of exercises appears at the **end of a chapter.** The starred problems are relatively difficult and suited to the more mathematically competent student.

4. Statistics in Context sections, in four of the beginning chapters, each describe an important statistical application where a statistical approach to understanding variation is vital. These extended examples reveal, early on in the course, the value of understanding the subject of statistics.

5. P–values are emphasized in examples concerning tests of hypotheses. Graphs giving the relevant normal or t density curve, rejection region, and P–value are presented.

6. **Regression analysis** is a primary statistical technique so we provide a more thorough coverage of the topic than is usual at this level. The basics of regression are introduced in Chapter 11, whereas Chapter 12 stretches the discussion to several issues of practical importance. These include methods of **model checking,** handling nonlinear relations, and multiple regression analysis. Complex formulas and calculations are judiciously replaced by computer output so the main ideas can be learned and appreciated with a minimum of stress.

7. **Computer Aided Statistical Analyses** use software packages that can remove much of the drudgery of hand calculation and plotting. They allow students to work with larger data sets where patterns are more pronounced and to make complicated calculations. Besides discussion of some computer output in the text, computer exercises are included in all chapters where relevant.

8. **Convenient Electronic Data Bank** at the end of the book contains a substantial collection of data. These data sets, together with numerous others throughout the book, allow for considerable flexibility in the choice between concept-orientated and applications-orientated exercises. The Data Bank and the other larger data sets are available for download on the accompanying website located at www.wiley.com/college/johnson.

9. **Technical Appendix A** presents a few statistical facts of a mathematical nature. These are separated from the main text so that they can be left out if the instructor so desires.

ABOUT THE FIFTH EDITION

The fifth edition of *STATISTICS—Principles and Methods* maintains the objectives and level of presentation of the earlier editions. The goals are the developing (i) of an understanding of the reasonings by which findings from sample data can be extended to general conclusions and (ii) a familiarity with some basic statistical methods. There are numerous data sets and computer outputs which give an appreciation of the role of the computer in modern data analysis.

Throughout, we have endeavored to give clear and concise explanations of the concepts and important statistical terminology and methods. Discussion of the statistical methods includes an explanation of their underlying assumptions and the dangers of ignoring them. Real-life settings are used to motivate the statistical ideas and well organized discussions proceed to cover statistical methods with heavy emphasis on examples. The fifth edition enhances these special features. More particularly, the major improvements are:

Using Statistics Wisely Feature. Provides important guidelines for the appropriate use of statistics. Included at the end of each chapter.

Integrated Technology. A summary of the steps for using MINITAB, EXCEL, and the TI-84 calculator is included at the end of most chapters. This concentrates the presentation of special purpose instructions so that, with few exceptions, only computer output is needed in the text.

More Data-Based Exercises. Some of the new exercises are keyed to new data based examples in the text. Others are based on the new grizzly bear data set added to the data bank.

New Exercises. Further new exercises providing practice on concepts, together with many new computational exercises, augment the already rich collection of exercises.

ORGANIZATION

This book is organized into fifteen chapters, an optional technical appendix (Appendix A), and a collection of tables (Appendix B). Although designed for a one-semester or a two-quarter course, it is enriched with ample additional material to allow the instructor some choices of topics. Beyond Chapter 1, which sets the theme of statistics, and distinguishes population and sample, the subject matter could be classified as follows:

Topic	Chapter
Descriptive study of data	2, 3
Probability and distributions	4, 5, 6
Sampling variability	7
Core ideas and methods of statistical inference	8, 9, 10
Special topics of statistical inference	11, 12, 13, 14, 15

We regard Chapters 1 to 10 as constituting the core material of an introductory statistics course, with the exception of the starred sections in Chapter 6. Although this material is just about enough for a one-semester course, many instructors may wish to eliminate some sections in order to cover the basics of regression analysis in Chapter 11. This is most conveniently done by initially skipping Chapter 3 and then taking up only those portions that are linked to Chapter 11. Also, instead of a thorough coverage of probability that is provided in Chapter 4, the later sections of that chapter may receive a lighter coverage.

SUPPLEMENTS

Instructor Solution Manual. (ISBN 0471-78869-4) This manual contains complete solutions to all exercises.

Test Bank. (Available on the accompanying website: www.wiley.com/college/johnson) Contains a large number of additional questions for each chapter.

Student Solutions Manual. (ISBN 0-471-71884-X) This manual contains complete solutions to all odd-numbered exercises.

Electronic Data Bank. (Available on the accompanying website: www.wiley.com/college/johnson) Contains data sets from varying sources that can be used to perform analyses with statistical software packages.

WileyPLUS. This powerful online tool provides a completely integrated suite of teaching and learning resources in one easy-to-use website. *WileyPLUS* offers an online assessment system with full gradebook capabilities and algorithmically generated skill building questions. To view a demo of *WileyPLUS*, contact your local Wiley Sales Representative or visit: www.wiley.com/college/wileyplus.

ACKNOWLEDGMENTS

We thank Minitab (State College, Pa.) and the SAS Institute (Cary, N.C.) for permission to include commands and output from their software packages. A special thanks to K. T. Wu and Kam Tsui for many helpful suggestions and comments. We also thank all those who have contributed the data sets which enrich the presentation and all those who reviewed the previous editions. The following people gave their careful attention to this edition:

Hongshik Ahn, SUNY at Stony Brook
Casey T. Cremins, University of Maryland
Cecil Hallum, Sam Houston State University
Robert F. Herrmann, Elmhurst College
Mark McKibben, Goucher College
Ameina Summerlin, University of South Alabama
Gwen Terwilliger, University of Toledo
Quoc-Phong Vu, Ohio University
Yimin Xiao, Michigan State University

Special thanks go to Paul Lorczak who gave his input on the final pages for this fifth edition of STATISTICS—*Principles and Methods*.

Richard A. Johnson
Gouri K. Bhattacharyya

Contents

6   THE NORMAL DISTRIBUTION   219

7   VARIATION IN REPEATED SAMPLES — SAMPLING DISTRIBUTIONS   259

8   DRAWING INFERENCES FROM LARGE SAMPLES   289

DATA BANK   637

ANSWERS TO SELECTED ODD-NUMBERED EXERCISES   651

INDEX   665

1

Introduction

Gallup Opinion Index

Trend in the popularity of the big three sports. Currently football is the favorite spectator sport in the United States. Earlier, baseball was the most popular.

What is your favorite spectator sport?

	1999	1990	1981	1972	1960	1948
Football	30%	35%	38%	36%	21%	17%
Baseball	14%	16%	16%	21%	34%	39%
Basketball	17%	15%	9%	8%	9%	10%
Other	39%	33%	37%	35%	36%	34%

Hometown fans attending today's game are but a sample of the population of all local baseball fans. A self-selected sample may not be entirely representative of the population on issues such as ticket price increases. © Chromosohn/John Sohm/Photo Researchers

1. WHAT IS STATISTICS?

The word statistics originated from the Latin word "status," meaning "state." For a long time, it was identified solely with the displays of data and charts pertaining to the economic, demographic, and political situations prevailing in a country. Even today, a major segment of the general public thinks of statistics as synonymous with forbidding arrays of numbers and myriad graphs. This image is enhanced by numerous government reports that contain a massive compilation of numbers and carry the word statistics in their titles: "Statistics of Farm Production," "Statistics of Trade and Shipping," "Labor Statistics," to name a few. However, gigantic advances during the twentieth century have enabled statistics to grow and assume its present importance as a discipline of data-based reasoning. Passive display of numbers and charts is now a minor aspect of statistics, and few, if any, of today's statisticians are engaged in the routine activities of tabulation and charting.

What, then, are the role and principal objectives of statistics as a scientific discipline? Stretching well beyond the confines of data display, statistics deals with collecting informative data, interpreting these data, and drawing conclusions about a phenomenon under study. The scope of this subject naturally extends to all processes of acquiring knowledge that involve fact finding through collection and examination of data. Opinion polls (surveys of households to study sociological, economic, or health-related issues), agricultural field experiments (with new seeds, pesticides, or farming equipment), clinical studies of vaccines, and cloud seeding for artificial rain production are just a few examples. The principles and methodology of statistics are useful in answering questions such as, What kind and how much data need to be collected? How should we organize and interpret the data? How can we analyze the data and draw conclusions? How do we assess the strength of the conclusions and gauge their uncertainty?

> Statistics as a subject provides a body of principles and methodology for designing the process of data collection, summarizing and interpreting the data, and drawing conclusions or generalities.

2. STATISTICS IN OUR EVERYDAY LIFE

Fact finding through the collection and interpretation of data is not confined to professional researchers. In our attempts to understand issues of environmental protection, the state of unemployment, or the performance of competing football teams, numerical facts and figures need to be reviewed and interpreted. In our day-to-day life, learning takes place through an often implicit analysis of factual information.

We are all familiar to some extent with reports in the news media on important statistics.

Employment. Monthly, as part of the Current Population Survey, the Bureau of Census collects information about employment status from a sample of about 65,000 households. Households are contacted on a rotating basis with three-fourths of the sample remaining the same for any two consecutive months.

The survey data are analyzed by the Bureau of Labor Statistics, which reports monthly unemployment rates. □

Cost of Living. The consumer price index (CPI) measures the cost of a fixed market basket of over 400 goods and services. Each month, prices are obtained from a sample of over 18,000 retail stores that are distributed over 85 metropolitan areas. These prices are then combined taking into account the relative quantity of goods and services required by a hypothetical "1967 urban wage earner." Let us not be concerned with the details of the sampling method and calculations as these are quite intricate. They are, however, under close scrutiny because of the importance to the hundreds of thousands of Americans whose earnings or retirement benefits are tied to the CPI. □

Election time brings the pollsters into the limelight.

Gallup Poll. This, the best known of the national polls, produces estimates of the percentage of popular vote for each candidate based on interviews with a minimum of 1500 adults. Beginning several months before the presidential election, results are regularly published. These reports help predict winners and track changes in voter preferences. □

Our sources of factual information range from individual experience to reports in news media, government records, and articles in professional journals. As consumers of these reports, citizens need some idea of statistical reasoning to properly interpret the data and evaluate the conclusions. Statistical reasoning provides criteria for determining which conclusions are supported by the data and which are not. The credibility of conclusions also depends greatly on the use of statistical methods at the data collection stage. Statistics provides a key ingredient for any systematic approach to improve any type of process from manufacturing to service.

Quality and Productivity Improvement. In the past 30 years, the United States has faced increasing competition in the world marketplace. An international revolution in quality and productivity improvement has heightened the pressure on the U.S. economy. The ideas and teaching of W. Edwards Deming helped rejuvenate Japan's industry in the late 1940s and 1950s. In the 1980s and 1990s, Deming stressed to American executives that, in order to survive, they must mobilize their work force to make a continuing commitment to quality improvement. His ideas have also been applied to government. The city of

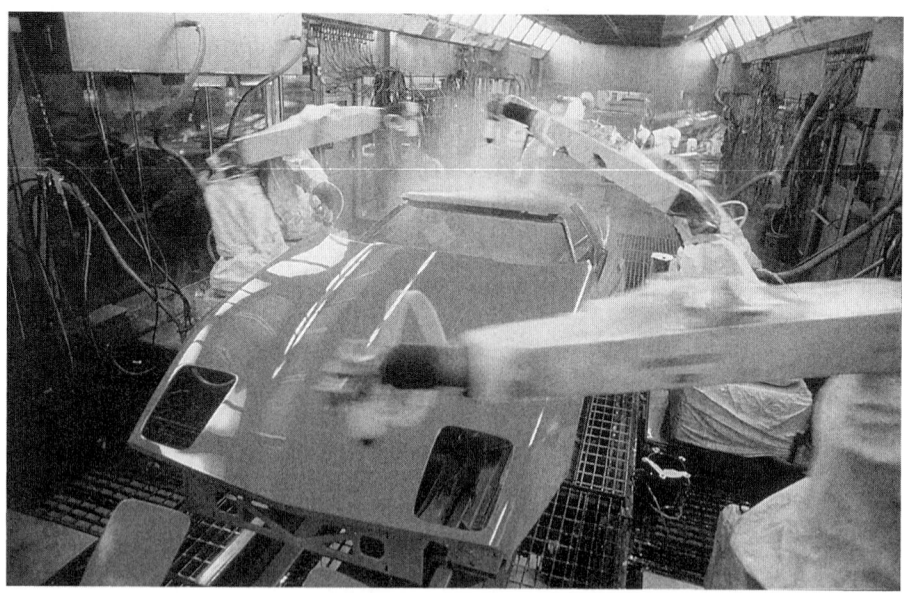

Statistical reasoning can guide the purposeful collection and analysis of data toward the continuous improvement of any process. © Andrew Sacks/Stone/Getty Images

Madison, WI, has implemented quality improvement projects in the police department and in bus repair and scheduling. In each case, the project goal was better service at less cost. Treating citizens as the customers of government services, the first step was to collect information from them in order to identify situations that needed improvement. One end result was the strategic placement of a new police substation and a subsequent increase in the number of foot patrol persons to interact with the community.

Once a candidate project is selected for improvement, data must be collected to assess the current status and then more data collected on the effects of possible changes. At this stage, statistical skills in the collection and presentation of summaries are not only valuable but necessary for all participants.

In an industrial setting, statistical training for all employees—production line and office workers, supervisors, and managers—is vital to the quality transformation of American industry. □

3. STATISTICS IN AID OF SCIENTIFIC INQUIRY

The phrase scientific inquiry refers to a systematic process of learning. A scientist sets the goal of an investigation, collects relevant factual information (or data), analyzes the data, draws conclusions, and decides further courses of action. We briefly outline a few illustrative scenarios.

Training Programs.   Training or teaching programs in many fields designed for a specific type of clientele (college students, industrial workers, minority groups, physically handicapped people, retarded children, etc.) are continually monitored, evaluated, and modified to improve their usefulness to society. To learn about the comparative effectiveness of different programs, it is essential to collect data on the achievement or growth of skill of the trainees at the completion of each program. □

Monitoring Advertising Claims.   The public is constantly bombarded with commercials that claim the superiority of one product brand in comparison to others. When such comparisons are founded on sound experimental evidence, they serve to educate the consumer. Not infrequently, however, misleading advertising claims are made due to insufficient experimentation, faulty analysis of data, or even blatant manipulation of experimental results. Government agencies and consumer groups must be prepared to verify the comparative quality of products by using adequate data collection procedures and proper methods of statistical analysis. □

Plant Breeding.   To increase food production, agricultural scientists develop new hybrids by cross-fertilizing different plant species. Promising new strains need to be compared with the current best ones. Their relative productivity is assessed by planting some of each variety at a number of sites. Yields are

Statistically designed experiments are needed to document the advantages of the new hybrid versus the old species. © Mitch Wojnarowicz/The Image Works

recorded and then analyzed for apparent differences. The strains may also be compared on the basis of disease resistance or fertilizer requirements.   □

Building Beams.   Wooden beams that support roofs on houses and public buildings must be strong. Most beams are constructed by laminating together several boards. Wood scientists have collected data that show stiffer boards are generally stronger. This relation can be used to predict the strength of candidate boards for laminating on the basis of their stiffness measurements.   □

Factual information is crucial to any investigation. The branch of statistics called **experimental design** can guide the investigator in planning the manner and extent of data collection.

After the data are collected, statistical methods are available that summarize and describe the prominent features of data. These are commonly known as descriptive statistics. Today, a major thrust of the subject is the evaluation of information present in data and the assessment of the new learning gained from this information. This is the area of inferential statistics and its associated methods are known as the methods of statistical inference.

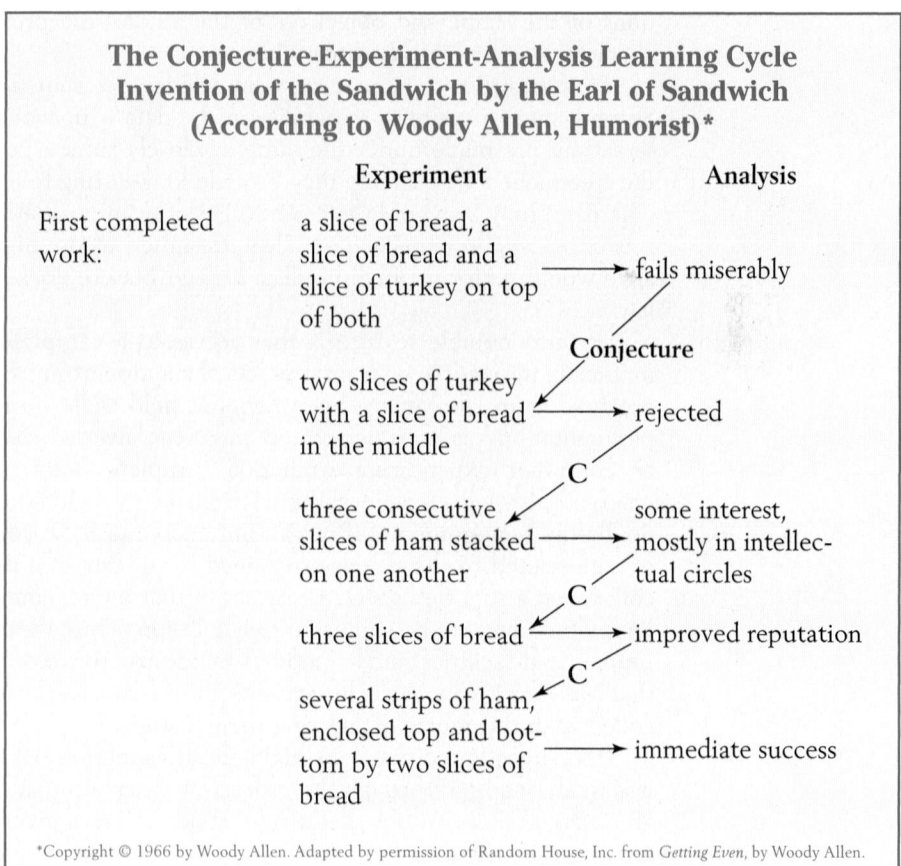

The Conjecture-Experiment-Analysis Learning Cycle
Invention of the Sandwich by the Earl of Sandwich
(According to Woody Allen, Humorist)*

	Experiment	Analysis
First completed work:	a slice of bread, a slice of bread and a slice of turkey on top of both	fails miserably
	Conjecture	
	two slices of turkey with a slice of bread in the middle	rejected
	C	
	three consecutive slices of ham stacked on one another	some interest, mostly in intellectual circles
	C	
	three slices of bread	improved reputation
	C	
	several strips of ham, enclosed top and bottom by two slices of bread	immediate success

*Copyright © 1966 by Woody Allen. Adapted by permission of Random House, Inc. from *Getting Even*, by Woody Allen.

It must be realized that a scientific investigation is typically a process of trial and error. Rarely, if ever, can a phenomenon be completely understood or a theory perfected by means of a single, definitive experiment. It is too much to expect to get it all right in one shot. Even after his first success with the electric light bulb, Thomas Edison had to continue to experiment with numerous materials for the filament before it was perfected. Data obtained from an experiment provide new knowledge. This knowledge often suggests a revision of an existing theory, and this itself may require further investigation through more experiments and analysis of data. Humorous as it may appear, the excerpt boxed above from a Woody Allen writing captures the vital point that a scientific process of learning is essentially iterative in nature.

4.   TWO BASIC CONCEPTS—POPULATION AND SAMPLE

In the preceding sections, we cited a few examples of situations where evaluation of factual information is essential for acquiring new knowledge. Although these examples are drawn from widely differing fields and only sketchy descriptions of the scope and objectives of the studies are provided, a few common characteristics are readily discernible.

First, in order to acquire new knowledge, relevant data must be collected. Second, some amount of variability in the data is unavoidable even though observations are made under the same or closely similar conditions. For instance, the treatment for an allergy may provide long-lasting relief for some individuals whereas it may bring only transient relief or even none at all to others. Likewise, it is unrealistic to expect that college freshmen whose high school records were alike would perform equally well in college. Nature does not follow such a rigid law.

A third notable feature is that access to a complete set of data is either physically impossible or from a practical standpoint not feasible. When data are obtained from laboratory experiments or field trials, no matter how much experimentation has been performed, more can always be done. In public opinion or consumer expenditure studies, a complete body of information would emerge only if data were gathered from every individual in the nation—undoubtedly a monumental if not an impossible task. To collect an exhaustive set of data related to the damage sustained by all cars of a particular model under collision at a specified speed, every car of that model coming off the production lines would have to be subjected to a collision! Thus, the limitations of time, resources, and facilities, and sometimes the destructive nature of the testing, mean that we must work with incomplete information—the data that are actually collected in the course of an experimental study.

The preceding discussions highlight a distinction between the data set that is actually acquired through the process of observation and the vast collection of all potential observations that can be conceived in a given context. The statistical name for the former is **sample**; for the latter, it is **population**, or **statistical** population. To further elucidate these concepts, we observe that each measurement in a data set originates from a distinct source which may be a patient, tree, farm, household, or some other entity depending on the object of a study. The source of each measurement is called a **sampling unit**, or simply, a **unit**.

To emphasize population as the entire collection of units, we term it the **population of units**.

A **unit** is a single entity, usually a person or an object, whose characteristics are of interest.

The **population of units** is the complete collection of units about which information is sought.

There is another aspect to any population and that is the value, for each unit, of a characteristic or variable of interest. There can be several characteristics of interest for a given population of units, as indicated in Table 1.

TABLE 1   Populations, Units, and Variables

Population	Unit	Variables/Characteristics
Registered voters in your state	Voter	Political party Voted or not in last election Age Sex Conservative/liberal
All rental apartments near campus	Apartment	Rent Size in square feet Number of bedrooms Number of bathrooms TV and Internet connections
All campus fast food restaurants	Restaurant	Number of employees Seating capacity Hiring/not hiring
All computers owned by students at your school	Computer	Speed of processor Size of hard disk Speed of Internet connection Screen size

For a given variable or characteristic of interest, we call the collection of values, evaluated for every unit in the population, the **statistical population** or just the population. We refer to the collection of units as the **population of units** when there is a need to differentiate it from the collection of values.

> A statistical **population** is the set of measurements (or record of some qualitative trait) corresponding to the entire collection of units about which information is sought.

The population represents the target of an investigation. We learn about the population by taking a sample from the population. A sample or sample data set then consists of measurements recorded for those units that are actually observed. It constitutes a part of a far larger collection about which we wish to make inferences—the set of measurements that would result if all the units in the population could be observed.

> A **sample** from a statistical population is the subset of measurements that are actually collected in the course of an investigation.

Example 1   Identifying the Population and Sample

Questions concerning the effect on health of two or fewer cups of coffee a day are still largely unresolved. Current studies seek to find physiological changes that could prove harmful. An article carried the headline CAFFEINE DECREASES CEREBRAL BLOOD FLOW. It describes a study[1] which establishes a physiological side effect—a substantial decrease in cerebral blood flow for persons drinking two to three cups of coffee daily.

The cerebral blood flow was measured twice on each of 20 subjects. It was measured once after taking an oral dose of caffeine equivalent to two to three cups of coffee and then, on another day, after taking a look-alike dose but without caffeine. The order of the two tests was random and subjects were not told which dose they received. The measured decrease in cerebral blood flow was significant.

Identify the population and sample.

SOLUTION   As the article implies, the conclusion should apply to you and me. The population could well be the potential decreases in cerebral blood flow for all adults living in the United States. It might even apply to all the decrease in blood flow for all caffeine users in the world, although the cultural customs

[1]A. Field et al. "Dietary Caffeine Consumption and Withdrawal: Confounding Variables in Quantitative Cerebral Perfusion Studies?" *Radiology* **227** (2003), pp. 129–135.

may vary the type of caffeine consumption from coffee breaks to tea time to kola nut chewing.

The sample consists of the decreases in blood flow for the 20 subjects who agreed to participate in the study.

Example 2   A Misleading Sample

A host of a radio music show announced that she wants to know which singer is the favorite among city residents. Listeners were then asked to call in and name their favorite singer.

Identify the population and sample. Comment on how to get a sample that is more representative of the city's population.

SOLUTION   The population is the collection of singer preferences of all city residents and the purported goal was to learn who was the favorite singer. Because it would be nearly impossible to question all the residents in a large city, one must necessarily settle for taking a sample.

Having residents make a local call is certainly a low-cost method of getting a sample. The sample would then consist of the singers named by each person who calls the radio station. Unfortunately, with this selection procedure, the sample is not very representative of the responses from all city residents. Those that listen to the particular radio station are already a special subgroup with similar listening tastes. Furthermore, those listeners who take the time and effort to call are usually those who feel strongest about their opinions. The resulting responses could well be much stronger in favor of a particular country western or rock singer than is the case for preference among the total population of city residents or even those who listen to the station.

If the purpose of asking the question is really to determine the favorite singer of the city's residents, we have to proceed otherwise. One procedure commonly employed is a phone survey where the phone numbers are chosen at random. For instance, one can imagine that the numbers 0, 1, 2, 3, 4, 5, 6, 7, 8, and 9 are written on separate pieces of paper and placed in a hat. Slips are then drawn one at a time and replaced between drawings. Later, we will see that computers can mimic this selection quickly and easily. Four draws will produce a random telephone number within a three-digit exchange. Telephone numbers chosen in this manner will certainly produce a much more representative sample than the self-selected sample of persons who call the station.

Self-selected samples consisting of responses to call-in or write-in requests will, in general, not be representative of the population. They arise primarily from subjects who feel strongly about the issue in question. To their credit, many TV news and entertainment programs now state that their call-in polls are nonscientific and merely reflect the opinions of those persons who responded.

USING A RANDOM NUMBER TABLE TO SELECT A SAMPLE

The choice of which population units to include in a sample must be impartial and objective. When the total number of units is finite, the name or number of each population unit could be written on a separate slip of paper and the slips placed in a box. Slips could be drawn one at a time without replacement and the corresponding units selected as the sample of units. Unfortunately, this simple and intuitive procedure is cumbersome to implement. Also, it is difficult to mix the slips well enough to ensure impartiality.

Alternatively, a better method is to take 10 identical marbles, number them 0 through 9, and place them in an urn. After shuffling, select 1 marble. After replacing the marble, shuffle and draw again. Continuing in this way, we create a sequence of random digits. Each digit has an equal chance of appearing in any given position, all pairs have the same chance of appearing in any two given positions, and so on. Further, any digit or collection of digits is unrelated to any other disjoint subset of digits. For convenience of use, these digits can be placed in a table called a **random number table**.

The digits in Table 1 of Appendix B were actually generated using computer software that closely mimics the drawing of marbles. A portion of this table is shown here as Table 2.

To obtain a random sample of units from a population of size N, we first number the units from 1 to N. Then numbers are read from the table of random digits until enough different numbers in the appropriate range are selected.

TABLE 2   Random Digits: A Portion of Table 1, Appendix B

Row									
1	0695	7741	8254	4297	0000	5277	6563	9265	1023	5925
2	0437	5434	8503	3928	6979	9393	8936	9088	5744	4790
3	6242	2998	0205	5469	3365	7950	7256	3716	8385	0253
4	7090	4074	1257	7175	3310	0712	4748	4226	0604	3804
5	0683	6999	4828	7888	0087	9288	7855	2678	3315	6718
6	7013	4300	3768	2572	6473	2411	6285	0069	5422	6175
7	8808	2786	5369	9571	3412	2465	6419	3990	0294	0896
8	9876	3602	5812	0124	1997	6445	3176	2682	1259	1728
9	1873	1065	8976	1295	9434	3178	0602	0732	6616	7972
10	2581	3075	4622	2974	7069	5605	0420	2949	4387	7679
11	3785	6401	0540	5077	7132	4135	4646	3834	6753	1593
12	8626	4017	1544	4202	8986	1432	2810	2418	8052	2710
13	6253	0726	9483	6753	4732	2284	0421	3010	7885	8436
14	0113	4546	2212	9829	2351	1370	2707	3329	6574	7002
15	4646	6474	9983	8738	1603	8671	0489	9588	3309	5860

Example 3   Using the Table of Random Digits to Select Items for a Price Check

One week, the advertisement for a large grocery store contains 72 special sale items. Five items will be selected with the intention of comparing the sales price with the scan price at the checkout counter. Select the five items at random to avoid partiality.

SOLUTION   The 72 sale items are first numbered from 1 to 72. Since the population size $N = 72$ has two digits, we will select random digits two at a time from Table 2. Arbitrarily, we decide to start in row 7 and columns 19 and 20. Starting with the two digits in columns 19 and 20 and reading down, we obtain

12   97   34   69   32   86   32   51

We ignore 97 and 86 because they are larger than the population size 72. We also ignore any number when it appears a second time as 32 does here. Consequently, the sale items numbered

12   34   69   32   51

are selected for the price check.

For large sample size situations or frequent applications, it is often more convenient to use computer software to choose the random numbers.

Example 4   Selecting a Sample by Random Digit Dialing

A major Internet service provider wants to learn about the proportion of people in one target area who are aware of its latest product. Suppose there is a single three-digit telephone exchange that covers the target area. Use Table 1, in Appendix B, to select six telephone numbers for a phone survey.

SOLUTION   We arbitrarily decide to start at row 26 and columns 25 to 28. Proceeding downward, we obtain

6435   1307   9320   1619   0766   7566

Together with the three-digit exchange, these six numbers form the phone numbers called in the survey. Every phone number, listed or unlisted, has the same chance of being selected. The same holds for every pair, every triplet, and so on. Commercial phones may have to be discarded and another four digits selected. If there are two exchanges in the area, separate selections could be done for each exchange.

For large sample sizes, it is better to use computer-generated random digits or even computer-dialed random phone numbers.

Data collected with a clear-cut purpose in mind are very different from **anecdotal data.** Most of us have heard people say they won money at a casino, but certainly most people cannot win most of the time as casinos are not in the business of giving away money. People tend to tell good things about themselves. In a similar vein, some drivers' lives are saved when they are thrown free of car wrecks because they were not wearing seat belts. Although such stories are told and retold, you must remember that there is really no opportunity to hear from those who would have lived if they had worn their seat belts. Anecdotal information is usually repeated because it has some striking feature that may not be representative of the mass of cases in the population. Consequently, it is not apt to provide reliable answers to questions.

5.   THE PURPOSEFUL COLLECTION OF DATA

Many poor decisions are made, in both business and everyday activities, because of the failure to understand and account for variability. Certainly, the purchasing habits of one person may not represent those of the population, or the reaction of one mouse, on exposure to a potentially toxic chemical compound, may not represent that of a large population of mice. However, despite diversity among the purchasing habits of individuals, we can obtain accurate information about the purchasing habits of the population by collecting data on a large number of persons. By the same token, much can be learned about the toxicity of a chemical if many mice are exposed.

Just making the decision to collect data to answer a question, to provide the basis for taking action, or to improve a process is a key step. Once that decision has been made, an important next step is to develop a **statement of purpose** that is both specific and unambiguous. If the subject of the study is public transportation being behind schedule, you must carefully specify what is meant by late. Is it 1 minute, 5 minutes, or more than 10 minutes behind scheduled times that should result in calling a bus or commuter train late? Words like soft or uncomfortable in a statement are even harder to quantify. One common approach, for a quality like comfort, is to ask passengers to rate the ride on public transportation on the five-point scale

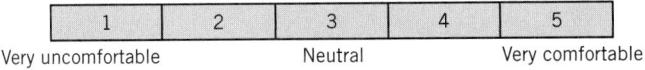

Very uncomfortable   Neutral   Very comfortable

where the numbers 1 through 5 are attached to the scale, with 1 for very uncomfortable and so on through 5 for very comfortable.

We might conclude that the ride is comfortable if the majority of persons in the sample check either of the top two boxes.

Example 5   A Clear Statement of Purpose Concerning Water Quality

Each day, a city must sample the lake water in and around a swimming beach to determine if the water is safe for swimming. During late summer, the primary difficulty is algae growth and the safe limit has been set in terms of water clarity.

SOLUTION   The problem is already well defined so the statement of purpose is straightforward.

PURPOSE: Determine whether or not the water clarity at the beach is below the safe limit.

The city has already decided to take three water samples and to analyze them separately. In Chapter 8, we will learn how to decide if the water is safe despite the variation in the three sample values.

The overall purpose can be quite general but a specific statement of purpose is required at each step to guide the collection of data. For instance:

GENERAL PURPOSE: Design a data collection and monitoring program at a completely automated plant that handles radioactive materials.

One issue is to ensure that the production plant will shut down quickly if materials start accumulating anywhere along the production line. More specifically, the weight of materials could be measured at critical positions. A quick shutdown will be implemented if any of these exceed a safe limit. For this step, a statement of purpose could be:

PURPOSE: Implement a fast shutdown if the weight at any critical position exceeds 1.2 kilograms.

The safe limit 1.2 kilograms should be obtained from experts; preferrably it would be a consensus of expert opinion.

There still remain statistical issues of how many critical positions to choose and how often to measure the weight. These are followed with questions on how to analyze data and specify a rule for implementing a fast shutdown.

A clearly specified statement of purpose will guide the choice of what data to collect and help ensure that it will be relevant to the purpose. Without a clearly specified purpose, or terms unambiguously defined, much effort can be wasted in collecting data that will not answer the question of interest.

6.   STATISTICS IN CONTEXT

A primary health facility became aware that sometimes it was taking too long to return patients' phone calls. That is, patients would phone in with requests for information. These requests, in turn, had to be turned over to doctors or nurses who would collect the information and return the call. The overall objective was to understand the current procedure and then improve on it. As a good first step, it was decided to find how long it was taking to return calls under the current procedure. Variation in times from call to call is expected, so the purpose of the initial investigation is to benchmark the variability with the current procedure by collecting a sample of times.

PURPOSE: Obtain a reference or benchmark for the current procedure by collecting a sample of times to return a patient's call under the current procedure.

For a sample of incoming calls collected during the week, the time received was noted along with the request. When the return call was completed, the elapsed time, in minutes, was recorded. Each of these times is represented as a dot in Figure 1. Notice that over one-third of the calls took over 120 minutes, or over two hours, to return. This could be a long time to wait for information if it concerns a child with a high fever or an adult with acute symptoms. If the purpose was to determine what proportion of calls took too long to return, we would need to agree on a more precise definition of "too long" in terms of number of minutes. Instead, these data clearly indicate that the process needs improvement and the next step is to proceed in that direction.

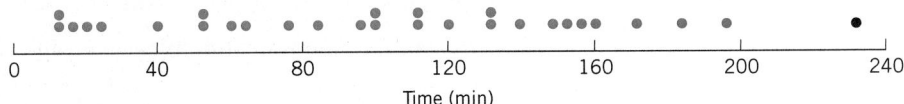

Figure 1   Time in minutes to return call.

In any context, to pursue potential improvements of a process, one needs to focus more closely on particulars. Three questions

When   Where   Who

should always be asked before gathering further data. More specifically, data should be sought that will answer the following questions.

When do the difficulties arise? Is it during certain hours, certain days of the week or month, or in coincidence with some other activities?

Where do the difficulties arise? Try to identify the locations of bottlenecks and unnecessary delays.

Who was performing the activity and who was supervising? The idea is not to pin blame, but to understand the roles of participants with the goal of making improvements.

It is often helpful to construct a **cause-and-effect diagram** or **fishbone diagram.** The main centerline represents the problem or the effect. A somewhat simplified fishbone chart is shown in Figure 2 for the *where* question regarding the location of delays when returning patients' phone calls. The main centerline represents the problem: Where are delays occurring? Calls come to the reception desk, but when these lines are busy, the calls go directly to nurses on the third or fourth floor. The main diagonal arms in Figure 2 represent the floors and the smaller horizontal lines more specific locations on the floor where the delay could occur. For instance, the horizontal line representing a delay in retrieving a patient's medical record connects to the second floor diagonal line. The resulting figure resembles the skeleton of a fish. Consideration of the diagram can help guide the choice of what new data to collect.

Fortunately, the quality team conducting this study had already given preliminary consideration to the *When, Where,* and *Who* questions and recorded not only the time of day but also the day and person receiving the call. That is, their current data gave them a start on determining if the time to return calls depends on when or where the call is received.

Although we go no further with this application here, the quality team next developed more detailed diagrams to study the flow of paper between the time the call is received and when it is returned. They then identified bottlenecks in the flow of information that were removed and the process was improved. In later chapters, you will learn how to compare and display data from two locations or old and new processes, but the key idea emphasized here is the purposeful collection of relevant data.

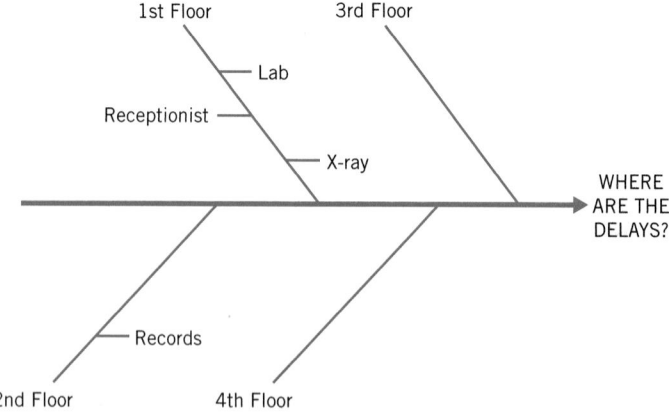

Figure 2   A cause-and-effect diagram for the location of delays.

7.   OBJECTIVES OF STATISTICS

The subject of statistics provides the methodology to make inferences about the population from the collection and analysis of sample data. These methods enable one to derive plausible generalizations and then assess the extent of uncertainty underlying these generalizations. Statistical concepts are also essential during the planning stage of an investigation when decisions must be made as to the mode and extent of the sampling process.

> The major objectives of statistics are:
> 1.   To make inferences about a population from an analysis of information contained in sample data. This includes assessments of the extent of uncertainty involved in these inferences.
> 2.   To design the process and the extent of sampling so that the observations form a basis for drawing valid inferences.

The design of the sampling process is an important step. A good design for the process of data collection permits efficient inferences to be made, often with a straightforward analysis. Unfortunately, even the most sophisticated methods of data analysis cannot, in themselves, salvage much information from data that are produced by a poorly planned experiment or survey.

The early use of statistics in the compilation and passive presentation of data has been largely superseded by the modern role of providing analytical tools with which data can be efficiently gathered, understood, and interpreted. Statistical concepts and methods make it possible to draw valid conclusions about the population on the basis of a sample. Given its extended goal, the subject of statistics has penetrated all fields of human endeavor in which the evaluation of information must be grounded in data-based evidence.

The basic statistical concepts and methods described in this book form the core in all areas of application. We present examples drawn from a wide range of applications to help develop an appreciation of various statistical methods, their potential uses, and their vulnerabilities to misuse.

Reference

1. American Statistical Association. *Careers in Statistics*. (A copy may be obtained by writing to the American Statistical Association, 1429 Duke Street, Alexandria, VA 22314-3402.)

USING STATISTICS WISELY

1. Compose a clear statement of purpose and use it to help decide upon which variables to observe.

2. Carefully define the population of interest.

3. Whenever possible, select samples using a random device or random number table.

4. Do not unquestionably accept conclusions based on self-selected samples.

5. Remember that conclusions reached in TV, magazine, or newspaper reports might not be as obvious as reported. When reading or listening to reports, you must be aware that the advocate, often a politician or advertiser, may only be presenting statistics that emphasize positive features.

KEY IDEAS

Before gathering data, on a characteristic of interest, identify a unit or sampling unit. This is usually a person or object. The population of units is the complete collection of units. In statistics we concentrate on the collection of values of the characteristic, or record of a qualitative trait, evaluated for each unit in the population. We call this the statistical population or just the population.

A sample or sample data set from the population is the subset of measurements that are actually collected.

Statistics is a body of principles that help to first design the process and extent of sampling and then guides the making of inferences about the population (inferential statistics). Descriptive statistics help summarize the sample and procedures for statistical inference allow us to make generalizations about the population from the information in the sample.

A statement of purpose is a key step in designing the data collection process.

8. REVIEW EXERCISES

1.1   A newspaper headline reads,

U.S. TEENS TRUST, FEAR THEIR PEERS

and the article explains that a telephone poll was conducted of 1055 persons 13 to 17 years old. Identify a statistical population and the sample.

1.2   Consider the population of all students at your college. You want to learn about total monthly entertainment expenses for a student.

(a) Specify the population unit.

(b) Specify the variable of interest.

(c) Specify the statistical population.

1.3   Consider the population of persons living in Chicago. You want to learn about the proportion which are illegal aliens.

(a) Specify the population unit.

(b) Specify the variable of interest.

(c) Specify the statistical population.

1.4   A student is asked to estimate the mean height of all male students on campus. She decides to use the heights of members of the basketball team because they are conveniently printed in the game program.

(a) Identify the statistical population and the sample.

(b) Comment on the selection of the sample.

(c) How should a sample of males be selected?

1.5   A student newspaper reports that 30% of undergraduates own a digital MP3 player. This statement, extracted from a survey about physical exercise, is based on the finding that among 10 students who jog for exercise, 3 own an MP3 player.

(a) Specify the population unit.

(b) Specify the statistical population and the sample.

(c) Comment on the representativeness of the sample.

1.6   A magazine that features the latest electronics and computer software for homes enclosed a short questionnaire on a postcard. Readers were asked to answer questions concerning their use and ownership of various software and hardware products, and to then send the card to the publisher. A summary of the results appeared in a later issue of the magazine that used the data to make statements such as 40% of readers have purchased program X. Identify a population and sample and comment on the representativeness of the sample. Are readers who have not purchased any new products mentioned in the questionnaire as likely to respond as those who have purchased?

1.7   Each year a local weekly newspaper gives out "Best of the City" awards in categories such as restaurant, deli, pastry shop, and so on. Readers are asked to fill in their favorites on a form enclosed in this free weekly paper and then send it to the publisher. The establishment receiving the most votes is declared the winner in its category. Identify the population and sample and comment on the representativeness of the sample.

1.8   Which of the following are anecdotal and which are based on sample?

(a) Out of 200 students questioned, 40 admitted they lied regularly?

(b) Bobbie says the produce at Market W is the freshest in the city.

(c) Out of 50 persons interviewed at a shopping mall, 18 had made a purchase that day.

1.9   Which of the following are anecdotal and which are based on a sample?

    (a)   Tom says he gets the best prices on electronics at the www.bestelc.com Internet site.

    (b)   Out of 22 students, 6 had multiple credit cards.

    (c)   Among 55 people checking in at the airport, 12 were going to destinations outside of the continental United States.

1.10   What is wrong with this statement of purpose?

    PURPOSE: *Determine if a newly designed rollerball pen is comfortable to hold when writing.*

    Give an improved statement of purpose.

1.11   What is wrong with this statement of purpose?

    PURPOSE: *Determine if it takes too long to get cash from the automated teller machine during the lunch hour.*

    Give an improved statement of purpose.

1.12   Give a statement of purpose for determining the amount of time it takes to make hotel reservations in San Francisco using the Internet.

1.13   Thirty-five classrooms on campus are equiped for multimedia instruction. Use Table 1, Appendix B, to select 4 of these classrooms to visit and check whether or not the instructor is using the equipment during that day's first hour lecture.

1.14   Fifty band members would like to ride the band bus to an out-of-town game. However, there is room for only 44. Use Table 1, Appendix B, to select the 44 persons that will go. Determine how to make your selection by taking only a few two-digit selections.

1.15   Eight young students need mentors. Of these, there are three that you enjoy being with while you are indifferent about the others. Two of the students will be randomly assigned to you. Label the students you like by 0, 1, and 2 and the others by 3, 4, 5, 6, and 7. Then, the process of assigning two students at random is equivalent to choosing two different digits from the table of random digits and ignoring any 8 or 9. Repeat the experiment of assigning two students 20 times by using the table of random digits. Record the pairs of digits you draw for each experiment.

    (a)   What is the proportion of the 20 experiments that give two students that you like?

    (b)   What is the proportion of the 20 experiments that give one of the students you like and one other?

    (c)   What is the proportion of the 20 experiments that give none of the students you like?

1.16   According to the cause-and-effect diagram on page 17, where are the possible delays on the first floor?

1.17   Refer to the cause-and-effect diagram on page 17. The workers have now noticed that a delay could occur:

    (a)   On the fourth floor at the pharmacy

    (b)   On the third floor at the practitioners' station

    Redraw the diagram and include this added information.

1.18   The front page of a national newspaper[1] contains the statement

    Each American creates 4.4 pounds of garbage a day

    (a)   Does this mean every single American produces the same amount of garbage? What do you think this statement means?

    (b)   Was the number 4.4 obtained from a sample? Explain.

    (c)   How would you select a sample?

1.19   As a very extreme case of self-selection, imagine a five-foot-high solid wood fence surrounding a collection of Great Danes and Miniature Poodles. You want to estimate the proportion of Great Danes inside and decide to collect your sample by observing the first seven dogs to jump out of the fence.

    (a)   Explain how this is a self-selected sample that is, of course, very misleading.

    (b)   How is this sample selection procedure like a call-in election poll?

[1]*Parade* Magazine, June 13, 1999.

2

Organization and
Description of Data

Acid Rain Is Killing Our Lakes

© SuperStock, Inc.

Acid precipitation is linked to the disappearance of sport fish and other organisms from lakes. Sources of air pollution, including automobile emissions and the burning of fossil fuels, add to the natural acidity of precipitation. The Wisconsin Department of Natural Resources initiated a precipitation monitoring program with the goal of developing appropriate air pollution controls to reduce the problem. The acidity of the first 50 rains monitored, measured on a pH scale from 1 (very acidic) to 7 (basic), are summarized by the histogram.

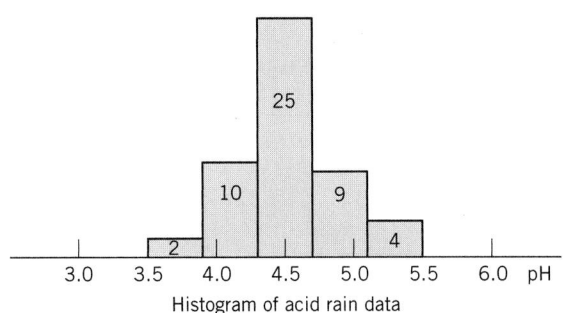

Histogram of acid rain data

Notice that all the rains are more acidic than normal rain, which has a pH of 5.6. (As a comparison, apples are about pH 3 and milk is about pH 6.)

Researchers in Canada have established that lake water with a pH below 5.6 may severely affect the reproduction of game fish. More research will undoubtedly improve our understanding of the acid rain problem and lead, it is hoped, to an improved environment.

1.   INTRODUCTION

In Chapter 1, we cited several examples of situations where the collection of data by appropriate processes of experimentation or observation is essential to acquire new knowledge. A data set may range in complexity from a few entries to hundreds or even thousands of them. Each entry corresponds to the observation of a specified characteristic of a sampling unit. For example, a nutritionist may provide an experimental diet to 30 undernourished children and record their weight gains after two months. Here, children are the sampling units, and the data set would consist of 30 measurements of weight gains. Once the data are collected, a primary step is to organize the information and extract a descriptive summary that highlights its salient features. In this chapter, we learn how to organize and describe a set of data by means of tables, graphs, and calculation of some numerical summary measures.

2.   MAIN TYPES OF DATA

In discussing the methods for providing summary descriptions of data, it helps to distinguish between the two basic types:

1.   Qualitative or categorical data
2.   Numerical or measurement data

When the characteristic under study concerns a qualitative trait that is only classified in categories and not numerically measured, the resulting data are called categorical data. Hair color (blond, brown, red, black), employment status (employed, unemployed), and blood type (O, A, B, AB) are but some examples. If, on the other hand, the characteristic is measured on a numerical scale, the resulting data consist of a set of numbers and are called measurement data. We will use the term numerical-valued variable or just variable to refer to a characteristic that is measured on a numerical scale. The word "variable" signifies that the measurements vary over different sampling units. In this terminology, observations of a numerical-valued variable yield measurement data. A few examples of numerical-valued variables are the shoe size of an adult male, daily number of traffic fatalities in a state, intensity of an earthquake, height of a 1-year-old pine seedling, the time in line at an automated teller, and the number of offspring in an animal litter.

Although in all these examples the stated characteristic can be numerically measured, a close scrutiny reveals two distinct types of underlying scale of measurement. Shoe sizes are numbers such as 6, $6\frac{1}{2}$, 7, $7\frac{1}{2}$, . . . , which proceed in steps of $\frac{1}{2}$. The count of traffic fatalities can only be an integer and so is the number of offspring in an animal litter. These are examples of discrete variables. The name discrete draws from the fact that the scale is made up of distinct numbers with gaps in between. On the other hand, some variables such as height, weight, and survival time can ideally take any value in an interval. Since the measurement scale does not have gaps, such variables are called continuous.

We must admit that a truly continuous scale of measurement is an idealization. Measurements actually recorded in a data set are always rounded either for the sake of simplicity or because the measuring device has a limited accuracy. Still, even though weights may be recorded in the nearest pounds or time recorded in the whole hours, their actual values occur on a continuous scale so the data are referred to as continuous. Counts are inherently discrete and treated as such, provided that they take relatively few distinct values (e.g., the number of children in a family or the number of traffic violations of a driver). But when a count spans a wide range of values, it is often treated as a continuous variable. For example, the count of white blood cells, number of insects in a colony, and number of shares of stock traded per day are strictly discrete, but for practical purposes, they are viewed as continuous.

A summary description of categorical data is discussed in Section 3.1. The remainder of this chapter is devoted to a descriptive study of measurement data, both discrete and continuous. As in the case of summarization and commentary on a long, wordy document, it is difficult to prescribe concrete steps for summary descriptions that work well for all types of measurement data. However, a few important aspects that deserve special attention are outlined here to provide general guidelines for this process.

Describing a Data Set of Measurements

1. Summarization and description of the overall pattern.
   (a) Presentation of tables and graphs.
   (b) Noting important features of the graphed data including symmetry or departures from it.
   (c) Scanning the graphed data to detect any observations that seem to stick far out from the major mass of the data—the outliers.
2. Computation of numerical measures.
   (a) A typical or representative value that indicates the center of the data.
   (b) The amount of spread or variation present in the data.

---

3. DESCRIBING DATA BY TABLES AND GRAPHS

3.1 CATEGORICAL DATA

When a qualitative trait is observed for a sample of units, each observation is recorded as a member of one of several categories. Such data are readily organized in the form of a frequency table that shows the counts (frequencies) of the individual categories. Our understanding of the data is further enhanced by calculation of the proportion (also called relative frequency) of observations in each category.

$$\frac{\text{Relative frequency}}{\text{of a category}} = \frac{\text{Frequency in the category}}{\text{Total number of observations}}$$

Example 1   Calculating Relative Frequencies to Summarize an Opinion Poll

A campus press polled a sample of 280 undergraduate students in order to study student attitude toward a proposed change in the dormitory regulations. Each student was to respond as support, oppose, or neutral in regard to the issue. The numbers were 152 support, 77 neutral, and 51 opposed. Tabulate the results and calculate the relative frequencies for the three response categories.

SOLUTION   Table 1 records the frequencies in the second column, and the relative frequencies are calculated in the third column. The relative frequencies show that about 54% of the polled students supported the change, 18% opposed, and 28% were neutral.

TABLE 1   Summary Results
of an Opinion Poll

Responses	Frequency	Relative Frequency	
Support	152	$\frac{152}{280} =$	.543
Neutral	77	$\frac{77}{280} =$	.275
Oppose	51	$\frac{51}{280} =$	.182
Total	280		1.000

Remark: The relative frequencies provide the most relevant information as to the pattern of the data. One should also state the sample size, which serves as an indicator of the credibility of the relative frequencies. (More on this in Chapter 8.)

Categorical data are often presented graphically as a **pie chart** in which the segments of a circle exhibit the relative frequencies of the categories. To obtain the angle for any category, we multiply the relative frequency by 360 degrees, which corresponds to the complete circle. Although laying out the angles by hand can be tedious, many software packages generate the chart with a single command. Figure 1 presents a pie chart for the data in Example 1.

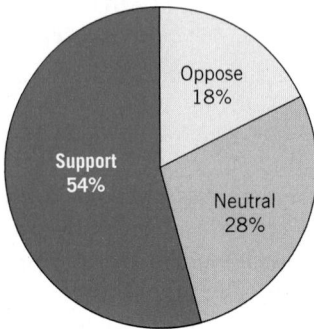

Figure 1   Pie chart of student opinion on change in dormitory regulations.

When questions arise that need answering but the decision makers lack precise knowledge of the state of nature or the full ramifications of their decisions, the best procedure is often to collect more data. In the context of quality improvement, if a problem is recognized, the first step is to collect data on the magnitude and possible causes. This information is most effectively communicated through graphical presentations.

A **Pareto diagram** is a powerful graphical technique for displaying events according to their frequency. According to Pareto's empirical law, any collection of events consists of only a few that are major in that they are the ones that occur most of the time.

Figure 2 gives a Pareto diagram for the type of defects found in a day's production of facial tissues. The cumulative frequency is 22 for the first cause and

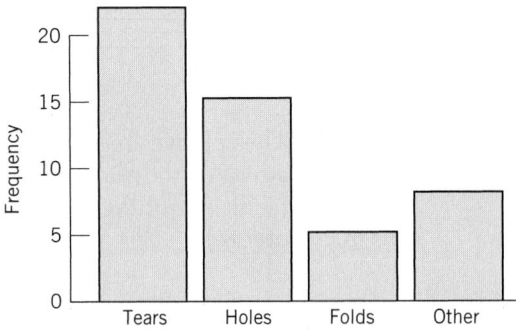

Figure 2   Pareto diagram of facial tissue defects.

22 + 15 = 37 for the first and second causes combined. This illustrates Pareto's rule, with two of the causes being responsible for 37 out of 50, or 74%, of the defects.

Example 2   A Pareto Diagram Clarifies Circumstances Needing Improvement

Graduate students in a counseling course were asked to choose one of their personal habits that needed improvement. In order to reduce the effect of this habit, they were asked to first gather data on the frequency of the occurrence and the circumstances. One student collected the following frequency data on fingernail biting over a two-week period.

Frequency	Activity
58	Watching television
21	Reading newspaper
14	Talking on phone
7	Driving a car
3	Grocery shopping
12	Other

Make a Pareto diagram showing the relationship between nail biting and type of activity.

SOLUTION   The cumulative frequencies are 58, 58 + 21 = 79, and so on, out of 115. The Pareto diagram is shown in Figure 3, where watching TV accounts for 50.4% of the instances.

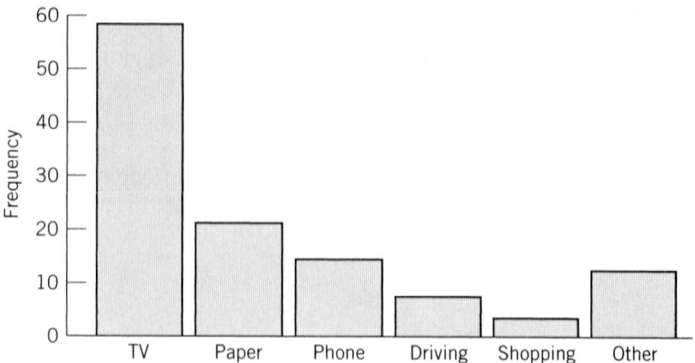

Figure 3   Pareto diagram for nail biting example.

The next step for this person would be to try and find a substitute for nail biting while watching television.

3.2 DISCRETE DATA

We next consider summary descriptions of measurement data and begin our discussion with discrete measurement scales. As explained in Section 2, a data set is identified as discrete when the underlying scale is discrete and the distinct values observed are not too numerous.

Similar to our description of categorical data, the information in a discrete data set can be summarized in a frequency table, or **frequency distribution** that includes a calculation of the relative frequencies. In place of the qualitative categories, we now list the distinct numerical measurements that appear in the data set and then count their frequencies.

Example 3   Creating a Frequency Distribution

The daily number of Internet system crashes are observed over 30 days at a university computing center, and the data of Table 2 are obtained. Determine the frequency distribution.

TABLE 2   Daily Number of Internet System Crashes

1	3	1	1	0	1	0	1	1	0
2	2	0	0	0	1	2	1	2	0
0	1	6	4	3	3	1	2	4	0

SOLUTION   The frequency distribution of this data set is presented in Table 3, where the last column shows the calculated relative frequencies.

TABLE 3   Frequency Distribution for Daily Number (x) of Internet System Crashes

Value x	Frequency	Relative Frequency
0	9	.300
1	10	.333
2	5	.167
3	3	.100
4	2	.067
5	0	.000
6	1	.033
Total	30	1.000

The frequency distribution of a discrete variable can be presented pictorially by drawing either lines or rectangles to represent the relative frequencies. First, the distinct values of the variable are located on the horizontal axis. For a line diagram, we draw a vertical line at each value and make the height of the line equal to the relative frequency. A **histogram** employs vertical rectangles instead of lines. These rectangles are centered at the values and their areas represent relative frequencies. Typically, the values proceed in equal steps so the rectangles are all of the same width and their heights are proportional to the relative frequencies as well as frequencies. Figure 4(a) shows the line diagram and 4(b) the histogram of the frequency distribution of Table 3.

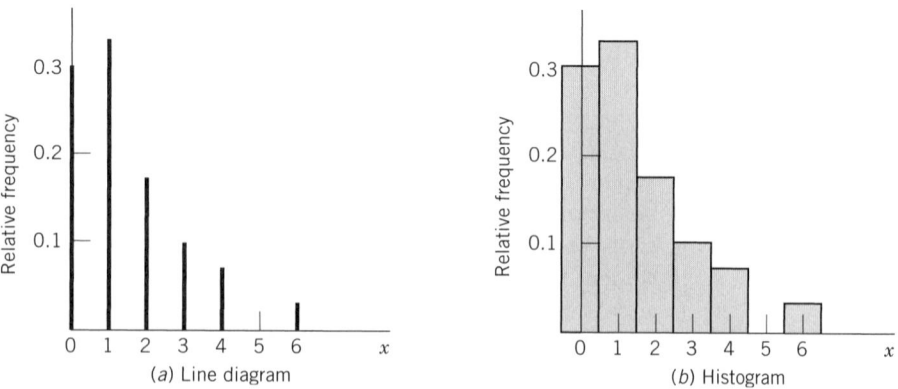

Figure 4   Graphic display of the frequency distribution of data in Table 3.

3.3 DATA ON A CONTINUOUS VARIABLE

We now consider tabular and graphical presentations of data sets that contain numerical measurements on a virtually continuous scale. Of course, the recorded measurements are always rounded. In contrast with the discrete case, a data set of measurements on a continuous variable may contain many distinct values. Then, a table or plot of all distinct values and their frequencies will not provide a condensed or informative summary of the data.

The two main graphical methods used to display a data set of measurements are the **dot diagram** and the **histogram**. Dot diagrams are employed when there are relatively few observations (say, less than 20 or 25); histograms are used with a larger number of observations.

Dot Diagram

When the data consist of a small set of numbers, they can be graphically represented by drawing a line with a scale covering the range of values of the measurements. Individual measurements are plotted above this line as prominent dots. The resulting diagram is called a **dot diagram**.

Example 4   A Dot Diagram Reveals an Unusual Observation

The number of days the first six heart transplant patients at Stanford survived after their operations were 15, 3, 46, 623, 126, 64. Make a dot diagram.

SOLUTION   These survival times extended from 3 to 623 days. Drawing a line segment from 0 to 700, we can plot the data as shown in Figure 5. This dot diagram shows a cluster of small survival times and a single, rather large value.

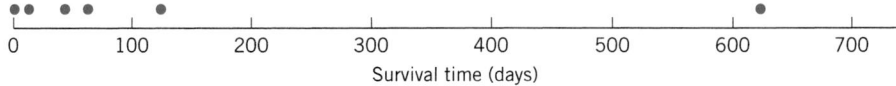

Figure 5   Dot diagram for the heart transplant data.

Frequency Distribution on Intervals

When the data consist of a large number of measurements, a dot diagram may be quite tedious to construct. More seriously, overcrowding of the dots will cause them to smear and mar the clarity of the diagram. In such cases, it is convenient to condense the data by grouping the observations according to intervals and recording the frequencies of the intervals. Unlike a discrete frequency distribution, where grouping naturally takes place on points, here we use intervals of values. The main steps in this process are outlined as follows.

Constructing a Frequency Distribution for a Continuous Variable

1. Find the minimum and the maximum values in the data set.
2. Choose intervals or cells of equal length that cover the range between the minimum and the maximum without overlapping. These are called class intervals, and their endpoints class boundaries.
3. Count the number of observations in the data that belong to each class interval. The count in each class is the **class frequency** or **cell frequency.**
4. Calculate the relative frequency of each class by dividing the class frequency by the total number of observations in the data:

$$\text{Relative frequency} = \frac{\text{Class frequency}}{\text{Total number of observations}}$$

The choice of the number and position of the class intervals is primarily a matter of judgment guided by the following considerations. The number of

Paying Attention

© Britt Erlanson/The Image Bank/Getty Images
Paying attention in class. Observations on 24 first-grade students.

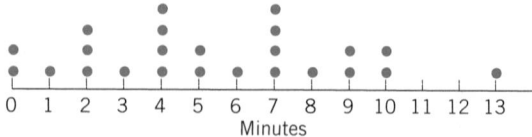

Figure 6   Time not concentrating on the mathematics assignment (out of 20 minutes).

First-grade teachers allot a portion of each day to mathematics. An educator, concerned about how students utilize this time, selected 24 students and observed them for a total of 20 minutes spread over several days. The number of minutes, out of 20, that the student was not on task was recorded (courtesy of T. Romberg). These lack-of-attention times are graphically portrayed in the dot diagram in Figure 6. The student with 13 out of 20 minutes off-task stands out enough to merit further consideration. Is this a student who finds the subject too difficult or might it be a very bright child who is bored?

classes usually ranges from 5 to 15, depending on the number of observations in the data. Grouping the observations sacrifices information concerning how the observations are distributed within each cell. With two few cells, the loss of information is serious. On the other hand, if one chooses too many cells and the data set is relatively small, the frequencies from one cell to the next would jump up and down in a chaotic manner and no overall pattern would emerge. As an initial step, frequencies may be determined with a large number of intervals that can later be combined as desired in order to obtain a smooth pattern of the distribution.

Computers conveniently order data from smallest to largest so that the observations in any cell can easily be counted. The construction of a frequency distribution is illustrated in Example 5.

Example 5   A Frequency Distribution for Bookstore Sales

Most students purchase their books during the registration period. University bookstore receipts from 40 students provided the sales data of Table 4 where the values have been ordered from smallest to largest.

TABLE 4   The Data of Forty Cash Register Receipts (in Dollars) at a University Bookstore

16.00	58.50	68.20	78.00	79.45	142.20	145.35	186.70
209.05	216.75	219.70	247.55	249.10	256.00	257.15	262.35
268.60	269.60	270.15	284.45	319.00	332.00	343.20	350.75
354.90	372.60	383.40	389.20	404.55	420.20	428.50	432.40
444.60	446.40	456.80	458.10	493.95	511.95	521.05	621.35

Construct a frequency distribution of the sales data.

SOLUTION   To construct a frequency distribution, we first notice that the minimum sale is $16.00 and the maximum sale $621.35. We choose class intervals of length 125 as a matter of convenience.

The selection of class boundaries is a bit of fussy work. Because the data have two decimal places, we could add a third decimal figure to avoid the possibility of any observation falling exactly on the boundary. For example, we could end the first class interval at 124.995. Alternatively, and more neatly, we could write 0–125 and make the **endpoint convention** that the left-hand limit is included but not the right.

The first interval contains 5 observations, so its frequency is 5 and its relative frequency $\frac{5}{40} = .125$. Table 5 gives the frequency distribution. The relative frequencies add to 1, as they should in any frequency distribution.

Remark: The rule requiring equal class intervals is inconvenient when the data are spread over a wide range but are highly concentrated in a small part of the range with relatively few numbers elsewhere. Using smaller intervals where the data are highly concentrated and larger intervals where the data are sparse helps to reduce the loss of information due to grouping.

TABLE 5   Frequency Distribution for Bookstore Sales
Data (left endpoints included, but right
endpoints excluded)

Class Interval	Frequency	Relative Frequency
$ 0–125	5	$\dfrac{5}{40} = .125$
125–250	8	$\dfrac{8}{40} = .200$
250–375	13	$\dfrac{13}{40} = .325$
375–500	11	$\dfrac{11}{40} = .275$
500–625	3	$\dfrac{3}{40} = .075$
Total	40	1.000

Tabulations of income, age, and other characteristics in official reports are often made with unequal class intervals.

Histogram

A frequency distribution can be graphically presented as a histogram. To draw a histogram, we first mark the class intervals on the horizontal axis. On each interval, we then draw a vertical rectangle whose area represents the relative frequency—that is, the proportion of the observations occurring in that class interval. The total area of all rectangles equals the sum of the relative frequencies, which is 1.

The total area of a histogram is 1.

The histogram for Table 5 is shown in Figure 7. For example, the rectangle drawn on the class interval 0–125 has its area = .001 × 125 = .125, which is the relative frequency of this class. Actually, we determined the height .001 as

$$\text{Height} = \frac{\text{Relative frequency}}{\text{Width of interval}} = \frac{.125}{125} = .001$$

The units on the vertical axis can be viewed as relative frequencies per unit of the horizontal scale. For instance, .0016 is the relative frequency per dollar for the interval $125–$250.

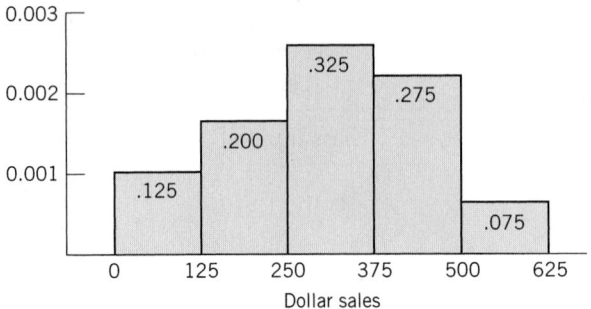

Figure 7   Histogram of the bookstore sales data of
Tables 4 and 5. Sample size = 40.

Visually, we note that the tallest block or most frequent class interval is
$250–$375. Also, proportion .275 + .075 = .35 of the sales are $375 or more.

Remark: When all class intervals have equal widths, the heights of the rectangles are proportional to the relative frequencies that the areas represent.
The formal calculation of height, as area divided by the width, is then re-
dundant. Instead, one can mark the vertical scale according to the relative
frequencies—that is, make the heights of the rectangles equal to the relative frequencies. The resulting picture also makes the areas represent the
relative frequencies if we read the vertical scale as if it is in units of the class
interval. This leeway when plotting the histogram is not permitted in the
case of unequal class intervals.

Figure 8 shows one ingenious way of displaying two histograms for comparison.
In spite of their complicated shapes, their back-to-back plot as a "tree" allows for
easy visual comparison of the male and female age distributions.

Stem-and-Leaf Display

A stem-and-leaf display provides a more efficient variant of the histogram for
displaying data, especially when the observations are two-digit numbers. This
plot is obtained by sorting the observations into rows according to their leading
digit. The stem-and-leaf display for the data of Table 6 is shown in Table 7. To
make this display:

1. List the digits 0 through 9 in a column and draw a vertical line. These
    correspond to the leading digit.
2. For each observation, record its second digit to the right of this vertical
    line in the row where the first digit appears.
3. Finally, arrange the second digits in each row so they are in increasing
    order.

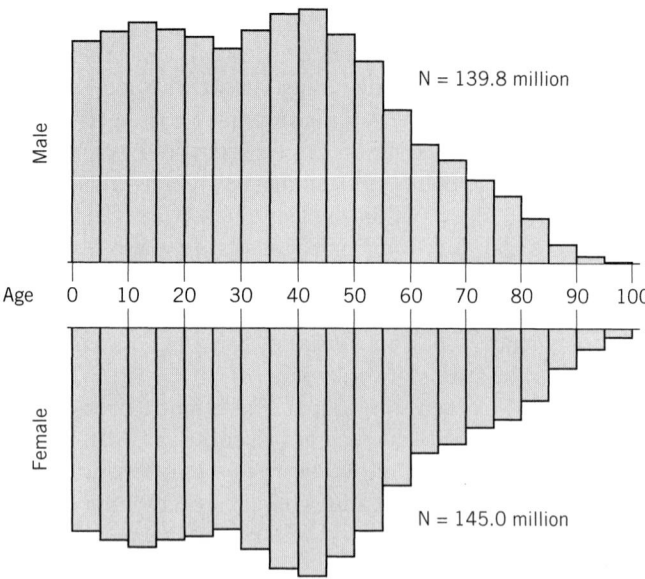

N = 139.8 million

N = 145.0 million

Figure 8   Population tree (histograms) of the male and female age distributions in the United States in 2001. (*Source:* U.S. Bureau of the Census.)

TABLE 6   Examination Scores of 50 Students

75	98	42	75	84	87	65	59	63
86	78	37	99	66	90	79	80	89
68	57	95	55	79	88	76	60	77
49	92	83	71	78	53	81	77	58
93	85	70	62	80	74	69	90	62
84	64	73	48	72				

TABLE 7   Stem-and-Leaf Display for the Examination Scores

0	
1	
2	
3	7
4	289
5	35789
6	022345689
7	01234556778899
8	00134456789
9	0023589

In the stem-and-leaf display, the column of first digits to the left of the vertical line is viewed as the stem, and the second digits as the leaves. Viewed sidewise, it looks like a histogram with a cell width equal to 10. However, it is more informative than a histogram because the actual data points are retained. In fact, every observation can be recovered exactly from the stem-and-leaf display.

A stem-and-leaf display retains all the information in the leading digits of the data. Within the leaf unit = 0.01, 3.5 | 0 2 3 7 8 presents the data 3.50, 3.52, 3.53, 3.57, and 3.58. Leaves may also be two-digit at times. Within the first leaf digit = 0.01, 0.4 | 07 13 82 90 presents the data 0.407, 0.413, 0.482, and 0.490.

Further variants of the stem-and-leaf display are described in Exercises 2.25 and 2.26. This versatile display is one of the most applicable techniques of exploratory data analysis.

When the sample size is small or moderate, no information is lost with the stem-and-leaf diagram because you can see every data point. The major disadvantage is that, when the sample size is large, diagrams with hundreds of numbers in a row cannot be constructed in a legible manner.

Exercises

2.1 Cities must find better ways to dispose of solid waste. According to the Environmental Protection Agency, the composition of the 217 million tons of solid municipal waste created in 1997 was

Paper and paperboard	38.6%
Yard waste	12.8%
Food waste	10.1%
Plastics	9.9%
Metals	7.7%
Other materials	

(a) Determine the percentage of other materials in the solid waste. This category includes glass, wood, rubber, and so on.

(b) Create a Pareto chart.

(c) What percentage of the total solid waste is paper or paperboard? What percentage is from the top two categories? What percentage is from the top five categories?

2.2 Solid waste disposal has become a major problem for most cities. It is usually fourth in cost behind education, police, and fire protection. On a statewide basis, the largest producers of solid municipal waste in 1997, in millions of tons per year, were

California	56.0
Texas	33.9
New York	30.2
Florida	23.8
Michigan	19.5
Illinois	13.3
North Carolina	12.6

The total amount of garbage created in all states is 217 million tons.

(a) Create a Pareto chart.

(b) What proportion of the total solid waste is created in California? What proportion by the top three states? What proportion by the top five states?

(c) Does this mean that a person in California, Texas, or New York creates more garbage than someone in another state? How would you adjust the figures to make that kind of comparison?

2.3 Recorded here are the blood types of 40 persons who have volunteered to donate blood at a plasma center. Summarize the data in a frequency table. Include calculations of the relative frequencies.

```
O  O  A  B  A  O  A  A  A  O
B  O  B  O  O  A  O  O  A  A
A  A  AB A  B  A  A  O  O  A
O  O  A  A  A  O  A  O  O  AB
```

2.4   In a study of the job hazards in the roofing industry in California, records of the disabling injuries were classified according to the accident types. Of the total number of 1132 injuries, 329 were due to falls, 256 from burns, 219 from overexertion, 202 from being struck, 40 from foreign substance in eye, and 86 from other miscellaneous reasons. (Source: U.S. Dept. of HEW Publication NIOSH 75–176.)

Present these data in a frequency table and also give the relative frequencies.

2.5   The following table shows how workers in one department get to work.

Mode of Transportation	Frequency
Drive alone	25
Car pool	3
Ride bus	7
Other	5

(a)   Calculate the relative frequency of each mode of transportation.

(b)   Construct a pie chart.

2.6   Of the $207 million raised by a major university's fund drive, $117 million came from individuals and bequests, $24 million from industry and business, and $66 million from foundations and associations. Present this information in the form of a pie chart.

2.7   Data from one campus dorm on the number of burglaries are collected each week of the semester. These data are to be grouped into the classes 0–1, 2–3, 3–5, 6 or more. Both endpoints included. Explain where a difficulty might arise.

2.8   Data from one campus dorm, on the number of complaints about the dorm food are collected each week of the semester. These weekly counts are to be grouped into the classes 0–1, 2–3, 4–5, 7 or more. Both endpoints are included. Explain where a difficulty might arise.

2.9   A sample of persons will each be asked to give the number of their close friends. The responses are to be grouped into the following classes: 0, 1–3, 3–5, 6 or more. Left endpoint is included. Explain where difficulties might arise.

2.10   The weights of the players on the university football team (to the nearest pound) are to be grouped into the following classes: 160–175, 175–190, 190–205, 205–220, 220–235, 235 or more. The left endpoint is included but not the right endpoint. Explain where difficulties might arise.

2.11   On flights from San Francisco to Chicago, the number of empty seats are to be grouped into the following classes: 0–4, 5–9, 10–14, 15–19, more than 19.

Is it possible to determine from this frequency distribution the exact number of flights on which there were:

(a)   Fewer than 10 empty seats?

(b)   More than 14 empty seats?

(c)   At least 5 empty seats?

(d)   Exactly 9 empty seats?

(e)   Between 5 and 15 empty seats inclusively?

2.12   A major West Coast power company surveyed 50 customers who were asked to respond to the statement, "People should rely mainly on themselves to solve problems caused by power outages" with one of the following responses.

1.   Definitely agree.

2.   Somewhat agree.

3.   Somewhat disagree.

4.   Definitely disagree.

The responses are as follows:

```
4  2  1  3  3  2  4  2  1  1  2  2  2  2  1  3  4
1  4  4  1  3  2  4  1  4  3  3  1  1  1  2  1  1
4  4  4  4  4  1  2  2  2  4  4  4  1  3  4  2
```

Construct a frequency table.

2.13   A sample of 50 departing airline passengers at the main check-in counter produced the following number of bags checked through to final destinations.

```
0  1  2  2  1  2  1  2  3  0  1  0
1  1  0  1  3  0  1  2  1  1  1  2
1  2  2  1  2  0  0  2  2  1  1  1
1  1  1  1  2  0  1  3  0  1  2  1
1  3
```

(a) Make a relative frequency line diagram.

(b) Comment on the pattern.

(c) What proportion of passengers who check in at the main counter fail to check any bags?

2.14 A person with asthma took measurements by blowing into a peak-flow meter on seven consecutive days.

429   425   471   422   432   444   454

Display the data in a dot diagram.

2.15 Before microwave ovens are sold, the manufacturer must check to ensure that the radiation coming through the door is below a specified safe limit. The amounts of radiation leakage (mW/cm^2) with the door closed from 25 ovens are as follows (courtesy of John Cryer):

```
15   9  18  10   5  12   8
 5   8  10   7   2   1
 5   3   5  15  10  15
 9   8  18   1   2  11
```

Display the data in a dot diagram.

2.16 Five students reported that the previous night they slept

7   8   7   10   9

hours. Display the data in a dot diagram.

2.17 The city of Madison regularly checks the water quality at swimming beaches located on area lakes. The concentration of fecal coliforms, in number of colony forming units (CFU) per 100 ml of water, was measured on fifteen days during the summer at one beach.

```
180  1600   90  140   50  260  400   90
380   110   10   60   20  340   80
```

(a) Make a dot diagram.

(b) Comment on the pattern and any unusual features.

(c) The city closes any swimming beach if a count is over 1350. What proportion of days, among the fifteen, was this beach closed?

2.18 Tornadoes kill many people every year in the United States. The yearly number of lives lost during the 54 years 1950 through 2003 are summarized in the following table.

Number of Deaths	Frequency
24 or less	2
25–49	17
50–74	17
75–99	6
100–149	5
150–199	2
200–249	1
250 or more	3
Total	54

(a) Calculate the relative frequency for the intervals [0, 25), [25, 50) and so on where the right-hand endpoint is excluded. Take the last interval to be [250, 550).

(b) Plot the relative frequency histogram. (*Hint:* Since the intervals have unequal widths, make the height of each rectangle equal to the relative frequency divided by the width of the interval.)

(c) What proportion of the years had 49 or fewer deaths due to tornadoes.

(d) Comment on the shape of the distribution.

2.19 A zoologist collected wild lizards in the Southwestern United States. Thirty lizards from the genus *Phrynosoma* were placed on a treadmill and their speed measured. The recorded speed (meters/second) is the fastest time to run a half meter. (Courtesy of K. Bonine.)

1.28 1.36 1.24 2.47 1.94 2.52 2.67 1.29
1.56 2.66 2.17 1.57 2.10 2.54 1.63 2.11
2.57 1.72 0.76 1.02 1.78 0.50 1.49 1.57
1.04 1.92 1.55 1.78 1.70 1.20

(a) Construct a frequency distribution using the class intervals 0.45–0.90, 0.90–1.35, and so on, with the endpoint convention that the left endpoint is included and the right endpoint is excluded. Calculate the relative frequencies.

(b) Make a histogram.

2.20 The California Seismic Safety Commission maintains a list of significant damaging California earthquakes. Their sizes, by row starting at the most recent quake, are

6.8 6.6 7.5 6.2 6.5 7.1 6.1
5.8 5.5 6.9 6.6 6.2 5.3 5.9
6.0 5.3 5.9 6.2 6.4 7.0 6.2
6.1 6.0 6.1 5.5 6.4 5.9 5.7
5.9 5.9 6.1 5.3 6.6 5.8 7.7
5.9 7.1 6.3 7.0 6.3 8.3 8.0
6.8 6.3 8.3 7.0 7.0 7.0 6.5

Construct a histogram using equal-length intervals starting with (5.1, 5.8] where the right-hand endpoint is included but not the left-hand endpoint.

2.21 Refer to Exercise 2.20, construct a density histogram using the intervals (5.2, 5.6], (5.6, 6.0], (6.0, 6.4], (6.4, 6.8], (6.8, 7.2], and (7.2, 8.4].

2.22 The following data represent the scores of 40 students on a college qualification test (courtesy of R. W. Johnson).

162 171 138 145 144 126 145 162 174 178
167   98 161 152 182 136 165 137 133 143
184 166 115 115   95 190 119 144 176 135
194 147 160 158 178 162 131 106 157 154

Make a stem-and-leaf display.

2.23 A federal government study of the oil reserves in Elk Hills, CA, included a study of the amount of iron present in the oil.

Amount of Iron (percent ash)

20	18	25	26	17
14	20	14	18	15
22	15	17	25	22
12	52	27	24	41
34	20	17	20	19
20	16	20	15	34
22	29	29	34	27
13	6	24	47	32
12	17	36	35	41
36	32	46	30	51

Make a stem-and-leaf display.

2.24 The following is a stem-and-leaf display with two-digit leaves. (The leading leaf digit = 10.0.)

```
1 |
2 | 46  68  93
3 | 19  44  71  82  97
4 | 05  26  43  90
5 | 04  68
6 | 13
```

List the corresponding measurements.

2.25 If there are too many leaves on some stems in a stem-and-leaf display, we might double the number of stems. The leaves 0–4 could hang on one stem and 5–9 on the repeated stem. For the observations

193 198 200 202 203 203 205 205 206 207
207 208 212 213 214 217 219 220 222 226 237

we would get the double-stem display

```
19 | 3
19 | 8
20 | 0233
20 | 556778
21 | 234
21 | 79
22 | 02
22 | 6
23 |
23 | 7
```

Construct a double-stem display with one-digit leaves for the data of Exercise 2.23.

2.26 If the double-stem display still has too few stems, we may wish to construct a stem-and-leaf display with a separate stem to hold leaves 0 and 1, 2 and 3, 4 and 5, 6 and 7, and a stem to hold 8 and 9. The resulting stem-and-leaf display is called a five-stem display. The following is a five-digit stem-and-leaf display. (Leaf unit = 1.0)

1	8
2	001
2	2233
2	444555
2	667
2	9
3	0

List the corresponding measurements.

2.27 The following table lists values of the Consumer Price Index for 25 selected areas both for 1992 and 2001. Construct a five-stem display for the consumer price index in 2001.

	1992	2001
Anchorage	128	155
Atlanta	139	176
Boston	149	191
Chicago	141	178
Cincinnati	134	168
Cleveland	137	173
Dallas	134	170
Denver	130	181
Detroit	136	174
Honolulu	155	178
Houston	129	159
Kansas City	134	172
Los Angeles	147	177
Miami	135	173
Milwaukee	137	172
Minneapolis	135	177
New York	150	187
Philadelphia	147	181
Pittsburgh	136	173
Portland	140	182
St. Louis	135	167
San Diego	147	191
San Francisco	143	190
Seattle	139	186

4. MEASURES OF CENTER

The graphic procedures described in Section 3 help us to visualize the pattern of a data set of measurements. To obtain a more objective summary description and a comparison of data sets, we must go one step further and obtain numerical values for the location or center of the data and the amount of variability present. Because data are normally obtained by sampling from a large population, our discussion of numerical measures is restricted to data arising in this context. Moreover, when the population is finite and completely sampled, the same arithmetic operations can be carried out to obtain numerical measures for the population.

To effectively present the ideas and associated calculations, it is convenient to represent a data set by symbols to prevent the discussion from becoming anchored to a specific set of numbers. A data set consists of a number of measurements which are symbolically represented by $x_1, x_2, \ldots, x_n$. The last subscript n denotes the number of measurements in the data, and $x_1, x_2, \ldots$ represent the first observation, the second observation, and so on. For instance, a data set consisting of the five measurements 2.1, 3.2, 4.1, 5.6, and 3.7 is represented in symbols by x_1, x_2, x_3, x_4, x_5, where $x_1 = 2.1, x_2 = 3.2, x_3 = 4.1, x_4 = 5.6,$ and $x_5 = 3.7$.

The most important aspect of studying the distribution of a sample of measurements is locating the position of a central value about which the measurements are distributed. The two most commonly used indicators of center are the mean and the median.

The **mean**, or **average**, of a set of measurements is the sum of the measurements divided by their number. For instance, the mean of the five measurements 2.1, 3.2, 4.1, 5.6, and 3.7 is

$$\frac{2.1 + 3.2 + 4.1 + 5.6 + 3.7}{5} = \frac{18.7}{5} = 3.74$$

To state this idea in general terms, we use symbols. If a sample consists of n measurements $x_1, x_2, \ldots, x_n$, the mean of the sample is

$$\frac{x_1 + x_2 + \cdots + x_n}{n} = \frac{\text{sum of the } n \text{ measurements}}{n}$$

The notation $\bar{x}$ will be used to represent a sample mean. To further simplify the writing of a sum, the Greek capital letter Σ (sigma) is used as a statistical shorthand. With this symbol:

The sum $x_1 + x_2 + \cdots + x_n$ is denoted as $\sum_{i=1}^{n} x_i$.

Read this as "the sum of all x_i with i ranging from 1 to n."

For example, $\sum_{i=1}^{5} x_i$ represents the sum $x_1 + x_2 + x_3 + x_4 + x_5$.

Remark: When the number of terms being summed is understood from the context, we often simplify to $\sum x_i$, instead of $\sum_{i=1}^{n} x_i$. Some further operations with the Σ notation are discussed in Appendix A1.

We are now ready to formally define the sample mean.

The **sample mean** of a set of n measurements $x_1, x_2, \ldots, x_n$ is the sum of these measurements divided by n. The sample mean is denoted by $\bar{x}$.

$$\bar{x} = \frac{\sum_{i=1}^{n} x_i}{n} \qquad \text{or} \qquad \frac{\sum x_i}{n}$$

According to the concept of "average," the mean represents a center of a data set. If we picture the dot diagram of a data set as a thin weightless horizontal bar on which balls of equal size and weight are placed at the positions of the data points, then the mean $\bar{x}$ represents the point on which the bar will balance. The computation of the sample mean and its physical interpretation are illustrated in Example 6.

Example 6   Calculating and Interpreting the Sample Mean

The birth weights in pounds of five babies born in a hospital on a certain day are 9.2, 6.4, 10.5, 8.1, and 7.8. Obtain the sample mean and create a dot diagram.

SOLUTION   The mean birth weight for these data is

$$\bar{x} = \frac{9.2 + 6.4 + 10.5 + 8.1 + 7.8}{5} = \frac{42.0}{5} = 8.4 \text{ pounds}$$

The dot diagram of the data appears in Figure 9, where the sample mean (marked by Δ) is the balancing point or center of the picture.

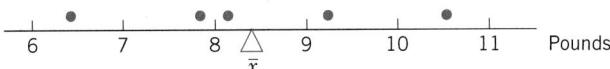

Figure 9   Dot diagram and the sample mean for the birth-weight data.

Another measure of center is the middle value.

> The sample median of a set of n measurements $x_1, \ldots, x_n$ is the middle value when the measurements are arranged from smallest to largest.

Roughly speaking, the median is the value that divides the data into two equal halves. In other words, 50% of the data lie below the median and 50% above it. If n is an odd number, there is a unique middle value and it is the median. If n is an even number, there are two middle values and the median is defined as their average. For instance, the data 3, 5, 7, 8 have two middle values 5 and 7, so the median = $(5 + 7)/2 = 6$.

Example 7   Calculating the Sample Median

Find the median of the birth-weight data given in Example 6.

SOLUTION   The measurements, ordered from smallest to largest, are

$$6.4 \quad 7.8 \quad \boxed{8.1} \quad 9.2 \quad 10.5$$

The middle value is 8.1, and the median is therefore 8.1 pounds.

Example 8   Choosing between the Mean and Median

Calculate the median of the survival times given in Example 4. Also calculate the mean and compare.

SOLUTION   To find the median, first we order the data. The ordered values are

$$3 \quad 15 \quad 46 \quad 64 \quad 126 \quad 623$$

There are two middle values, so

$$\text{Median} \;=\; \frac{46 \,+\, 64}{2} \;=\; 55 \text{ days}$$

The sample mean is

$$\bar{x} \;=\; \frac{3 \,+\, 15 \,+\, 46 \,+\, 64 \,+\, 126 \,+\, 623}{6} \;=\; \frac{877}{6} \;=\; 146.2 \text{ days}$$

Note that one large survival time greatly inflates the mean. Only 1 out of the 6 patients survived longer than $\bar{x} = 146.2$ days. Here the median of 55 days appears to be a better indicator of the center than the mean.

Example 8 demonstrates that the median is not affected by a few very small or very large observations, whereas the presence of such extremes can have a considerable effect on the mean. For extremely asymmetrical distributions, the median is likely to be a more sensible measure of center than the mean. That is why government reports on income distribution quote the median income as a summary, rather than the mean. A relatively small number of very highly paid persons can have a great effect on the mean salary.

If the number of observations is quite large (greater than, say, 25 or 30), it is sometimes useful to extend the notion of the median and divide the ordered data set into quarters. Just as the point for division into halves is called the median, the points for division into quarters are called quartiles. The points of division into more general fractions are called percentiles.

> **The sample 100 p-th percentile is a value such that after the data are ordered from smallest to largest, at least 100 p % of the observations are at or below this value and at least 100 (1 − p) % are at or above this value.**

If we take $p = .5$, the above conceptual description of the sample $100(.5) = $ 50th percentile specifies that at least half the observations are equal or smaller and at least half are equal or larger. If we take $p = .25$, the sample $100(.25) = 25$th percentile has proportion one-fourth of the observations that are the same or smaller and proportion three-fourths that are the same or larger.

We adopt the convention of taking an observed value for the sample percentile except when two adjacent values satisfy the definition, in which case their average is taken as the percentile. This coincides with the way the median is defined when the sample size is even. When all values in an interval satisfy the definition of a percentile, the particular convention used to locate a point in the interval does not appreciably alter the results in large data sets, except perhaps for the determination of extreme percentiles (those before the 5th or after the 95th percentile).

The following operating rule will simplify the calculation of the sample percentile.

Calculating the Sample 100p-th Percentile

1. Order the data from smallest to largest.
2. Determine the product (*sample size*) $\times$ (*proportion*) $= np$.

If np is not an integer, round it up to the next integer and find the corresponding ordered value.

If np is an integer, say k, calculate the average of the kth and $(k + 1)$st ordered values.

The quartiles are simply the 25th, 50th, and 75th percentiles.

Sample Quartiles

Lower (first) quartile	Q_1 = 25th percentile
Second quartile (or median)	Q_2 = 50th percentile
Upper (third) quartile	Q_3 = 75th percentile

Example 9 Calculating Quartiles to Summarize Length of Phone Calls

An administrator wanted to study the utilization of long-distance telephone service by a department. One variable of interest is the length, in minutes, of long-distance calls made during one month. There were 38 calls that resulted in a connection. The lengths of calls, already ordered from smallest to largest, are presented in Table 8. Locate the quartiles and also determine the 90th percentile.

Table 8   The Lengths of Long-Distance Phone Calls in Minutes

1.6	1.7	1.8	1.8	1.9	2.1	2.5	3.0	3.0	4.4
4.5	4.5	5.9	7.1	7.4	7.5	7.7	8.6	9.3	9.5
12.7	15.3	15.5	15.9	15.9	16.1	16.5	17.3	17.5	19.0
19.4	22.5	23.5	24.0	31.7	32.8	43.5	53.3		

SOLUTION   To determine the first quartile, we take $p = .25$ and calculate the product 38 $\times .25 = 9.5$. Because 9.5 is not an integer, we take the next largest integer, 10. In Table 8, we see that the 10th ordered observation is 4.4 so the first quartile is $Q_1 = 4.4$ minutes.

We confirm that this observation has 10 values *at or below* it and 29 values *at or above* so that it does satisfy the conceptual definition of the first quartile.

For the median, we take $p = .5$ and calculate $38 \times .5 = 19$. Because this is an integer, we average the 19th and 20th smallest observations to obtain the median, $(9.3 + 9.5)/2 = 9.4$ minutes.

Next, to determine the third quartile, we take $p = .75$ and calculate 38 $\times .75 = 28.5$. The next largest integer is 29, so the 29th ordered observation is the third quartile $Q_3 = 17.5$ minutes. More simply, we could mimic the calculation of the first quartile but now count down 10 observations starting with the largest value.

For the 90th percentile, we determine $38 \times .90 = 34.2$, which we increase to 35. The 90th percentile is 31.7 minutes. Only 10% of calls last 31.7 minutes or longer.

Exercises

2.28   Calculate the mean and median for each of the following data sets.
   (a) 2   6   3   10   4
   (b) 4   2   8   4   2

2.29   Calculate the mean and median for each of the following data sets.
   (a) 2   5   1   4   3
   (b) 26   30   38   32   26   31
   (c) −1   2   0   1   4   −1   2

2.30   The height that bread rises may be one indicator of how light it will be. As a first step, before modifying her existing recipe, a student cook measured the raise height (cm) on eight occasions:

   6.3   6.9   5.7   5.4   5.6   5.5   6.6   6.5

   Find the mean and median of the raised heights.

2.31   With reference to the water quality in Exercise 2.17:
   (a) Find the sample mean.
   (b) Does the sample mean or the median give a better indication of the water quality of a "typical" day? Why?

2.32   The monthly income in dollars for seven sales persons at a car dealership are

2450   2275   2425   4700   2650   2350   2475

(a)   Calculate the mean and median salary.
(b)   Which of the two is preferable as a measure of center and why?

2.33   Records show that in Las Vegas, NV, the normal daily maximum temperature (°F) for each month starting in January is

56   62   68   77   87   99   105   102   95   82   66   57

Verify that the mean of these figures is 79.67. Comment on the claim that the daily maximum temperature in Las Vegas averages a pleasant 79.67.

2.34   A major wine producer reported sales (in hundreds of cases) for two-week periods during one summer:

85   82   77   83   80   77   94

Obtain the sample mean and median.

2.35   With reference to the radiation leakage data given in Exercise 2.15:

(a)   Calculate the sample mean.
(b)   Which gives a better indication of the amount of radiation leakage, the sample mean or the median?

2.36   Recent crime reports on the number of aggravated assaults at each of the 27 largest universities reporting for the year are summarized in the computer output

Descriptive Statistics: AggAslt

```
Variable   N    Mean   Median   StDev
AggAslt   27   10.30   10.00    7.61
```

Locate two measures of center tendency, or location, and interpret the values.

2.37   The weights (oz) of nineteen babies born in Madison, Wisconsin, are summarized in the computer output

Descriptive Statistics: Weight

```
Variable   N    Mean   Median   StDev
Weight    19   118.05  117.00   15.47
```

Locate two measures of center tendency, or location, and interpret the values.

2.38   The following table gives the number of days in which selected metropolitan areas failed to meet air-quality standards in 1997 and 2002.

	2002	1997
Atlanta	24	26
Baltimore	42	30
Boston	1	8
Chicago	2	9
Dallas	15	15
Detroit	26	12
Houston	23	47
Los Angeles	80	63
Miami	1	3
New York	31	23
Philadelphia	33	32
Pittsburgh	53	20
Sacramento	69	2
St. Louis	34	15
San Diego	20	14
Washington, DC	34	28

(a)   Find the sample mean for 2002.
(b)   Comment on the effect of a large observation.

2.39   With reference to Exercise 2.14, find the sample mean peak-flow reading.

2.40   Old Faithful, the most famous geyser in Yellowstone Park, had the following durations (measured in seconds) in six consecutive eruptions:

240   248   113   268   117   253

(a)   Find the sample median.
(b)   Find the sample mean.

2.41   Loss of calcium is a serious problem for older women. To investigate the amount of loss, a researcher measured the initial amount of bone mineral content in the radius bone of the dominant hand of elderly women and then the amount remaining after one year. The differences, representing the loss of bone mineral content, are given in the following table (courtesy of E. Smith).

8	7	13	3	6
4	8	6	3	4
0	1	11	7	1
8	6	12	13	10
9	11	3	2	9
7	1	16	3	2
10	15	2	5	8
17	8	2	5	5

(a) Find the sample mean.

(b) Does the sample mean or the median give a better indication of the amount of mineral loss?

2.42 Physical education researchers interested in the development of the overarm throw measured the horizontal velocity of a thrown ball at the time of release. The results for first-grade children (in feet/second) (courtesy of L. Halverson and M. Roberton) are

Males

54.2	39.6	52.3	48.4	35.9	30.4	25.2	45.4	48.9	48.9
45.8	44.0	52.5	48.3	59.9	51.7	38.6	39.1	49.9	38.3

Females

30.3	43.0	25.7	26.7	27.3	31.9	53.7	32.9	19.4	23.7
23.3	23.3	37.8	39.5	33.5	30.4	28.5			

(a) Find the sample median for males.

(b) Find the sample median for females.

(c) Find the sample median for the combined set of males and females.

2.43 On opening day one season, 10 major league baseball games were played and they lasted the following numbers of minutes.

167 211 187 176 170 158 198 218 145 232

Find the sample median.

2.44 If you were to use the data on the length of major league baseball games in Exercise 2.43 to estimate the total amount of videotape needed to film another 10 major league baseball games, which is the more meaningful de-

scription, the sample mean or the sample median? Explain.

2.45 The following measurements of the diameters (in feet) of Indian mounds in southern Wisconsin were gathered by examining reports in the *Wisconsin Archeologist* (courtesy of J. Williams).

22 24 24 30 22 20 28 30 24 34 36 15 37

(a) Create a dot diagram.

(b) Calculate the mean and median and then mark these on the dot diagram.

(c) Calculate the quartiles.

2.46 With reference to Exercise 2.38, calculate the quartiles for 2002.

2.47 Refer to the data of college qualification test scores given in Exercise 2.22.

(a) Find the median.

(b) Find Q_1 and Q_3.

2.48 A large mail-order firm employs numerous persons to take phone orders. Computers on which orders are entered also automatically collect data on phone activity. One variable useful for planning staffing levels is the number of calls per shift handled by each employee. From the data collected on 25 workers, calls per shift were (courtesy of Land's End)

118	118	57	92	127	109	96	68	73
69	106	91	93	94	102	105	100	104
80	50	96	82	72	108	73		

Calculate the sample mean.

2.49 With reference to Exercise 2.48, calculate the quartiles.

2.50 The speedy lizard data, from Exercise 2.19, are

1.28	1.36	1.24	2.47	1.94	2.52	2.67	1.29
1.56	2.66	2.17	1.57	2.10	2.54	1.63	2.11
2.57	1.72	0.76	1.02	1.78	0.50	1.49	1.57
1.04	1.92	1.55	1.78	1.70	1.20		

(a) Find the sample median, first quartile, and third quartile.

(b) Find the sample 90th percentile.

2.51 With reference to the water quality data in Exercise 2.17:

(a) Find the sample median, first quartile, and third quartile.

(b) Find the sample 90th percentile.

2.52 *Some properties of the mean and median.*

  1. If a fixed number c is added to all measurements in a data set, then the mean of the new measurements is

  $c +$ (the original mean).

  2. If all measurements in a data set are multiplied by a fixed number d, then the mean of the new measurements is

  $d \times$ (the original mean).

  (a) Verify these properties for the data set

  4 8 8 7 9 6

  taking $c = 4$ in property (1) and $d = 2$ in (2).

  (b) The same properties also hold for the median. Verify these for the data set and the numbers c and d given in part (a).

2.53 On a day, the noon temperature measurements (in °F) reported by five weather stations in a state were

  74 80 76 76 73

  (a) Find the mean and median temperature in °F.

  (b) The Celsius (°C) scale is related to the Farenheit (°F) scale by $C = \frac{5}{9}(F - 32)$. What are the mean and median temperatures in °C? (Answer without converting each temperature measurement to °C. Use the properties stated in Exercise 2.52.)

2.54 Given here are the mean and median salaries of machinists employed by two competing companies A and B.

	Company	
	A	B
Mean salary	$60,000	$55,500
Median salary	$46,000	$49,000

Assume that the salaries are set in accordance with job competence and the overall quality of workers is about the same in the two companies.

(a) Which company offers a better prospect to a machinist having superior ability? Explain your answer.

(b) Where can a medium-quality machinist expect to earn more? Explain your answer.

2.55 Refer to the alligator data in Table D.11 of the Data Bank. Using the data on testosterone x_4 for male alligators:

(a) Make separate dot plots for the Lake Apopka and Lake Woodruff alligators.

(b) Calculate the sample means for each group.

(c) Do the concentrations of testosterone appear to differ between the two groups? What does this suggest the contamination has done to male alligators in the Lake Apopka habitat?

2.56 Refer to the alligator data in Table D.11 of the Data Bank. Using the data on testosterone x_4 from Lake Apopka:

(a) Make separate dot plots for the male and female alligators.

(b) Calculate the sample means for each group.

(c) Do the concentrations of testosterone appear to differ between the two groups? We would expect differences. What does your graph suggest the contamination has done to alligators in the Lake Apopka habitat?

5. MEASURES OF VARIATION

Besides locating the center of the data, any descriptive study of data must numerically measure the extent of variation around the center. Two data sets may exhibit similar positions of center but may be remarkably different with respect to variability. For example, the dots in Figure 10b are more scattered than those in Figure 10a.

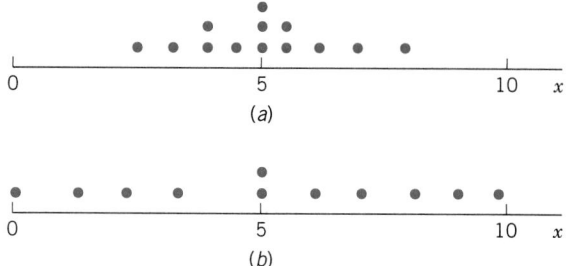

Figure 10   Dot diagrams with similar center values but different amounts of variation.

Because the sample mean $\bar{x}$ is a measure of center, the variation of the individual data points about this center is reflected in their deviation from the mean

$$\text{Deviation} = \text{Observation} - (\text{Sample mean})$$
$$= x - \bar{x}$$

For instance, the data set 3, 5, 7, 7, 8 has mean $\bar{x} = (3 + 5 + 7 + 7 + 8)/5 = 30/5 = 6$, so the deviations are calculated by subtracting 6 from each observation. See Table 9.

TABLE 9   Calculation of Deviations

Observation x	Deviation $x - \bar{x}$
3	−3
5	−1
7	1
7	1
8	2

One might feel that the average of the deviations would provide a numerical measure of spread. However, some deviations are positive and some negative, and the total of the positive deviations exactly cancels the total of the negative ones. In the foregoing example, we see that the positive deviations add to 4 and the negative ones add to −4, so the total deviation is 0. With a little reflection on the definition of the sample mean, the reader will realize that this was not just an accident. For any data set, the total deviation is 0 (for a formal proof of this fact, see Appendix A1).

$$\sum (\text{Deviations}) = \sum (x_i - \bar{x}) = 0$$

To obtain a measure of spread, we must eliminate the signs of the deviations before averaging. One way of removing the interference of signs is to square the numbers. A measure of spread, called the sample variance, is constructed by adding the squared deviations and dividing the total by the number of observations minus one.

Sample variance of n observations:

$$s^2 = \frac{\text{sum of squared deviations}}{n - 1}$$

$$= \frac{\sum\limits_{i=1}^{n} (x_i - \bar{x})^2}{n - 1}$$

Example 10   Calculating Sample Variance

Calculate the sample variance of the data 3 5 7 7 8.

SOLUTION   For this data set, $n = 5$. To find the variance, we first calculate the mean, then the deviations and the squared deviations. See Table 10.

TABLE 10   Calculation of Variance

Observation x	Deviation $x - \bar{x}$	(Deviation)^2 $(x - \bar{x})^2$
3	-3	9
5	-1	1
7	1	1
7	1	1
8	2	4
Total 30	0	16
$\sum x$	$\sum (x - \bar{x})$	$\sum (x - \bar{x})^2$

$$\bar{x} = \frac{30}{5} = 6$$

$$\text{Sample variance} \quad s^2 = \frac{16}{5 - 1} = 4$$

Remark: Although the sample variance is conceptualized as the **average squared deviation,** notice that the divisor is $n - 1$ rather than n. The divisor, $n - 1$, is called the degrees of freedom[1] associated with s^2.

Because the variance involves a sum of squares, its unit is the square of the unit in which the measurements are expressed. For example, if the data pertain to measurements of weight in pounds, the variance is expressed in (pounds)^2. To obtain a measure of variability in the same unit as the data, we take the positive square root of the variance, called the sample standard deviation. The standard deviation rather than the variance serves as a basic measure of variability.

Sample Standard Deviation

$$s = \sqrt{\text{Variance}} = \sqrt{\frac{\sum\limits_{i=1}^{n} (x_i - \bar{x})^2}{n - 1}}$$

Example 11   Calculating the Sample Standard Deviation

Calculate the standard deviation for the data of Example 10.

SOLUTION   We already calculated the variance $s^2 = 4$ so the standard deviation is $s = \sqrt{4} = 2$.

To show that a larger spread of the data does indeed result in a larger numerical value of the standard deviation, we consider another data set in Example 12.

Example 12   Using Standard Deviations to Compare Variation in Two Data Sets

Calculate the standard deviation for the data 1, 4, 5, 9, 11. Plot the dot diagram of this data set and also the data set of Example 10.

SOLUTION   The standard deviation is calculated in Table 11. The dot diagrams, given in Figure 11, show that the data points of Example 10 have less spread than those of Example 12. This visual comparison is confirmed by a smaller value of s for the first data set.

---

[1]The deviations add to 0 so a specification of any $n - 1$ deviations allows us to recover the one that is left out. For instance, the first four deviations in Example 10 add to -2, so to make the total 0, the last one must be $+2$, as it really is. In the definition of s^2, the divisor $n - 1$ represents the number of deviations that can be viewed as free quantities.

TABLE 11 Calculation of s

x	$(x - \bar{x})$	$(x - \bar{x})^2$
1	−5	25
4	−2	4
5	−1	1
9	3	9
11	5	25
Total 30	0	64

$$\bar{x} = 6 \qquad\qquad s^2 = \frac{64}{4} = 16$$

$$s = \sqrt{16} = 4$$

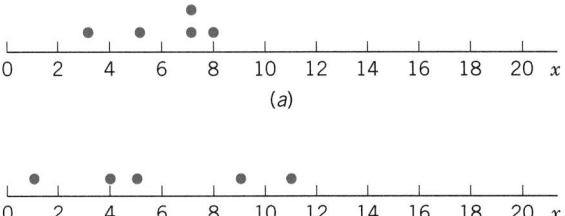

Figure 11   Dot diagrams of two data sets.

An alternative formula for the sample variance is

$$s^2 = \frac{1}{n - 1}\left[\sum x_i^2 - \frac{\left(\sum x_i\right)^2}{n}\right]$$

It does not require the calculation of the individual deviations. In hand calculation, the use of this alternative formula often reduces the arithmetic work, especially when $\bar{x}$ turns out to be a number with many decimal places. The equivalence of the two formulas is shown in Appendix A1.2.

Example 13   Calculating Sample Variance Using the Alternative Formula

In a psychological experiment a stimulating signal of fixed intensity was used on six experimental subjects. Their reaction times, recorded in seconds, were 4, 2, 3, 3, 6, 3. Calculate the standard deviation for the data by using the alternative formula.

SOLUTION   These calculations can be conveniently carried out in tabular form:

	x	x^2
	4	16
	2	4
	3	9
	3	9
	6	36
	3	9
Total	21	83
	$= \sum x$	$= \sum x^2$

$$s^2 = \frac{1}{n-1}\left[\sum x^2 - \frac{\left(\sum x\right)^2}{n}\right] = \frac{83 - (21)^2/6}{5} = \frac{83 - 73.5}{5}$$

$$= \frac{9.5}{5} = 1.9$$

$$s = \sqrt{1.9} = 1.38 \text{ seconds}$$

The reader may do the calculations with the first formula and verify that the same result is obtained.

In Example 12, we have seen that one data set with a visibly greater amount of variation yields a larger numerical value of s. The issue there surrounds a comparison between different data sets. In the context of a single data set, can we relate the numerical value of s to the physical closeness of the data points to the center $\bar{x}$? To this end, we view one standard deviation as a benchmark distance from the mean $\bar{x}$. For bell-shaped distributions, an empirical rule relates the standard deviation to the proportion of the data that lie in an interval around $\bar{x}$.

Empirical Guidelines for Symmetric Bell-Shaped Distributions

Approximately   68%   of the data lie within $\bar{x} \pm s$
95%   of the data lie within $\bar{x} \pm 2s$
99.7%   of the data lie within $\bar{x} \pm 3s$

Example 14   Comparing Bookstore Sales with the Empirical Guidelines

Examine the 40 bookstore sales receipts in Table 4 in the context of the empirical guideline.

SOLUTION   Using a computer (see, for instance, Exercise 2.126), we obtain

$$\bar{x} = \$306.744$$
$$s = \$143.289 \qquad 2s = 2(143.289) = \$286.578$$

Going two standard deviations either side of $\bar{x}$ results in the interval

$$\$306.744 - 286.578 = \$20.166 \qquad \text{to} \qquad \$593.32 = 306.744 + 286.578$$

By actual count, all the observations except $16.00 and $621.35 fall in this interval. We find that $38/40 = .95$, or 95% of the observations lie within two standard deviations of $\bar{x}$. This example is too good! Ordinarily, we would not find exactly 95% but something close to it.

Other Measures of Variation

Another measure of variation that is sometimes employed is

Sample range = Largest observation − Smallest observation

The range gives the length of the interval spanned by the observations.

Example 15 Calculating the Sample Range

Calculate the range for the length of phone call data given in Example 9.

SOLUTION   The data given in Table 8 contained

$$\text{Smallest observation} = 1.6$$
$$\text{Largest observation} = 53.3$$

Therefore, the length of the interval covered by these observations is

$$\text{Sample range} = 53.3 - 1.6 = 51.7 \text{ minutes}$$

As a measure of spread, the range has two attractive features: It is extremely simple to compute and interpret. However, it suffers from the serious disadvantage that it is much too sensitive to the existence of a very large or very small observation in the data set. Also, it ignores the information present in the scatter of the intermediate points.

To circumvent the problem of using a measure that may be thrown far off the mark by one or two wild or unusual observations, a compromise is made by measuring the interval between the first and third quartiles.

Sample interquartile range = Third quartile − First quartile

The sample interquartile range represents the length of the interval covered by the center half of the observations. This measure of the amount of variation is not disturbed if a small fraction of the observations are very large or very small. The sample interquartile range is usually quoted in government reports on income and other distributions that have long tails in one direction, in preference to standard deviation as the measure of spread.

Example 16   Calculating the Interquartile Range

Calculate the sample interquartile range for the length of long distance phone calls data given in Table 8.

SOLUTION   In Example 9, the quartiles were found to be $Q_1 = 4.4$ and $Q_3 = 17.5$. Therefore,

$$\text{Sample interquartile range} = Q_3 - Q_1$$
$$= 17.5 - 4.4$$
$$= 13.1 \text{ minutes}$$

Boxplots

A recently created graphic display, called a boxplot, highlights the summary information in the quartiles. Begin with the

Five-number summary: minimum, Q_1, Q_2, Q_3, maximum.

The center half of the data, from the first to the third quartile, is represented by a rectangle (box) with the median indicated by a bar. A line extends from Q_3 to the maximum value and another from Q_1 to the minimum. Figure 12 gives the boxplot for the length of phone calls data in Table 8. The long line to the right is a consequence of the largest value, 53.3 minutes, and, to some extent, the second largest value, 43.5 minutes.

Boxplots are particularly effective for displaying several samples alongside each other for the purpose of visual comparison.

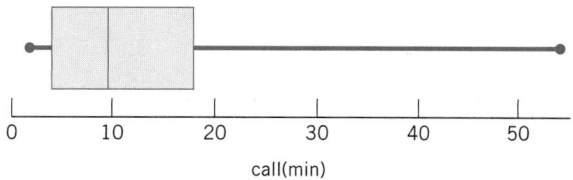

Figure 12   Boxplot of the length of phone call data in Table 8.

Figure 13 displays the amount of reflected light in the near-infrared band as recorded by satellite when flying over forest areas and urban areas, respectively. Because high readings tend to correspond to forest and low readings to urban areas, the readings have proven useful in classifying unknown areas.

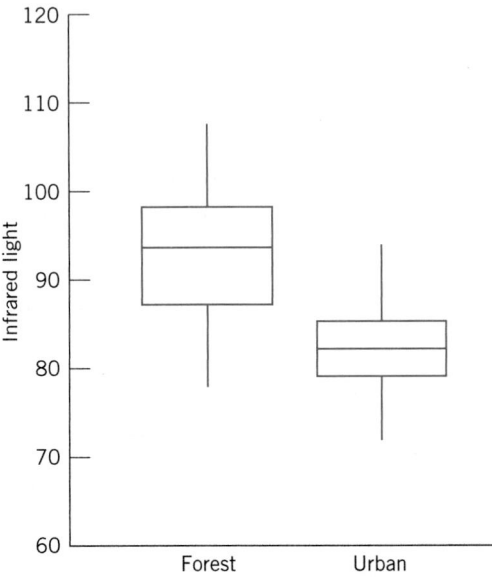

Figure 13   Boxplots of near-infrared light reflected from forest and urban areas.

Exercises

2.57   For the data set

$$8 \quad 3 \quad 4$$

(a) Calculate the deviations $(x - \bar{x})$ and check to see that they add up to 0.

(b) Calculate the sample variance and the standard deviation.

2.58   Repeat (a) and (b) of Exercise 2.57 for the data set

$$4 \quad 9 \quad 2$$

2.59   For the data set   8   6   14   4:

(a) Calculate the deviations $(x - \bar{x})$ and check to see that they add up to 0.

(b) Calculate the variance and the standard deviation.

2.60   Repeat (a) and (b) of Exercise 2.59 for the data set

$$2.5 \quad 1.7 \quad 2.1 \quad 1.5 \quad 1.7$$

2.61   For the data of Exercise 2.57, calculate s^2 by using the alternative formula.

2.62   For the data of Exercise 2.59, calculate s^2 by using the alternative formula.

2.63   For each data set, calculate s^2.

(a) 1   4   3   2   2

(b) −2   1   −1   −3   0   −2

(c) 9   8   8   9   8   8   9

2.64   The monthly rents for 7 one-bedroom apartments located in one area of the city, are

$$625 \quad 740 \quad 805 \quad 670 \quad 705 \quad 740 \quad 870$$

(a) Give two possible factors that may contribute to variation in the monthly rents.

Calculate

(b) The sample variance.

(c) The sample standard deviation.

2.65 Find the standard deviation of the measurements of diameters given in Exercise 2.45.

2.66 Five students reported that the previous night they slept

$$7 \quad 8 \quad 7 \quad 10 \quad 9$$

hours. Calculate:

(a) The sample variance.

(b) The sample standard deviation.

2.67 The city of Madison regularly checks the quality of water at swimming beaches located on area lakes. Fifteen times the concentration of fecal coliforms, in number of colony forming units (CFU) per 100 ml of water, was measured during the summer at one beach.

180 1600 90 140 50 260 400 90
380 110 10 60 20 340 80

(a) Calculate the sample variance.

(b) Calculate the sample standard deviation.

(c) One day, the water quality was bad—the reading was 1600 CFU—and the beach was closed. Drop this value and calculate the sample standard deviation for the days where the water quality was suitable for swimming. Comment on the change.

2.68 With reference to the radiation leakage data given in Exercise 2.15, calculate:

(a) The sample variance.

(b) The sample standard deviation.

2.69 With reference to the data on the length of 10 major league baseball games in Exercise 2.43:

(a) Find the sample mean.

(b) Find the sample variance.

(c) Find the sample standard deviation.

2.70 With reference to Exercise 2.13,

(a) Find the sample mean.

(b) Find the sample standard deviation.

2.71 A sample of seven compact discs at the music store stated the performance times as lasting the following numbers of minutes for Beethoven's Ninth Symphony.

66.9 66.2 71.0 68.6 65.4 68.4 71.9

(a) Find the sample median.

(b) Find the sample mean.

(c) Find the sample standard deviation.

2.72 Recent crime reports on the number of aggravated assaults at each of the 27 largest universities reporting for the year are summarized in the computer output.

Descriptive Statistics: AggAslt

Variable	N	Mean	Median	StDev
AggAslt	27	10.30	10.00	7.61

Variable	Minimum	Maximum	Q1	Q3
AggAslt	0.00	29.00	5.00	14.00

(a) Locate a measure of variation and also calculate the sample variance.

(b) Calculate the interquartile range and interpret this value.

(c) Give a value for a standard deviation that would correspond to greater variation in the numbers of aggravated assaults.

2.73 The weights (oz) of nineteen babies born in Madison, Wisconsin, are summarized in the computer output.

Descriptive Statistics: Weight

Variable	N	Mean	Median	StDev
Weight	19	118.05	117.00	15.47

Variable	Minimum	Maximum	Q1	Q3
Weight	89.00	144.00	106.00	131.00

(a) Locate a measure of variation and also calculate the sample variance.

(b) Calculate the interquartile range and interpret this value.

(c) Give a value for a standard deviation that would correspond to variation in the weights.

2.74 *Some properties of the standard deviation.*

1. If a fixed number c is added to all measurements in a data set, the deviations $(x - \bar{x})$ remain unchanged (see Exercise 2.52). Consequently, s^2 and s remain unchanged.

2. If all measurements in a data set are multiplied by a fixed number d, the deviations $(x - \bar{x})$ get multiplied by d. Consequently, s^2 gets multiplied by d^2, and s by $|d|$. (*Note:* The standard deviation is never negative.)

Verify these properties for the data set

$$5 \quad 9 \quad 9 \quad 8 \quad 10 \quad 7$$

taking $c = 4$ in property (1) and $d = 2$ in (2).

2.75 For the data set of Exercise 2.22, calculate the interquartile range.

2.76 For the data set of Exercise 2.38, calculate the interquartile range for 2002.

2.77 Should you be surprised if the range is larger than twice the interquartile range? Explain.

2.78 Calculations with the test scores data of Exercise 2.22 give $\bar{x} = 150.125$ and $s = 24.677$.

(a) Find the proportion of the observations in the intervals $\bar{x} \pm 2s$ and $\bar{x} \pm 3s$.

(b) Compare your findings in part (a) with those suggested by the empirical guidelines for bell-shaped distributions.

2.79 Refer to the data on bone mineral content in Exercise 2.41.

(a) Calculate $\bar{x}$ and s.

(b) Find the proportion of the observations that are in the intervals $\bar{x} \pm s$, $\bar{x} \pm 2s$, and $\bar{x} \pm 3s$.

(c) Compare the results of part (b) with the empirical guidelines.

2.80 Refer to the data on lizards in Exercise 2.19.

(a) Calculate $\bar{x}$ and s.

(b) Find the proportion of the observations that are in the intervals $\bar{x} \pm s$, $\bar{x} \pm 2s$, and $\bar{x} \pm 3s$.

(c) Compare the results of part (b) with the empirical guidelines.

2.81 Refer to the data on iron content in Exercise 2.23.

(a) Calculate $\bar{x}$ and s.

(b) Find the proportions of the observations that are in the intervals $\bar{x} \pm s$, $\bar{x} \pm 2s$, and $\bar{x} \pm 3s$.

(c) Compare the results of part (b) with the empirical guidelines.

2.82 *Sample z score.* The *z scale* (or *standard scale*) measures the position of a data point relative to the mean and in units of the standard deviation. Specifically,

$$z \text{ value of a measurement} = \frac{\text{Measurement} - \bar{x}}{s}$$

When two measurements originate from different sources, converting them to the z scale helps to draw a sensible interpretation of their relative magnitudes. For instance, suppose a student scored 65 in a math course and 72 in a history course. These (raw) scores tell little about the student's performance. If the class averages and standard deviations were $\bar{x} = 60$, $s = 20$ in math and $\bar{x} = 78$, $s = 10$ in history, this student's

$$z \text{ score in math} = \frac{65 - 60}{20} = .25$$

$$z \text{ score in history} = \frac{72 - 78}{10} = -.60$$

Thus, the student was .25 standard deviations above the average in math and .6 standard deviations below the average in history.

(a) If $\bar{x} = 490$ and $s = 120$, find the z scores of 350 and 620.

(b) For a z score of 2.4, what is the raw score if $\bar{x} = 210$ and $s = 50$?

2.83 The weights (oz) of nineteen babies born in Madison, Wisconsin, are summarized in the computer output.

Descriptive Statistics: Weight

Variable	N	Mean	Median	StDev
Weight	19	118.05	117.00	15.47

Referring to Exercise 2.82 obtain the z score for a baby weighing

(a) 102 oz

(b) 144 oz

2.84 Two cities provided the following information on public school teachers' salaries.

Minimum	Q_1	Median	Q_3	Maximum
City A 28,400	34,000	38,300	40,400	46,300
City B 29,600	36,500	41,200	45,700	51,800

(a) Construct a boxplot for the salaries in City A.

(b) Construct a boxplot, on the same graph, for the salaries in City B.

(c) Are there larger differences at the lower or the higher salary levels? Explain.

2.85 Refer to the data on throwing speed in Exercise 2.42. Make separate boxplots to compare males and females.

2.86 Refer to Exercise 2.27 and the data on the consumer price index for various cities. Find the increase, for each city, by subtracting the 1992 value from the 2001 value.

(a) Obtain the five-number summary: minimum, Q_1, Q_2, Q_3, and maximum. Which city had the largest increase? Were there any decreases?

(b) Make a boxplot of the increases.

2.87 Refer to Exercise 2.27 and the data on the consumer price index for various cities. Find the increase, for each city, by subtracting the 1992 value from the 2001 value.

(a) Find the sample mean and standard deviation of these differences.

(b) What proportion of the increases lie between $\bar{x} \pm 2s$?

2.88 Refer to Exercise 2.38 and the data on the number of days that a metropolitan area failed to meet the air-quality standard. Find the decrease, for each area, by subtracting the 2002 value from the 1997 value.

(a) Obtain the five-number summary: minimum, Q_1, Q_2, Q_3, and maximum. Which metropolitan area had the largest decrease? Were there any increases?

(b) Make a boxplot of the decreases.

2.89 Refer to Exercise 2.38 and the data on the number of days that a metropolitan area failed to meet the air-quality standard. Find the decrease, for each area, by subtracting the 2002 value from the 1997 value.

(a) Find the sample mean and standard deviation of these differences.

(b) Comment on how a large observation can influence the value of the mean.

2.90 Presidents also take midterms! After two years of the President's term, members of Congress are up for election. The following table gives the number of net seats lost, by the party of the President, in the House of Representatives since the end of World War II.

Net House Seats Lost in Midterm Elections

1950	Truman (D)	55
1954	Eisenhower (R)	16
1962	Kennedy (D)	4
1966	Johnson (D)	47
1970	Nixon (R)	12
1974	Nixon/Ford (R)	43
1978	Carter (D)	11
1982	Reagan (R)	26
1986	Reagan (R)	5
1990	Bush (R)	8
1994	Clinton (D)	52
1998	Clinton (D)	−5 (gain)

For the data on the number of House seats lost:

(a) Calculate the sample mean.

(b) Calculate the standard deviation.

(c) Make a dot plot.

(d) What is one striking feature of these data that could be useful in predicting future midterm election results? (*Hint:* Would you have expected more elections to result in net gains?)

2.91 With reference to Exercise 2.90:

(a) Calculate the median number of lost House seats.

(b) Find the maximum and minimum losses and identify these with a President.

(c) Determine the range for the number of House seats lost.

6.   CHECKING THE STABILITY OF THE OBSERVATIONS OVER TIME

The calculations for the sample mean and sample variance treat all the observations alike. The presumption is that there are no apparent trends in data over time and there are no unusual observations. Another way of saying this is that the process producing the observations is in **statistical control.** The concept of statistical control allows for variability in the observations but requires that the pattern of variability be the same over time. Variability should not increase or decrease with time and the center of the pattern should not change.

To check on the stability of the observations over time, observations should be plotted versus time, or at least the order in which they were taken. The resulting plot is called a **time plot** or sometimes a **time series plot.**

Example 17   A Time Plot of Overtime Hours

The Madison Police Department charts several important variables, one of which is the number of overtime hours due to extraordinary events. These events would include murders, major robberies, and so forth. Although any one event is not very predictable, there is some constancy when data are grouped into six-month periods.

The values of overtime hours for extraordinary events for eight recent years, beginning with 2200, 875, . . ., through 1223, are

$$\begin{array}{cccccccc} 2200 & 875 & 957 & 1758 & 868 & 398 & 1603 & 523 \\ 2034 & 1136 & 5326 & 1658 & 1945 & 344 & 807 & 1223 \end{array}$$

Is the extraordinary event overtime hours process in control? Construct a time plot and comment.

SOLUTION   The time plot is shown in Figure 14. There does not appear to be any trend, but there is one large value of 5326 hours.

Example 18   A Time Plot of the Yen/Dollar Exchange Rate

The exchange rate between the United States and Japan can be stated as the number of yen that can be purchased with $1. Although this rate changes daily, we quote the official value for the year:

Year	1985	1986	1987	1988	1989	1990	1991	1992	1993	1994
Exchange rate	238.5	168.4	144.6	128.2	138.1	145.0	134.6	126.8	111.1	102.2

	1995	1996	1997	1998	1999	2000	2001	2002	2003	
	94.0	108.8	121.1	130.9	113.7	107.8	121.6	125.1	116.8	

Is this exchange rate in statistical control? Make a time plot and comment.

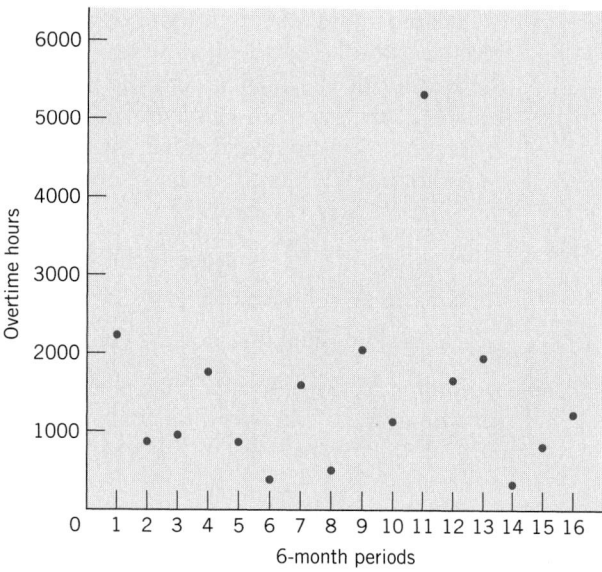

Figure 14   Time plot of extraordinary event hours versus time order.

SOLUTION   The time plot is shown in Figure 15. There is a rather strong downhill trend over most of the time period so the exchange rate is definitely not in statistical control. A dollar has purchased fewer and fewer yen over the years. It is the downward trend that is the primary feature and a cause for serious concern with regard to trade deficits. The upturn or leveling off in the last few years suggests a change in trend.

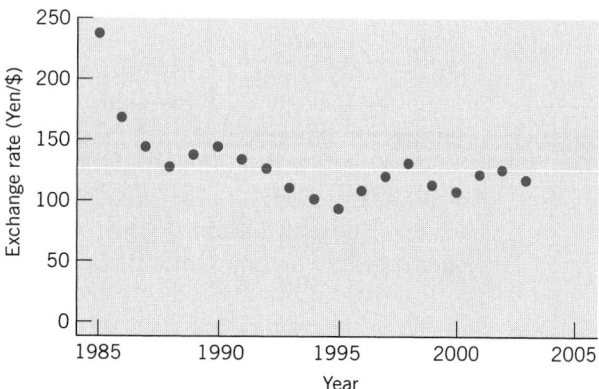

Figure 15   Time plot of the exchange rate.

Manufacturers need to monitor critical dimensions, temperatures, and other variables so that, although they vary, the variation is kept small enough so that the quality of the final product is maintained. A graphical approach, called a **control chart,** is recommended for this purpose because it allows for the visual inspection of outliers and trends. It adds a **centerline** and **control limits** to the time plot to help identify unusual observations.

To construct a control chart:

1. Plot the observations versus time order.

2. Add a solid centerline at the level of the sample mean $\bar{x}$.

3. Add dashed lines for the control limits at $\bar{x} - 2s$ and $\bar{x} + 2s$.

According to the empirical rule, if the process is in statistical control so the observations are stable over time, only about 5% of the observations will fall outside of the control limits. In some very critical applications, manufacturers sometimes use $\bar{x} - 3s$ and $\bar{x} + 3s$ as the control limits.

The upper and lower control limits on the control charts help to identify unusually low or unusually high observations. It allows us to distinguish between typical variation and variation that is especially large and could be due to special or assignable causes. Any time an observation falls outside of the control limits, an effort should be made to search for the reason.

| **Example 19** | A Control Chart for Overtime Hours |

Manufacturing processes are not the only ones that can benefit from control charting. Refer to the data in Example 17 on the number of overtime hours for police due to extraordinary events. Is the extraordinary event overtime hours process in control? Construct a control chart and comment.

SOLUTION   A computer calculation gives $\bar{x} = 1478$ and $s = 1183$ so the centerline is drawn at the sample mean 1478 and the upper control limit is $\bar{x} + 2s = 1478 + 2 \times 1183 = 3844$. The lower control limit is negative; we replace it by 0. Figure 16 gives the resulting control chart for extraordinary event overtime hours.

There is no discernible trend, but one point does exceed the upper control limit. By checking more detailed records, it was learned that the point outside of the control limits occurred when protests took place on campus in response to the bombing of a foreign capital. These events required city police to serve 1773 extraordinary overtime hours in just one 2-week period and 683 in the next period. That is, there was really one exceptional event, or special cause, that could be identified with the unusual point.

The one large value, 5326 hours, not only affects the centerline by inflating the mean, but it also increases the variance and that raises the upper control limit. In Exercise 2.98, you are asked to remove this outlier and redo the control chart.

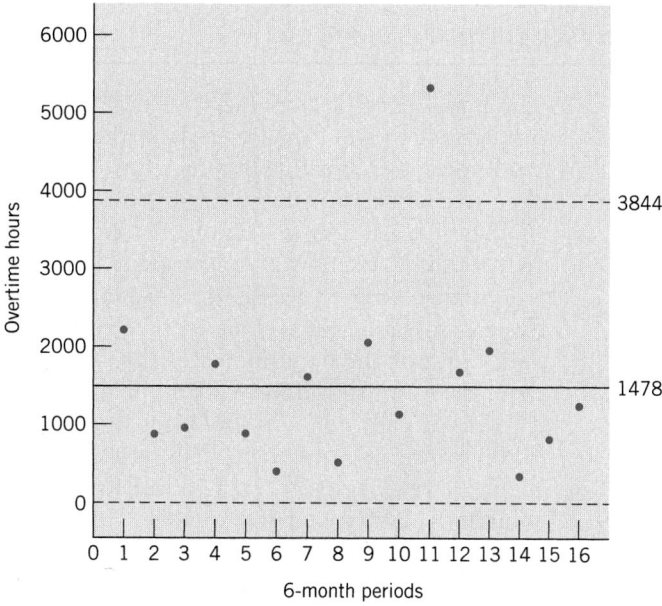

Figure 16   Control chart for extraordinary event overtime hours.

Exercises

2.92 Make a time plot of the phone call data in Exercise 2.48 and comment on the statistical control.

2.93 A city department has introduced a quality improvement program and has allowed employees to get credit for overtime hours when attending meetings of their quality groups. The total number of overtime meeting hours for each of the 26 pay periods in one year by row were

30	215	162	97	194	163	60	41	100
43	96	69	80	42	162	75	95	65
57	131	54	114	64	114	38	140

Make a time plot of the overtime meeting hours data.

2.94 Make a control chart for the data referred to in Exercise 2.92 and comment.

2.95 Make a control chart for the data in Exercise 2.93 and comment.

2.96 The exchange rate between the United States and Canada can be stated as the number of Canadian dollars that can be purchased with $1. The official values for the year are

Year	1992	1993	1994	1995	1996	1997
Exchange rate	1.21	2.29	1.37	1.37	1.36	1.38

	1998	1999	2000	2001	2002	2003
	1.48	1.49	1.48	1.55	1.57	1.41

Is this exchange rate in statistical control? Make a time plot and comment.

2.97 Make a control chart of the data in Exercise 2.96 and comment.

2.98 Make a control chart for the extraordinary event overtime data in Example 17 after removing the outlier identified in that example. You need to recalculate the mean and standard deviation.

7.   MORE ON GRAPHICS

The importance of graphing your data cannot be overemphasized. If a feature you expect to see is not present in the plots, statistical analyses will be of no avail. Moreover, creative graphics can often highlight features in the data and even give new insights.

The devastation of Napoleon's Grand Army during his ill-fated attempt to capture Russia was vividly depicted by Charles Minard. The 422,000 troops that entered Russia near Kaunas are shown as a wide (shaded) river flowing toward Moscow and the retreating army as a small (black) stream. The width of the band indicates the size of the army at each location on the map. Even the simplified version of the original graphic, appearing in Figure 17, dramatically conveys the losses that reduced the army of 422,000 men to 10,000 returning members. The temperature scale at the bottom, pertaining to the retreat, helps to explain the loss of life, including the incident where thousands died trying to cross the Berezina River in subzero temperatures. (For a copy of Minard's more detailed map and additional discussion, see E. R. Tufte, *The Visual Display of Quantitative Information*, Cheshire, CT: Graphics Press, 1983.)

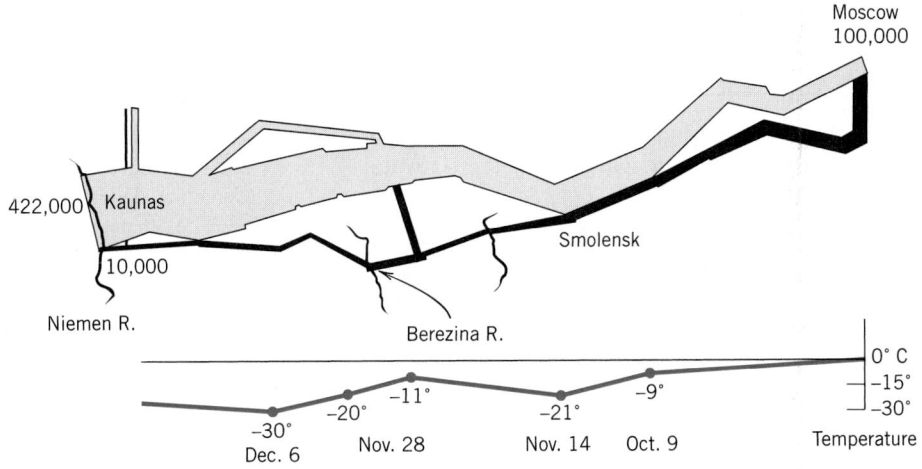

Figure 17   The demise of Napoleon's army in Russia, 1812–1813, based on Charles Minard.

Another informative graphic, made possible with modern software, is the display of nitrogen oxides by states in Figure 18. This figure illustrates the trend in emissions following the implementation of phase I of the Acid Rain Program.

1990 1999

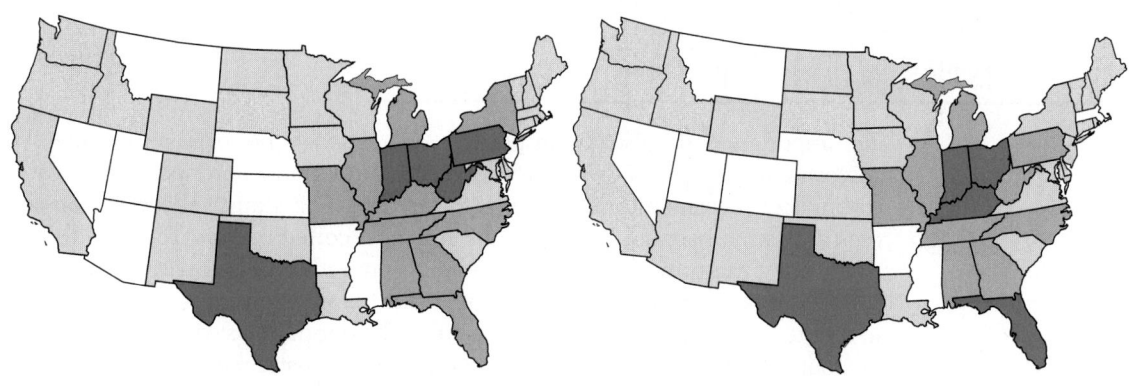

Total Utility NOx Emissions (thousand tons)

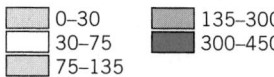

0–30   135–300
30–75   300–450
75–135

Figure 18   Total state-level nitrogen oxides, Environmental Protection Agency.

Graphs can give a vivid overall picture. © Datamation

8. STATISTICS IN CONTEXT

The importance of visually inspecting data cannot be overemphasized. We present a mini-case study[1] that shows the importance of first appropriately plotting and then monitoring manufacturing data. This statistical application concerns a ceramic part used in a popular brand of coffee makers. To make this ceramic part, a mixture of clay, water, and oil is poured into the cavity between two dies of a pressing machine. After pressing but before the part is dried to a hardened state, critical dimensions are measured. The depth of a slot is of interest here.

Sources of variation in the slot depth abound: the natural uncontrolled variation in the clay–water–oil mixture, the condition of the press, differences in operators, and so on. Some variation in the depth of the slot is inevitable. Even so, for the part to fit when assembled, the slot depth needs to be controlled within certain limits.

Every half hour during the first shift, slot depth is measured on three ceramic parts selected from production. Table 12 gives the data obtained on a Friday. The sample mean for the first sample of 218, 217, and 219 (thousandths of inch) is $(218 + 217 + 219)/3 = 654/3 = 218$, and so on.

TABLE 12   Slot Depth (thousandths of an inch)

Time	7:00	7:30	8:00	8:30	9:00	9:30	10:00	
1	218	218	216	217	218	218	219	
2	217	218	218	220	219	217	219	
3	219	217	219	221	216	217	218	
SUM	654	653	653	658	653	652	656	
$\bar{x}$	218.0	217.7	217.7	219.3	217.7	217.3	218.7	

Time	10:30	11:00	11:30	12:30	1:00	1:30	2:00	2:30
1	216	216	218	219	217	219	217	215
2	219	218	219	220	220	219	220	215
3	218	217	220	221	216	220	218	214
SUM	653	651	657	660	653	658	655	644
$\bar{x}$	217.7	217.0	219.0	220.0	217.7	219.3	218.3	214.7

An x-bar chart will indicate when changes have occurred and there is a need for corrective actions. Because there are 3 slot measurements at each time, it is the 15 sample means that are plotted versus time order. We will take the centerline to be the mean of the 15 sample means, or

$$\text{Centerline: } \bar{\bar{x}} = \frac{218.0 + \cdots + 214.7}{15} = 218.0$$

[1]Courtesy of Don Ermer.

When each plotted mean is based on several observations, the variance can be estimated by combining the variances from each sample. The first sample has variance $s_1^2 = [(218 - 218)^2 + (217 - 218)^2 + (219 - 218)^2]/(3 - 1) = 1.000$ and so on. The details are not important, but a computer calculation of the variance used to set control limits first determines the average of the 15 individual sample variances,

$$\frac{1.000 + 0.333 + \cdots + 0.333}{15} = 1.58$$

and, for reasons given in Chapter 7, divides by 3 to give the variance of a single sample mean. That is, $1.58/3 = .527$ is the appropriate s^2. The control limits are set at three times the estimated standard deviation $s = \sqrt{.527} = .726$, or $3 \times .726 = 2.2$ units from the centerline.

Lower control limit: LCL $= 218.0 - 2.2 = 215.8$
Upper control limit: UCL $= 218.0 + 2.2 = 220.2$

The x-bar chart is shown in Figure 19. What does the chart tell us?

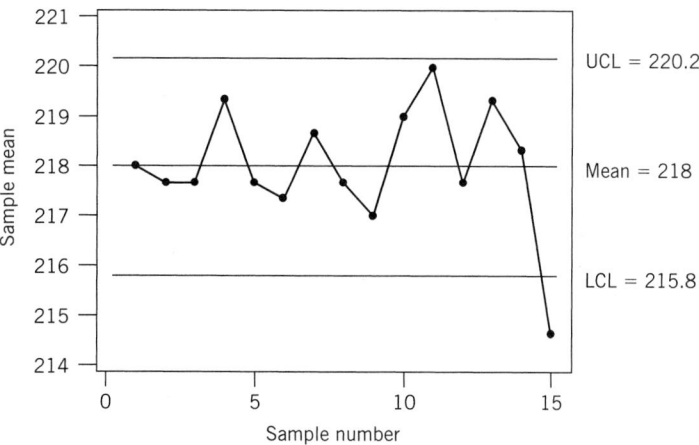

Figure 19   X-bar chart for depth.

The x-bar chart shows that the process was stable throughout the day and no points were out of control except the last sample. It was then that an unfortunate oversight occurred. Because it was near the end of her shift and the start of the weekend, the operator did not report the out-of-control value to either the setup person or the foreman. She knew the setup person was already cleaning up for the end of the shift and that the foreman was likely thinking about going across the street to the Legion Bar for some refreshments as soon as the shift ended. The operator did not want to ruin anyone's weekend plans so she kept quiet.

When the pressing machine was started up on Monday morning, one of the dies broke. The cost of the die was over a thousand dollars. But, when a customer was called and told there would be a delay in delivering the ceramic parts, he canceled the order. Certainly the loss of a customer is an even more expensive item.

Later, it was concluded that the clay had likely dried and stuck to the die leading to the break. A problem was predicted by the chart on Friday. Although the chart correctly indicated a problem at that time, someone had to act for the monitoring procedure to work.

USING STATISTICS WISELY

1. As a first step, always graph the data as a dot diagram or histogram to assess the overall pattern of data.

2. When comparing histograms based on different class intervals, be sure to create histograms whose height is relative frequency divided by width of interval.

3. Calculate summary statistics to describe the data set. Always determine the sample mean and standard deviation. The five-number summary

   minimum   first quartile   median   third quartile   maximum

   provides an additional summary when the sample sizes are moderately large. It helps describe cases where the dot diagram or histogram has a single long tail.

4. Use the median to describe the center when a small sample contains an extreme observation. The sample median is not influenced by a few very large or very small observations that may even be incorrectly recorded.

5. Do not routinely calculate summary statistics without identifying unusual observations (outliers) which may have undue influence on the value of a summary statistic.

KEY IDEAS AND FORMULAS

Qualitative data refer to frequency counts in categories. These are summarized by calculating the

$$\text{Relative frequency} \; = \; \frac{\text{Frequency}}{\text{Total number of observations}}$$

for the individual categories.

The term **numerical-valued variable** or just **variable** refers to a characteristic that varies over units and is measured on a numerical scale. **Discrete variables** are usually counts and all discrete variables have gaps in their scale of measurement. Continuous variables, like height or weight, can conceptually take any value in an interval. Data resulting from measurements of a variable are either discrete or continuous data.

For a discrete data set, the frequency is the count of the number of observations having a distinct value. The relative frequency is the proportion of sample units having this property.

$$\text{Relative frequency} = \frac{\text{Frequency}}{\text{Total number of observations}}$$

The discrete data set is summarized by a frequency distribution that lists the distinct data points and their corresponding relative frequencies. Either a line diagram or a histogram can be used for a graphical display.

Continuous measurement data should be graphed as a dot diagram when the data set is small, say, fewer than 20 or 25 observations. Larger data sets are first summarized in a frequency table. This is constructed by grouping the observations in class intervals, preferably of equal lengths. The class intervals are non-overlapping and cover the range of the data set from smallest to largest. We recommend specifying an endpoint convention that tells which of the class boundaries, or endpoints of the class intervals, to include and which to exclude from each class interval. A list of the class intervals along with the corresponding relative frequencies provides a frequency distribution which can graphically be displayed as a histogram. The histogram is constructed to have total area 1, equal to total relative frequency. That is, for each class interval, we draw a rectangle whose area represents the relative frequency of the class interval.

A stem-and-leaf display is another effective means of display when the data set is not too large. It is more informative than a histogram because it retains the individual observations in each class interval instead of lumping them into a frequency count. Two variants are the double-stem display and five-stem display.

Pareto diagrams display events according to their frequency in order to highlight the most important few that occur most of the time.

A summary of measurement data (discrete or continuous) should also include numerical measures of center and spread.

Two important measures of center are

$$\text{Sample mean} \qquad \bar{x} = \frac{\sum x}{n}$$

$$\text{Sample median} = \text{ middle most value of the ordered data set}$$

The quartiles and, more generally, percentiles are other useful locators of the distribution of a data set. The second quartile is the same as the median. The sample quartiles divide the ordered data into nearly four equal parts. The $100p$-th percentile has least proportion p at or below and proportion $1 - p$ at or above.

The amount of variation, or spread, of a data set is measured by the sample standard deviation s. The sample variance s^2 is given by

$$s^2 = \frac{\sum (x - \bar{x})^2}{n - 1}$$

Also, $s^2 = \dfrac{1}{n-1}\left[\sum x^2 - \dfrac{\left(\sum x\right)^2}{n}\right]$ (convenient for hand calculation)

Sample standard deviation $s = +\sqrt{s^2}$

The standard deviation indicates the amount of spread of the data points around the mean $\bar{x}$. If the histogram appears symmetric and bell-shaped, then the interval

$\bar{x} \pm s$ includes approximately 68% of the data

$\bar{x} \pm 2s$ includes approximately 95% of the data

$\bar{x} \pm 3s$ includes approximately 99.7% of the data

Two other measures of variation are

Sample range $=$ Largest observation $-$ Smallest observation

and

Sample interquartile range $=$ Third quartile $-$ First quartile

The five-number summary, namely, the median, the first and third quartiles, the smallest observation, and the largest observation, together serve as useful indicators of the distribution of a data set. These are displayed in a boxplot.

TECHNOLOGY

Creating graphs and computing statistical summaries have become considerably easier because of recent developments in software. In all our professional applications of statistics, we begin by entering the data in a worksheet. We then read from the worksheet while another person checks against the original. This procedure has eliminated many errors in data entry and allowed us to proceed knowing that the computer software is using the correct data.

In the technology sections of this text, we give the essential details for using MINITAB, EXCEL, and the TI-84/-83 Plus graphing calculator. The first two use the worksheet format.

MINITAB

The MINITAB screen is split with the bottom part being the worksheet. In the example here, we have typed sales in the top entry and then in row 1 of the first column we have typed the smallest number in the book sales in Table 4. The rest of the dollar sales are typed in the other cells in column 1.

Alternatively, the data sets in the book are stored as MINITAB worksheets on the accompanying CD. For instance, the book sales are in C2T4.MTW, indicating Table 4 of Chapter 2. To open this worksheet:

Under **File** choose **Open Worksheet** and arrow up to the drive that contains the data CD.
Click on the MINITAB folder and then click on the file name C2T4. Click **OK.**

This will activate the worksheet in MINITAB and you do not have to manually enter the numbers.

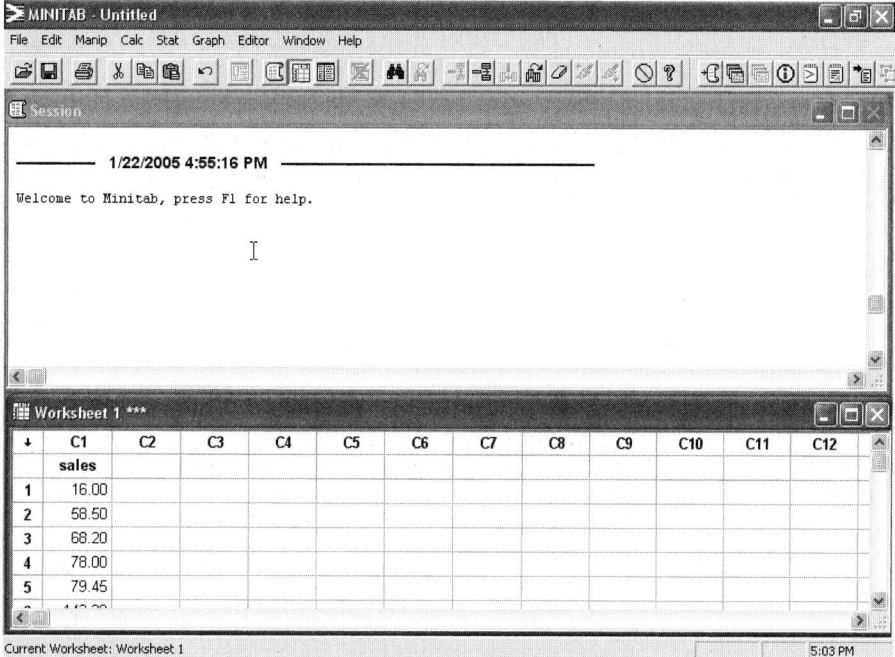

The summary statistics can be obtained by pulling down the menu in the top bar under **Stat,** then choosing **Basic Statistics** and then **Graphic Summary.** More specifically,

Data in C1

Dialog box:
Stat > Basic Statistics > Graphical Summary.
Type *C1* in **Variables.** Click **OK.**

MINITAB uses a slightly different definition of the quartiles and their values may slightly differ from those calculated by the method in this book.

MINITAB will also create a histogram with the choices

Data in *C1*

Dialog box:
Graph > Histogram.
Select *Simple* in **Variables.** Click **OK.**
Type *C1* in **X.** Click **OK.**

MINITAB will also create boxplots, dot plots, and stem-and-leaf displays. With the data in *C1*,

Dialog box:

Graph > Dotplot.
Select *Simple.* Click **OK.**
Type *C1* in **Graph variables.** Click **OK.**

produces a dot plot. To obtain a boxplot, replace the first step by

Graph > Boxplot.

and to obtain a stem-and-leaf display, replace the first two steps by the single step

Graph > Stem-and-Leaf.

Clicking on **Labels** before the last **OK** will allow you to put titles on your graph.

EXCEL

Begin with the data in column A. For the book sales from Table 4, the spread sheet is given on page 73.

Alternatively, the data sets in the book are stored in EXCEL workbooks on the accompanying CD. For instance, the book sales are in C2T4.xls, indicating Table 4 of Chapter 2. Go to the drive containing the data CD and click on the file name C2T4. The workbook having the data from Table 4 will then open.

Most of the statistical procedures we will use start with

Select **Tools** and then **Data Analysis.**
If **Data Analysis** is not listed, then it must be added once. To do so, select **Tools** then **Add-Ins.** Check **Analysis Toolpak** and click **OK.**

To obtain the summary statistics,

Select **Tools,** then **Data Analysis,** and then **Descriptive Statistics.**
Click **OK.** Place cursor in the **Input Range** window and use the mouse to highlight the data in column A. Check **Summary Statistics** and click **OK.**

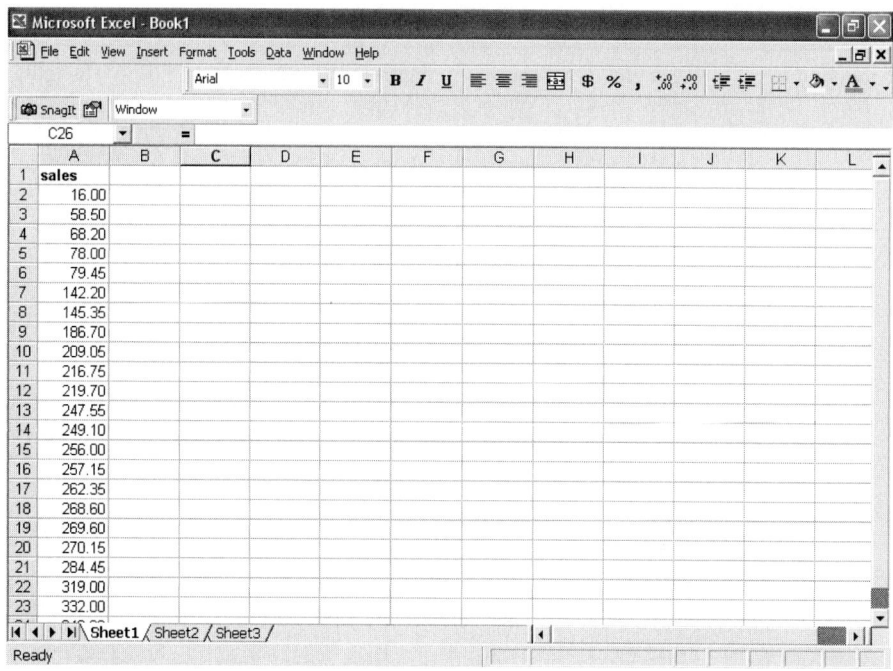

TI-84/83 PLUS

Press **STAT,** select **1:Edit,** and then enter the data in **L**₁.

Press **STAT,** highlight **Calc,** and select **1:1–Var Stats.**

With **1:1–Var Stats** in the Home screen, press **2nd 1:1–Var Stats 1** to insert **L**₁ on the Home screen.

Then press **ENTER.**

9. REVIEW EXERCISES

2.99 Recorded here are the numbers of civilians employed in the United States by major occupation

	Number of Workers in Millions	
	1980	2001
Goods producing	25.7	25.1
Service (private)	48.5	86.2
Government	16.2	20.9
Total	90.4	132.2

groups for the years 1980 and 2001. (*Source: Statistical Abstract of the United States, 2004.*)

(a) For each year, calculate the relative frequencies of the occupation groups.

(b) Comment on changes in the occupation pattern between 1980 and 2001.

2.100 Table 13 gives data collected from the students attending an elementary statistics course at the University of Wisconsin. These data include sex, height, number of years in college, and the general area of intended major [Humanities (H); Social Science (S); Biological Science (B); Physical Science (P)].

TABLE 13   Class Data

Student No.	Sex	Height in Inches	Year in College	Intended Major	Student No.	Sex	Height in Inches	Year in College	Intended Major
1	F	67	3	S	26	M	67	1	B
2	M	72	3	P	27	M	68	3	P
3	M	70	4	S	28	M	72	4	B
4	M	70	1	B	29	F	68	3	P
5	F	61	4	P	30	F	66	2	B
6	F	66	3	B	31	F	65	2	B
7	M	71	3	H	32	M	64	4	B
8	M	67	4	B	33	M	72	1	H
9	M	65	3	S	34	M	67	4	B
10	F	67	3	B	35	M	73	3	S
11	M	74	3	H	36	F	71	4	B
12	M	68	3	S	37	M	71	3	B
13	M	74	2	P	38	M	69	2	S
14	F	64	4	P	39	F	69	4	P
15	M	69	3	S	40	M	74	4	S
16	M	64	3	B	41	M	73	3	B
17	M	72	4	P	42	M	68	3	B
18	M	71	3	B	43	F	66	2	S
19	F	67	2	S	44	M	73	2	P
20	M	70	4	S	45	M	73	2	S
21	M	66	4	S	46	M	67	4	S
22	F	67	2	B	47	F	62	3	S
23	M	68	4	S	48	M	68	2	B
24	M	71	3	H	49	M	71	3	S
25	M	75	1	S					

(a) Summarize the data of "intended major" in a frequency table.

(b) Summarize the data of "year in college" in a frequency table and draw either a line diagram or a histogram.

2.101 Referring to Exercise 2.100, plot the dot diagrams of heights separately for the male and female students and compare.

2.102 Refer to the data on power outages in Table D.1 in the Data Bank. Make a Pareto chart for the cause of the outage.

2.103 The dollar amounts claimed by businessmen for their lunches are to be grouped into the following classes: 0–5, 5–10, 10–15, 15–20, 20 or more. The left endpoint is included. Is it possible to determine from this frequency dis-

tribution the exact number of lunches for which the amount claimed was:

(a) Less than 15?

(b) 10 or more?

(c) 30 or more?

2.104 Mung bean sprouts are more widely used in Asian cooking than the beans themselves. To study their growth, an experimenter presoaked some beans until they sprouted about 1 millimeter. Five were randomly selected and placed in individual petri dishes. After 96 hours, their lengths (mm)

143   131   101   143   111

were obtained. Find the mean and standard deviation.

2.105 The weights of twenty adult grizzly bears captured and released are summarized in the computer output

Descriptive Statistics: bearwt

Variable	N	Mean	Median	StDev
Bearwt	20	227.4	232.5	82.7

(a) Locate two measures of center tendency, or location, and interpret the values.

(b) Locate the standard deviation.

(c) Calculate the z score for a grizzly bear that weighs 320 pounds.

2.106 The stem-and-leaf display given here shows the final examination scores of students in a sociology course. (Leaf unit $= 1.0$)

Stem-and-Leaf
Display of Scores

2	57
3	244
4	1179
5	03368
6	012447
7	223556899
8	00457
9	0036

(a) Find the median score.

(b) Find the quartiles Q_1 and Q_3.

(c) What proportion of the students scored below 70? 80 and over?

2.107 The following are the numbers of passengers on the minibus tour of Hollywood.

9 12 10 11 11 7 12 6 11 4 10 10 11 9 10
7 10 8 8 9 8 9 11 9 8 6 10 6 8 11

(a) Find the sample median.

(b) Find the sample mean.

(c) Find the sample variance.

2.108 The following table shows the age at inauguration of each U.S. president.

(a) Make a stem-and-leaf display with a double stem.

(b) Find the median, Q_1 and Q_3.

Name	Age at Inauguration
1. Washington	57
2. J. Adams	61
3. Jefferson	57
4. Madison	57
5. Monroe	58
6. J. Q. Adams	57
7. Jackson	61
8. Van Buren	54
9. W. H. Harrison	68
10. Tyler	51
11. Polk	49
12. Taylor	64
13. Fillmore	50
14. Pierce	48
15. Buchanan	65
16. Lincoln	52
17. A. Johnson	56
18. Grant	46
19. Hayes	54
20. Garfield	49
21. Arthur	50
22. Cleveland	47
23. B. Harrison	55
24. Cleveland	55
25. McKinley	54
26. T. Roosevelt	42
27. Taft	51
28. Wilson	56
29. Harding	55
30. Coolidge	51
31. Hoover	54
32. F. D. Roosevelt	51
33. Truman	60
34. Eisenhower	62
35. Kennedy	43
36. L. Johnson	55
37. Nixon	56
38. Ford	61
39. Carter	52
40. Reagan	69
41. G. Bush	64
42. Clinton	46
43. G. W. Bush	54

2.109 (a) Calculate $\bar{x}$ and s for the data 6, 8, 4, 9, 8.

 (b) Consider the data set 106, 108, 104, 109, 108, which is obtained by adding 100 to each number given in part (a). Use your results of part (a) and the properties stated in Exercises 2.52 and 2.74 to obtain the $\bar{x}$ and s for this modified data set. Verify your results by direct calculations with this new data set.

 (c) Consider the data set -18, -24, -12, -27, -24, which is obtained by multiplying each number of part (a) by -3. Repeat the problem given in part (b) for this new data set.

2.110 Refer to the class data in Exercise 2.100. Calculate the following.

 (a) $\bar{x}$ and s for the heights of males.

 (b) $\bar{x}$ and s for the heights of females.

 (c) Median and the quartiles for the heights of males.

 (d) Median and the quartiles for the heights of females.

2.111 In a genetic study, a regular food was placed in each of 20 vials and the number of flies of a particular genotype feeding on each vial recorded. The counts of flies were also recorded for another set of 20 vials that contained grape juice. The following data sets were obtained (courtesy of C. Denniston and J. Mitchell).

No. of Flies (Regular Food)

15 20 31 16 22 22 23 33 38 28
25 20 21 23 29 26 40 20 19 31

No. of Flies (Grape Juice)

6 19 0 2 11 12 13 12 5 16
2 7 13 20 18 19 19 9 9 9

 (a) Plot separate dot diagrams for the two data sets.

 (b) Make a visual comparison of the two distributions with respect to their centers and spreads.

 (c) Calculate $\bar{x}$ and s for each data set.

2.112 The data below were obtained from a detailed record of purchases of toothpaste over several years (courtesy of A. Banerjee). The usage times (in weeks) per ounce of toothpaste for a household taken from a consumer panel were

.74 .45 .80 .95 .84 .82 .78 .82 .89 .75 .76 .81
.85 .75 .89 .76 .89 .99 .71 .77 .55 .85 .77 .87

 (a) Plot a dot diagram of the data.

 (b) Find the relative frequency of the usage times that do not exceed .80.

 (c) Calculate the mean and the standard deviation.

 (d) Calculate the median and the quartiles.

2.113 To study how first-grade students utilize their time when assigned to a math task, a researcher observes 24 students and records their times off-task out of 20 minutes (courtesy of T. Romberg).

Times Off-Task (minutes)					
4	0	2	2	4	1
4	6	9	7	2	7
5	4	13	7	7	10
10	0	5	3	9	8

For this data set, find:

 (a) Mean and standard deviation.

 (b) Median.

 (c) Range.

2.114 Tornadoes cause many deaths each year in the United States. The ordered values for the yearly number of deaths for the 54 years 1950 through 2003 are

15   24   25   27   28   30   30   31   32   33
34   34   36   39   39   39   40   43   44   46
50   51   52   53   53   55   58   59   60   64
66   67   67   69   70   73   73   83   84   89
94   94   98   114 122 129 130 131 159 193
230 301 366 519

Calculate:

 (a) $\bar{x}$ and s.

 (b) Median and the other quartiles.

 (c) Range and interquartile range.

2.115 Refer to Exercise 2.114.

(a) Determine the intervals $\bar{x} \pm s$, $\bar{x} \pm 2s$, and $\bar{x} \pm 3s$.

(b) Find the proportion of the measurements that lie in each of these intervals.

(c) Compare your findings with the empirical guidelines for bell-shaped distributions.

2.116 The following summary statistics were obtained from a data set.

$$\bar{x} = 80.5 \qquad \text{Median} = 84.0$$
$$s = 10.5 \qquad Q_1 = 75.5$$
$$Q_3 = 96.0$$

Approximately what proportion of the observations are:

(a) Below 96.0?

(b) Above 84.0?

(c) In the interval 59.5 to 101.5?

(d) In the interval 75.5 to 96.0?

(e) In the interval 49.0 to 112.0?

State which of your answers are based on the assumption of a bell-shaped distribution.

2.117 The 50 measurements of acid rain in Wisconsin, whose histogram is given on the cover page of the chapter, are

3.58	3.80	4.01	4.01	4.05	4.05
4.12	4.18	4.20	4.21	4.27	4.28
4.30	4.32	4.33	4.35	4.35	4.41
4.42	4.45	4.45	4.50	4.50	4.50
4.50	4.51	4.52	4.52	4.52	4.57
4.58	4.60	4.61	4.61	4.62	4.62
4.65	4.70	4.70	4.70	4.70	4.72
4.78	4.78	4.80	5.07	5.20	5.26
5.41	5.48				

(a) Calculate the median and quartiles.

(b) Find the 90th percentile.

(c) Determine the mean and standard deviation.

(d) Display the data in the form of a boxplot.

2.118 Refer to Exercise 2.117.

(a) Determine the intervals $\bar{x} \pm s$, $\bar{x} \pm 2s$, and $\bar{x} \pm 3s$.

(b) What proportions of the measurements lie in those intervals?

(c) Compare your findings with the empirical guidelines for bell-shaped distributions.

2.119 Refer to the earthquake size data in Exercise 2.20.

(a) Calculate the median and quartiles.

(b) Calculate the mean and standard deviation.

(c) Display the data in the form of a boxplot.

2.120 The Dow Jones average provides an indication of overall market level. The changes in this average, from year end, from 1969 to 1970 through 2002 to 2003 are summarized in the following frequency table, where the left-hand endpoint is excluded.

Yearly Changes in the Dow Jones Average

Change in DJ Average	Frequency
$(-1600, -850]$	1
$(-850, -250]$	2
$(-250, 0]$	7
$(0, 250]$	13
$(250, 850]$	5
$(850, 1650]$	4
$(1650, 2450]$	2
Total	34

(a) Calculate the relative frequency for the intervals.

(b) Plot the relative frequency histogram. (*Hint:* Since the intervals have unequal widths, make the height of each rectangle equal to the relative frequency divided by the width of the interval.)

(c) What proportion of the changes were negative.

(d) Comment on the location and shape of the distribution.

2.121 The winning times of the men's 400-meter freestyle swimming in the Olympics (1908 to 2000) appear in the following table.

Winning Times in Minutes and Seconds			
Year	Time	Year	Time
1908	5:36.8	1964	4:12.2
1912	5:24.4	1968	4:09.0
1920	5:26.8	1972	4:00.27
1924	5:04.2	1976	3:51.93
1928	5:01.6	1980	3:51.31
1932	4:48.4	1984	3:51.23
1936	4:44.5	1988	3:46.95
1948	4:41.0	1992	3:45.00
1952	4:30.7	1996	3:47.97
1956	4:27.3	2000	3:40.59
1960	4:18.3		

(a) Draw a dot diagram and label the points according to time order.

(b) Explain why it is not reasonable to group the data into a frequency distribution.

2.122 The *mode* of a collection of observations is defined as the observed value with largest relative frequency. The mode is sometimes used as a center value. There can be more than one mode in a data set. Find the mode for the data given in Exercise 2.13.

2.123 Refer to the earthquake size data in Exercise 2.20.

(a) Make a time plot of the data.

(b) What earthquake sizes, if any, are outliers?

2.124 With reference to Exercise 2.123:

(a) Make a control chart of the data earthquake size.

(b) Identify any earthquake sizes that were out of statistical control.

The Following Exercises Require a Computer.

Calculations of the descriptive statistics such as $\bar{x}$ and s are increasingly tedious with larger data sets. Current computer software programs alleviate the drudgery of hand calculations. Use MINITAB or some other package program.

2.125 Find $\bar{x}$ and s for:

(a) The lizard data in Exercise 2.19.

(b) The acid rain data in Exercise 2.117.

2.126 Lumber intended for building houses and other structures must be monitored for strength. The measurement of strength (pounds per square inch) for 61 specimens of Southern Pine (*Source:* U.S. Forest Products Laboratory) yielded

4001	3927	3048	4298	4000	3445
4949	3530	3075	4012	3797	3550
4027	3571	3738	5157	3598	4749
4263	3894	4262	4232	3852	4256
3271	4315	3078	3607	3889	3147
3421	3531	3987	4120	4349	4071
3686	3332	3285	3739	3544	
4103	3401	3601	3717	4846	
5005	3991	2866	3561	4003	
4387	3510	2884	3819	3173	
3470	3340	3214	3670	3694	

Using MINITAB, the sequence of choices

> **Data**(in 2.126.DAT):
> Strength
> **Dialog box:**
> **Stat > Basic Statistics > Graphical Summary.**
> Type *Strength* in **Variables.** *Click* **OK.**

produces a rather complete summary of the data. It includes the output on page 79.

(a) Use this output to identify a departure from a bell-shaped pattern.

(b) MINITAB uses a slightly different scheme to determine the first and third quartiles, but the difference is not of practical importance with large samples. Calculate the first quartile using the definition in this book and compare with the value in the output.

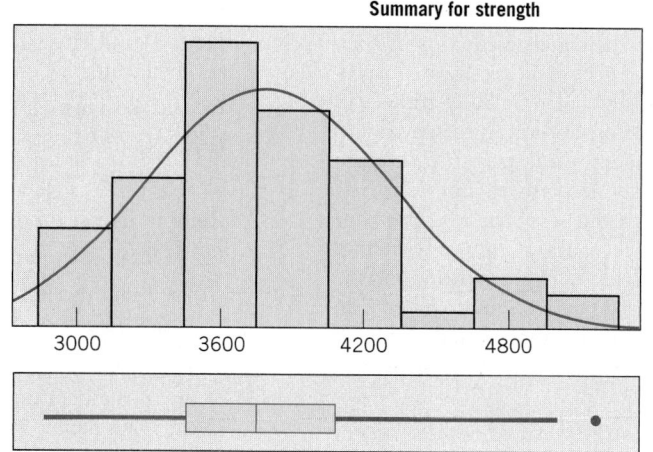

Summary for strength

Mean	3801.0
StDev	513.5
Variance	263724.4
Skewness	0.577366
Kurtosis	0.265812
N	61

Minimum	2866.0
1st Quartile	3457.5
Median	3738.0
3rd Quartile	4087.0
Maximum	5157.0

2.127 Find $\bar{x}$ and s for the data set in Table 4.

2.128 Find $\bar{x}$ and s for the final times to run 1.5 miles in Table D.5 in the Data Bank.

2.129 The SAS computer software package produced the following output. Compare the mean and standard deviation with that of the MINITAB output in Exercise 2.126. Which output gives more digits?

UNIVARIATE PROCEDURE

VARIABLE = X1

MOMENTS

N	61	VARIANCE	263724.4
MEAN	3800.951		
STD DEV	513.5411		

QUANTILES (DEF = 5)

100% MAX	5157	99%	5157
75% Q3	4071	95%	4846
50% MED	3738	90%	4349
25% Q1	3470	10%	3173
0% MIN	2866	5%	3075
		1%	2866

RANGE	2291
Q3-Q1	601

EXTREMES

LOWEST	OBS	HIGHEST	OBS
2866(	31)	4749(	28)
2884(	42)	4846(	22)
3048(	3)	4949(	12)
3075(	14)	5005(	29)
3078(	47)	5157(	26)

2.130 The salmon fisheries support a primary industry in Alaska and their management is of high priority. Salmon are born in freshwater rivers and streams but then swim out into the ocean for a few years before returning to spawn and die. In order to identify the origins of mature fish, researchers studied growth rings on their scales. The growth the first year in freshwater is measured by the width of the growth rings for that period of life. The growth ring for the first year in the ocean environment will give an indication of growth for that period. A set of these measurements are given in Table D.7 in the Data Bank.

(a) Describe the freshwater growth for males by making a histogram and calculating the mean, standard deviation, and quartiles.

(b) Describe the freshwater growth for females by making a histogram and calculating the mean, standard deviation, and quartiles.

(c) Construct boxplots to compare the growth of males and females.

2.131 Refer to the alligator data in Table D.11 of the Data Bank. Using the data on x_5 for thirty-seven alligators:

(a) Make a histogram.

(b) Obtain the sample mean and standard deviation.

2.132 Refer to Exercise 2.131.

(a) Obtain the quartiles.

(b) Obtain the 90th percentile. How many of the alligators above the 90th percentile are female?

2.133 Refer to the data on malt extract in Table D.8 of the Data Bank.

(a) Obtain sample mean and standard deviation.

(b) Obtain quartiles.

(c) Check conformity with the empirical rule.

3

Descriptive Study of Bivariate Data

Hydrogen–Carbon Association in Moon Rocks

© Photo Researchers.

In their quest for clues to the origin and composition of the planets, scientists performed chemical analyses of rock specimens collected by astronauts and unmanned space probes. The Apollo moon landings made it possible to study firsthand the geology of the moon. Eleven lunar rocks were analyzed for carbon and hydrogen content.

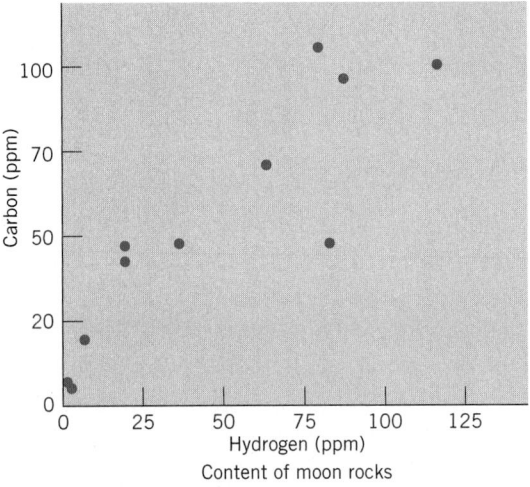

Rocks with large amounts of hydrogen tend to have large amounts of carbon. Other rocks tend to have small amounts of both elements, indicating a positive association between hydrogen and carbon content.

1.   INTRODUCTION

In Chapter 2, we discussed the organization and summary description of data concerning a single variable. Observations on *two or more* variables are often recorded for the individual sampling units; the height and weight of individuals, or the number of goals scored by and against a team. By studying such **bivariate** or **multivariate** data, one typically wishes to discover if any relationships exist between the variables, how strong the relationships appear to be, and whether one variable of primary interest can be effectively predicted from information on the values of the other variables. To illustrate the concepts, we restrict our attention to the simplest case where only two characteristics are observed on the individual sampling units. Some examples are:

Gender and the type of occupation of college graduates.

Smoking habit and lung capacity of adult males.

Average daily carbohydrate intake and protein intake of 10-year-old children.

The age of an aircraft and the time required for repairs.

The two characteristics observed may both be qualitative traits, both numerical variables, or one of each kind. For brevity, we will only deal with situations where the characteristics observed are either both categorical or both numerical. Summarization of bivariate categorical data is discussed in Section 2. Sections 4, 5, and 6 are concerned with bivariate measurement data and treat such issues as graphical presentations, examination of relationship, and prediction of one variable from another.

2.   SUMMARIZATION OF BIVARIATE CATEGORICAL DATA

When two traits are observed for the individual sampling units and each trait is recorded in some qualitative categories, the resulting data can be summarized in the form of a two-way frequency table. The categories for one trait are marked along the left margin, those for the other along the upper margin, and the frequency counts recorded in the cells. The total frequency for any row is given in the right-hand margin and those for any column given at the bottom margin. Both are called **marginal totals.**

Data in this summary form are commonly called cross-classified or cross-tabulated data. In statistical terminology, they are also called contingency tables.

Example 1   Calculation of Relative Frequencies Aids Interpretation

A survey was conducted by sampling 400 persons who were questioned regarding union membership and attitude toward decreased national spending on social welfare programs. The cross-tabulated frequency counts are presented in Table 1.

TABLE 1   Cross-Tabulated Frequency Counts

	Support	Indifferent	Opposed	Total
Union	112	36	28	176
Nonunion	84	68	72	224
Total	196	104	100	400

The entries of this table are self-explanatory. For instance, of the 400 persons polled, there were 176 union members. Among these union members, 112 expressed support, 36 were indifferent, and 28 were opposed. To gain further understanding of how the responses are distributed, calculate the relative frequencies of the cells.

SOLUTION   For this purpose, we divide each cell frequency by the sample size 400. The relative frequencies (for instance, $84/400 = .21$) are shown in Table 2.

TABLE 2   Relative Frequencies for the Data of Table 1

	Support	Indifferent	Opposed	Total
Union	.28	.09	.07	.44
Nonunion	.21	.17	.18	.56
Total	.49	.26	.25	1.00

Depending on the specific context of a cross-tabulation, one may also wish to examine the cell frequencies relative to a marginal total. In Example 1, you may wish to compare the attitude patterns of the union members with those of the nonmembers. This is accomplished by calculating the relative frequencies separately for the two groups (for instance, $84/224 = .375$), as Table 3 shows.

From the calculations in Table 3, it appears that the attitude patterns are different between the two groups—support seems to be stronger among union members than nonmembers.

Now the pertinent question is: Can these observed differences be explained by chance or are there real differences of attitude between the populations of members and nonmembers? We will pursue this aspect of statistical inference in Chapter 13.

TABLE 3   Relative Frequencies by Group

	Support	Indifferent	Opposed	Total
Union	.636	.205	.159	1.00
Nonunion	.375	.304	.321	1.00

SIMPSON'S PARADOX

Quite surprising and misleading conclusions can occur when data from different sources are combined into a single table. We illustrate this reversal of implications with graduate school admissions data.

Example 2 Combining Tables Can Produce Misleading Summaries

We consider graduate school admissions at a large midwestern university but, to simplify, we use only two departments as the whole school. We are interested in comparing admission rates by gender and obtain the data in Table 4 for the school.

TABLE 4   School Admission Rates

	Admit	Not Admit	Total Applicants
Male	233	324	557
Female	88	194	282
Total	321	518	839

Does there appear to be a gender bias?

SOLUTION   It is clear from these admissions statistics that the proportion of males admitted, $233/557 = .418$, is greater than the proportion of females admitted, $88/282 = .312$.

Does this imply some type of discrimination? Not necessarily. By checking the admission records, we were able to further categorize the cases according to department in Table 5. Table 4 is the aggregate of these two sets of data.

TABLE 5   Admission Rates by Department

Mechanical Engineering				History			
	Admit	Not Admit	Total		Admit	Not Admit	Total
Male	151	35	186	Male	82	289	371
Female	16	2	18	Female	72	192	264
Total	167	37	204	Total	154	481	635

One of the two departments, mechanical engineering, has mostly male applicants. Even so the proportion of males admitted, $151/186 = .812$, is smaller than the proportion of females admitted, $16/18 = .889$. The same is true for the history department where the proportion of males admitted, $82/371 = .221$, is again smaller than the proportion of females admitted, $72/264 = .273$. When the data are studied department by department, the reverse but correct conclusion holds; females have a higher admission rate in both cases!

To obtain the correct interpretation, these data need to be presented as the full three-way table of gender-admission action-department as given above. If department is ignored and the data aggregated across this variable, "department" can act as an unrecorded or lurking variable. In this example, it has reversed the direction of possible gender bias and led to the erroneous conclusion that males have a higher admission rate than females.

The reversal of the comparison, such as in Example 2, when data are combined from several groups is called Simpson's paradox.

When data from several sources are aggregated into a single table, there is always the danger that unreported variables may cause a reversal of the findings. In practical applications, there is not always agreement on how much effort to expend following up on unreported variables. When comparing two medical treatments, the results often need to be adjusted for the age, gender, and sometimes current health of the subjects and other variables.

Exercises

3.1 Nausea from air sickness affects some travelers. A drug company, wanting to establish the effectiveness of its motion sickness pill, randomly gives either its pill or a look-alike sugar pill (placebo) to 200 passengers.

	Degree of Nausea				
	None	Slight	Moderate	Severe	Total
Pill	43	36	18	3	100
Placebo	19	33	36	12	100
Total					

(a) Complete the marginal totals.
(b) Calculate the relative frequencies separately for each row.
(c) Comment on any apparent differences in response between the pill and the placebo.

3.2 Breakfast cereals from three leading manufacturers can be classified either above average or below average in sugar content. Data for ten cereals from each manufacturer are given below:

	Below Average	Above Average	Total
General Mills	3	7	10
Kellogg's	4	6	10
Quaker	6	4	10
Total			

(a) Complete the marginal totals.
(b) Calculate the relative frequencies separately for each row.
(c) Comment on any apparent differences between the cereals produced by the three companies.

3.3 The aging of commercial aircraft can make them more vulnerable to "skin-cracking" rivets. A major manufacturer collected data on its three most popular models in active use to determine the magnitude of the problem.

Number of Aircraft Still in Service

Model	≤20 Years	>20 Years
B7	90	123
B27	1214	435
B37	1042	9

Compare the aging of the three types of planes by calculating the relative frequencies.

3.4 A survey was conducted to study the attitudes of the faculty, academic staff, and students in regard to a proposed measure for reducing the heating and air-conditioning expenses on campus.

	Favor	Indifferent	Opposed	Total
Faculty	36	42	122	200
Academic staff	44	77	129	250
Students	106	178	116	400

Compare the attitude patterns of the three groups by computing the relative frequencies.

3.5 Groundwater from 19 wells was classified as low or high in alkalinity and low or high in dissolved iron. There were 9 wells with high alkalinity, 7 that were high in iron, and 5 that were high in both.

(a) Based on these data, complete the following two-way frequency table.

(b) Calculate the relative frequencies of the cells.

(c) Calculate the relative frequencies separately for each row.

	Iron	
Alkalinity	Low	High
Low		
High		

3.6 Interviews with 150 persons engaged in a stressful occupation revealed that 57 were alcoholics, 64 were mentally depressed, and 42 were both.

(a) Based on these records, complete the following two-way frequency table.

(b) Calculate the relative frequencies.

	Alcoholic	Not Alcoholic	Total
Depressed			
Not depressed			
Total			

3.7 Cross-tabulate the "Class data" of Exercise 2.100 according to gender (M, F) and the general areas of intended major (H, S, B, P). Calculate the relative frequencies.

3.8 A psychologist interested in obese children gathered data on a group of children and their parents.

	Child	
Parent	Obese	Not Obese
At least one obese	12	24
Neither obese	8	36

(a) Calculate the marginal totals.

(b) Convert the frequencies to relative frequencies.

(c) Calculate the relative frequencies separately for each row.

3.9 Typically, there is a gender unbalance among tenured faculty, especially in the sciences. At a large university, tenured faculty members in two departments, English and Computer Science, were categorized according to gender.

	Male	Female
English	23	19
Computer Science	27	5

(a) Calculate relative frequencies separately for each row.

(b) Comment on major differences in the patterns for the two rows.

3.10   A large research hospital and a community hospital are located in your area. The surgery records for the two hospitals are:

	Died	Survived	Total
Research hospital	90	2110	2200
Community hospital	23	677	700

The outcome is "survived" if the patient lives at least six weeks.

(a)   Calculate the proportion of patients that survive surgery at each of the hospitals.

(b)   Which hospital do these data suggest you should choose for surgery?

3.11   Refer to Exercise 3.10. Not all surgery cases, even of the same type, are equally serious. Large research hospitals tend to get the most serious surgery cases, whereas community hospitals tend to get more of the routine cases. Suppose that patients can be classified as being in either "Good" or "Poor" condition and the outcomes of surgery are as shown in table below.

(a)   Calculate the proportions that survive for each hospital and each condition.

(b)   From these data, which hospital would you choose if you were in good condition? If you were in bad condition?

(c)   Compare your answer with that to Exercise 3.10. Explain this reversal as an example of Simpson's paradox and identify the lurking variable in Exercise 3.10.

Survival Rates by Condition

Good Condition	Died	Survived	Total
Research hospital	15	685	700
Community hospital	16	584	600
Total	31	1269	1300

Poor Condition	Died	Survived	Total
Research hospital	75	1425	1500
Community hospital	7	93	100
Total	82	1518	1600

3.   A DESIGNED EXPERIMENT FOR MAKING A COMPARISON

We regularly encounter claims that, as a group, smokers have worse health records than nonsmokers with respect to one disease or another or that a new medical treatment is better than the former one. Properly designed experiments can often provide data that are so conclusive that a comparison is clear-cut. An example of a comparative study will illustrate the design issue and how to conduct an experiment.

During the early development of a medicated skin patch to help smokers break the habit, a test was conducted with 112 volunteers. Because of the so-called placebo effect, of people tending to respond positively just because attention is paid to them, half the volunteers were given an unmedicated skin patch. The data will consist of a count of the number of persons who are abstinent at the end of the study.

Purpose: To determine the effectiveness of a medicated nicotine patch for smoking cessation based on the end-of-therapy numbers of abstinent persons in the medicated and unmedicated groups.

What is involved in comparing two approaches or methods for doing something? First the subjects, or experimental units, must be assigned to the two groups in such a manner that neither method is favored. One approach is to list the subjects' names on a piece of paper, cut the paper into strips, each with one name on it, and then draw one at a time until half the names are drawn. Ideally, we like to have groups of equal size, so if there is an odd number of subjects, draw just over one-half. These subjects are assigned to the first approach. The other subjects are assigned to the second approach. This step, called random assignment, helps guarantee a valid comparison. Any subject likely to respond positively has the same chance of supporting the first approach as supporting the second approach. When subjects cannot be randomly assigned, we will never know if observed differences in the number of abstinent smokers is due to the approaches or some other variables associated with the particular persons assigned to the two groups.

Subjects were randomly assigned to the medicated or unmedicated (placebo) groups. They were not told which group. As with many medical trials, this was a double blind trial. That is, the medical staff in contact with the patients was also kept unaware of which patients were getting the treated patch and which were not. At the end of the study, the number of persons in each group who were abstinent and who were smoking were recorded.

The data[1] collected from this experiment are summarized in Table 6.

TABLE 6   Quitting Smoking

	Abstinent	Smoking	
Medicated patch	21	36	57
Unmedicated patch	11	44	55
	32	80	112

The proportion abstinent is $21/57 = .368$ for the medicated skin patch group and only $11/55 = .200$ for the control. The medicated patch seems to work. Later, in Chapter 10, we verify that the difference $.368 - .200 = .168$ is greater than can be explained by chance variation.

In any application where the subjects might learn from the subjects before them, it would be a poor idea to perform all the trials for treatment 1 and then all those for treatment 2. Learning or other uncontrolled variables must not be given the opportunity to systematically affect the experiment. We could number the subjects 1 to 112 and write each of these numbers on a separate slip of paper. The 112 slips of paper should be mixed and then drawn one at a time to determine the sequence in which the trials are conducted.

Researchers continue to investigate the effectiveness of patches. One study involved over 200 pregnant women in their first trimester.[2] Typically, these sub-

[1] M. Fiore, S. Kenford, D. Jorenby, D. Wetter, S. Smith, and T. Baker. "Two Studies of the Clinical Effectiveness of the Nicotine Patch with Different Counseling Treatments." *Chest* **105** (1994), pp. 524–533.
[2] K. Wisborg, T. Henriksen, L. Jespersen, and N. Secher. "Nicotine Patches for Pregnant Smokers: Randomized Controlled Study." *Obstetrics & Gynecology*, **96** (2000), pp. 967–971.

jects were highly motivated to quit smoking. Subjects were randomly assigned either a placebo or medicated patch but all received counseling. They were not told which patch they received so this was a blind trial. It was found that about one-fourth of subjects were not smoking six months later but there was no significant difference between the placebo and medicated patch groups.

Exercises

3.12 With reference to the quit-smoking experiment, suppose two new subjects are available. Explain how you would assign one subject to receive the placebo and one to receive the medicated patch.

3.13 With reference to the quit-smoking experiment:

(a) Suppose the placebo trials were ignored and you were only told that 21 of 57 were abstinent after using the medicated patches. Would this now appear to be stronger evidence in favor of the patches?

(b) Explain why the placebo trials provide a more valid reference for results of the medicated patch trials.

4. SCATTER DIAGRAM OF BIVARIATE MEASUREMENT DATA

We now turn to a description of data sets concerning two variables, each measured on a numerical scale. For ease of reference, we will label one variable x and the other y. Thus, two numerical observations (x, y) are recorded for each sampling unit. These observations are *paired* in the sense that an (x, y) pair arises from the same sampling unit. An x observation from one pair and an x or y from another are unrelated. For n sampling units, we can write the measurement pairs as

$$(x_1, y_1), (x_2, y_2), \ldots, (x_n, y_n)$$

The set of x measurements alone, if we disregard the y measurements, constitutes a data set for one variable. The methods of Chapter 2 can be employed for descriptive purposes, including graphical presentation of the pattern of distribution of the measurements, and, calculation of the mean, standard deviation, and other quantities. Likewise, the y measurements can be studied disregarding the x measurements. However, a major purpose of collecting bivariate data is to answer such questions as:

Are the variables related?

What form of relationship is indicated by the data?

Can we quantify the strength of their relationship?

Can we predict one variable from the other?

Studying either the x measurements by themselves or the y measurements by themselves would not help answer these questions.

An important first step in studying the relationship between two variables is to graph the data. To this end, the variable x is marked along the horizontal axis and y on the vertical axis on a graph paper. The pairs (x, y) of observations are then plotted as dots on the graph. The resulting diagram is called a scatter diagram or scatter plot. By looking at the scatter diagram, a visual impression can be formed about the relation between the variables. For instance, we can observe whether the points band around a line or a curve or if they form a patternless cluster.

Example 3   A Scatter Diagram Provides a Visual Display of a Relationship
Recorded in Table 7 are the data of

$$x = \text{Undergraduate GPA}$$

and

$$y = \text{Score in the Graduate Management Aptitude Test (GMAT)}$$

for applicants seeking admission to an MBA program.
Construct a scatter diagram.

TABLE 7   Data of Undergraduate GPA
x and GMAT Score y

x	y	x	y	x	y
3.63	447	2.36	399	2.80	444
3.59	588	2.36	482	3.13	416
3.30	563	2.66	420	3.01	471
3.40	553	2.68	414	2.79	490
3.50	572	2.48	533	2.89	431
3.78	591	2.46	509	2.91	446
3.44	692	2.63	504	2.75	546
3.48	528	2.44	336	2.73	467
3.47	552	2.13	408	3.12	463
3.35	520	2.41	469	3.08	440
3.39	543	2.55	538	3.03	419
				3.00	509

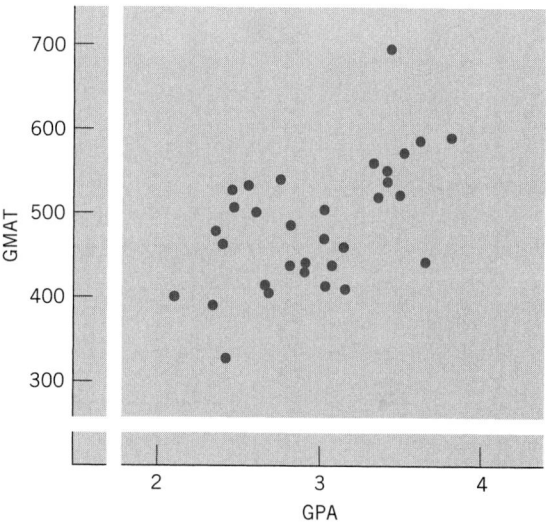

Figure 1   Scatter diagram of applicants' scores.

SOLUTION   The scatter diagram is plotted in Figure 1. The southwest-to-northeast pattern of the points indicates a positive relation between x and y. That is, the applicants with a high GPA tend to have a high GMAT. Evidently, the relation is far from a perfect mathematical relation.

When the two measurements are made on two or more groups, visual comparisons between groups are made by plotting the points on the same scatter plot. A different symbol is used for each group. The resulting graph is called a **multiple scatter plot** where "multiple" refers to groups.

Example 4   A Multiple Scatter Diagram for Visually Comparing Relationships

Concern was raised by environmentalists that spills of contaminants were affecting wildlife in and around an adjacent lake. Estrogenic contaminants in the environment can have grave consequences on the ability of living things to reproduce. Researchers examined the reproductive development of young male alligators hatched from eggs taken from around (1) Lake Apopka, the lake which was contaminated, and (2) Lake Woodruff, which acted as a control. The contaminants were thought to influence sex steroid concentrations.

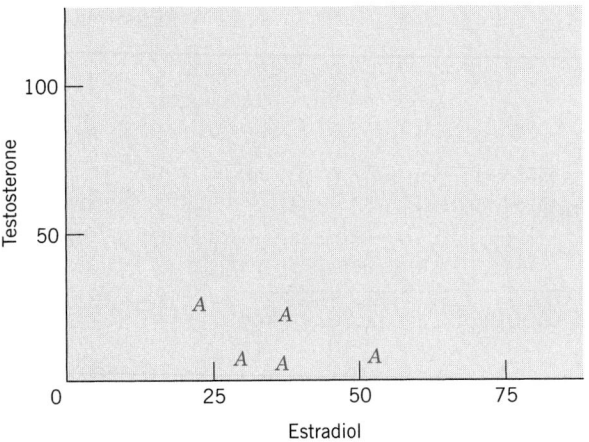

(*a*) Scatter diagram for Lake Apopka

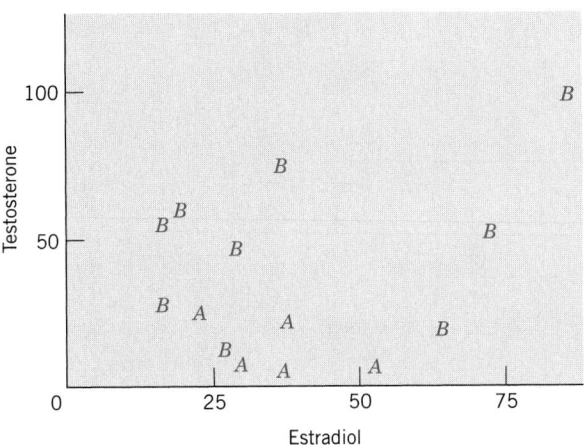

(*b*) Multiple scatter diagram

Figure 2   Scatter diagrams. *A* = Lake Apopka. *B* = Lake Woodruff.

The concentrations of two steroids, estradiol and testosterone, were determined by radioimmunoassay.

Lake Apopka

Estradiol	38	23	53	37	30
Testosterone	22	24	8	6	7

Lake Woodruff

Estradiol	29	64	19	36	27	16	15	72	85
Testosterone	47	20	60	75	12	54	33	53	100

(a) Make a scatter diagram of the two concentrations for the Lake Apopka alligators.

(b) Create a multiple scatter diagram by adding to the same plot the pairs of concentrations for the Lake Woodruff male alligators. Use a different symbol for the two lakes.

(c) Comment on any major differences between the two groups of male alligators.

SOLUTION (a) Figure 2a gives the scatter diagram for the Lake Apopka alligators.

(b) Figure 2b is the multiple scatter diagram with the points for Lake Woodruff marked as B.

(c) The most prominent feature of the data is that the male alligators from the contaminated lake have, generally, much lower levels of testosterone than those from the nearly pollution-free control lake. (The A's are at the bottom third of the scatter diagram.) Low testosterone levels in males have grave consequences regarding reproduction.

5. THE CORRELATION COEFFICIENT—A MEASURE OF LINEAR RELATION

The scatter diagram provides a visual impression of the nature of the relation between the x and y values in a bivariate data set. In a great many cases, the points appear to band around a straight line. Our visual impression of the closeness of the scatter to a linear relation can be quantified by calculating a numerical measure, called the correlation coefficient.

The correlation coefficient, denoted by r, is a measure of strength of the linear relation between the x and y variables. Before introducing its formula, we outline some important features of the correlation coefficient and discuss the manner in which it serves to measure the strength of a linear relation.

1. The value of r is always between -1 and $+1$.

2. The magnitude of r indicates the strength of a linear relation, whereas its sign indicates the direction. More specifically,

$r > 0$ if the pattern of (x, y) values is a band that runs from lower left to upper right.

$r < 0$ if the pattern of (x, y) values is a band that runs from upper left to lower right.

$r = +1$ if all (x, y) values lie exactly on a straight line with a positive slope (perfect positive linear relation).

$r = -1$ if all (x, y) values lie exactly on a straight line with a negative slope (perfect negative linear relation).

A high numerical value of r, that is, a value close to $+1$ or -1, represents a strong linear relation.

3. A value of r close to zero means that the linear association is very weak.

The correlation coefficient is close to zero when there is no visible pattern of relation; that is, the y values do not change in any direction as the x values change. A value of r near zero could also happen because the points band

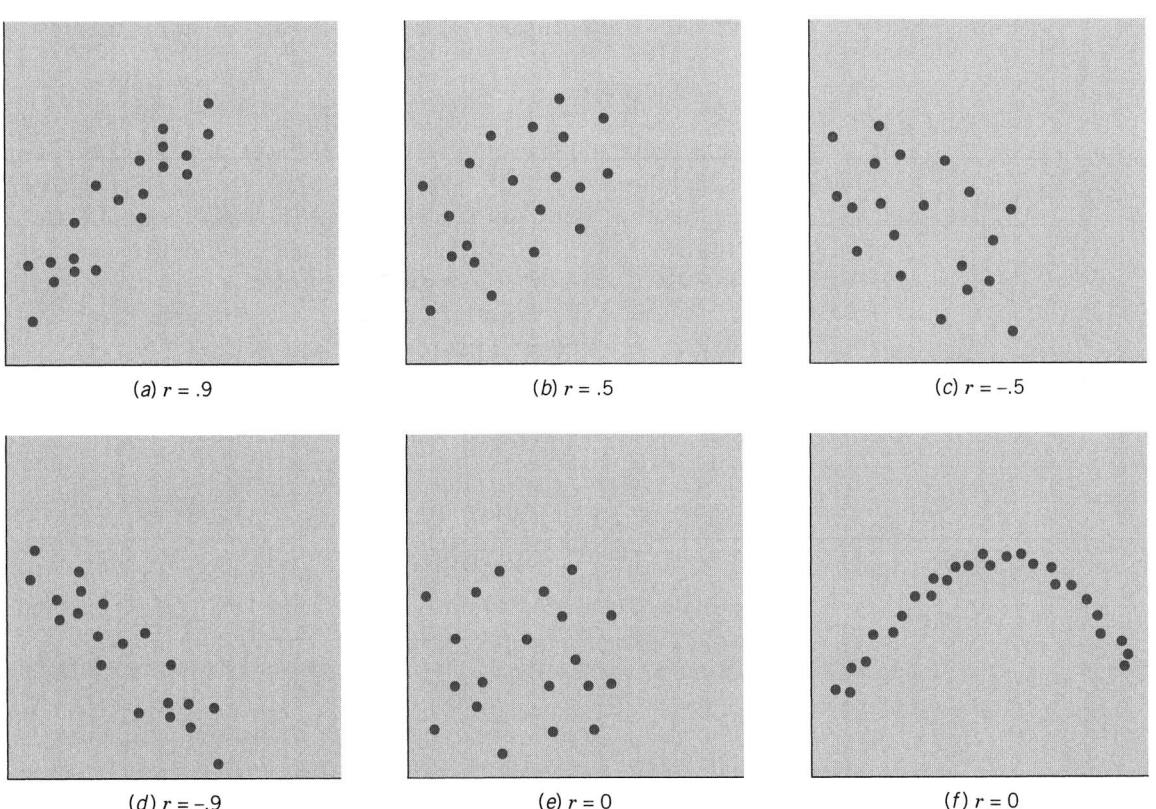

(a) r = .9   (b) r = .5   (c) r = −.5

(d) r = −.9   (e) r = 0   (f) r = 0

Figure 3   Correspondence between the values of r and the amount of scatter.

around a curve that is far from linear. After all, r measures linear association, and a markedly bent curve is far from linear.

Figure 3 shows the correspondence between the appearance of a scatter diagram and the value of r. Observe that (e) and (f) correspond to situations where $r = 0$. The zero correlation in (e) is due to the absence of any relation between x and y, whereas in (f) it is due to a relation following a curve that is far from linear.

CALCULATION OF r

The sample correlation coefficient quantifies the association between two numerically valued characteristics. It is calculated from n pairs of observations on the two characteristics

$$(x_1, y_1), (x_2, y_2), \ldots, (x_n, y_n)$$

The correlation coefficient is best interpreted in terms of the **standardized observations**

$$\frac{\text{Observation} \;-\; \text{Sample mean}}{\text{Sample standard deviation}} = \frac{x_i - \bar{x}}{s_x}$$

where $s_x = \sqrt{\sum_{i=1}^{n} (x_i - \bar{x})^2/(n - 1)}$ and the subscript x on s distinguishes the sample standard deviation of the x observations from the sample standard deviation s_y of the y observations.

Since the difference $x_i - \bar{x}$ has the units of x and the sample standard deviation s_x also has the same units, the standardized observation is free of the units of measurements. The sample correlation coefficient is the sum of the products of the standardized x observation times the standardized y observations divided by $n - 1$.

Sample Correlation Coefficient

$$r = \frac{1}{n - 1} \sum_{i=1}^{n} \left(\frac{x_i - \bar{x}}{s_x} \right) \left(\frac{y_i - \bar{y}}{s_y} \right)$$

When the pair (x_i, y_i) has both components above their sample means or both below their sample means, the product of standardized observations will be positive; otherwise it will be negative. Consequently, if most pairs have both components simultaneously above or simultaneously below their means, r will be positive.

An alternative formula for r is used for calculation. It is obtained by canceling the common term $n - 1$.

Calculation Formula for the Sample Correlation Coefficient

$$r = \frac{S_{xy}}{\sqrt{S_{xx}}\sqrt{S_{yy}}}$$

where

$$S_{xy} = \Sigma\,(x - \bar{x})(y - \bar{y})$$

$$S_{xx} = \Sigma\,(x - \bar{x})^2 \qquad S_{yy} = \Sigma\,(y - \bar{y})^2$$

The quantities S_{xx} and S_{yy} are the sums of squared deviations of the x observations and the y observations, respectively. S_{xy} is the sum of cross products of the x deviations with the y deviations. This formula will be examined in more detail in Chapter 11.

Example 5   Calculation of Sample Correlation

Calculate r for the $n = 4$ pairs of observations

$$(2, 5), \quad (1, 3), \quad (5, 6), \quad (0, 2)$$

SOLUTION   We first determine the mean $\bar{x}$ and deviations $x - \bar{x}$ and then $\bar{y}$ and the deviations $y - \bar{y}$. See Table 8.

TABLE 8   Calculation of r

x	y	$x - \bar{x}$	$y - \bar{y}$	$(x - \bar{x})^2$	$(y - \bar{y})^2$	$(x - \bar{x})(y - \bar{y})$
2	5	0	1	0	1	0
1	3	−1	−1	1	1	1
5	6	3	2	9	4	6
0	2	−2	−2	4	4	4
Total 8	16	0	0	14	10	11
$\bar{x} = 2$	$\bar{y} = 4$			S_{xx}	S_{yy}	S_{xy}

Consequently,

$$r = \frac{S_{xy}}{\sqrt{S_{xx}}\sqrt{S_{yy}}} = \frac{11}{\sqrt{14}\sqrt{10}} = .930$$

The value .930 is large and it implies a strong association where both x and y tend to be small or both tend to be large.

It is sometimes convenient, when using hand-held calculators, to evaluate r using the alternative formulas for S_{xx}, S_{yy}, and S_{xy}.

$$S_{xx} = \Sigma x^2 - \frac{\left(\Sigma x\right)^2}{n} \qquad S_{yy} = \Sigma y^2 - \frac{\left(\Sigma y\right)^2}{n}$$

$$S_{xy} = \Sigma xy - \frac{\left(\Sigma x\right)\left(\Sigma y\right)}{n}$$

This calculation is illustrated in Table 9.

TABLE 9 Alternate Calculation of r

x	y	x^2	y^2	xy
2	5	4	25	10
1	3	1	9	3
5	6	25	36	30
0	2	0	4	0
Total 8	16	30	74	43
Σx	Σy	Σx^2	Σy^2	Σxy

$$r = \frac{43 - \dfrac{8 \times 16}{4}}{\sqrt{30 - \dfrac{8^2}{4}}\,\sqrt{74 - \dfrac{(16)^2}{4}}} = .930$$

We remind the reader that r measures the closeness of the pattern of scatter to a line. Figure 3f presents a strong relationship between x and y, but one that is not linear. The small value of r for these data does not properly reflect the strength of the relation. Clearly, r is not an appropriate summary of a curved pattern. Another situation where the sample correlation coefficient r is not appropriate occurs when the scatter plot breaks into two clusters. Faced with separate clusters as depicted in Figure 4, it is best to try and determine the underlying cause. It may be that a part of the sample has come from one population and a part from another.

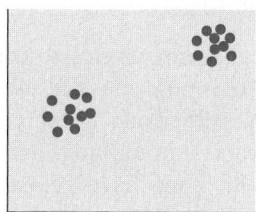

Figure 4 r is not appropriate—samples from two populations.

CORRELATION AND CAUSATION

Data analysts often jump to unjustified conclusions by mistaking an observed correlation for a cause-and-effect relationship. A high sample correlation coefficient does not necessarily signify a causal relation between two variables. A classic example concerns an observed high positive correlation between the number of storks sighted and the number of births in a European city. It is hoped no one would use this evidence to conclude that storks bring babies or, worse yet, that killing storks would control population growth.

The observation that two variables tend to simultaneously vary in a certain direction does not imply the presence of a direct relationship between them. If we record the monthly number of homicides x and the monthly number of religious meetings y for several cities of widely varying sizes, the data will probably indicate a high positive correlation. It is the fluctuation of a third variable (namely, the city population) that causes x and y to vary in the same direction, despite the fact that x and y may be unrelated or even negatively related. Picturesquely, the third variable, which in this example is actually causing the observed correlation between crime and religious meetings, is referred to as a **lurking variable**. The false correlation that it produces is called a **spurious correlation**. It is more a matter of common sense than of statistical reasoning to determine if an observed correlation has a practical interpretation or is spurious.

> An observed correlation between two variables may be **spurious**. That is, it may be caused by the influence of a third variable.

When using the correlation coefficient as a measure of relationship, we must be careful to avoid the possibility that a lurking variable is affecting any of the variables under consideration.

Example 6 Spurious Correlation Caused by Lurking Variables

Figure 5 gives a scatter diagram of the number of persons in prison versus the number of cell phone subscribers in each of 7 years. This plot exhibits a pattern of strong positive correlation. Would restricting the number of cell phones result in fewer persons in prison?

SOLUTION The scatter diagram reveals a strong positive correlation, but common sense suggests there is no cause-and-effect relation to tie an increase in the number of cell phone subscribers to an increase in the prison population. Realistically, the two variables should not have a causal relationship.

In Figure 6 we have repeated the scatter diagram but have labeled each point according to the year. For example, 94 stands for 1994. The years increase exactly in the same order as the points from lower left to upper right in the scatter diagram. More things change over time or from year to year. Year is just a stand-in, or proxy, for all of them. Since the

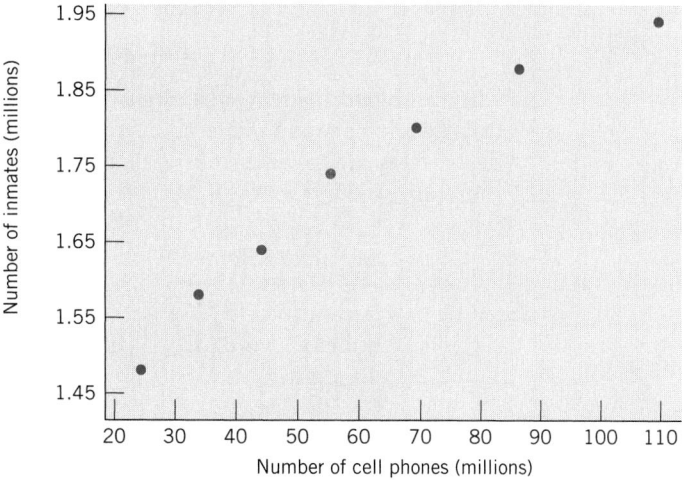

Figure 5   Scatter diagram reveals pattern of strong positive correlation.

population of the United States increased over these years, population size could be one lurking variable.

Once the year notation is added to the graph, it is clear that other variables are leading to the observed correlation. A graph, with time order added, can often help discredit false claims of causal relations. If the order of the years had been scrambled, we could not discredit the suggestion of a causal relation.

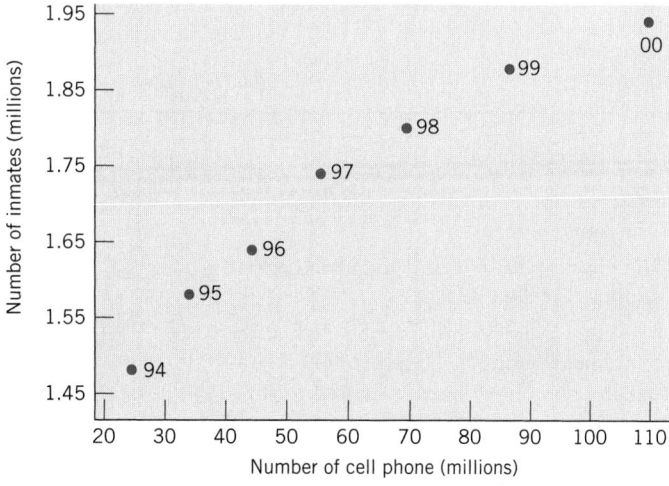

Figure 6   Scatter diagram pattern has strong relation to year.

Lurking Variables

The Insurance Institute for Highway Safety announced the safest model cars for the calendar years 1995 to 1998 in terms of fewest fatalities. The death rates are given in terms of one million cars that are registered for the year.

Lowest Fatality Rates	Highest Fatality Rates
Infinity J30 (20)	Chevrolet Camaro (308)
Cadillac Seville (33)	Chevrolet Camaro convertible (290)
Buick Riviera (36)	Pontiac Firebird (267)
Toyota Camry (37)	Geo Metro, four-door (212)
Volvo 850 (39)	Ford Aspire, four-door (210)
Lincoln Continental (39)	

Although it must be acknowledged there is truth in the statement that larger cars are generally safer than small cars, there is a big lurking variable here—the driver. How often does the teenager cruise in the luxury car? There is a strong correlation between the age of the driver and the type of car driven and also between the age of the driver and driver behavior.

WARNING. Don't confuse the existence of a high correlation between two variables with the situation where one is the cause of the other. Recall Example 6, where the number of cell phones and the number of persons in prison have a high correlation. There is no commonsense connection—no causal relation.

A high correlation between two variables can sometimes be caused when there is a third, "lurking" variable that strongly influences both of them.

Exercises

3.14 Would you expect a positive, negative, or nearly zero correlation for each of the following? Give reasons for your answers.

  (a) The physical fitness of a dog and the physical fitness of the owner.

  (b) For each person, the number of songs downloaded from the Internet last month and the number of hours listening to MP3 format music.

  (c) For a student, the number of close friends and the number of hours spent studying alone last week.

3.15 In each of the following instances, would you expect a positive, negative, or zero correlation?

  (a) Number of salespersons and total dollar sales for real estate firms.

  (b) Total payroll and percent of wins of national league baseball teams.

  (c) The amount spent on a week of TV advertising and sales of a cola.

  (d) Age of adults and their ability to maintain a strenuous exercise program.

3.16 Would you expect a positive, negative, or nearly zero correlation for each of the following? Give reasons for your answers.

(a) The number of swim suits owned and the distance from a residence to the nearest outdoor public swimming area.

(b) The times to a first flower for a white daisy and yellow daisy planted two feet apart in several locations around the state.

(c) Weight and the amount of physical activity for a thirty year old.

(d) A person's height and the number of mystery novels they read last year.

3.17 If the value of r is small, can we conclude that there is not a strong relationship between the two variables?

3.18 For the data set

x	1	2	7	4	6
y	5	4	1	3	2

(a) Construct a scatter diagram.

(b) Guess the sign and value of the correlation coefficient.

(c) Calculate the correlation coefficient.

3.19 Refer to the alligator data in Table D.11 of the Data Bank. Using the data on x_3 and x_4 for male and female alligators from Lake Apopka:

(a) Make a scatter diagram of the pairs of concentrations for the male alligators. Calculate the sample correlation coefficient.

(b) Create a multiple scatter diagram by adding, on the same plot, the pairs of concentrations for the female alligators. Use a different symbol for females. Calculate the sample correlation coefficient for this latter group.

(c) Comment on any major differences between the male and female alligators.

3.20 (a) Construct scatter diagrams of the data sets

(i)
x	0	4	2	6	3
y	3	5	1	7	4

(ii)
x	0	4	2	6	3
y	7	1	4	3	5

(b) Calculate r for the data set (i).

(c) Guess the value of r for the data set (ii) and then calculate r. (Note: The x and y values are the same for both sets, but they are paired differently in the two cases.)

3.21 Match the following values of r with the correct diagrams (Figure 7).

(a) $r = -.3$ (b) $r = .1$ (c) $r = .9$

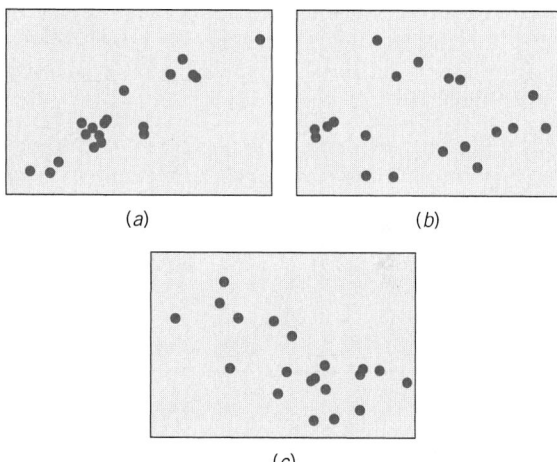

(a)   (b)

(c)

Figure 7   Scatter diagrams.

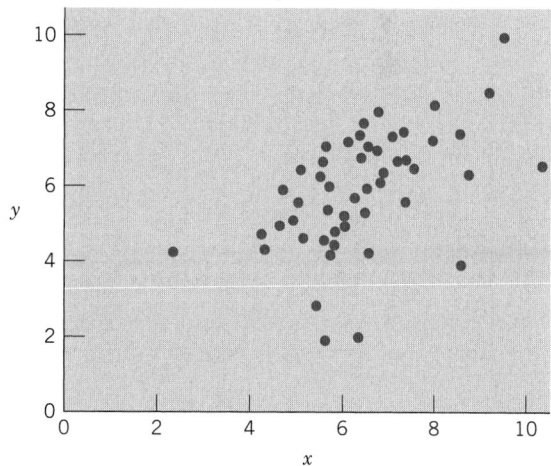

Figure 8   Scatter diagram.

3.22 Is the correlation in Figure 8 about (a) .1, (b) .5, (c) .9, or (d) −.7?

3.23 Calculations from a data set of $n = 36$ pairs of (x, y) values have provided the following results.

$$\sum (x - \bar{x})^2 = 530.7 \quad \sum (y - \bar{y})^2 = 215.2$$
$$\sum (x - \bar{x})(y - \bar{y}) = -204.3$$

Obtain the correlation coefficient.

3.24 Over the years, a traffic officer noticed that cars with fuzzy dice hanging on the rear-view mirror always seemed to be speeding. Perhaps tongue in cheek, he suggested that outlawing the sale of fuzzy dice would reduce the number of cars exceeding the speed limit. Comment on lurking variables.

3.25 Heating and combustion analyses were performed in order to study the composition of moon rocks. Recorded here are the determinations of hydrogen (H) and carbon (C) in parts per million (ppm) for 11 specimens.

Hydrogen (ppm)	120	82	90	8	38	20	2.8	66	2.0	20	85
Carbon (ppm)	105	110	99	22	50	50	7.3	74	7.7	45	51

Calculate r.

3.26 A zoologist collected 20 wild lizards in the southwestern United States. After measuring their total length (mm), they were placed on a treadmill and their speed (m/sec) recorded.

Speed	1.28	1.36	1.24	2.47	1.94	2.52	2.67
Length	179	157	169	146	143	131	159

Speed	1.29	1.56	2.66	2.17	1.57	2.10	2.54
Length	142	141	130	142	116	130	140

Speed	1.63	2.11	2.57	1.72	0.76	1.02
Length	138	137	134	114	90	114

(a) Create a scatter plot. Comment on any unusual observations.

(b) Calculate the sample correlation coefficient.

3.27 An ongoing study of wolves is being conducted at the Yukon-Charley Rivers National Preserve.

Table D.9 in the Data Bank gives the physical characteristics of wolves that were captured.

Males

Body length (cm)	134	143	148	127	136	146
Weight (lb)		71	93	101	84	88	117

Body length (cm)	142	139	140	133	123
Weight (lb)		86	86	93	86	106

(a) Plot length versus weight for the male wolves. From your visual inspection, estimate the value of the correlation coefficient.

(b) Calculate the sample correlation coefficient for male wolves.

(c) Create a multiple scatter diagram by adding the points for female wolves from Table D.9 to your plot in part (a). Do the patterns of correlation for males and females appear to be similar or different?

3.28 An ongoing study of wolves is being conducted at the Yukon-Charley Rivers National Preserve. Table D.9 in the Data Bank gives the physical characteristics of wolves that were captured.

Females

Body length (cm)	123	129	143	124	125	122	125	122
Weight (lb)		57	84	90	71	71	77	68	73

(a) Plot length versus weight for the female wolves. From your visual inspection, estimate the value of the correlation coefficient.

(b) Calculate the sample correlation coefficient for female wolves.

3.29 Refer to Example 6 concerning spurious correlation. Replace number of cell phone subscribers with attendance at broadway shows (in millions).

Cell phones	24.1	33.8	40.0	55.3	69.2	86.0	109.4
Inmates	1.48	1.58	1.64	1.74	1.80	1.88	1.94
Attendance	8.1	9.0	9.5	10.6	11.5	11.7	11.4
Year	1994	1995	1996	1997	1998	1999	2000

(a) Create a scatter diagram and identify the kind of association.

(b) Comment on possible lurking variables.

3.30 *A further property of r.* Suppose all x measurements are changed to $x' = ax + b$ and all y measurements to $y' = cy + d$, where $a, b, c,$ and d are fixed numbers $(a \neq 0, c \neq 0)$. Then the correlation coefficient remains unchanged if a and c have the same signs; it changes sign but not numerical value if a and c are of opposite signs.

This property of r can be verified along the lines of Exercise 2.74 in Chapter 2. In particular, the deviations $x - \bar{x}$ change to $a(x - \bar{x})$ and the deviations $y - \bar{y}$ change to $c(y - \bar{y})$. Consequently, $\sqrt{S_{xx}}$, $\sqrt{S_{yy}}$, and S_{xy} change to $|a|\sqrt{S_{xx}}, |c|\sqrt{S_{yy}}$, and acS_{xy}, respectively (recall that we must take the positive square root of a sum of squares of the deviations). Therefore, r changes to

$$\frac{ac}{|a||c|} r = \begin{cases} r \text{ if } a \text{ and } c \text{ have the same signs} \\ -r \text{ if } a \text{ and } c \text{ have opposite signs} \end{cases}$$

(a) For a numerical verification of this property of r, consider the data of Exercise 3.18. Change the x and y measurements according to

$$x' = 3x - 2$$
$$y' = -y + 10$$

Calculate r from the (x', y') measurements and compare with the result of Exercise 3.18.

(b) Suppose from a data set of height measurements in inches and weight measurements in pounds, the value of r is found to be .86. What would the value of r be if the heights were measured in centimeters and weights in kilograms?

3.31 The amount of municipal solid waste created has become a major problem. According to the Environmental Protection Agency, the yearly amounts (millions of tons) are:

Year	1960	1970	1980	1990	2000
Garbage (millions of tons)	88	121	152	205	232
Population (millions)	179	203	227	249	281

(a) Plot the amount of garbage (millions of tons) versus year.

(b) Visually, does there appear to be a strong correlation? Explain.

(c) Give one possible lurking variable.

3.32 Refer to the data on garbage in Exercise 3.31.

(a) Plot the amount of garbage (millions of tons) versus population (millions).

(b) Does there appear to be a strong correlation? Explain.

(c) How does your interpretation of the association differ from that to Exercise 3.31, parts (b) and (c)?

3.33 Refer to the data on garbage in Exercises 3.31 and 3.30.

(a) Replace year by (year—1960). Calculate the correlation coefficient between (year—1960) and amount of garbage in millions of tons.

(b) Based on your calculation in part (a), what is the correlation between year itself and the amount of garbage? Explain.

3.34 Refer to the data on garbage in Exercises 3.31 and 3.30.

(a) Calculate the correlation coefficient between the amount of garbage in millions of tons and the population size in millions.

(b) Based on your calculation in part (a), give the correlation coefficient between the amount of garbage in pounds and population size in number of persons. Explain your answer. [*Hint:* Recall that there are 2000 pounds in a ton so (number of pounds) = 2000 × (number of tons).]

6. PREDICTION OF ONE VARIABLE FROM ANOTHER (LINEAR REGRESSION)

An experimental study of the relation between two variables is often motivated by a need to predict one from the other. The administrator of a job training program may wish to study the relation between the duration of training and the score of the trainee on a subsequent skill test. A forester may wish to estimate the timber volume of a tree from the measurement of the trunk diameter a few feet above the ground. A medical technologist may be interested in predicting the blood alcohol measurement from the read-out of a newly devised breath analyzer.

In such contexts as these, the predictor or input variable is denoted by x, and the response or output variable is labeled y. The object is to find the nature of relation between x and y from experimental data and use the relation to predict the response variable y from the input variable x. Naturally, the first step in such a study is to plot and examine the scatter diagram. If a linear relation emerges, the calculation of the numerical value of r will confirm the strength of the linear relation. Its value indicates how effectively y can be predicted from x by fitting a straight line to the data.

A line is determined by two constants: its height above the origin (intercept) and the amount that y increases whenever x is increased by one unit (slope). (See Figure 9.) Chapter 11 explains an objective method of best fitting a straight line, called the method of least squares. This best fit line, or least squares line, is close to the points graphed in the scatter plot in terms of minimizing the amount of vertical distance.

Equation of the Line Fitted by Least Squares

$$\hat{y} = \hat{\beta}_0 + \hat{\beta}_1 x$$

where

$$\text{Slope } \hat{\beta}_1 = \frac{S_{xy}}{S_{xx}} = \frac{\sum (x - \bar{x})(y - \bar{y})}{\sum (x - \bar{x})^2}$$

$$\text{Intercept } \hat{\beta}_0 = \bar{y} - \hat{\beta}_1 \bar{x}$$

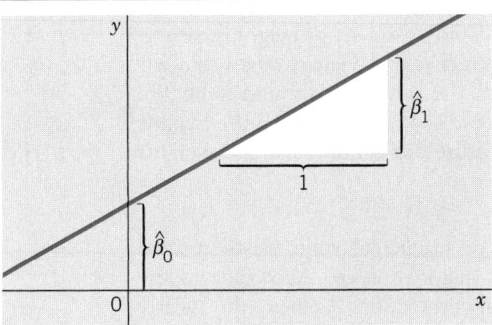

Figure 9   The line $\hat{y} = \hat{\beta}_0 + \hat{\beta}_1 x$.

Besides the sample mean $\bar{x}$ and $\bar{y}$, the fitted line involves the sum of the squared deviations of the x observations, S_{xx}, and the sum of the cross products of the x observations and the y deviations, S_{xy}. The formulas will be examined in more detail in Chapter 11.

Example 7   Calculation of the Line Fitted by Least Squares

A chemist wishes to study the relation between the drying time of a paint and the concentration of a base solvent that facilitates a smooth application. The data of concentration setting x and the observed drying times y are recorded in the first two columns of Table 10. Plot the data, calculate r, and obtain the fitted line.

TABLE 10   Data of Concentration x and
Drying Time y (in minutes) and
the Basic Calculations

Concentration x	Drying Time y	x^2	y^2	xy
0	1	0	1	0
1	5	1	25	5
2	3	4	9	6
3	9	9	81	27
4	7	16	49	28
Total 10	25	30	165	66

SOLUTION   The scatter diagram in Figure 10 gives the appearance of a linear relation. To calculate r and determine the equation of the fitted line, we first calculate the basic quantities $\bar{x}$, $\bar{y}$, S_{xx}, S_{yy}, and S_{xy} using the totals in Table 10.

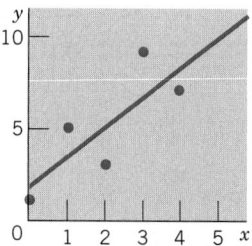

Figure 10   Scatter diagram.

$$\bar{x} = \frac{10}{5} = 2 \qquad \bar{y} = \frac{25}{5} = 5$$

$$S_{xx} = 30 - \frac{(10)^2}{5} = 10$$

$$S_{yy} = 165 - \frac{(25)^2}{5} = 40$$

$$S_{xy} = 66 - \frac{10 \times 25}{5} = 16$$

$$r = \frac{16}{\sqrt{40 \times 10}} = \frac{16}{20} = .8$$

$$\hat{\beta}_1 = \frac{16}{10} = 1.6$$

$$\hat{\beta}_0 = 5 - (1.6)2 = 1.8$$

The equation of the fitted line is

$$\hat{y} = 1.8 + 1.6x$$

The line is shown on the scatter diagram in Figure 10.

   If we are to predict the drying time y corresponding to the concentration 2.5, we substitute $x = 2.5$ in our prediction equation and get the result.

   At $x = 2.5$, the predicted drying time $= 1.8 + 1.6(2.5) = 5.8$ minutes.

Graphically, this amounts to reading the ordinate of the fitted line at $x = 2.5$.

   Software programs greatly simplify the calculation and plotting of the fitted line. The MINITAB calculations for Example 7 are shown in Figure 11. Column 1 is named x and column 2, y.

Data:

C1: 0 1 2 3 4
C2: 1 5 3 9 7
Stat > **Regression** > **Fitted Line Plot.**
Type *C2* in **Response.** Type *C1* in Predictors.
Under **Type of Regression Model** choose **Linear.** Click **OK.**

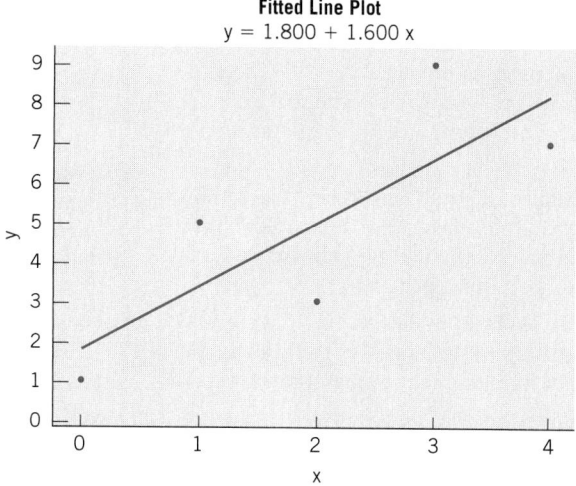

Figure 11 MINITAB output for fitted line in Example 7.

The sample correlation r was introduced as a measure of association between two variables. When r is near 1 or -1, points in the scatter plot are closely clustered about a straight line and the association is high. In these circumstances, the value of one variable can be accurately predicted from the value of the other. Said another way, when the value of r^2 is near 1, we can predict the value of y from its corresponding x value. In all cases, the slope of the least squares line $\hat{\beta}_1$ and the sample correlation r are related since $\hat{\beta}_1 = r\sqrt{S_{yy}}/\sqrt{S_{xx}}$. If the sample correlation is positive, then the slope of the least squares line is positive. Otherwise, both are negative or both zero.

Here we have only outlined the basic ideas concerning the prediction of one variable from another in the context of a linear relation. Chapter 11 expands on these ideas and treats statistical inferences associated with the prediction equation.

Exercises

3.35 Plot the line $y = 2 + 3x$ on graph paper by locating points for $x = 1$ and $x = 4$. What is its intercept? Its slope?

3.36 Plot the line $y = 8 - 2x$ on graph paper by locating the points for $x = 0$ and $x = 3$. What is its intercept? Its slope?

3.37 A store manager has determined that the monthly profit y realized from selling a particular brand of car battery is given by

$$y = 10x - 155$$

where x denotes the number of these batteries sold in a month.

(a) If 41 batteries were sold in a month, what was the profit?

(b) At least how many batteries must be sold in a month in order to make a profit?

3.38 Identify the predictor variable x and the response variable y in each of the following situations.

(a) A training director wishes to study the relationship between the duration of training for new recruits and their performance in a skilled job.

(b) The aim of a study is to relate the carbon monoxide level in blood samples from smokers with the average number of cigarettes they smoke per day.

(c) An agronomist wishes to investigate the growth rate of a fungus in relation to the level of humidity in the environment.

(d) A market analyst wishes to relate the expenditures incurred in promoting a product in test markets and the subsequent amount of product sales.

3.39 Given these five pairs of (x, y) values

x	1	2	3	4	5
y	1	2.2	2.6	3.4	3.9

(a) Plot the points on graph paper.

(b) From a visual inspection, draw a straight line that appears to fit the data well.

(c) Compute the least squares estimates $\hat{\beta}_0$ and $\hat{\beta}_1$ and draw the fitted line.

3.40 For the data set

x	1	2	7	4	6
y	4	3	0	2	1

(a) Construct a scatter diagram.

(b) From a visual inspection, draw a straight line that appears to fit the data well.

(c) Calculate the least squares estimates and draw the least squares fitted line on your plot.

3.41 In an experiment to study the relation between the time waiting in line, y (minutes), to get to the head of the checkout line at her favorite grocery store and the number of persons ahead in line, x, a student collected the following statistics:

$$n \ = \ 9 \quad \Sigma x \ = \ 19 \quad \Sigma y \ = \ 39.9$$
$$S_{xx} \ = \ 9.4 \quad S_{yy} \ = \ 17.8 \quad S_{xy} \ = \ 10.2$$

(a) Find the equation of the least squares fitted line.

(b) Using the fitted line, predict the time waiting in line when 3 persons are already in line.

3.42 Wolves used to range over much of Michigan, Minnesota, and Wisconsin. They were reintroduced several years ago, but counts over the winter showed that the populations are expanding rapidly over the past few years.

Total Number of Wolves in Wisconsin and Michigan

Year	1994	1995	1996	1997	1998
No. wolves	114	163	215	257	318

Year	1999	2000	2001	2002
No. wolves	379	482	500	600

(a) Plot the number of wolves versus the year the count was taken.

(b) Fit a least squares line to summarize the growth. To simplify the calculation, code 1994 as $x = 1$, 1995 as $x = 2$, and so on.

(c) Does your fitted straight line summarize the growth in the wolf population over this period of time? If so, what numerical value summarizes the change in population size from one winter to the next?

3.43 The amount of municipal solid waste created has become a major problem. According to the Environmental Protection Agency, the yearly amount (millions of tons) are:

Year	1960	1970	1980	1990	2000
Garbage (million tons)	88	121	152	205	232
Population (millions)	179	203	227	249	281

(a) Plot the amount of garbage (millions of tons) versus population (millions).

(b) Fit a straight line using x = population in millions.

(c) According to the fitted line, how much garbage is created by a typical person?

USING STATISTICS WISELY

1. To study the association between two variables, you need to collect the pair of values obtained from each unit in the sample. There is no information about association in the summaries of the observations on individual variables.

2. To study association when both variables are categorical, cross-tabulate the frequencies in a two-way table. Calculate relative frequencies based on the total number.

3. To look for association between any pair of variables whose values are numerical, create a scatter diagram and look for a pattern of association.

4. Never confuse a strong association with a causal relationship. The relation may be due to a lurking variable.

5. Remember that the correlation coefficient measures the clustering of points about a straight line. It is not appropriate for a relationship on a curve or disjoint groups of points.

6. Before using a fitted line to predict one variable from another, create a scatter plot and inspect the pattern to see if a straight-line relationship is appropriate.

KEY IDEAS AND FORMULAS

Comparative trials often have a placebo, or inactive treatment, which serves as a control. This eliminates from the comparison a placebo effect where some subjects in the control group responded positively just because they were given the attention of being participants. An experiment has double blind trials when neither the subject nor the person administering the treatments knows which treatment is given.

A random assignment of treatments helps prevent uncontrolled sources of variation from systematically influencing the responses.

Data on two traits can be summarized in a two-way table of frequencies where the categories for one trait are in the left margin and categories for the other trait along the upper margin. These are said to be cross-classified or cross-tabulated data and the summary tables of frequencies are called contingency tables.

The combining of two contingency tables that pertain to the same two traits, but are based on very different populations, can lead to very misleading conclusions if the two tables are combined. This is called Simpson's paradox when there is a third variable that strongly influences the other two.

A scatter plot or scatter diagram shows all the values (x_i, y_i) of a pair of variables as points in two dimensions. This plot can be visually inspected for the strength of association between the two variables.

The correlation coefficient r measures how closely the scatter approximates a straight-line pattern.

A positive value of correlation indicates a tendency of large values of x to occur with large values of y, and also for small values of both to occur together.

A negative value of correlation indicates a tendency of large values of x to occur with small values of y and vice versa.

A high correlation does not necessarily imply a causal relation.

In fact, a high value of correlation between two variables may be spurious. That is, the two variables may not be connected but the apparent correlation is caused by a third lurking variable that strongly influences both of the original two variables.

A least squares fit of a straight line helps describe the relation of the response or output variable y to the predictor or input variable x.

A y value may be predicted for a known x value by reading from the fitted line $\hat{y} = \hat{\beta}_0 + \hat{\beta}_1 x$.

For pairs of measurements (x, y)

$$\text{Sample correlation} \quad r = \frac{S_{xy}}{\sqrt{S_{xx}}\sqrt{S_{yy}}}$$

where $S_{xx} = \sum (x - \bar{x})^2$, $S_{yy} = \sum (y - \bar{y})^2$, and $S_{xy} = \sum (x - \bar{x})(y - \bar{y})$.

$$\text{Fitted line} \quad \hat{y} = \hat{\beta}_0 + \hat{\beta}_1 x$$

where

$$\text{Slope} \quad \hat{\beta}_1 = \frac{S_{xy}}{S_{xx}} \quad \text{and} \quad \text{Intercept} \quad \hat{\beta}_0 = \bar{y} - \hat{\beta}_1 \bar{x}$$

TECHNOLOGY

Fitting a straight line and calculating the correlation coefficient

MINITAB

Fitting a straight line

Begin with the values for the predictor variable x in $C1$ and the response variable y in $C2$.

Stat > **Regression** > **Regression.**
Type C2 in **Response.** Type $C1$ in **Predictors.**
Click **OK.**

To calculate the correlation coefficient, start as above with data in *C1* and *C2*.

> **Stat > Basic Statistics > Correlation.**
> Type *C1 C2* in **Variables.** Click **OK.**

EXCEL

Fitting a straight line

Begin with the values of the predictor variable in column A and the values of the response variable in column B.

> Highlight the data and go to **Insert,** then **Chart.**
> Select **XY (Scatter)** and click **Finish.**
> Go to **Chart,** then **Add Trendline.**
> Click on the **Options** tab and check **Display equation on chart.**
> Click **OK.**

To calculate the correlation coefficient, begin with the predictor variable in column A and the response variable in column B.

> Click on a blank cell. Select **Insert** and then **Function** (or click on the f_x icon).
> Select **Statistical** and then **CORREL.** Click **OK.**
> **Highlight** the data in Column A for **Array1** and **Highlight** the data in Column B for **Array2.** Then, click **OK.**

TI-84/-83 PLUS

Fitting a straight line

Enter the values of the predictor variable in **L1** and those of the response variable in **L2.**

> Select **STAT** then **Calc** and then **4: LinReg (ax+b).**
> With **LinReg** on the Home screen press **Enter.**

The calculator will return the intercept *a*, slope *b*, correlation coefficient *r*. (If *r* is not shown, go to the 2nd **O: CATALOG** and select **DiagnosticON.** Press **ENTER** twice. Then go back to **LinREg.**)

---

7. REVIEW EXERCISES

3.44 Applicants for welfare are allowed an appeals process when they feel they have been unfairly treated. At the hearing, the applicant may choose self-representation or representation by an attorney. The appeal may result in an increase, decrease, or no change in benefit recommendation. Court records of 320 appeals cases provided the data at the top of the next page. Calculate the relative frequencies for each row and compare the patterns of the appeals decisions between the two types of representation.

Type of Representation	Amount of Aid			Total
	Increased	Unchanged	Decreased	
Self	59	108	17	
Attorney	70	63	3	
Total				

3.45 Sugar content (g) and carbohydrate content (g) are obtained from the package of the breakfast cereals referred to in Exercise 3.2.

General Mills		Kellogg's		Quaker	
Sugar	Carb.	Sugar	Carb.	Sugar	Carb.
13	12	13	15	9	14
1	18	4	18	6	17
13	11	14	14	10	12
13	11	12	15	0	12
12	12	3	21	14	29
5	16	18	21	13	23
19	19	15	8	9	31
16	22	16	13	12	23
14	26	15	31	16	15
16	28	4	20	13	15

(a) Calculate the sample mean carbohydrates for all 30 cereals.

(b) Construct a table like the one in Exercise 3.2 but using carbohydrates rather than sugar.

(c) Calculate the relative frequencies separately for each row. Comment on any pattern.

3.46 Refer to Exercise 3.45.

(a) Make a scatter plot for the cereals made by General Mills.

(b) Calculate r for the cereals made by General Mills. Do sugar content and carbohydrate content seem to be associated or unrelated?

3.47 A dealer's recent records of 75 truck sales provided the following frequency information on size of truck and type of drive.

Truck Size	2-Wheel Drive	4-Wheel Drive
Small	12	21
Full	27	15

(a) Determine the marginal totals.

(b) Obtain the table of relative frequencies.

(c) Calculate the relative frequencies separately for each row.

(d) Does there appear to be a difference in the choice of drive for purchasers of small- and full-size trucks?

3.48 A high-risk group of 1083 male volunteers was included in a major clinical trial for testing a new vaccine for type B hepatitis. The vaccine was given to 549 persons randomly selected from the group, and the others were injected with a neutral substance (placebo). Eleven of the vaccinated people and 70 of the nonvaccinated ones later got the disease.

(a) Present these data in the following two-way frequency table.

(b) Compare the rates of incidence of hepatitis between the two subgroups.

	Hepatitis	No Hepatitis	Total
Vaccinated Not vaccinated			
Total			

3.49 Would you expect a positive, negative, or nearly zero correlation for each of the following? Give reasons for your answers.

(a) The time a student spends playing computer games each week and the time they spend talking with friends in a group.

(b) The number of finals taken by undergraduates and their number of hours of sleep during finals week.

(c) A person's height and the number of movies he or she watched last month.

(d) The temperature at a baseball game and beer sales.

3.50 Examine each of the following situations and state whether you would expect to find a high correlation between the variables. Give reasons why an observed correlation cannot be interpreted as a direct relationship between the variables and indicate at least one possible lurking variable.

(a) The correlation between the number of Internet users and truck sales in cities of varying sizes.

(b) The correlation between yearly sales of satellite TV receivers and portable CD players over the past 10 years.

(c) The correlation between yearly sales of cell phones and number of new automated teller machines over the past 10 years.

(d) Correlation between the concentration x of air pollutants and the number of riders y on public transportation facilities when the data are collected from several cities that vary greatly in size.

(e) Correlation between the wholesale price index x and the average speed y of winning cars in the Indianapolis 500 during the last 10 years.

3.51 The tar yield of cigarettes is often assayed by the following method: A motorized smoking machine takes a two-second puff once every minute until a fixed butt length remains. The total tar yield is determined by laboratory analysis of the pool of smoke taken by the machine. Of course, the process is repeated on several cigarettes of a brand to determine the average tar yield. Given here are the data of average tar yield and the average number of puffs for six brands of filter cigarettes.

Average tar (milligrams)	12.2	14.3	15.7	12.6	13.5	14.0
Average no. of puffs	8.5	9.9	10.7	9.0	9.3	9.5

(a) Plot the scatter diagram.

(b) Calculate r.

Remark: Fewer puffs taken by the smoking machine mean a faster burn time. The amount of tar inhaled by a human smoker depends largely on how often the smoker puffs.

3.52 As part of a study of the psychobiological correlates of success in athletes, the following measurements (courtesy of W. Morgan) are obtained from members of the U.S. Olympic wrestling team.

Anger x	6	7	5	21	13	5	13	14
Vigor y	28	23	29	22	20	19	28	19

(a) Plot the scatter diagram.

(b) Calculate r.

(c) Obtain the least squares line.

(d) Predict the vigor score y when the anger score is $x = 8$.

3.53 Refer to Exercise 3.45.

(a) Make a scatter plot for the cereals made by Kellogg's.

(b) Calculate r for the cereals made by Kellogg's. Do sugar content and carbohydrate content seem to be associated or unrelated?

3.54 Given the following (x, y) values

x	0	2	5	4	1	6
y	4	3	3	1	6	1

(a) Make a scatter plot.

(b) Calculate r.

3.55 Given these five pairs of values

x	0	3	5	8	9
y	2	3	5	4	6

(a) Plot the scatter diagram.

(b) From a visual inspection, draw a straight line that appears to fit the data well.

(c) Compute the least squares estimates $\hat{\beta}_0$, $\hat{\beta}_1$ and draw the fitted line.

3.56 Given the following six pairs of values

x	1	2	3	5	6	7
y	3	5	2	0	2	0

(a) Obtain the least squares estimates $\hat{\beta}_0$, $\hat{\beta}_1$ and the fitted line.

(b) Predict the y value for $x = 6$.

3.57 Identify the predictor variable x and the response variable y in each of the following situations.

(a) The state highway department wants to study the relationship between road roughness and a car's gas consumption.

(b) A concession salesperson at football games wants to relate total fall sales to the number of games the home team wins.

(c) A sociologist wants to investigate the number of weekends a college student goes home in relation to the trip distance.

The Following Exercises Require a Computer

3.58 In Figure 11, we have illustrated the output from MINITAB commands for fitting a straight line. To create the scatter plot, without the fitted line, choose:

Graph > Scatter plot. Choose Simple. Click OK. Type **C2** in **Y variables** and **C1** in **X variables.** Click **OK.**

Use MINITAB (or another package program) to obtain the scatter diagram, correlation coefficient, and regression line for:

(a) The GPA and GMAT scores data of Table 7 in Example 3.

(b) The hydrogen x and carbon y data in Exercise 3.25.

3.59 For fitting body length to weight for all wolves given in Table D.9 in the Data Bank, use MINITAB or some other computer package to obtain:

(a) The scatter diagram.

(b) The correlation coefficient.

(c) The regression line.

3.60 Use MINITAB or some other computer package to obtain the scatter diagram, correlation coefficient, and regression line of:

(a) The final on the initial times to row given in Table D.4 in the Data Bank.

(b) Drop one unusual pair and repeat part(a). Comment on any major differences.

3.61 A director of student counseling is interested in the relationship between the numerical score x and the social science score y on college qualification tests. The following data (courtesy of R. W. Johnson) are recorded.

x	41	39	53	67	61	67
y	29	19	30	27	28	27
x	46	50	55	72	63	59
y	22	29	24	33	25	20
x	53	62	65	48	32	64
y	28	22	27	22	27	28
x	59	54	52	64	51	62
y	30	29	21	36	20	29
x	56	38	52	40	65	61
y	34	21	25	24	32	29
x	64	64	53	51	58	65
y	27	26	24	25	34	28

(a) Plot the scatter diagram.

(b) Calculate r.

4

Probability

Uncertainty of Weather Forecasts

Today's forecast: Increasing cloudiness with a 25% chance of snow.

Probabilities express the chance of events that cannot be predicted with certainty. Even unlikely events sometimes occur. © Patti McConville/The Image Bank/Getty Images.

1.   INTRODUCTION

In Chapter 1, we introduced the notions of *sample* and *statistical population* in the context of investigations where the outcomes exhibit variation. Although complete knowledge of the statistical population remains the target of an investigation, we typically have available only the partial information contained in a sample. Chapter 2 focused on some methods for describing the salient features of a data set by graphical presentations and calculation of the mean, standard deviation, and other summary statistics. When the data set represents a sample from a statistical population, its description is only a preliminary part of a statistical analysis. Our major goal is to make generalizations or inferences about the target population on the basis of information obtained from the sample data. An acquaintance with the subject of probability is essential for understanding the reasoning that leads to such generalizations.

In everyday conversations, we all use expressions of the kind:

"Most likely our team will win this Saturday."

"It is unlikely that the weekend will be cold."

"I have a 50–50 chance of getting a summer job at the camp."

The phrases "most likely," "probable," "quite likely," and so on are used qualitatively to indicate the chance that an event will occur. Probability, as a subject, provides a means of quantifying uncertainty. In general terms, the probability of an event is a numerical value that gauges how likely it is that the event will occur. We assign probability on a scale from 0 to 1 with a very low value indicating extremely unlikely, a value close to 1 indicating very likely, and the intermediate values interpreted accordingly. A full appreciation for the concept of a numerical measure of uncertainty and its role in statistical inference can be gained only after the concept has been pursued to a reasonable extent. We can, however, preview the role of probability in one kind of statistical reasoning.

> *Suppose it has been observed that in 50% of the cases a certain type of muscular pain goes away by itself. A hypnotist claims that her method is effective in relieving the pain. For experimental evidence, she hypnotizes 15 patients and 12 get relief from the pain. Does this demonstrate that hypnotism is effective in stopping the pain?*
>
> *Let us scrutinize the claim from a statistical point of view. If indeed the method had nothing to offer, there would still be a 50–50 chance that a patient is cured. Observing 12 cures out of 15 amounts to obtaining 12 heads in 15 tosses of a coin. We will see later that the probability of at least 12 heads in 15 tosses of a fair coin is .018, indicating that the event is not likely to happen. Thus, if we tentatively assume the model (or hypothesis) that the method is ineffective, 12 or more cures are very unlikely. Rather than agree that an unlikely*

event has occurred, we conclude that the experimental evidence strongly supports the hypnotist's claim.

This kind of reasoning, called *testing a statistical hypothesis*, will be explored in greater detail later. For now, we will be concerned with introducing the ideas that lead to assigned values for probabilities.

2.   PROBABILITY OF AN EVENT

The probability of an event is viewed as a numerical measure of the chance that the event will occur. The idea is naturally relevant to situations where the outcome of an experiment or observation exhibits variation.

Although we have already used the terms "experiment" and "event," a more specific explanation is now in order. In the present context, the term experiment is not limited to the studies conducted in a laboratory. Rather, it is used in a broad sense to include any operation of data collection or observation where the outcomes are subject to variation. Rolling a die, drawing a card from a shuffled deck, sampling a number of customers for an opinion survey, and quality inspection of items from a production line are just a few examples.

> An experiment is the process of observing a phenomenon that has variation in its outcomes.

Before attempting to assign probabilities, it is essential to consider all the eventualities of the experiment. Pertinent to their description, we introduce the following terminologies and explain them through examples.

> The sample space associated with an experiment is the collection of all possible distinct outcomes of the experiment.
>
> Each outcome is called an elementary outcome, a simple event, or an element of the sample space.
>
> An event is the set of elementary outcomes possessing a designated feature.

The elementary outcomes, which together comprise the sample space, constitute the ultimate breakdown of the potential results of an experiment. For instance, in rolling a die, the elementary outcomes are the points 1, 2, 3, 4, 5, and 6, which together constitute the sample space. The outcome of a football game would be either a win, loss, or tie for the home team. Each time the experiment is performed, one and only one elementary outcome can occur. A sample space can be specified by either listing all the elementary outcomes, using convenient symbols to identify them, or making a descriptive statement that characterizes the entire collection. For general discussion, we denote:

The sample space by S

The elementary outcomes by $e_1, e_2, e_3, \ldots$

Events by $A, B,$ and so on.

In specific applications, the elementary outcomes may be given other labels that provide a more vivid identification.

An event A occurs when any one of the elementary outcomes in A occurs.

Example 1   A Tree Diagram and Events for Coin Tossing

Toss a coin twice and record the outcome head (H) or tail (T) for each toss. Let A denote the event of getting exactly one head and B the event of getting no heads at all. List the sample space and give the compositions of A and B.

SOLUTION   For two tosses of a coin, the elementary outcomes can be conveniently identified by means of a **tree diagram**.

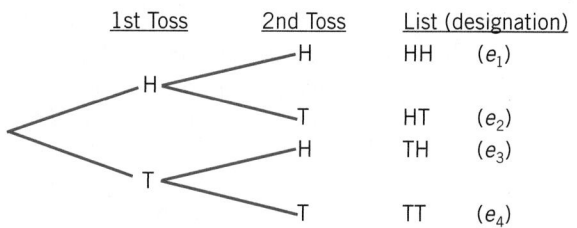

The sample space can then be listed as $S = \{HH, HT, TH, TT\}$. With the designation given above, we can also write

$$S = \{e_1, e_2, e_3, e_4\}$$

The order in which the elements of S are listed is inconsequential. It is the collection that matters.

Consider the event A of getting exactly one head. Scanning the above list, we see that only the elements HT (e_2) and TH (e_3) satisfy this requirement. Therefore, the event A has the composition

$$A = \{e_2, e_3\}$$

which is, of course, a subset of S. The event B of getting no heads at all consists of the single element e_4 so $B = \{e_4\}$. That is, B is a simple event as well as an event. The term "event" is a general term that includes simple events.

Example 2   A Sample Space and an Event Based on a Count

On a Saturday afternoon, 135 customers will be observed during check-out and the number paying by card, credit or debit, will be recorded. Identify (a) the sample space and (b) the event that more than 50% of purchases are made with a card.

SOLUTION   (a)   Since the number of customers who purchase with a card could be any of the numbers 0, 1, 2, . . . , 135, the sample space can be listed simply as

$$S = \{0, 1, 2, \ldots , 135\}$$

Using the notation e for elementary outcome, one can also describe this sample space as $S = \{e_0, e_1, e_2, \ldots , e_{135}\}$.

(b)   Let A stand for the event that more than 50% of the customers purchase with a card. Calculating $.5 \times 135 = 67.5$, we identify

$$A = \{68, 69, \ldots , 135\}$$

Both Examples 1 and 2 illustrate sample spaces that have a finite number of elements. There are also sample spaces with infinitely many elements. For instance, suppose a gambler at a casino will continue pulling the handle of a slot machine until he hits the first jackpot. The conceivable number of attempts does not have a natural upper limit so the list never terminates. That is, $S = \{1, 2, 3, \ldots \}$ has an infinite number of elements. However, we notice that the elements could be arranged one after another in a sequence. An infinite sample space where such an arrangement is possible is called "countably infinite." Either of these two types of sample spaces is called a discrete sample space.

Another type of infinite sample space is also important. Suppose a car with a full tank of gasoline is driven until its fuel runs out and the distance traveled recorded. Since distance is measured on a continuous scale, any nonnegative number is a possible outcome. Denoting the distance traveled by d, we can describe this sample space as $S = \{d; d \geq 0\}$, that is, the set of all real numbers greater than or equal to zero. Here the elements of S form a continuum and cannot be arranged in a sequence. Any S that is an interval is called a continuous sample space.

To avoid unnecessary complications, we will develop the basic principles of probability in the context of finite sample spaces. We first elaborate on the notion of the probability of an event as a numerical measure of the chance that it will occur. The most intuitive interpretation of this quantification is to consider the fraction of times the event would occur in many repeated trials of the experiment.

> The probability of an event is a numerical value that represents the pro-
> portion of times the event is expected to occur when the experiment is
> repeated many times under identical conditions.
>
> The probability of event A is denoted by $P(A)$.

Since a proportion must lie between 0 and 1, the probability of an event is
a number between 0 and 1. To explore a few other important properties of
probability, let us refer to the experiment in Example 1 of tossing a coin
twice. The event A of getting exactly one head consists of the elementary outcomes HT (e_2) and TH (e_3). Consequently, A occurs if either of these occurs.
Because

$$\begin{bmatrix} \text{Proportion of times} \\ A \text{ occurs} \end{bmatrix} = \begin{bmatrix} \text{Proportion of times} \\ \text{HT occurs} \end{bmatrix} + \begin{bmatrix} \text{Proportion of times} \\ \text{TH occurs} \end{bmatrix}$$

the number that we assign as $P(A)$ must be the sum of the two numbers $P(\text{HT})$
and $P(\text{TH})$. Guided by this example, we state some general properties of probability.

**The probability of an event is the sum of the probabilities assigned to all
the elementary outcomes contained in the event.**

Next, since the sample space S includes all conceivable outcomes, in every trial
of the experiment some element of S must occur. Viewed as an event, S is certain to occur, and therefore its probability is 1.

The sum of the probabilities of all the elements of S must be 1.

In summary:

Probability must satisfy:

1. $0 \leq P(A) \leq 1$ for all events A

2. $P(A) = \displaystyle\sum_{\text{all } e \text{ in } A} P(e)$

3. $P(S) = \displaystyle\sum_{\text{all } e \text{ in } S} P(e) = 1$

We have deduced these basic properties of probability by reasoning from
the definition that the probability of an event is the proportion of times the
event is expected to occur in many repeated trials of the experiment.

An assessment of the probabilities of events and their consequences can help to guide decisions. Calvin and Hobbes © 1990 Universal Press Syndicate. Reprinted with permission.

Exercises

4.1   Match the proposed probability of *A* with the appropriate verbal description. (More than one description may apply.)

Probability		Verbal Description
(a)	.95	(i) No chance of happening
(b)	.02	(ii) Very likely to happen
(c)	3.0	(iii) As much chance of occurring as not
(d)	−.1	(iv) Very little chance of happening
(e)	.3	(v) May occur but by no means certain
(f)	.5	(vi) An incorrect assignment
(g)	0	

4.2   For each numerical value assigned to the probability of an event, identify the verbal statements that are appropriate.

(a)   1.1   (b) $\dfrac{1}{1.1}$ (c) $\dfrac{1}{2}$ (d) $\dfrac{45}{47}$

(e) $\dfrac{1}{89}$ (f) 1.0

Verbal statements: (i) cannot be a probability, (ii) the event is very unlikely to happen, (iii) 50–50 chance of happening, (iv) sure to happen, (v) more likely to happen than not.

4.3   Identify the statement that best describes each $P(A)$.

(a) $P(A) = 2.0$ (i) $P(A)$ is incorrect.

(b) $P(A) = .35$ (ii) *A* rarely occurs.

(c) $P(A) = .04$ (iii) *A* occurs moderately often.

4.4   Construct a sample space for each of the following experiments.

(a) Someone claims to be able to taste the difference between the same brand of bottled, tap, and canned draft beer. A glass of each is poured and given to the subject in an unknown order. The subject is asked to identify the contents of each glass. The number of correct identifications will be recorded.

(b) Record the number of traffic fatalities in a state next year.

(c) Observe the length of time a new video recorder will continue to work satisfactorily without service.

Which of these sample spaces are discrete and which are continuous?

4.5 Identify these events in Exercise 4.4.

(a) Not more than one correct identification.

(b) Less accidents than last year.
(*Note*: If you don't know last year's value, use 345.)

(c) Longer than the 90-day warranty but less than 425.4 days.

4.6 When bidding on two projects, the president and vice president of a construction company make the following probability assessments for winning the contracts.

President	Vice President
P (win none) $= .1$	P (win none) $= .1$
P (win only one) $= .5$	P (win Project 1) $= .4$
P (win both) $= .4$	P (win Project 2) $= .2$
	P (win both) $= .3$

For both cases, examine whether or not the probability assignment is permissible.

4.7 Bob, John, Linda, and Sue are the finalists in the spelling contest of a local school district. The winner and the first runner-up will be sent to a statewide competition.

(a) List the sample space concerning the outcomes of the local contest.

(b) Give the composition of each of the following events.

A = Linda wins the local contest
B = Bob does not go to the state contest

4.8 Consider the following experiment: A coin will be tossed twice. If both tosses show heads, the experiment will stop. If one head is obtained in the two tosses, the coin will be tossed one more time, and in the case of both tails in the two tosses, the coin will be tossed two more times.

(a) Make a tree diagram and list the sample space.

(b) Give the composition of the following events.

A = [Two heads] B = [Two tails]

4.9 There are four elementary outcomes in a sample space. If $P(e_1) = .2$, $P(e_2) = .5$, and $P(e_3) = .1$, what is the probability of e_4?

4.10 Suppose $S = \{e_1, e_2, e_3\}$. If the simple events e_1, e_2, and e_3 are all equally likely, what are the numerical values of $P(e_1)$, $P(e_2)$, and $P(e_3)$?

4.11 The sample space for the response of a single person's attitude toward a political issue consists of the three elementary outcomes e_1 = {Unfavorable}, e_2 = {Favorable}, and e_3 = {Undecided}. Are the following assignments of probabilities permissible?

(a) $P(e_1) = .8,$ $P(e_2) = .1,$ $P(e_3) = .1$
(b) $P(e_1) = .4,$ $P(e_2) = .4,$ $P(e_3) = .4$
(c) $P(e_1) = .5,$ $P(e_2) = .5,$ $P(e_3) = 0$

4.12 A campus organization will select one day of the week for an end-of-year picnic. Assume that the weekdays, Monday through Friday, are equally likely and that each weekend day, Saturday and Sunday, is twice as likely as a weekday to be selected.

(a) Assign probabilities to the seven outcomes.

(b) Find the probability a weekday will be selected.

4.13 The month in which the year's highest temperature occurs in a city has probabilities in the ratio 1:3:6:10 for May, June, July, and August, respectively. Find the probability that the highest temperature occurs in either May or June.

4.14 *Probability and odds.* The probability of an event is often expressed in terms of odds. Specifically, when we say that the odds are k to m that an event will occur, we mean that the probability of the event is $k/(k + m)$. For instance, "the odds are 4 to 1 that candidate Jones will win" means that $P(\text{Jones will win}) = \frac{4}{5} = .8$. Express the following statements in terms of probability.

(a) The odds are 3 to 1 that there will be good weather tomorrow.

(b) The odds are 7 to 2 that the city council will delay the funding of a new sports arena.

3. METHODS OF ASSIGNING PROBABILITY

An assignment of probabilities to all the events in a sample space determines a probability model. In order to be a valid probability model, the probability assignment must satisfy the properties 1, 2, and 3 stated in the previous section. Any assignment of numbers $P(e_i)$ to the elementary outcomes will satisfy the three conditions of probability provided these numbers are nonnegative and their sum over all the outcomes e_i in S is 1. However, to be of any practical import, the probability assigned to an event must also be in agreement with the concept of probability as the proportion of times the event is expected to occur. Here we discuss the implementation of this concept in two important situations.

3.1. EQUALLY LIKELY ELEMENTARY OUTCOMES— THE UNIFORM PROBABILITY MODEL

Often, the description of an experiment ensures that each elementary outcome is as likely to occur as any other. For example, consider the experiment of rolling a fair die and recording the top face. The sample space can be listed as

$$S = \{e_1, e_2, e_3, e_4, e_5, e_6\}$$

where e_1 stands for the elementary outcome of getting the face 1, and similarly, $e_2, \ldots, e_6$. Without actually rolling a die, we can deduce the probabilities. Because a fair die is a symmetric cube, each of its six faces is as likely to appear as any other. In other words, each face is expected to occur one-sixth of the time. The probability assignments should therefore be

$$P(e_1) = P(e_2) = \cdots = P(e_6) = \frac{1}{6}$$

and any other assignment would contradict the statement that the die is fair. We say that rolling a fair die conforms to a uniform probability model because the total probability 1 is evenly apportioned to all the elementary outcomes.

What is the probability of getting a number higher than 4? Letting A denote this event, we have the composition $A = \{e_5, e_6\}$, so

$$P(A) = P(e_5) + P(e_6) = \frac{1}{6} + \frac{1}{6} = \frac{1}{3}$$

When the elementary outcomes are modeled as equally likely, we have a uniform probability model. If there are k elementary outcomes in S, each is assigned the probability of $1/k$.

An event A consisting of m elementary outcomes is then assigned

$$P(A) \ = \ \frac{m}{k} \ = \ \frac{\text{No. of elementary outcomes in } A}{\text{No. of elementary outcomes in } S}$$

Gregor Mendel, pioneer geneticist, perceived a pattern in the characteristics of generations of pea plants and conceived a theory of heredity to explain them. According to Mendel, inherited characteristics are transmitted from one generation to another by genes. Genes occur in pairs and the offspring obtain their pair by taking one gene from each parent. A simple uniform probability model lies at the heart of Mendel's explanation of the selection mechanism.

One experiment that illustrated Mendel's theory consists of cross fertilizing a pure strain of red flowers with a pure strain of white flowers. This produces hybrids having one gene of each type that are pink-flowered. Crossing these hybrids leads to one of four possible gene pairs. Under Mendel's laws, these four are equally likely. Consequently, $P[\text{Pink}] = \frac{1}{2}$ and $P[\text{White}] = P[\text{Red}] = \frac{1}{4}$. (Compare with the experiment of tossing two coins.)

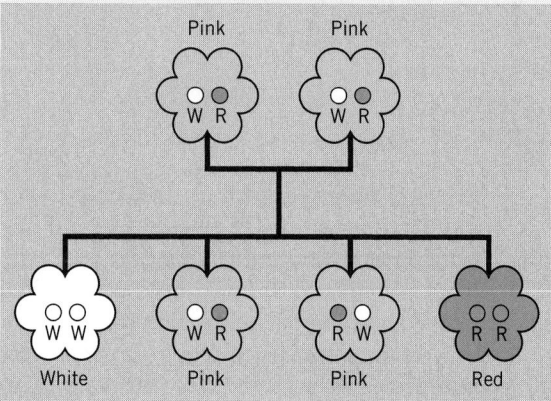

An experiment carried out by Correns, one of Mendel's followers, resulted in the frequencies 141, 291, and 132 for the white, pink, and red flowers, respectively. These numbers are nearly in the ratio $1:2:1$. (*Source:* W. Johannsen, *Elements of the Precise Theory of Heredity*, Jena: G. Fischer, 1909.)

Example 3   The Uniform Probability Model for Tossing a Fair Coin

Find the probability of getting exactly one head in two tosses of a fair coin.

SOLUTION   As listed in Example 1, there are four elementary outcomes in the sample space: $S = \{HH, HT, TH, TT\}$. The very concept of a fair coin implies that the four elementary outcomes in S are equally likely. We therefore assign the probability $\frac{1}{4}$ to each of them. The event $A = [\text{One head}]$ has two elementary outcomes—namely, HT and TH. Hence, $P(A) = \frac{2}{4} = .5$.

Example 4   Random Selection and the Uniform Probability Model

Suppose that among 50 students in a class, 42 are right-handed and 8 left-handed. If one student is randomly selected from the class, what is the probability that the selected student is left-handed?

SOLUTION   The intuitive notion of random selection is that each student is as likely to be selected as any other. If we view the selection of each individual student as an elementary outcome, the sample space consists of 50 e's of which 8 are in the event "left-handed." Consequently, $P[\text{Left-handed}] = \frac{8}{50} = .16$.

> *Note:*   Considering that the selected student will be either left-handed (L) or right-handed (R), we can write the sample space as $S = \{L, R\}$, but we should be aware that the two elements L and R are not equally likely.

3.2.   PROBABILITY AS THE LONG-RUN RELATIVE FREQUENCY

In many situations, it is not possible to construct a sample space where the elementary outcomes are equally likely. If one corner of a die is cut off, it would be unreasonable to assume that the faces remain equally likely and the assignments of probability to various faces can no longer be made by deductive reasoning. When speaking of the probability (or risk) that a man will die in his thirties, one may choose to identify the occurrence of death at each decade or even each year of age as an elementary outcome. However, no sound reasoning can be provided in favor of a uniform probability model. In fact, from extensive mortality studies, demographers have found considerable disparity in the risk of death for different age groups.

When the assumption of equally likely elementary outcomes is not tenable, how do we assess the probability of an event? The only recourse is to repeat the experiment many times and observe the proportion of times the event occurs. Letting N denote the number of repetitions (or trials) of an experiment, we set

$$\text{Relative frequency of event } A \text{ in } N \text{ trials} = \frac{\text{No. of times } A \text{ occurs in } N \text{ trials}}{N}$$

For instance, let A be the event of getting a 6 when rolling a die. If the die is rolled 100 times and 6 comes up 23 times, the observed relative frequency of A would be $\frac{23}{100} = .23$. In the next 100 tosses, 6 may come up 18 times. Collecting these two sets together, we have $N = 200$ trials with the observed relative frequency

$$\frac{23 + 18}{200} = \frac{41}{200} = .205$$

Imagine that this process is continued by recording the results from more and more tosses of the die and updating the calculations of relative frequency. Figure 1 shows a typical plot of the relative frequency versus the number N of trials of the experiment. We see that the relative frequencies fluctuate as N changes, but the fluctuations become damped with increasing N. Two persons separately performing the same experiment N times are not going to get exactly the same graph. However, the numerical value at which the relative frequency stabilizes, in the long run, will be the same. This concept, called **long-run stability of relative frequency**, is illustrated in Figure 1a.

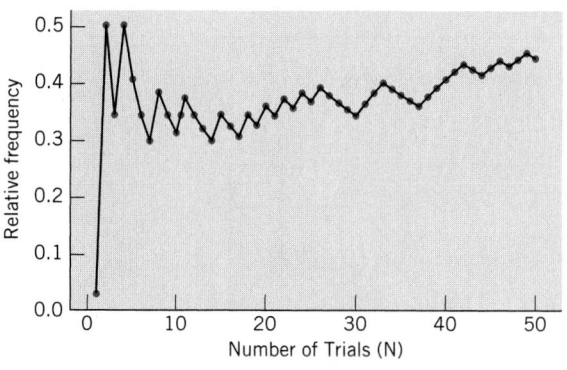

(a) Relative frequency versus number of trials. First 1–50.

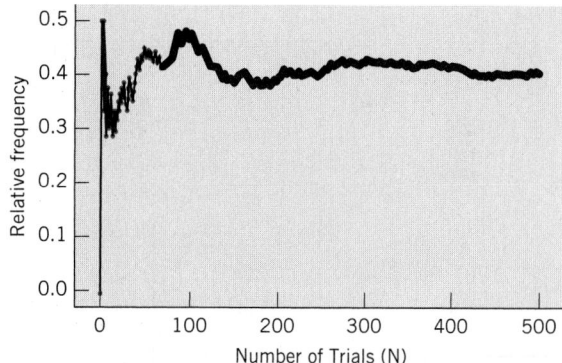

(b) Relative frequency versus number of trials. First 500 trials.

Figure 1   Stabilization of relative frequency.

Figure 1a graphically displays the considerable fluctuations present in the relative frequency as the number of trials increases from 1 to 50. Figure 1b displays the relative frequencies for the first 500 trials. In Figure 1b, the stabilization of relative frequency is evident, although the results for the first 50 trials are a little hard to discern in this view.

Probability as Long-Run Relative Frequency

We define $P(A)$, the probability of an event A, as the value to which the relative frequency stabilizes with increasing number of trials.

Although we will never know $P(A)$ exactly, it can be estimated accurately by repeating the experiment many times.

The property of the long-run stabilization of relative frequencies is based on the findings of experimenters in many fields who have undertaken the strain of studying the behavior of the relative frequencies under prolonged repetitions of their experiments. French gamblers, who provided much of the early impetus for the study of probability, performed experiments tossing dice and coins, drawing cards, and playing other games of chance thousands and thousands of times. They observed the stabilization property of relative frequency and applied this knowledge to achieve an understanding of the uncertainty involved in these games. Demographers have compiled and studied volumes of mortality data to examine the relative frequency of the occurrence of such events as death in particular age groups. In each

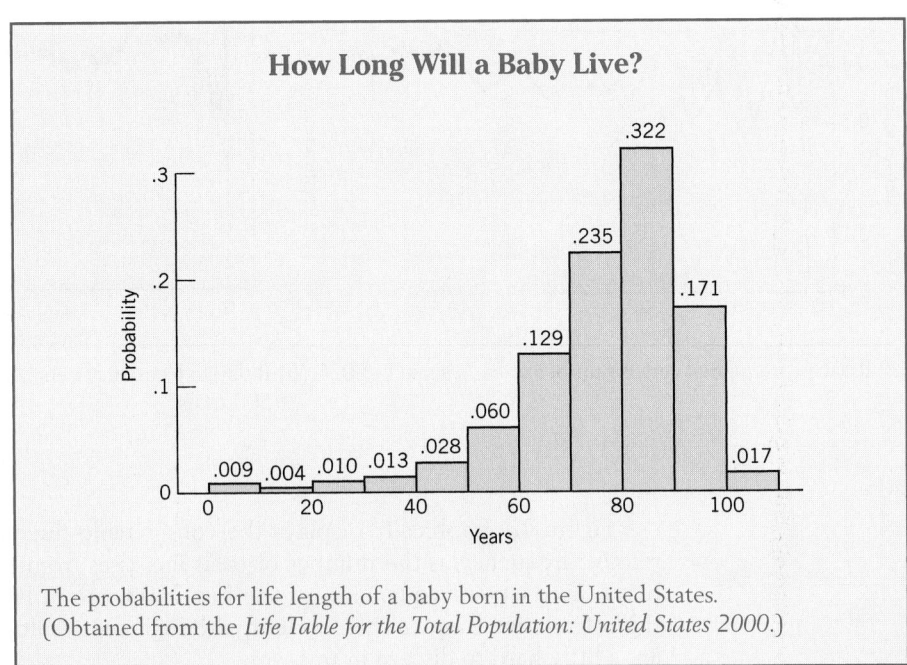

How Long Will a Baby Live?

The probabilities for life length of a baby born in the United States. (Obtained from the *Life Table for the Total Population: United States 2000.*)

context, the relative frequencies were found to stabilize at specific numerical values as the number of cases studied increased. Life and accident insurance companies actually depend on the stability property of relative frequencies.

As another example of an idealized model, consider the assignment of probabilities to the day of the week a child will be born. We may tentatively assume the simple model that all seven days of the week are equally likely. Each day is then assigned the probability $\frac{1}{7}$. If A denotes the event of a birth on the weekend (Saturday or Sunday), our model leads to the probability $P(A) = \frac{2}{7}$. The plausibility of the uniform probability model can only be ascertained from an extensive set of birth records.

Each newborn in the United States can be considered as a trial of the experiment where the day of birth determines whether or not the event A occurs. One year[1], the outcomes for 3563.1 thousand newborns constitute a very large number of replications.

The resulting proportion of babies born on either Saturday or Sunday is $\frac{915.7}{3565.1} = .257$. Although this may appear to be close to the value $\frac{2}{7} = .286$ predicted by the uniform model, the difference $.286 - .257 = .029$ is still larger than would ordinarily occur by chance. A possible explanation for the difference in probabilities is the prevalence of elective induction of labor on weekdays.

Exercises

4.15 Refer to the day of birth data in the preceding text above. Assuming conditions are the same today, estimate the probability that a baby will be born during a weekday. That is, not on Saturday or Sunday.

4.16 Among 41,131 turkey permit holders for a recent hunting season in Wisconsin, 8845 harvested a bird. Assuming conditions are the same today, estimate the probability that a turkey will be harvested for a single permit.

4.17 Consider the experiment of tossing a coin three times.

(a) List the sample space by drawing a tree diagram.

(b) Assign probabilities to the elementary outcomes.

(c) Find the probability of getting exactly one head.

[1]*Source:* R. Rindfuss et al., Convenience and the Occurrence of Births Induction of Labor in the United States and Canada, *International Journal of Health Services*, 9:3 (1979), p. 441, Table 1. Copyright ©1979, Baywood Publishing Company, Inc.

4.18   A letter is chosen at random from the word "FRIEND." What is the probability that it is a vowel?

4.19   A stack contains eight tickets numbered 1, 1, 2, 2, 2, 3, 3, 3. One ticket will be drawn at random and its number will be noted.

    (a)   List the sample space and assign probabilities to the elementary outcomes.

    (b)   What is the probability of drawing an odd-numbered ticket?

4.20   Suppose you are eating at a pizza parlor with two friends. You have agreed to the following rule to decide who will pay the bill. Each person will toss a coin. The person who gets a result that is different from the other two will pay the bill. If all three tosses yield the same result, the bill will be shared by all. Find the probability that:

    (a)   Only you will have to pay.

    (b)   All three will share.

4.21   A white and a colored die are tossed. The possible outcomes are shown in the illustration below.

    (a)   Identify the events A = [Sum = 6], B = [Sum = 7], C = [Sum is even], D = [Same number on each die].

    (b)   If both die are "fair," assign probability to each elementary outcome.

    (c)   Obtain $P(A)$, $P(B)$, $P(C)$, $P(D)$.

4.22   A roulette wheel has 34 slots, 2 of which are green, 16 are red, and 16 are black. A successful bet on black or red doubles the money, whereas one on green fetches 30 times as much. If you play the game once by betting $2 on the black, what is the probability that:

    (a)   You will lose your $2?

    (b)   You will win $2?

*4.23   One part of a quiz consists of two multiple-choice questions with the suggested answers: True (T), False (F), or Insufficient Data to Answer (I). An unprepared student randomly marks one of the three answers to each question.

    (a)   Make a tree diagram to list the sample space, that is, all possible pairs of answers the student might mark.

    (b)   What is the probability of exactly one correct answer?

4.24   Based on the data of the Center for Health Statistics, the 2001 birth rates in 50 states are grouped in the following frequency table.

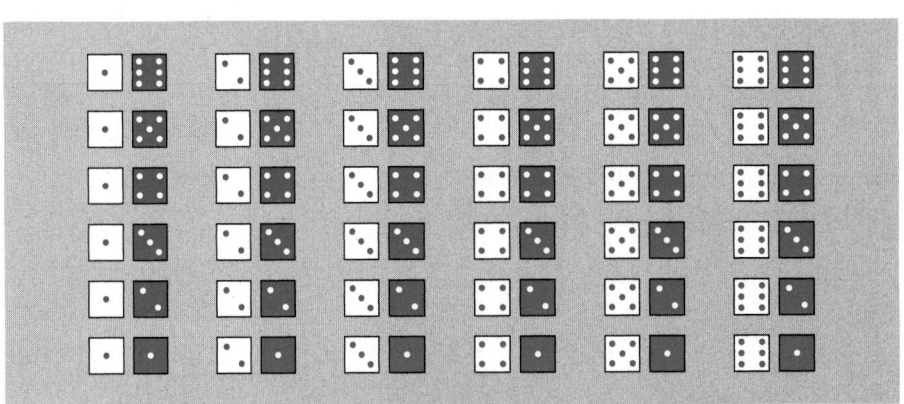

Birth rate (per thousand)	10–12	12–14	14–16
No. of states	4	21	18

Birth rate (per thousand)	16–18	18 and over	Total
No. of states	6	1	50

(Endpoint convention: Lower point is included, upper is not.)

If one state is selected at random, what is the probability that the birth rate there is:

(a)   Under 16?

(b)   Under 18 but not under 14?

(c)   16 or over?

4.25   Fifteen persons reporting to a Red Cross center one day are typed for blood, and the following counts are found:

Blood group	O	A	B	AB	Total
No. of persons	3	5	6	1	15

If one person is randomly selected, what is the probability that this person's blood group is:

(a)   AB?

(b)   Either A or B?

(c)   Not O?

4.26   Friends will be called, one after another, and asked to go on a weekend trip with you. You will call until one agrees to go (A) or four friends are asked, whichever occurs first. List the sample space for this experiment.

4.27   Campers arriving at a summer camp will be asked one after another whether they have protection against Lyme disease (Y) or not (N). The inspection will continue until one camper is found to be not protected or until five campers are checked, whichever occurs first. List the sample space for this experiment.

4.28   (a)   Consider the simplistic model that human births are evenly distributed over the 12 calendar months. If a person is randomly selected, say, from a phone directory, what is the probability that his or her birthday would be in November or December?

(b)   The following record shows a classification of 41,208 births in Wisconsin (courtesy of Professor Jerome Klotz). Calculate the relative frequency of births for each month and comment on the plausibility of the uniform probability model.

Jan.	3,478	July	3,476
Feb.	3,333	Aug.	3,495
March	3,771	Sept.	3,490
April	3,542	Oct.	3,331
May	3,479	Nov.	3,188
June	3,304	Dec.	3,321
		Total	41,208

4.29   A government agency will randomly select one of the 14 paper mills in a state to investigate its compliance with federal safety standards. Suppose, unknown to the agency, 9 of these mills are in compliance, 3 are borderline cases, and 2 are in gross violation.

(a)   Formulate the sample space in such a way that a uniform probablility model holds.

(b)   Find the probability that a gross violator will be detected.

4.30   A plant geneticist crosses two parent strains, each with gene pairs of type aA. An offspring receives one gene from each parent.

(a)   Construct the sample space for the genetic type of the offspring.

(b)   Assign probabilities assuming that the selection of genes is random.

(c)   If A is dominant and the aa offspring are short while all the others are tall, find $P[\text{short offspring}]$.

4.31   Explain why the long-run relative frequency interpretation of probability does not apply to the following situations.

(a) The proportion of days when the home loan rate at your bank is above its value at the start of the year.

(b) The proportion of cars that do not meet emission standards if the data are collected from service stations where the mechanics have been asked to check emissions while attending to other requested services.

(c) The proportion of days during a year that a store sells a room air conditioner if the data are only collected during the winter.

4.32 A local bookstore intended to award three gift certificates in the amounts $100, $50, and $25 to the first, second, and third customer to identify a mystery author. Unfortunately, a careless clerk in charge of mailing forgot the order and just randomly placed the gift certificates in the already addressed envelopes.

(a) List the sample space using F, S, and T for the three persons.

(b) State the compositions of the events

A = [exactly one certificate is sent to the correct person]

B = [all of the certificates are sent to incorrect persons]

4.33 Refer to Exercise 4.32.

(a) Assign probabilities to the elementary outcomes.

(b) Find $P(A)$ and $P(B)$.

4.34 Refer to Exercise 4.28. Using relative frequencies to estimate probabilities, find which 3 consecutive months have the lowest probability of a new birth.

4. EVENT RELATIONS AND TWO LAWS OF PROBABILITY

Later, when making probability calculations to support generalizations from the actual sample to the complete population, we will need to calculate probabilities of combined events, such as whether the count of no shows for a flight is either large or low.

Recall that the probability of an event A is the sum of the probabilities of all the elementary outcomes that are in A. It often turns out, however, that the event of interest has a complex structure that requires tedious enumeration of its elementary outcomes. On the other hand, this event may be related to other events that can be handled more easily. The purpose of this section is to first introduce the three most basic event relations: complement, union, and intersection. These event relations will then motivate some laws of probability.

The event operations are conveniently described in graphical terms. We first represent the sample space as a collection of points in a diagram, each identified with a specific elementary outcome. The geometric pattern of the plotted points is irrelevant. What is important is that each point is clearly tagged to indicate which elementary outcome it represents and to watch that no elementary outcome is missed or duplicated in the diagram. To represent an event A, identify the points that correspond to the elementary outcomes in A, enclose them in a boundary line, and attach the tag A. This representation, called a Venn diagram, is illustrated in Figure 2.

Example 5 Venn Diagram for Coin Tossing

Make a Venn diagram for the experiment of tossing a coin twice and indicate the following events.

A: Tail at the second toss

B: At least one head

SOLUTION   Here the sample space is $S = \{HH, HT, TH, TT\}$, and the two events have the compositions $A = \{HT, TT\}$, $B = \{HH, HT, TH\}$. Figure 2 shows the Venn diagram.

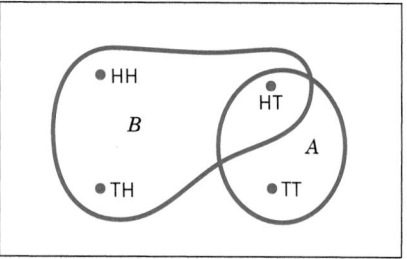

Figure 2   Venn diagram of the events in Example 5.

Example 6 A Venn Diagram for the Selection of Monkeys

Four monkeys, procured by a laboratory for a drug trial, are the types and ages

Monkey	Type	Age
1	Rhesus	6
2	Rhesus	8
3	Spider	6
4	Spider	6

Suppose two monkeys will be selected by lottery and assigned to an experimental drug. Considering all possible choices of two monkeys, make a Venn diagram and show the following events.

A: The selected monkeys are of the same type.

B: The selected monkeys are of the same age.

SOLUTION   Here the elementary outcomes are the possible choices of a pair of numbers from {1, 2, 3, 4}. These pairs are listed and labeled as e_1, e_2, e_3, e_4, e_5, e_6 for ease of reference.

$$\begin{array}{llll} \{1, 2\} & (e_1) & \{2, 3\} & (e_4) \\ \{1, 3\} & (e_2) & \{2, 4\} & (e_5) \\ \{1, 4\} & (e_3) & \{3, 4\} & (e_6) \end{array}$$

The pair {1, 2} has both monkeys of the same type, and so does the pair {3, 4}. Consequently, $A = \{e_1, e_6\}$. Those with the same ages are {1, 3}, {1, 4}, and {3, 4}, so $B = \{e_2, e_3, e_6\}$. Figure 3 shows the Venn diagram.

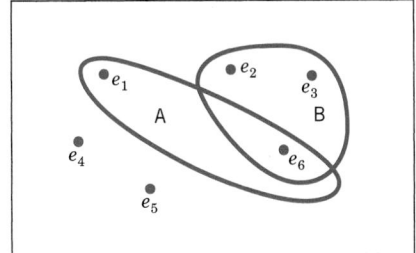

Figure 3   Venn diagram of the events in Example 6.

We now proceed to define the three basic event operations and introduce the corresponding symbols.

The complement of an event A, denoted by $\overline{A}$, is the set of all elementary outcomes that are not in A. The occurrence of $\overline{A}$ means that A *does not occur.*

The union of two events A and B, denoted by $A \cup B$, is the set of all elementary outcomes that are in A, B, or both. The occurrence of $A \cup B$ means that *either A or B or both occur.*

The intersection of two events A and B, denoted by AB, is the set of all elementary outcomes that are in A and B. The occurrence of AB means that *both A and B occur.*

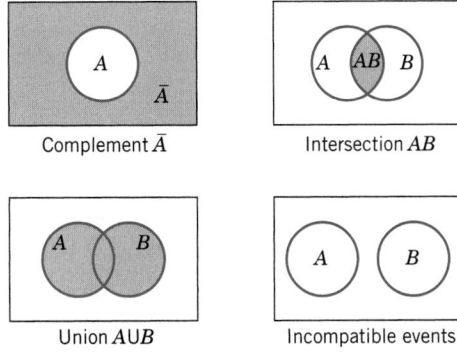

Note that $A \cup B$ is a larger set containing A as well as B, whereas AB is the common part of the sets A and B. Also it is evident from the definitions that $A \cup B$ and $B \cup A$ represent the same event, while AB and BA are both expressions for the intersection of A and B. The operations of union and intersection can be extended to more than two events. For instance, $A \cup B \cup C$ stands for the set of all points that are in *at least one* of A, B, and C, whereas ABC represents the *simultaneous occurrence* of all three events.

Two events A and B are called **incompatible** or **mutually exclusive** if their intersection AB is empty. Because incompatible events have no elementary outcomes in common, they cannot occur simultaneously.

Example 7   Determining the Composition of Events Defined by Complement, Union, or Intersection

Refer to the experiment in Example 6 of selecting two monkeys out of four. Let A = [Same type], B = [Same age], and C = [Different types]. Give the compositions of the events

$$C, \quad \overline{A}, \quad A \cup B, \quad AB, \quad BC$$

SOLUTION   The pairs consisting of different types are $\{1, 3\}$, $\{1, 4\}$, $\{2, 3\}$, and $\{2, 4\}$, so $C = \{e_2, e_3, e_4, e_5\}$. The event $\overline{A}$ is the same as the event C. Employing the definitions of union and intersection, we obtain

$$A \cup B = \{e_1, e_2, e_3, e_6\}$$
$$AB = \{e_6\}$$
$$BC = \{e_2, e_3\}$$

Let us now examine how probabilities behave as the operations of complementation, union, and intersection are applied to events. It would be worthwhile for the reader to review the properties of probability listed in Section 2. In particular, recall that $P(A)$ is the sum of probabilities of the elementary outcomes that are in A, and $P(S) = 1$.

First, let us examine how $P(\overline{A})$ is related to $P(A)$. The sum $P(A) + P(\overline{A})$ is the sum of the probabilities of all elementary outcomes that are in A plus the sum of the probabilities of elementary outcomes not in A. Together, these two sets comprise S and we must have $P(S) = 1$. Consequently, $P(A) + P(\overline{A}) = 1$, and we arrive at the following law.

Law of Complement

$$P(A) = 1 - P(\overline{A})$$

This law or formula is useful in calculating $P(A)$ when $\overline{A}$ is of a simpler form than A so that $P(\overline{A})$ is easier to calculate.

Turning to the operation of union, recall that $A \cup B$ is composed of points (or elementary outcomes) that are in A, in B, or in both A and B. Consequently, $P(A \cup B)$ is the sum of the probabilities assigned to these elementary outcomes, each probability taken *just once*. Now, the sum $P(A) + P(B)$ includes contributions from all these points, but it double counts those in the region AB (see the figure of $A \cup B$). To adjust for this double counting, we must therefore subtract $P(AB)$ from $P(A) + P(B)$. This results in the following law.

Addition Law

$$P(A \cup B) = P(A) + P(B) - P(AB)$$

If the events A and B are incompatible, their intersection AB is empty, so $P(AB) = 0$, and we obtain

---

Special Addition Law for Incompatible Events

$$P(A \cup B) = P(A) + P(B)$$

The addition law expresses the probability of a larger event $A \cup B$ in terms of the probabilities of the smaller events A, B, and AB. Some applications of these two laws are given in the following examples.

Example 8   Using the Law of Complement for Probability

A child is presented with three word-association problems. With each problem, two answers are suggested—one is correct and the other wrong. If the child has no understanding of the words whatsoever and answers the problems by guessing, what is the probability of getting at least one correct answer?

SOLUTION   Let us denote a correct answer by C and a wrong answer by W. The elementary outcomes can be conveniently enumerated by means of a tree diagram.

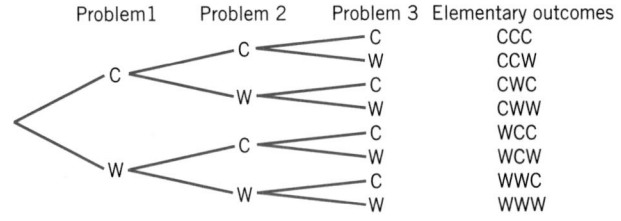

There are 8 elementary outcomes in the sample space and, because they are equally likely, each has the probability $\frac{1}{8}$. Let A denote the event of getting at least one correct answer. Scanning our list, we see that A contains 7 elementary outcomes, all except WWW. Our direct calculation yields $P(A) = \frac{7}{8}$.

Now let us see how this probability calculation could be considerably simplified. First, making a complete list of the sample space is not necessary. Since the elementary outcomes are equally likely, we need only determine that there are a total of 8 elements in S. How can we obtain this count without making a list? Note that an outcome is represented by three letters. There are 2 choices for each letter—namely, C or W. We then have $2 \times 2 \times 2 = 8$ ways of filling the three slots. The tree diagram explains this multiplication rule of counting. Evidently, the event A contains many elementary outcomes. On the other hand, $\overline{A}$ is the event of getting all answers wrong. It consists of the single elementary outcome WWW, so $P(\overline{A}) = \frac{1}{8}$. According to the law of complement,

$$P(A) = 1 - P(\overline{A})$$
$$= 1 - \frac{1}{8} = \frac{7}{8}$$

Example 9   Using the Addition Law for Probability

Refer to Example 6 where two monkeys are selected from four by lottery. What is the probability that the selected monkeys are either of the same type or the same age?

SOLUTION   In Example 6, we already enumerated the six elementary outcomes that comprise the sample space. The lottery selection makes all choices equally likely and the uniform probability model applies. The two events of interest are

$$A = [\text{Same type}] = \{e_1, e_6\}$$
$$B = [\text{Same age}] = \{e_2, e_3, e_6\}$$

Because A consists of two elementary outcomes and B consists of three,

$$P(A) = \frac{2}{6} \quad \text{and} \quad P(B) = \frac{3}{6}$$

Here we are to calculate $P(A \cup B)$. To employ the addition law, we also need to calculate $P(AB)$. In Figure 3, we see $AB = \{e_6\}$, so $P(AB) = \frac{1}{6}$. Therefore,

$$P(A \cup B) = P(A) + P(B) - P(AB)$$
$$= \frac{2}{6} + \frac{3}{6} - \frac{1}{6} = \frac{4}{6} = \frac{2}{3}$$

which is confirmed by the observation that $A \cup B = \{e_1, e_2, e_3, e_6\}$ indeed has four outcomes.

Example 10 Determining Probabilities from Those Given in a Venn Diagram

The accompanying Venn diagram shows three events A, B, and C and also the probabilities of the various intersections. [For instance, $P(AB) = .07$, $P(A\overline{B}) = .13$.] Determine:

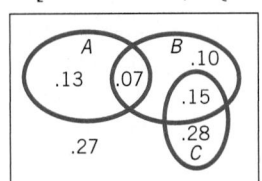

(a) $P(A)$

(b) $P(B\overline{C})$

(c) $P(A \cup B)$

SOLUTION   To calculate a probability, first identify the set in the Venn diagram. Then add the probabilities of those intersections that together comprise the stated event. We obtain

(a) $P(A) = .13 + .07 = .20$

(b) $P(B\overline{C}) = .10 + .07 = .17$

(c) $P(A \cup B) = .13 + .07 + .10 + .15 = .45$

Example 11 Expressing Relations between Events in Set Notation

Refer to Example 10. Express the following events in set notation and find their probabilities.

(a) Both B and C occur.

(b) C occurs and B does not.

(c) Exactly one of the three events occurs.

SOLUTION   The stated events and their probabilities are

(a) BC $P(BC) = .15$

(b) $\overline{B}C$ $P(\overline{B}C) = .28$

(c) $(A\overline{B}\,\overline{C}) \cup (\overline{A}B\overline{C}) \cup (\overline{A}\,\overline{B}C)$
        The probability $= .13 + .10 + .28 = .51$

Exercises

4.35   A day of the week will be selected to hold an all-day club picnic. The sample space has seven elementary outcomes $e_1, e_2, \ldots, e_7$ where e_1 represents Sunday, e_2 Monday, and so on. Two events are given as $A = \{e_4, e_5, e_6, e_7\}$ and $B = \{e_1, e_6, e_7\}$.

(a) Draw a Venn diagram and show the events A and B.

(b) Determine the composition of the following events: (i) AB (ii) $\overline{B}$ (iii) $A\overline{B}$ (iv) $A \cup B$.

4.36   A sample space consists of 8 elementary outcomes with the following probabilities.

$$P(e_1) = .04 \qquad P(e_2) = P(e_3) = P(e_4) = .12$$
$$P(e_5) = P(e_6) = P(e_7) = P(e_8) = .15$$

Three events are given as
$A = \{e_1, e_2, e_5, e_6, e_7\}$, $B = \{e_2, e_3, e_6, e_7\}$,
and $C = \{e_6, e_8\}$.

(a) Draw a Venn diagram and show these events.

(b) Give the composition and determine the probability of (i) $\bar{B}$ (ii) BC (iii) $A \cup C$ (iv) $\bar{A} \cup C$.

4.37 Refer to Exercise 4.36 Corresponding to each verbal description given here, write the event in set notation, give its composition, and find its probability.

(a) C does not occur.

(b) Both A and B occur.

(c) A occurs and B does not occur.

(d) Neither A nor C occurs.

4.38 Suppose you have had interviews for summer jobs at a grocery store, a discount store, and a movie theater. Let G, D, and M denote the events of your getting an offer from the grocery store, the discount store, and the movie theater, respectively. Express the following events in set notation.

(a) You get offers from the discount store and the movie theater.

(b) You get offers from the discount store and the movie theater but fail to get an offer from the grocery store.

(c) You do not get offers from the grocery store and the movie theater.

4.39 Four applicants will be interviewed for an administrative position with an environmental lobby. They have the following characteristics.

1. Psychology major, male, GPA 3.5
2. Chemistry major, female, GPA 3.3
3. Journalism major, female, GPA 3.7
4. Mathematics major, male, GPA 3.8

One of the candidates will be hired.

(a) Draw a Venn diagram and exhibit these events:

A: A social science major is hired.
B: The GPA of the selected candidate is higher than 3.6.
C: A male candidate is hired.

(b) Give the composition of the events $A \cup B$ and AB.

4.40 For the experiment of Exercise 4.39, give a verbal description of each of the following events and also state the composition of the event.

(a) $\bar{C}$

(b) $C\bar{A}$

(c) $A \cup \bar{C}$

4.41 A sample space consists of 9 elementary outcomes $e_1, e_2, \ldots, e_9$ whose probabilities are

$P(e_1) = P(e_2) = .06$ $P(e_3) = P(e_4) = P(e_5) = .1$
$P(e_6) = P(e_7) = .2$ $P(e_8) = P(e_9) = .09$

Suppose
$A = \{e_1, e_5, e_8\}$, $B = \{e_2, e_5, e_8, e_9\}$.

(a) Calculate $P(A)$, $P(B)$, and $P(AB)$.

(b) Using the addition law of probability, calculate $P(A \cup B)$.

(c) List the composition of the event $A \cup B$ and calculate $P(A \cup B)$ by adding the probabilities of the elementary outcomes.

(d) Calculate $P(\bar{B})$ from $P(B)$ and also by listing the composition of $\bar{B}$.

4.42 Refer to Exercise 4.35. Suppose the elementary outcomes are assigned these probabilities.

$P(e_1) = P(e_2) = P(e_3) = .15$ $P(e_4) = P(e_5) = .05$
$P(e_6) = .2$ $P(e_7) = .25$

(a) Find $P(A)$, $P(B)$, and $P(AB)$.

(b) Employing the laws of probability and the results of part (a), calculate $P(\bar{A})$ and $P(A \cup B)$.

(c) Verify your answers to part (b) by adding the probabilities of the elementary outcomes in each of $\bar{A}$ and $A \cup B$.

4.43 For two events A and B, the following probabilities are specified.

$P(A) = .52$ $P(B) = .36$ $P(AB) = .20$

(a) Enter these probabilities in the following table.

(b) Determine the probabilities of $A\overline{B}$, $\overline{A}B$, and $\overline{A}\,\overline{B}$ and fill in the table.

	B	$\overline{B}$
A		
$\overline{A}$		

4.44   Refer to Exercise 4.43. Express the following events in set notation and find their probabilities.

(a) B occurs and A does not occur.

(b) Neither A nor B occurs.

(c) Either A occurs or B does not occur.

4.45   The following table shows the probabilities concerning two events A and B.

	B	$\overline{B}$
A	.20	.12
$\overline{A}$	.35	

(a) Determine the missing entries.

(b) What is the probability that A occurs and B does not occur?

(c) Find the probability that either A or B occurs.

(d) Find the probability that one of these events occurs and the other does not.

4.46   If $P(A) = .2$ and $P(B) = .9$, can A and B be mutually exclusive? Why or why not?

4.47   From the probabilities shown in this Venn diagram, determine the probabilities of the following events.

(a) A does not occur.

(b) A occurs and B does not occur.

(c) Exactly one of the events A and B occurs.

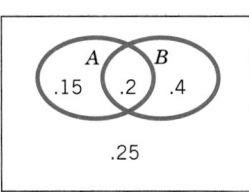

4.48   In a class of 32 seniors and graduate students, 20 are men and 12 are graduate students of whom 8 are women. If a student is randomly selected from this class, what is the probability that the selected student is (a) a senior? (b) a male graduate student?

4.49   Of 18 fast food restaurants in a city, 7 are in violation of sanitary standards, 8 are in violation of safety standards, and 4 are in violation of both. If a fast food restaurant is chosen at random, what is the probability that it is in compliance of both safety and sanitary standards?

4.50   Given that the probability that A occurs is .3, the probability that B does not occur is .6, and the probability that either A or B occurs is .5, find:

(a) The probability that A does not occur.

(b) The probability that both A and B occur.

(c) The probability that A occurs and B does not occur.

4.51   The medical records of the male diabetic patients reporting to a clinic during one year provide the following percentages.

	Light Case		Serious Case	
	Diabetes in Parents		Diabetes in Parents	
Age of Patient	Yes	No	Yes	No
Below 40	15	10	8	2
Above 40	15	20	20	10

Suppose a patient is chosen at random from this group, and the events A, B, and C are defined as follows.

A: He has a serious case.

B: He is below 40.

C: His parents are diabetic.

(a) Find the probabilities $P(A)$, $P(B)$, $P(BC)$, $P(ABC)$.

(b) Describe the following events verbally and find their probabilities: (i) $\overline{A}\,\overline{B}$ (ii) $\overline{A} \cup \overline{C}$ (iii) $\overline{A}B\overline{C}$.

4.52   The following frequency table shows the classification of 58 landfills in a state according to their concentration of the three hazardous chemicals arsenic, barium, and mercury.

	Barium			
Arsenic	High		Low	
	Mercury		Mercury	
	High	Low	High	Low
High	1	3	5	9
Low	4	8	10	18

If a landfill is selected at random, find the probability that it has:

(a) A high concentration of barium.

(b) A high concentration of mercury and low concentrations of both arsenic and barium.

(c) High concentrations of any two of the chemicals and low concentration of the third.

(d) A high concentration of any one of the chemicals and low concentrations of the other two.

4.53   A bank rewards its employees by giving awards to any employee that is cited by a customer for giving special service. Each award consists of two gift certificates contained in a sealed envelope. Each envelope contains certificates for one of the five following combinations of items.

1. Dinner and box of candy.

2. Round of golf and flowers.

3. Lunch and flowers.

4. Box of candy and lunch.

5. Music CD and lunch.

(a) An employee, cited twice for service, first selects one envelope from a collection of five and then the second from the full collection of five choices. List the sample space and assign probabilities to the simple events.

(b) State the compositions of the events

A = {The employee gets flowers}

B = {The employee gets lunch}

AB = {The employee gets flowers and lunch}

and give their probabilities.

4.54   Refer to Exercise 4.53. Let C denote the event that the employee gets either lunch or flowers or both.

(a) Relate C to the events A and B, and calculate $P(C)$ using a law of probability.

(b) State the composition of C and calculate its probability by adding the probabilities of the simple events.

5. CONDITIONAL PROBABILITY AND INDEPENDENCE

The probability of an event A must often be modified after information is obtained as to whether or not a related event B has taken place. Information about some aspect of the experimental results may therefore necessitate a revision of the probability of an event concerning some other aspect of the results. The revised probability of A when it is known that B has occurred is called the conditional probability of A given B and is denoted by $P(A|B)$. To illustrate how such modification is made, we consider an example that will lead us to the formula for conditional probability.

Example 12   Conditional Probability of Hypertension Given Body Weight

A group of executives is classified according to the status of body weight and incidence of hypertension. The proportions in the various categories appear in Table 1.

TABLE 1   Body Weight and Hypertension

	Overweight	Normal Weight	Underweight	Total
Hypertensive	.10	.08	.02	.20
Not hypertensive	.15	.45	.20	.80
Total	.25	.53	.22	1.00

(a) What is the probability that a person selected at random from this group will have hypertension?

(b) A person selected at random from this group is found to be overweight. What is the probability that this person is also hypertensive?

SOLUTION   Let A denote the event that a person is hypertensive, and let B denote the event that a person is overweight.

(a) Because 20% of the group is hypertensive and the individual is selected at random from this group, we conclude that $P(A) = .2$. This is the unconditional probability of A.

(b) When we are given the information that the selected person is overweight, the categories in the second and third columns of Table 1 are not relevant to this person. The first column shows that among the subgroup of overweight persons, the proportion having hypertension is .10/.25. Therefore, given the information that the person is in this subgroup, the probability that he or she is hypertensive is

$$P(A|B) = \frac{.10}{.25} = .4$$

Noting that $P(AB) = .10$ and $P(B) = .25$, we have derived $P(A|B)$ by taking the ratio $P(AB)/P(B)$. In other words, $P(A|B)$ is the proportion of the population having the characteristic A among all those having the characteristic B.

> The conditional probability of A given B is denoted by $P(A|B)$ and defined by the formula
>
> $$P(A|B) = \frac{P(AB)}{P(B)}$$
>
> Equivalently, this formula can be written
>
> $$P(AB) = P(B)P(A|B)$$
>
> This latter version is called the multiplication law of probability.

Similarly, the conditional probability of B given A can be expressed

$$P(B|A) = \frac{P(AB)}{P(A)}$$

which gives the relation $P(AB) = P(A)P(B|A)$. Thus, the multiplication law of probability states that the conditional probability of an event multiplied by the probability of the conditioning event gives the probability of the intersection.

The multiplication law can be used in one of two ways, depending on convenience. When it is easy to compute $P(A)$ and $P(AB)$ directly, these values can be used to compute $P(A|B)$, as in Example 12. On the other hand, if it is easy to calculate $P(B)$ and $P(A|B)$ directly, these values can be used to compute $P(AB)$.

Example 13   Conditional Probability of Survival

Refer to the box "How Long Will a Baby Live?" in Section 3.2. It shows the probabilities of death within 10-year age groups.

(a)   What is the probability that a newborn child will survive beyond age 90?

(b)   What is the probability that a person who has just turned 80 will survive beyond age 90?

SOLUTION

(a)   Let A denote the event "Survive beyond 90." Adding the probabilities of death in the age groups 90–100 and beyond, we find

$$P(A) = .171 + .017 = .188$$

(b)   Letting B denote the event "Survive beyond 80," we see that the required probability is the conditional probability $P(A|B)$. Because $AB = A$, $P(A) = .188$, and

$$P(B) = .322 + .171 + .017 = .510$$

we obtain

$$P(A|B) = \frac{P(AB)}{P(B)} = \frac{.188}{.510} = .369$$

Example 14   Using the Multiplication Law of Probability

There are 25 pens in a container on your desk. Among them, 20 will write well but 5 have defective ink cartridges. You will select 2 pens to take to a business appointment. Calculate the probability that:

(a) Both pens are defective.

(b) One pen is defective but the other will write well.

SOLUTION   We will use the symbols D for "defective" and G for "writes well", and attach subscripts to identify the order of the selection. For instance, $G_1 D_2$ will represent the event that the first pen checked will write well and the second is defective.

(a) Here the problem is to calculate $P(D_1 D_2)$. Evidently, $D_1 D_2$ is the intersection of the two events D_1 and D_2. Using the multiplication law, we write

$$P(D_1 D_2) = P(D_1)P(D_2 | D_1)$$

In order to calculate $P(D_1)$, we need only consider selecting one pen at random from 20 good and 5 defective pens. Clearly, $P(D_1) = \frac{5}{25}$. The next step is to calculate $P(D_2 | D_1)$. Given that D_1 has occurred, there will remain 20 good and 4 defective pens at the time the second selection is made. Therefore, the conditional probability of D_2 given D_1 is $P(D_2 | D_1) = \frac{4}{24}$. Multiplying these two probabilities, we get

$$P(\text{both defective}) = P(D_1 D_2) = \frac{5}{25} \times \frac{4}{24} = \frac{1}{30} = .033$$

(b) The event [exactly one defective] is the union of the two incompatible events $G_1 D_2$ and $D_1 G_2$. The probability of each of these can be calculated by the multiplication law as in part (a). Specifically,

$$P(G_1 D_2) = P(G_1)P(D_2 | G_1) = \frac{20}{25} \times \frac{5}{24} = \frac{1}{6}$$

$$P(D_1 G_2) = P(D_1)P(G_2 | D_1) = \frac{5}{25} \times \frac{20}{24} = \frac{1}{6}$$

The required probability is $P(G_1 D_2) + P(D_1 G_2) = \frac{2}{6} = .333$.

Remark: In solving the problems of Example 14, we have avoided listing the sample space corresponding to the selection of two pens from a collection of 25. A judicious use of the multiplication law has made it possible to focus attention on one draw at a time, thus simplifying the probability calculations.

A situation that merits special attention occurs when the conditional probability $P(A | B)$ turns out to be the same as the unconditional probability $P(A)$.

Information about the occurrence of B then has no bearing on the assessment of the probability of A. Therefore, when we have the equality $P(A|B) = P(A)$, we say the events A and B are independent.

Two events A and B are independent if

$$P(A|B) = P(A)$$

Equivalent conditions are

$$P(B|A) = P(B)$$

or

$$P(AB) = P(A)P(B)$$

The last form follows by recalling that $P(A|B) = P(AB)/P(B)$, so that the condition $P(A|B) = P(A)$ is equivalent to

$$P(AB) = P(A)P(B)$$

which may be used as an alternative definition of independence. The other equivalent form is obtained from

$$P(B|A) = \frac{P(AB)}{P(A)} = \frac{P(A)P(B)}{P(A)} = P(B)$$

The form $P(AB) = P(A)P(B)$ shows that the definition of independence is symmetric in A and B.

Example 15   Demonstrating Dependence between Hypertension and Overweight

Are the two events $A = $ [Hypertensive] and $B = $ [Overweight] independent for the population in Example 12?

SOLUTION   Referring to that example, we have

$$P(A) = .2$$
$$P(A|B) = \frac{P(AB)}{P(B)} = \frac{.10}{.25} = .4$$

Because these two probabilities are different, the two events A and B are not independent.

Caution: Do not confuse the terms "incompatible events" and "independent events." We say A and B are incompatible when their intersection AB is empty, so $P(AB) = 0$. On the other hand, if A and B are independent, $P(AB) = P(A)P(B)$. Both these properties cannot hold as long as A and B have nonzero probabilities.

We introduced the condition of independence in the context of checking a given assignment of probability to see if $P(A|B) = P(A)$. A second use of this condition is in the assignment of probability when the experiment consists of two physically unrelated parts. When events A and B refer to unrelated parts of an experiment, AB is assigned the probability $P(AB) = P(A)P(B)$.

Example 16   Using Independence to Assign Probability

Engineers use the term "reliability" as an alternative name for the probability that a device does not fail. Suppose a mechanical system consists of two components that function independently. From extensive testing, it is known that component 1 has reliability .98 and component 2 has reliability .95. If the system can function only if both components function, what is the reliability of the system?

SOLUTION   Consider the events

A_1:   Component 1 functions
A_2:   Component 2 functions
S: System functions

Here we have the event relation $S = A_1A_2$. Given that the components operate independently, we take the events A_1 and A_2 to be independent. Consequently, the multiplication law assigns

$$P(S) = P(A_1)P(A_2) = .98 \times .95 = .931$$

and the system reliability is .931.

In this system, the components are said to be connected in series, and the system is called a series system. A two-battery flashlight is an example. The conventional diagram for a series system is shown in the illustration:

Example 17   Independence and Assigning Probabilities When Sampling with Replacement

In the context of Example 14, suppose that a box contains 25 cards identifying the pens and their ability to write. One card is drawn at random. It is returned to the box and then another card is drawn at random. What is the probability that both draws produce pens that will not write?

SOLUTION   As before, we will use the letter D for defective and G for a pen that will write. By returning the first card to the box, the contents of the box remain unchanged. Hence, with each draw, $P(D) = \frac{5}{25}$, and the results of the two draws are independent. Instead of working with conditional probability as we did in Example 11, we can use the property of independence to calculate

$$P(D_1D_2) = P(D_1)P(D_2) = \frac{5}{25} \times \frac{5}{25} = .04$$

Remark 1: Evidently, this method of probability calculation extends to any number of draws if after each draw the selected card is returned to the box. For instance, the probability that the first draw produces a D and the next two draws produce G's is

$$P(D_1G_2G_3) = \frac{5}{25} \times \frac{20}{25} \times \frac{20}{25} = .128$$

Remark 2: Sampling with replacement is seldom used in practice, but it serves as a conceptual frame for simple probability calculations when a problem concerns sampling from a large population. For example, consider drawing 3 cards from a box containing 2500 cards, of which 2000 are G's and 500 D's. Whether or not a selected card is returned to the box before the next draw makes little difference in the probabilities. The model of independence serves as a reasonable approximation.

The connection between dependent trials and the size of the population merits further emphasis.

Example 18 Dependence and Sampling without Replacement

If the outcome of a single trial of any experiment is restricted to just two possible outcomes, it can be modeled as drawing a single ball from an urn containing only red (R) and white (W) balls. In the previous example, these two possible outcomes were good and defective. Consider taking a sample of size 3, without replacement, from each of two populations:

1.   Small population where the urn contains 7 W and 3 R.
2.   Large population where the urn contains 7000 W and 3000 R.

Compare with a sample of size 3 generated from a spinner having a probability .7 of white, where R or W is determined by a separate spin for each trial.

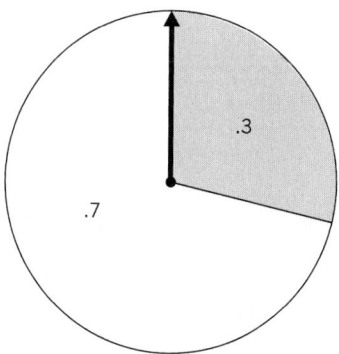

(a)   Calculate the probability assigned to each possible sample.
(b)   Let D = [At least one W]. Calculate the probability of D.

TABLE 2   A Comparison of Finite Populations and the Spinner Model

Draw 3 balls without replacement	Small population	Large population	Spinner
	$P(ABC) =$ $P(A)P(B\|A)P(C\|AB)$	$P(ABC) \approx$ $P(A)P(B)P(C)$	$P(ABC) =$ $P(A)P(B)P(C)$
Outcome	Not independent	Approximately independent	Independent

Outcome	Small population	Large population	Spinner
RRR	$\dfrac{3}{10} \times \dfrac{2}{9} \times \dfrac{1}{8} = \dfrac{6}{720}$	$\dfrac{3000}{10,000} \times \dfrac{2999}{9999} \times \dfrac{2998}{9998} \approx (.3)(.3)(.3)$	$(.3)(.3)(.3)$
RRW	$\dfrac{3}{10} \times \dfrac{2}{9} \times \dfrac{7}{8} = \dfrac{42}{720}$	$\approx (.3)(.3)(.7)$	$(.3)^2(.7)$
RWR	$\dfrac{3}{10} \times \dfrac{7}{9} \times \dfrac{2}{8} = \dfrac{42}{720}$	$\approx (.3)(.7)(.3)$	$(.3)^2(.7)$
WRR	$\dfrac{7}{10} \times \dfrac{3}{9} \times \dfrac{2}{8} = \dfrac{42}{720}$	$\approx (.7)(.3)(.3)$	$(.3)^2(.7)$
RWW	$\dfrac{3}{10} \times \dfrac{7}{9} \times \dfrac{6}{8} = \dfrac{126}{720}$	$\approx (.3)(.7)(.7)$	$(.3)(.7)^2$
WRW	$\dfrac{7}{10} \times \dfrac{3}{9} \times \dfrac{6}{8} = \dfrac{126}{720}$	$\approx (.7)(.3)(.7)$	$(.3)(.7)^2$
WWR	$\dfrac{7}{10} \times \dfrac{6}{9} \times \dfrac{3}{8} = \dfrac{126}{720}$	$\approx (.7)(.7)(.3)$	$(.3)(.7)^2$
WWW	$\dfrac{7}{10} \times \dfrac{6}{9} \times \dfrac{5}{8} = \dfrac{210}{720}$	$\approx (.7)(.7)(.7)$	$(.7)^3$

$P(D) = 1 - P(\overline{D}) = 1 - \dfrac{6}{720}$ $\approx 1 - (.3)^3$ $= 1 - (.3)^3$

If $A = 1^{st}$ is R, $B = 2^{nd}$ is R $C = 3^{rd}$ is R $D = $ at least one W

then $ABC = \{RRR\} = \overline{D}$ $D = \{RRW, RWR, WRR, RWW, WRW, WWR, WWW\}$

SOLUTION   (a)   We will write RWR for the outcome where the first and third draws are R and the second is W. Applying the general multiplication rule $P(ABC) = P(A)P(B|A)P(C|AB)$, when sampling the small population, we get

$$P(RWR) = P(R)P(W|R)P(R|RW) = \frac{3}{10} \times \frac{7}{9} \times \frac{2}{8} = \frac{42}{720}$$

For the larger population,

$$P(RWR) = \frac{3000}{10,000} \times \frac{7000}{9999} \times \frac{2999}{9998} \approx (.3) \times (.7) \times (.3) = (.3)^2(.7)$$

When the population size is large, the assumption of independence produces a very good approximation.

Under the spinner model, the probability of R is .3 for the first trial and this probability is the same for all trials. A spinner is a classic representation of a device with no memory, so that the outcome of the current trial is independent of the outcomes of all the previous trials. According to the product rule for independence, we assign

$$P(RWR) = (.3) \times (.7) \times (.3)$$

Notice that the spinner model is equivalent to sampling with replacement from either of the two finite populations.

The results for all eight possible samples are shown in Table 2.

(b) The event D is complicated, whereas $\overline{D} = \{RRR\}$, a single outcome. By the law of the complement,

$$P(D) = 1 - P(\overline{D}) = 1 - \frac{3}{10} \times \frac{2}{9} \times \frac{1}{8} = 1 - \frac{6}{720}$$

In the second case, $P(D)$ is approximately $1 - (.3) \times (.3) \times (.3)$ and this answer is exact for the spinner model.

Table 2 summarizes sampling from a small finite population, a large but finite population, and the spinner model. Dependence does matter when sampling without replacement from a small population.

Exercises

4.55 A person is randomly selected from persons working in your state. Consider the two events

A = [Earned over $60,000 last year]
B = [College graduate]

Given that the person is a college graduate, would you expect the probability of A to be larger, the same, or smaller than the unconditional probability $P(A)$? Explain your answer. Are A and B independent according to your reasoning?

4.56 A person is randomly selected from persons working in your state. Consider the two events

A = [Lawyer]
B = [Driving a new luxury car]

Given that the person selected drives a new luxury car, would you expect the probability of A to be larger, the same, or smaller than the unconditional probability $P(A)$? Explain your answer. Are A and B independent according to your reasoning?

4.57 Suppose that $P(A)$ = .68, $P(B)$ = .55, and $P(AB)$ = .35. Find

(a) The conditional probability that B occurs given that A occurs.

(b) The conditional probability that B does not occur given that A occurs.

(c) The conditional probability that B occurs given that A does not occur.

4.58 Suppose that $P(A)$ = .4, $P(B)$ = .6 and the probability that either A or B occurs is .7. Find

(a) The conditional probability that A occurs given that B occurs.

(b) The conditional probability that B occurs given that A does not occur.

4.59 The following data relate to the proportions in a population of drivers.

A = Defensive driver training last year
B = Accident in current year

The probabilities are given in the accompanying Venn diagram. Find $P(B|A)$. Are A and B independent?

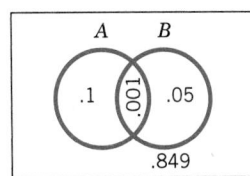

4.60 Suppose $P(A)$ = .55, $P(B)$ = .32, and $P(\overline{A}B)$ = .20.

(a) Determine all the probabilities needed to fill in the accompanying table.

	B	$\overline{B}$	
A			.55
$\overline{A}$	.20		
	.32		

(b) Find the conditional probability of A given that B does not occur.

4.61 For two events A and B, the following probabilities are given.

$P(A)$ = .4 $P(B)$ = .25 $P(A|B)$ = .7

Use the appropriate laws of probability to calculate

(a) $P(\overline{A})$

(b) $P(AB)$

(c) $P(A \cup B)$

4.62 Records of student patients at a dentist's office concerning fear of visiting the dentist suggest the following proportions.

	School		
	Elementary	Middle	High
Fear	.12	.08	.05
Do not fear	.28	.25	.22

For a student selected at random, consider the events

A = [Fear] M = [Middle school]

(a) Find the probabilities

$P(A)$ $P(AM)$
$P(M)$ $P(A \cup M)$

(b) Are A and M independent?

4.63 An urn contains two green balls and three red balls. Suppose two balls will be drawn at random one after another and *without replacement* (i.e., the first ball drawn is *not* returned to the urn before the second one is drawn).

(a) Find the probabilities of the events

A = [Green ball appears in the first draw]
B = [Green ball appears in the second draw]

(b) Are the two events independent? Why or why not?

4.64 Refer to Exercise 4.63. Now suppose two balls will be drawn *with replacement* (i.e., the first ball drawn will be returned to the urn before the second draw). Repeat parts (a) and (b).

4.65 In a county, men constitute 60% of the labor force. The rates of unemployment are 5.1% and 4.3% among males and females, respectively.

(a) In the context of selecting a worker at random from the country labor force, state what probabilities the foregoing percentages represent. (Use symbols such as M for male, E for employed.)

(b) What is the overall rate of unemployment in the county?

(c) If a worker selected at random is found to be unemployed, what is the probability that the worker is a woman?

4.66 Assume $P(A) = .4$, $P(B) = .4$ and $P(A \cup B) = .7$.

(a) Are A and B independent? Why or why not?

(b) Can A and B be mutually exclusive? Why or why not?

4.67 Suppose $P(A) = .6$ and $P(B) = .22$.

(a) Determine $P(A \cup B)$ if A and B are independent.

(b) Determine $P(A \cup B)$ if A and B are mutually exclusive.

(c) Find $P(A|\bar{B})$ if A and B are mutually exclusive.

4.68 Refer to Exercise 4.49.

(a) If a fast food restaurant selected at random is found to comply with safety standards, what is the probability that it violates sanitary standards?

(b) If a restaurant selected at random is found to violate at least one of the two standards, what is the probability that it complies with safety standards?

4.69 In a shipment of 15 room air conditioners, there are 3 with defective thermostats. Two air conditioners will be selected at random and inspected one after another. Find the probability that:

(a) The first is defective.

(b) The first is defective and the second good.

(c) Both are defective.

(d) The second air conditioner is defective.

(e) Exactly one is defective.

4.70 Refer to Exercise 4.69. Now suppose 3 air conditioners will be selected at random and checked one after another. Find the probability that:

(a) All 3 are good.

(b) The first 2 are good and the third defective.

(c) Two are good and 1 defective.

4.71 Of 20 rats in a cage, 12 are males and 9 are infected with a virus that causes hemorrhagic fever. Of the 12 male rats, 7 are infected with the virus. One rat is randomly selected from the cage.

(a) If the selected rat is found to be infected, what is the probability that it is a female?

(b) If the selected rat is found to be a male, what is the probability that it is infected?

(c) Are the events "the selected rat is infected" and "the selected rat is male" independent? Why or why not?

4.72 The following probabilities are given for two events A and B.

$$P(A) = \frac{1}{4} \qquad P(B) = \frac{1}{3} \qquad P(A|B) = \frac{3}{5}$$

(a) Using the appropriate laws of probability, calculate $P(\bar{A})$, $P(AB)$, and $P(A \cup B)$.

(b) Draw a Venn diagram to determine $P(\bar{A}B)$.

4.73 Of three events, A, B, and C, suppose events A and B are independent and events B and C are mutually exclusive. Their probabilities are $P(A) = .7$, $P(B) = .2$, and $P(C) = .3$. Express the following events in set notation and calculate their probabilities.

(a) Both B and C occur.

(b) At least one of A and B occurs.

(c) B does not occur.

(d) All three events occur.

4.74 Approximately 40% of the Wisconsin population have type O blood. If 4 persons are selected at random to be donors, find P[at least one type O].

4.75 The primary cooling unit in a nuclear power plant has reliability .999. There is also a back-up cooling unit to substitute for the primary unit when it fails. The reliability of the back-up unit is .910. Find the reliability of the cooling system of the power plant. Assume independence.

4.76 An accountant screens large batches of bills according to the following sampling inspection plan. She inspects 4 bills chosen at random from each batch and passes the batch if, among the 4, none is irregular. Find the probability that a batch will be passed if, in fact:

(a) 5% of its bills are irregular.

(b) 20% of its bills are irregular.

4.77 An electronic scanner is successful in detecting flaws in a material in 80% of the cases. Three material specimens containing flaws will be tested with the scanner. Assume that the tests are independent.

(a) List the sample space and assign probabilities to the simple events.

(b) Find the probability that the scanner is successful in at least two of the three cases.

4.78 Refer to Exercise 4.52. Given that a landfill selected at random is found to have a high concentration of mercury, what is the probability that its concentration is:

(a) High in barium?

(b) Low in both arsenic and barium?

(c) High in either arsenic or barium?

*4.79 *Imperfect clinical test.* Suppose a disease is present in 3% of a population. A diagnostic test is available but is yet to be perfected. The test shows 10% false positives and 5% false negatives. That is, for a patient having the disease, the test shows positive (+) with probability .95 and negative (−) with probability .05. For a patient not having the disease, the test shows "+" with probability .10 and "−" with probability .90. (*Note:* These are conditional probabilities.) Denote the events D = diseased, N = not diseased.

(a) Use the multiplication law to calculate the probabilities of the joint events $D+$,

$D-$, $N+$, and $N-$ and fill in the following probability table.

	+	−
D		
N		

(b) If Jeff's test shows "+," what is the probability of his having the disease? (Assume that Jeff can be considered a randomly selected person.)

*4.80 Consider tossing two fair coins and the events

A: Head in the first toss

B: Head in the second toss

C: Both heads or both tails in the two tosses

(a) Verify that the property of independence holds for all event pairs.

(b) Show that $P(ABC)$ is different from the product $P(A)P(B)P(C)$. (This illustrates the fact that pairwise independence does not ensure complete independence.)

4.81 Of the patients reporting to a clinic with the symptoms of sore throat and fever, 25% have strep throat, 40% have an allergy, and 10% have both.

(a) What is the probability that a patient selected at random has strep throat, an allergy, or both?

(b) Are the events "strep throat" and "allergy" independent?

4.82 Items coming off a production line are categorized as good (G), slightly blemished (B), and defective (D), and the percentages are 80%, 15%, and 5%, respectively. Suppose that two items will be randomly selected for inspection and the selections are independent.

(a) List all outcomes and assign probabilities.

(b) Find the probability that at least one of the items is slightly blemished.

(c) Find the probability that neither of the items is good.

6. RANDOM SAMPLING FROM A FINITE POPULATION

In our earlier examples of probability calculations, we have used the phrase "randomly selected" to mean that all possible selections are equally likely. It usually is not difficult to enumerate all the elementary outcomes when both the population size and sample size are small numbers. With larger numbers, making a list of all the possible choices becomes a tedious job. However, a counting rule is available that enables us to solve many probability problems.

We begin with an example where the population size and the sample size are both small numbers so all possible samples can be conveniently listed.

Example 19   Selecting a Random Sample of Size 2 from a Population of Size 5

There are five qualified applicants for two editorial positions on a college newspaper. Two of these applicants are men and three women. If the positions are filled by randomly selecting two of the five applicants, what is the probability that neither of the men is selected?

SOLUTION   Suppose the three women applicants are identified as a, b, and c and the two men as d and e. Two members are selected at random from the population:

$$\{a, \quad b, \quad c, \quad d, \quad e\}$$

women   men

The possible samples may be listed as

$$\{a, b\} \quad \{b, c\} \quad \{c, d\} \quad \{d, e\}$$
$$\{a, c\} \quad \{b, d\} \quad \{c, e\}$$
$$\{a, d\} \quad \{b, e\}$$
$$\{a, e\}$$

As the list shows, our sample space has 10 elementary outcomes. The notion of random selection entails that these are all equally likely, so each is assigned the probability $\frac{1}{10}$. Let A represent the event that two women are selected. Scanning our list, we see that A consists of the three elementary outcomes

$$\{a, b\} \quad \{a, c\} \quad \{b, c\}$$

Consequently,

$$P(A) = \frac{\text{No. of elements in } A}{\text{No. of elements in } S} = \frac{3}{10} = .3$$

Note that our probability calculation in Example 19 only requires knowledge of the two counts: the number of elements in S and the number of elements in A. Can we arrive at these counts without formally listing the sample space? An important counting rule comes to our aid.

The Rule of Combinations

Notation: The number of possible choices of r objects from a group of N distinct objects is denoted by $\binom{N}{r}$, which reads as "N choose r."

Formula:

$$\binom{N}{r} = \frac{N \times (N - 1) \times \cdots \times (N - r + 1)}{r \times (r - 1) \times \cdots \times 2 \times 1}$$

More specifically, the numerator of the formula $\binom{N}{r}$ is the product of r consecutive integers starting with N and proceeding downward. The denominator is also the product of r consecutive integers, but starting with r and proceeding down to 1.

To motivate the formula, let us consider the number of possible choices (or collections) of three letters from the seven letters $\{a, b, c, d, e, f, g\}$. This count is denoted by $\binom{7}{3}$.

It is easier to arrive at a formula for the number of ordered selections. The first choice can be any of the 7 letters, the second can be any of the remaining 6, and the third can be any of the remaining 5. Thinking in terms of a tree diagram (without actually drawing one), we arrive at the following count.

The number of ordered selections of 3 letters from 7 is given by the product $7 \times 6 \times 5$.

Next, note that a particular collection, say $\{a, b, c\}$, can produce $3 \times 2 \times 1$ orderings, as one can also verify by a tree diagram. The $\binom{7}{3}$ number of collections, each producing $3 \times 2 \times 1$ orderings, generate a total of $\binom{7}{3} \times 3 \times 2 \times 1$ orderings. Because this count must equal $7 \times 6 \times 5$, we get

$$\binom{7}{3} = \frac{7 \times 6 \times 5}{3 \times 2 \times 1}$$

This explains the formula of $\binom{N}{r}$ for the case $N = 7$ and $r = 3$.

Although not immediately apparent, there is a certain symmetry in the counts $\binom{N}{r}$. The process of selecting r objects is the same as choosing $N - r$

objects to leave behind. Because every choice of r objects corresponds to a choice of $N - r$ objects,

$$\binom{N}{r} = \binom{N}{N - r}$$

This relation often simplifies calculations. Since $\binom{N}{N} = 1$, we take $\binom{N}{0} = 1$.

Example 20 Evaluating Some Combinations

Calculate the values of $\binom{5}{2}$, $\binom{15}{4}$, and $\binom{15}{11}$.

SOLUTION

$$\binom{5}{2} = \frac{5 \times 4}{2 \times 1} = 10 \qquad \binom{15}{4} = \frac{15 \times 14 \times 13 \times 12}{4 \times 3 \times 2 \times 1} = 1365$$

Using the relation $\binom{N}{r} = \binom{N}{N - r}$, we have

$$\binom{15}{11} = \binom{15}{4} = 1365$$

Example 21 Calculating a Probability Using Combinations

Refer to Example 19 concerning a random selection of two persons from a group of two men and three women. Calculate the required probability without listing the sample space.

SOLUTION   The number of ways two persons can be selected out of five is given by

$$\binom{5}{2} = \frac{5 \times 4}{2 \times 1} = 10$$

Random selection means that the 10 outcomes are equally likely. Next, we are to count the outcomes that are favorable to the event A that both selected persons are women. Two women can be selected out of three in

$$\binom{3}{2} = \frac{3 \times 2}{2 \times 1} = 3 \text{ ways}$$

Taking the ratio, we obtain the result

$$P(A) = \frac{3}{10} = .3$$

Example 22   Probabilities of Being Selected under Random Selection

After some initial challenges, there remain 16 potential jurors of which 10 are male and 6 female. The defense attorney can dismiss 4 additional persons on the basis of answers to her questions.

(a)   How many ways can the 4 additional jurors be selected for dismissal?

(b)   How many selections are possible that result in 1 male and 3 females being dismissed?

(c)   If the selection process were random, what is the probability that 1 male and 3 females would be dismissed?

SOLUTION

(a)   According to the counting rule $\binom{N}{r}$, the number of ways 4 jurors can be selected out of 16 is

$$\binom{16}{4} = \frac{16 \times 15 \times 14 \times 13}{4 \times 3 \times 2 \times 1} = 1820$$

(b)   One male can be chosen from 10 in $\binom{10}{1} = 10$ ways. Also, 3 females can be chosen from 6 in

$$\binom{6}{3} = \frac{6 \times 5 \times 4}{3 \times 2 \times 1} = 20 \text{ ways}$$

Each of the 10 choices of a male can accompany each of the 20 choices of 3 females. Reasoning from the tree diagram, we conclude that the number of possible samples with the stated composition is

$$\binom{10}{1} \times \binom{6}{3} = 10 \times 20 = 200$$

(c)   Random sampling requires that the 1820 possible samples are all equally likely. Of these, 200 are favorable to the event $A = [1 \text{ male}$ and 3 females]. Consequently,

$$P(A) = \frac{200}{1820} = .110$$

The notion of a random sample from a finite population is crucial to statistical inference. In order to generalize from a sample to the population, it is imperative that the sampling process be impartial. This criterion is evidently met if we allow the selection process to be such that all possible samples are given equal opportunity to be selected. This is precisely the idea behind the term random sampling, and a formal definition can be phrased as follows.

A sample of size n selected from a population of N distinct objects is said to be a **random sample** if each collection of size n has the same probability $1/\binom{N}{n}$ of being selected.

Note that this is a conceptual rather than an operational definition of a random sample. On the surface, it might seem that a haphazard selection by the experimenter would result in a random sample. Unfortunately, a seemingly haphazard selection may have hidden bias. For instance, when asked to name a random integer between 1 and 9, more persons respond with 7 than any other number. Also, odd numbers are more popular than even numbers. Therefore, the selection of objects must be entrusted to some device that cannot think; in other words, some sort of mechanization of the selection process is needed to make it truly haphazard!

To accomplish the goal of a random selection, one may make a card for each of the N members of the population, shuffle, and then draw n cards. This method is easy to understand but awkward to apply to large-size populations. It is best to use random numbers as described in Chapter 1. Random numbers are also conveniently generated on a computer (see Exercise 4.102).

At the beginning of this chapter, we stated that probability constitutes the major vehicle of statistical inference. In the context of random sampling from a population, the tools of probability enable us to gauge the likelihood of various potential outcomes of the sampling process. Ingrained in our probability calculations lies the artificial assumption that the composition of the population is *known*. The route of statistical inference is exactly in the opposite direction, as depicted in Figure 4. It is the composition of the population that is unknown

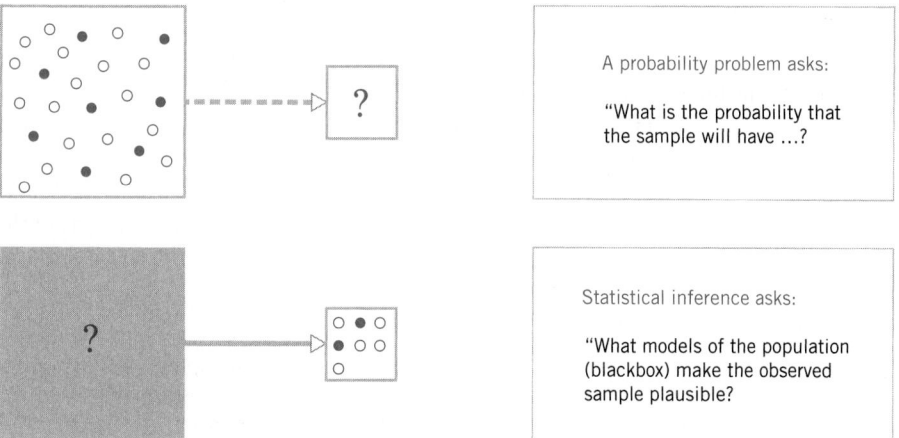

Figure 4   Probability versus statistical inference.

while we have at hand the observations (data) resulting from a random sample. Our object of inference is to ascertain what compositions (or models) of the population are compatible with the observed sample data. We view a model as plausible unless probability calculations based on this model make the sample outcome seem unlikely.

Exercises

4.83 Evaluate:

(a) $\binom{6}{3}$ (b) $\binom{10}{3}$ (c) $\binom{22}{2}$

(d) $\binom{22}{20}$ (e) $\binom{30}{3}$ (f) $\binom{30}{27}$

4.84 List all the samples from $\{a, b, c, d, e\}$ when (a) 2 out of 5 are selected, (b) 3 out of 5 are selected. Count the number of samples in each case.

4.85 Of 10 available candidates for membership in a university committee, 6 are men and 4 are women. The committee is to consist of 4 persons.

(a) How many different selections of the committee are possible?

(b) How many selections are possible if the committee must have 2 men and 2 women?

4.86 If a coin is tossed 11 times, the outcome can be recorded as a 11-character sequence of H's and T's according to the results of the successive tosses. In how many ways can there be 4 H's and 7 T's? (Put differently, in how many ways can one choose 4 positions out of 11 to put the letter H?)

4.87 A psychologist will select 5 preschool children from a class of 11 students in order to try out new abuse awareness material.

(a) How many different selections are possible?

(b) Suppose 4 of the 11 children are males. If the 5 selected children were to consist of 2 males and 3 females, how many different selections are possible?

4.88 Out of 11 people applying for an assembly job, 3 cannot do the work. Suppose two persons will be hired.

(a) How many distinct pairs are possible?

(b) In how many of the pairs will 0 or 1 people not be able to do the work?

(c) If two persons are chosen in a random manner, what is the probability that neither will be able to do the job?

4.89 After a preliminary screening, the list of qualified jurors consists of 10 males and 7 females. The 5 jurors the judge selects from this list are all males. Did the selection process seem to discriminate against females? Answer this by computing the probability of having no female members in the jury if the selection is random.

4.90 Suppose you participate in a lottery conducted by a local store to give away four prizes. Each customer is allowed to place 2 cards in the barrel. Suppose the barrel contains 5000 cards from which the 4 winning cards will be chosen at random. What is the probability that at least one of your cards will be drawn?

4.91 A batch of 20 used automobile alternators contains 4 defectives. If 3 alternators are sampled at random, find the probability of the event

(a) A = [None of the defectives appear]

(b) B = [Exactly two defectives appear]

4.92 *Ordered sampling versus unordered sampling.* Refer to Exercise 4.91. Suppose the sampling of 3 alternators is done by randomly choosing one after another and without replacement. The event A can then be described as $G_1G_2G_3$, where G denotes "good" and the suffixes refer to the order of the draws. Use the method of Example 14 to calculate $P(A)$ and $P(B)$. Verify that you get the same results as in Exercise 4.91.

This illustrates the following fact: To arrive at a random sample, we may randomly draw one object at a time without replacement and then disregard the order of the draws.

4.93 A college senior is selected at random from each state. Next, one senior is selected at random from the group of 50. Does this procedure produce a senior selected at random from those in the United States?

4.94 An instructor will choose 3 problems from a set of 7 containing 3 hard and 4 easy problems. If the selection is made at random, what is the probability that only the hard problems are chosen?

4.95 Nine agricultural plots for an experiment are laid out in a square grid as shown. Three plots are to be selected at random.

(a) Find the probability that all 3 are in the same row.

(b) Find the probability that all 3 are in different rows.

1	2	3
4	5	6
7	8	9

4.96 In one area of an orchard, there are 17 trees, of which 10 are bushy and 7 lean. If 4 trees are randomly selected for testing a new spray, what is the probability that exactly 2 bushy trees are selected?

*4.97 Referring to Exercise 4.96, now suppose that the trees are located in two rows: Row A has 8 trees of which 4 are bushy, and row B has 9 trees of which 6 are bushy. Two trees are to be randomly selected from each row for testing the spray, and the selections are independent for the two rows.

(a) Find the probability that the trees selected in row A are both bushy and those selected in row B are both lean.

(b) Find the probability that of the total of 4 trees selected in the manner described above, exactly 2 are bushy.

4.98 Are the following methods of selection likely to produce a random sample of 5 students from your school? Explain.

(a) Pick 5 students throwing flying discs on the mall.

(b) Pick 5 students who are studying in the library on Friday night.

(c) Select 5 students sitting near you in your statistics course.

4.99 An advertisement seeking volunteers for a clinical research draws 12 respondents. Of these respondents, 5 are below age 30 and 7 are over 30. The researcher will randomly select 4 persons to assign to a particular treatment regimen.

(a) How many selections are possible?

(b) What is the probability that 3 of the selected persons are below 30?

*4.100 Refer to Exercise 4.99, and further suppose that the 5 respondents who are below 30 consist of 2 males and 3 females, whereas those above 30 consist of 4 males and 3 females. Now, the researcher wants to randomly select 2 males and 2 females to be assigned to the treatment regimen. (The random selections from the different sexes are, of course, independent.)

(a) How many selections are possible?

(b) What is the probability that both selected males are over 30 and both selected females are under 30?

4.101 A box of tulip bulbs contains six bulbs that produce yellow flowers and five bulbs that produce red flowers. Four bulbs are to be randomly selected without replacement. Find the probability that:

(a) Exactly two of the selected bulbs produce red flowers.

(b) At least two of the selected bulbs produce red flowers.

(c) All four selected bulbs produce flowers of an identical color.

4.102 A file cabinet has eight student folders arranged alphabetically according to last name. Three files are selected at random.

(a) How many different selections are possible?

(b) Find the probability that the selected folders are all adjacent.

(*Hint*: Enumerate the selections of adjacent folders.)

USING STATISTICS WISELY

1. Begin by creating a sample space S which specifies all possible outcomes for the experiment.

2. Always assign probabilities to events that satisfy the axioms of probability. In the discrete case, the possible outcomes can be arranged in a sequence. The axioms are then automatically satisfied when probability $P(e_i)$ is assigned to the ith outcome e_i, where

$$0 \leq P(e_i) \quad \text{and} \quad \sum_{\text{all } e_i \text{ in } S} P(e_i) = 1$$

and then the probability of any event A is defined as

$$P(A) = \sum_{\text{all } e_i \text{ in } A} P(e_i)$$

3. Always use the rules of probability when combining the probabilities of events.

4. Do not confuse independent events with mutually exclusive events. When A and B are mutually exclusive, only one of them can occur. Their intersection is empty and so has probability 0.

5. Do not apply probability to $A\,B$ according to the special product rule

$$P(AB) = P(A)P(B)$$

unless the conditions for independence hold. Independence may be plausible when the events A and B pertain to physically unrelated parts of a large system and there are no common causes that jointly affect the occurrence of both events.

KEY IDEAS AND FORMULAS

An **experiment** is any process of observing a phenomenon that has variation in its outcomes. Each possible outcome is called an **elementary outcome**, a **simple event**, or an **element of the sample space**. The **sample space** is the collection of all of these outcomes. A **discrete sample space** has outcomes that can be arranged in a, possibly infinite, sequence. In contrast, a **continuous sample space** is an interval of possible outcomes.

A **tree diagram**, with separate sets of branches for each stage of an experiment, can help identify the elementary outcomes.

If an experiment is repeated a large number of times, experimentally we observe that the relative frequency of an event A.

$$\frac{\text{Number of times } A \text{ occurs}}{\text{Number of times experiment is performed}}$$

will stabilize at a numerical value. This long-run stability of relative frequency motivates us to assign a number $P(A)$ between 0 and 1 as the probability of the event A. In the other direction, we can approximate the probability of any event by repeating an experiment many times.

When the sample space is discrete, probability is then expressed as any assignment of non-negative numbers to the elementary outcomes so that probability one is assigned to the whole sample space. The probability model of an experiment is described by:

1. The sample space, a list or statement of all possible distinct outcomes.
2. Assignment of probabilities to all the elementary outcomes. $P(e) \geq 0$ and $\Sigma P(e) = 1$, where the sum extends over all e in S.

The probability of an event A is the sum of the probabilities of all the elementary outcomes that are in A.

$$P(A) = \sum_{\text{all } e \text{ in } A} P(e)$$

A uniform probability model holds when all the elementary outcomes in S are equiprobable. With a uniform probability model,

$$P(A) = \frac{\text{No. of } e \text{ in } A}{\text{No. of } e \text{ in } S}$$

In all cases, $P(A)$, viewed as the long-run relative frequency of A, can be approximately determined by repeating the experiment a large number of times.

Elementary outcomes and events can be portrayed in a Venn diagram. The event operations union, intersection, and complement can be depicted as well as the result of combining several operations.

The three basic laws of probability are

Law of complement $\qquad P(A) = 1 - P(\overline{A})$

Addition law $\qquad P(A \cup B) = P(A) + P(B) - P(AB)$

Multiplication law $\qquad P(AB) = P(B)P(A|B)$

These are useful in probability calculations when events are formed with the operations of complement, union, and intersection.

Two events are incompatible or mutually exclusive if their intersection is empty. In that case we have the special addition law for incompatible events

$$P(A \cup B) = P(A) + P(B)$$

The concept of conditional probability is useful to determine how the probability of an event A must be revised when another event B has occurred. It forms the basis of the multiplication law of probability and the notion of independence of events.

Conditional probability of A given B

$$P(A|B) = \frac{P(AB)}{P(B)}$$

Two events A and B are said to be **independent** if $P(A|B) = P(A)$. An equivalent condition for independence is that $P(A \cap B) = P(A) P(A)$.

The notion of **random sampling** is formalized by requiring that all possible samples are equally likely to be selected. The rule of combinations facilitates the calculation of probabilities in the context of random sampling from N distinct units.

TECHNOLOGY

Generating random digits

MINITAB

The following commands illustrate the generation of 5 random digits between 1 and 237 inclusive. As with random-digit tables, it is possible to get repeated values. It is prudent to generate a few more digits than you need in order to get enough unique numbers.

> **Dialog box:**
>
> **Calc > Random Data > Integer.** Type *C1* in **Store.**
> Type *5* in **Generate,** *1* in **Minimum,** and *237* in **Maximum.**
> Click **OK.**

EXCEL

The following commands illustrate the generation of 5 random digits between 1 and 237 inclusive. As with random-digit tables, it is possible to get repeated values.

> Select **Tools,** then **Data Analysis,** and then **Random Number Generation.**
> Click **OK.** Type *1* in **Number of Variables,** *5* in **Number of Random Numbers.**
> Select *Uniform* for **Distribution,** type *1* for **Between** and *238* after **and**
> (238 is 1 larger than the desired limit 237)
> Type (any positive number) *743* in **Random Seed.**
> Click **OK.**

The random numbers appear in the first column of the spreadsheet. You just ignore the decimal part of each entry to obtain random digits between 1 and 237 inclusive.

TI-84 PLUS

The following commands show the generation of 5 random digits between 1 and 237 inclusive. As with random-digit tables, it is possible to get repeated values.

Enter any nonzero number on the Home screen.

Press the **STO**→ button. Press the **MATH** button.

Select the **PRB** menu and then select **1: rand.**

From the Home screen press **ENTER.**

Press the **MATH** button. Select the **PRB** menu and then **5: randInt(.**

With **randInt(** on the Home screen, enter 1 and 237 so that the following appears

randInt (1,237)

Press **ENTER** to obtain the first random digit. Continue to press.

ENTER until the 5 random digits are obtained.

7. REVIEW EXERCISES

4.103 Describe the sample space for each of the following experiments.

(a) The number of different words used in a sentence containing 24 words.

(b) The air pressure (psi) in the right front tire of a car.

(c) In a survey, 50 students are asked to respond "yes" or "no" to the question "Do you hold at least a part-time job while attending school?" Only the number answering "yes" will be recorded.

(d) The time a TV satellite remains in operation.

4.104 For the experiments in Exercise 4.103, which sample spaces are discrete and which are continuous?

4.105 Identify these events in the corresponding parts of Exercise 4.103.

(a) More than 22 words.

(b) Air pressure less than or equal to 28 psi.

(c) At most 25% hold jobs.

(d) Less than 500.5 days.

4.106 Examine each of these probability assignments and state what makes it improper.

(a) Concerning tomorrow's weather,

$$P(\text{Rain}) = .4$$
$$P(\text{Cloudy but no rain}) = .4$$
$$P(\text{Sunny}) = .3$$

(b) Concerning your passing of the statistics course,

$$P(\text{Pass}) = 1.1 \qquad P(\text{Fail}) = .1$$

(c) Concerning your grades in statistics and economics courses,

$$P(A \text{ in statistics}) = .3$$
$$P(A \text{ in economics}) = .7$$
$$P(A\text{'s in both statistics and economics}) = .4$$

4.107 A driver is stopped for erratic driving, and the alcohol content of his blood is checked. Specify the sample space and the event $A =$ [level exceeds legal limit] if the legal limit is .10%.

4.108 The Wimbledon men's tennis championship ends when one player wins three sets.

(a) How many elementary outcomes end in three sets? In four?

*(b) If the players are evenly matched, what is the probability that the tennis match ends in four sets?

4.109 There are four tickets numbered 1, 2, 3, and 4. Suppose a two-digit number will be formed by first drawing one ticket at random and then drawing a second ticket at random from the remaining three. (For instance, if the first ticket drawn shows 3 and the second shows 1, the number recorded is 31.) List the sample space and determine the following probabilities.

(a) An even number.

(b) A number larger than 20.

(c) A number between 22 and 30.

4.110 To compare two varieties of wheat, say, a and b, a field trial will be conducted on four square plots located in two rows and two columns. Each variety will be planted on two of these plots.

Plot arrangement

(a) List all possible assignments for variety a.

(b) If the assignments are made completely at random, find the probability that the plots receiving variety a are:

(i) In the same column.

(ii) In different rows and different columns.

4.111 Refer to Exercise 4.110. Instead of a completely random choice, suppose a plot is chosen at random from each row and assigned to variety a. Find the probability that the plots receiving a are in the same column.

4.112 Chevalier de Méré, a French nobleman of the seventeenth century, reasoned that in a single throw of a fair die, $P(1) = \frac{1}{6}$, so in two throws, $P(1$ appears at least once$) = \frac{1}{6} + \frac{1}{6} = \frac{1}{3}$. What is wrong with the above reasoning? Use the sample space of Exercise 4.21 to obtain the correct answer.

4.113 A letter is chosen at random from the word *"STATISTICIAN."*

(a) What is the probability that it is a vowel?

(b) What is the probability that it is a T?

4.114 Does the uniform model apply to the following observations? Explain.

(a) Day of week on which the most persons depart by airplane from Chicago.

(b) Day of week on which the monthly low temperature occurs.

(c) Day of week on which the maximum amount of ozone is recorded.

(d) Month of year when a department store has the maximum sales revenues.

4.115 A three-digit number is formed by arranging the digits 1, 2, and 5 in a random order.

(a) List the sample space.

(b) Find the probability of getting a number less than 400.

(c) What is the probability that an even number is obtained?

4.116 A late shopper for Valentine's flowers calls by phone to have a flower wrapped. The store has only 5 roses, of which 3 will open by the next day, and 6 tulips, of which 2 will open by the next day.

(a) Construct a Venn diagram and show the events $A = $ [Rose], and $B = $ [Will open next day].

(b) If the store selects one flower at random, find the probability that it will not open by the next day.

4.117 In checking the conditions of a used car, let A denote the event that the car has a faulty transmission, B the event that it has faulty brakes, and C the event that it has a faulty exhaust system. Describe in words what the following events represent:

(a) $A \cup B$ (b) ABC

(c) $\overline{A}\,\overline{B}\,\overline{C}$ (d) $\overline{A} \cup \overline{B}$

4.118 Express the following statements in the notations of the event operations.

(a) *A* occurs and *B* does not.

(b) Neither *A* nor *B* occurs.

(c) Exactly one of the events *A* and *B* occurs.

4.119 Suppose each of the numbers .1, .28, and .52 represents the probability of one of the events *A*, *AB*, and *A* ∪ *B*. Connect the probabilities to the appropriate events.

4.120 From the probabilities exhibited in this Venn diagram, find $P(\overline{A})$, $P(AB)$, $P(B \cup C)$, and $P(BC)$.

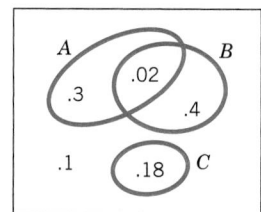

4.121 Using event relations, express the following events in terms of the three events *A*, *B*, and *C*.

(a) All three events occur.

(b) At least one of the three events occurs.

(c) *A* and *B* occur and *C* does not.

(d) Only *B* occurs.

4.122 Concerning three events *A*, *B*, and *C*, the probabilities of the various intersections are given in the accompanying table. [For instance, $P(AB\overline{C}) = .10$.]

	B		**$\overline{B}$**	
	C	$\overline{C}$	*C*	$\overline{C}$
A	.05	.10	.05	.17
$\overline{A}$	.20	.15	.18	.10

(a) Draw a Venn diagram, identify the intersections, and mark the probabilities.

(b) Determine the probabilities

$$P(AB) \qquad P(A\overline{C}) \qquad P(C)$$

(c) Fill in the accompanying probability table concerning the events *A* and *B*.

	B	**$\overline{B}$**
A		
$\overline{A}$		

4.123 Referring to Exercise 4.122, calculate the probabilities of the following events.

(a) Both *B* and *C* occur.

(b) Either *B* or *C* occurs.

(c) *B* occurs and *C* does not occur.

(d) Only one of the three events *A*, *B*, and *C* occurs.

4.124 Concerning three events A, B, and C, the following probabilities are specified.

$$P(A) = .51 \quad P(AB) = .17 \quad P(ABC) = .12$$
$$P(B) = .45 \quad P(BC) = .20$$
$$P(C) = .50 \quad P(AC) = .33$$

Draw a Venn diagram and determine the probabilities of all the intersections that appear in the diagram. Also, make a probability table like the one given in Exercise 4.122.

4.125 Referring to Exercise 4.124, find the probability that:

(a) *B* occurs and *C* does not occur.

(b) At least one of the events *A* and *B* occurs.

(c) Exactly two of the events *A*, *B*, and *C* occur.

4.126 Suppose a fair die has its even-numbered faces painted red and the odd-numbered faces are white. Consider the experiment of rolling the die once and the events

$$A = [2 \text{ or } 3 \text{ shows up}]$$
$$B = [\text{Red face shows up}]$$

Find the following probabilities:

(a) $P(A)$ (b) $P(B)$ (c) $P(AB)$

(d) $P(A|B)$ (e) $P(A \cup B)$

4.127 Given $P(AB) = .3$ and $P(B) = .5$, find $P(A|B)$. If, further, $P(A) = .6$, are A and B independent?

4.128 Suppose three events A, B, and C are such that B and C are mutually exclusive and

$$P(A) = .6 \qquad P(B) = .3 \quad P(C) = .25$$

$$P(A|B) = \frac{2}{3} \qquad P(\overline{A}C) = .1$$

(a) Show the events in a Venn diagram.

(b) Determine the probabilities of all the intersections and mark them in the Venn diagram.

(c) Find the probability that only one of the three events occurs.

4.129 Refer to Exercise 4.128. For each pair of events given below, determine whether or not the events are independent.

(a) A, C

(b) $A\overline{B}$, C

4.130 Concerning the events A and B, the following probabilities are given.

$$P(B) = \frac{1}{4} \qquad P(A|B) = \frac{1}{2} \qquad P(A|\overline{B}) = \frac{1}{3}$$

Determine (a) $P(A\overline{B})$, (b) $P(A)$, and (c) $P(\overline{B}|A)$.

4.131 Refer to the probability table given in Exercise 4.122 concerning three events A, B, and C.

(a) Find the conditional probability of A given that B does not occur.

(b) Find the conditional probability of B given that both A and C occur.

(c) Determine whether or not the events A and C are independent.

4.132 Mr. Hope, a character apprehended by Sherlock Holmes, was driven by revenge to commit two murders. He presented two seemingly identical pills, one containing a deadly poison, to an adversary who selected one while Mr. Hope took the other. The entire procedure was then to be repeated with the second victim. Mr. Hope felt that Providence would protect him, but what is the probability of the success of his endeavor?

4.133 A bowl contains 15 marbles, of which 10 are numbered 1 and 5 are numbered 2. Two marbles are to be randomly drawn from the bowl one after another and without replacement, and a two-digit number will be recorded according to the results. (For instance, if the first marble drawn shows 2 and the second shows 1, the number recorded is 21.)

(a) List the sample space and determine the probability of each outcome.

(b) Find the probability of getting an even number.

(c) Find the probability that the number is larger than 15.

*4.134 Three production lines contribute to the total pool of a company's product. Line 1 provides 20% to the pool and 10% of its products are defective; Line 2 provides 50% to the pool and 5% of its products are defective; Line 3 contributes 30% to the pool and 6% of its products are defective.

(a) What percent of the items in the pool are defective?

(b) Suppose an item is randomly selected from the pool and found to be defective. What is the probability that it came from Line 1?

4.135 In an optical sensory experiment, a subject shows a fast response (F), a delayed response (D), or no response at all (N). The experiment will be performed on two subjects.

(a) Using a tree diagram, list the sample space.

(b) Suppose, for each subject, $P(F) = .4$, $P(D) = .3$, $P(N) = .3$, and the responses of different subjects are independent. Assign probabilities to the elementary outcomes.

(i) Find the probability that at least one of the subjects shows a fast response.

(ii) Find the probability that both of the subjects respond.

4.136 Four border collie puppies from different litters are available for a new endurance training program.

Dog	Sex	Age (weeks)
1	F	8
2	M	8
3	M	12
4	F	8

Two puppies will be selected at random to receive the training. Let

A = [Selected puppies are of the same sex]

B = [Selected puppies are of the same age]

(a) Make a Venn diagram showing the two events.

(b) Find the probability of $A \cup B$.

(c) Are A and B independent? Explain why or why not.

(d) Find the probability of $\overline{AB}$.

4.137 A local moving company owns 11 trucks. Three are randomly selected for compliance with emission standards and all are found to be noncompliant. The company argues that these are the only three which do not meet the standards. Calculate the probability that, if only three are noncompliant, all three would be in the sample. Comment on the veracity of the company's claim.

4.138 In all of William Shakespeare's works, he used 884,647 different words. Of these, 14,376 appeared only once and 4343 appeared twice. If one word is randomly selected from a list of these 884,647 different words:

(a) What is the probability that the selected word appears only once?

(b) What is the probability that the selected word appears exactly twice?

(c) What is the probability that the selected word appears more than twice?

(d) Suppose that, instead of randomly selecting a word from the list of different words, you randomly select a word from a book of Shakespeare's complete works by selecting a page, line, and word number from a random-digit table. Is the probability of selecting a word that appears exactly once larger, smaller, or the same as your answer to part (a)?

4.139 An IRS agent receives a batch of 18 tax returns that were flagged by computer for possible tax evasions. Suppose, unknown to the agent, 7 of these returns have illegal deductions and the other 11 are in good standing. If the agent randomly selects 4 of these returns for audit, what is the probability that:

(a) None of the returns that contain illegal deductions are selected?

(b) At least 2 have illegal deductions?

*4.140 *Polya's urn scheme.* An urn contains 4 red and 6 green balls. One ball is drawn at random and its color is observed. The ball is then returned to the urn, and 3 new balls of the same color are added to the urn. A second ball is then randomly drawn from the urn that now contains 13 balls.

(a) List all outcomes of this experiment (use symbols such as $R_1 G_2$ to denote the outcome of the first ball red and the second green).

(b) What is the probability that the first ball drawn is green?

(c) What is the conditional probability of getting a red ball in the second draw given that a green ball appears in the first?

(d) What is the (unconditional) probability of getting a green ball in the second draw?

*4.141 *Birthdays.* It is somewhat surprising to learn the probability that 2 persons in a class share the same birthday. As an approximation, assume that the 365 days are equally likely birthdays.

(a) What is the probability that, among 3 persons, at least 2 have the same birthday? (*Hint:* The reasoning associated with a tree diagram shows that there are $365 \times 365 \times 365$ possible birthday outcomes. Of these, $365 \times 364 \times 363$ correspond to no common birthday.)

(b) Generalize the above reasoning to N persons. Show that

$$P[\text{No common birthday}] = \frac{365 \times 364 \times \cdots \times (365 - N + 1)}{(365)^N}$$

Some numerical values are

N	5	9	18	22	23
$P[\text{No common birthday}]$	.973	.905	.653	.524	.493

We see that with $N = 23$ persons, the probability is greater than $\frac{1}{2}$ that at least two share a common birthday.)

5

Probability Distributions

Rescue Service on a Lake

Student sailors and other boaters on Lake Mendota are protected by a boating rescue service. A long record for summer days provides a distribution of the number of rescues per day.

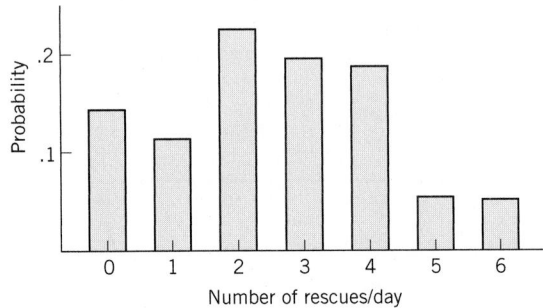

© John Terrence Turner/FPG International/
Getty Images

The distribution describes the randomness of daily rescue activity. For instance, on any given day, the most probable number of rescues is 2. The distribution can be the basis for decisions concerning manpower and the need for additional rescue boats.

1. INTRODUCTION

A prescription for the probability model of an experiment contains two basic ingredients: the sample space and the assignment of probability to each elementary outcome. In Chapter 4, we encountered several examples where the elementary outcomes had only qualitative descriptions rather than numerical values. For instance, with two tosses of a coin, the outcomes HH, HT, TH, and TT are pairs of letters that identify the occurrences of heads or tails. If a new vaccine is studied for the possible side effects of nausea, the response of each subject may be severe, moderate, or no feeling of nausea. These are qualitative outcomes rather than measurements on a numerical scale.

Often, the outcomes of an experiment are numerical values: for example, the daily number of burglaries in a city, the hourly wages of students on summer jobs, and scores on a college placement examination. Even in the former situation where the elementary outcomes are only qualitatively described, interest frequently centers on some related numerical aspects. If a new vaccine is tested on 100 individuals, the information relevant for an evaluation of the vaccine may be the numbers of responses in the categories—severe, moderate, or no nausea. The detailed record of 100 responses can be dispensed with once we have extracted this summary. Likewise, for an opinion poll conducted on 500 residents to determine support for a proposed city ordinance, the information of particular interest is how many residents are in favor of the ordinance, and how many are opposed. In these examples, the individual observations are not numerical, yet a numerical summary of a collection of observations forms the natural basis for drawing inferences. In this chapter, we concentrate on the numerical aspects of experimental outcomes.

2. RANDOM VARIABLES

Focusing our attention on the numerical features of the outcomes, we introduce the idea of a random variable.

> A **random variable** X associates a numerical value with each outcome of an experiment.

Corresponding to every elementary outcome of an experiment, a random variable assumes a numerical value, determined from some characteristic pertaining to the outcome. (In mathematical language, we say that a random variable X is a real-valued function defined on a sample space.) The word "random" serves as a reminder of the fact that, beforehand, we do not know the outcome of an experiment or its associated value of X.

Example 1   The Number of Heads as a Random Variable

Consider X to be the number of heads obtained in three tosses of a coin. List the numerical values of X and the corresponding elementary outcomes.

SOLUTION   First, X is a variable since the number of heads in three tosses of a coin can have any of the values 0, 1, 2, or 3. Second, this variable is random in the sense that the value that would occur in a given instance cannot be predicted with certainty. We can, though, make a list of the elementary outcomes and the associated values of X.

Outcome	Value of X
HHH	3
HHT	2
HTH	2
HTT	1
THH	2
THT	1
TTH	1
TTT	0

Note that, for each elementary outcome there is only one value of X. However, several elementary outcomes may yield the same value. Scanning our list, we now identify the events (i.e., the collections of the elementary outcomes) that correspond to the distinct values of X.

Numerical Value of X as an Event		Composition of the Event
$[X = 0]$	$=$	$\{TTT\}$
$[X = 1]$	$=$	$\{HTT, THT, TTH\}$
$[X = 2]$	$=$	$\{HHT, HTH, THH\}$
$[X = 3]$	$=$	$\{HHH\}$

Guided by this example, we observe the following general facts.

The events corresponding to the distinct values of X are incompatible.

The union of these events is the entire sample space.

Typically, the possible values of a random variable X can be determined directly from the description of the random variable without listing the sample space. However, to assign probabilities to these values, treated as events, it is sometimes helpful to refer to the sample space.

Example 2 A Random Variable That Is a Count with a Finite Maximum Value

Fifty cars are entered in a 100-mile road race. Let X be the number of cars that actually finish the race. Here X could conceivably take any of the values $0, 1, \ldots, 50$.

Example 3 A Random Variable That Is a Count with No Upper Limit

Once a week, a student buys a single lottery ticket. Let X be the number of tickets she purchases before she wins at least $1000 on a ticket. The possible values of X are then $1, 2, 3, \ldots$, where the list never terminates.

A random variable is said to be **discrete** if it has either a finite number of values or infinitely many values that can be arranged in a sequence. All the preceding examples are of this type. On the other hand, if a random variable represents some measurement on a continuous scale and is therefore capable of assuming all values in an interval, it is called a **continuous** random variable. Of course, any measuring device has a limited accuracy and, therefore, a continuous scale must be interpreted as an abstraction. Some examples of continuous random variables are the height of an adult male, the daily milk yield of a holstein, and the survival time of a patient following a heart attack.

Probability distributions of discrete random variables are explored in this chapter. As we shall see, the developments stem directly from the concepts of probability introduced in Chapter 4. A somewhat different outlook is involved in the process of conceptualizing the distribution of a continuous random variable. Details for the continuous case are postponed until Chapter 6.

Exercises

5.1 Identify each of the following as a discrete or continuous random variable.

(a) Number of empty seats on a flight from Atlanta to London.

(b) Yearly low temperature in your city.

(c) Yearly maximum daily amount of ozone in Los Angeles.

(d) Time it takes for a plumber to fix a bathroom faucet.

(e) Number of cars ticketed for illegal parking on campus today.

5.2 Identify the variable as a discrete or a continuous random variable.

(a) The loss of weight following a diet program.

(b) The magnitude of an earthquake as measured on the open-ended Richter scale.

(c) The seating capacity of an airplane.

(d) The number of cars sold at a dealership on one day.

(e) The percentage of fruit juice in a drink mix.

5.3 Two of the integers {1, 3, 5, 6, 7} are chosen at random without replacement. Let X denote the difference larger minus smaller number.

(a) List all choices and the corresponding values of X.

(b) List the distinct values of X and determine their probabilities.

5.4 The three finalists for an award are A, B, and C. They will be rated by two judges. Each judge assigns the ratings 1 for best, 2 for intermediate, and 3 for worst. Let X denote the total score for finalist A (the sum of the ratings received from the two judges).

(a) List all pairs of ratings that finalist A can receive.

(b) List the distinct values of X.

5.5 Refer to Exercise 5.4. Suppose instead there are two finalists A and B and four judges. Each judge assigns the ratings 1 for the best and 2 for the worst finalists.

(a) List all possible assignments of ratings to finalist A by the four judges.

(b) List the distinct values of X, the total score of A.

5.6 Two brands of beverages, B and M, are popular with students. The owner of one campus establishment will observe sales and, for each of three weekends, record which brand has the highest sales. List the possible outcomes, and for each outcome record the number of weekends X that the sales of B are highest. (Assume there are no ties.)

5.7 Each week a grocery shopper buys either canned (C) or bottled (B) soft drinks. The type of soft drink purchased in 3 consecutive weeks is to be recorded.

(a) List the sample space.

(b) If a different type of soft drink is purchased than in the previous week, we say that there is a switch. Let X denote the number of switches. Determine the value of X for each elementary outcome. (*Example:* For BBB, $X = 0$; for BCB, $X = 2$.)

5.8 A child psychologist interested in how friends are selected studies groups of three children. For one group, Ann, Barb, and Carol, each is asked which of the other two she likes best.

(a) Make a list of the outcomes. (Use A, B, and C to denote the three children.)

(b) Let X be the number of times Carol is chosen. List the values of X.

3. PROBABILITY DISTRIBUTION OF A DISCRETE RANDOM VARIABLE

The list of possible values of a random variable X makes us aware of all the eventualities of an experiment as far as the realization of X is concerned. By employing the concepts of probability, we can ascertain the chances of observing the various values. To this end, we introduce the notion of a probability distribution.

> The **probability distribution** or, simply the **distribution**, of a discrete random variable X is a list of the distinct numerical values of X along with their associated probabilities.
>   Often, a formula can be used in place of a detailed list.

Example 4   The Probability Distribution for Tossing a Fair Coin

If X represents the number of heads obtained in three tosses of a fair coin, find the probability distribution of X.

SOLUTION   In Example 1, we have already listed the eight elementary outcomes and the associated values of X. The distinct values of X are 0, 1, 2, and 3. We now calculate their probabilities.

The model of a fair coin entails that the eight elementary outcomes are equally likely, so each is assigned the probability $\frac{1}{8}$. The event $[X = 0]$ has the single outcome TTT, so its probability is $\frac{1}{8}$. Similarly, the probabilities of $[X = 1]$, $[X = 2]$, and $[X = 3]$ are found to be $\frac{3}{8}$, $\frac{3}{8}$, and $\frac{1}{8}$, respectively. Collecting these results, we obtain the probability distribution of X displayed in Table 1.

TABLE 1   The Probability Distribution of X, the Number of Heads in Three Tosses of a Coin

Value of X	Probability
0	$\frac{1}{8}$
1	$\frac{3}{8}$
2	$\frac{3}{8}$
3	$\frac{1}{8}$
Total	1

For general discussion, we will use the notation x_1, x_2, and so on, to designate the distinct values of a random variable X. The probability that a particular value x_i occurs will be denoted by $f(x_i)$. As in Example 4, if X can take k possible values x_1, . . . , x_k with the corresponding probabilities $f(x_1)$, . . . , $f(x_k)$, the probability distribution of X can be displayed in the format of Table 2. Since the quantities $f(x_i)$ represent probabilities, they must all be numbers between 0 and 1. Furthermore, when summed over all possible values of X, these probabilities must add up to 1.

TABLE 2   Form of a Discrete Probability Distribution

Value of x	Probability $f(x)$
x_1	$f(x_1)$
x_2	$f(x_2)$
.	.
.	.
.	.
x_k	$f(x_k)$
Total	1

The **probability distribution** of a discrete random variable X is described as the function

$$f(x_i) = P[X = x_i]$$

which gives the probability for each value and satisfies:

1. $f(x_i) \geq 0$ for each value x_i of X
2. $\sum_{i=1}^{k} f(x_i) = 1$

A probability distribution or the probability function describes the manner in which the total probability 1 gets apportioned to the individual values of the random variable.

A graphical presentation of a probability distribution helps reveal any pattern in the distribution of probabilities. Is there symmetry about some value or a long tail to one side? Is the distribution peaked with a few values having high probabilities or is it uniform?

We consider a display similar in form to a relative frequency histogram, discussed in Chapter 2. It will also facilitate the building of the concept of a continuous distribution. To draw a **probability histogram**, we first mark the values of X on the horizontal axis. With each value x_i as center, a vertical rectangle is drawn whose area equals the probability $f(x_i)$. The probability histogram for the distribution of Example 4 is shown in Figure 1.

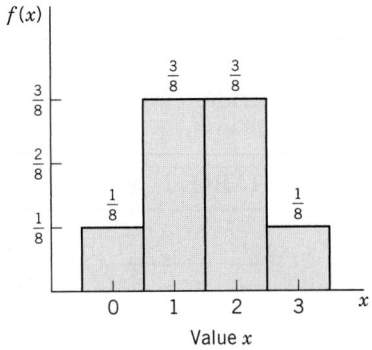

Figure 1   The probability histogram of X, the number of heads in three tosses of a coin.

Example 5 Probability Distribution for News Source Preference

Suppose 30% of the students at a large university prefer getting their daily news from the Internet as opposed to television. Four students are randomly selected. Let X be the number of students sampled that prefer news from the Internet. Obtain the probability distribution of X and plot the probability histogram.

SOLUTION Because each student will prefer either Internet (I) news or television (T), the number of elementary outcomes concerning a sample of four students is $2 \times 2 \times 2 \times 2 = 16$. These can be conveniently enumerated in the scheme of Example 8, Chapter 4 (called a tree diagram). However, we list them here according to the count X.

$X = 0$	$X = 1$	$X = 2$	$X = 3$	$X = 4$
TTTT	TTTI	TTII	TIII	IIII
	TTIT	TITI	ITII	
	TITT	TIIT	IITI	
	ITTT	ITTI	IIIT	
		ITIT		
		IITT		

Our objective here is to calculate the probability of each value of X. To this end, we first reflect on the assignment of probabilities to the elementary outcomes. For one student selected at random, we obviously have $P(I) = .3$ and $P(T) = .7$ because 30% of the students prefer Internet news. Moreover, as the population is vast while the sample size is very small, the observations on four students can, for all practical purposes, be treated as independent. That is, knowledge that the first student selected prefers Internet news does not change the probability that the second will prefer Internet news and so on.

Invoking independence and the multiplication law of probability, we calculate $P(TTTT) = .7 \times .7 \times .7 \times .7 = .2401$ so $P(X = 0) = .2401$. The event $[X = 1]$ has four elementary outcomes, each containing three T's and one I. Since $P(TTTI) = (.7)^3 \times (.3) = .1029$ and the same result holds for each of these 4 elementary outcomes, we get $P[X = 1] = 4 \times .1029 = .4116$. In the same manner,

$$P[X = 2] = 6 \times (.7)^2 \times (.3)^2 = .2646$$

$$P[X = 3] = 4 \times (.7) \times (.3)^3 = .0756$$

$$P[X = 4] = (.3)^4 = .0081$$

Collecting these results, we obtain the probability distribution of X presented in Table 3 and the probability histogram plotted in Figure 2.

TABLE 3   The Probability
Distribution of X
in Example 5

x	$f(x)$
0	.2401
1	.4116
2	.2646
3	.0756
4	.0081
Total	1.0000

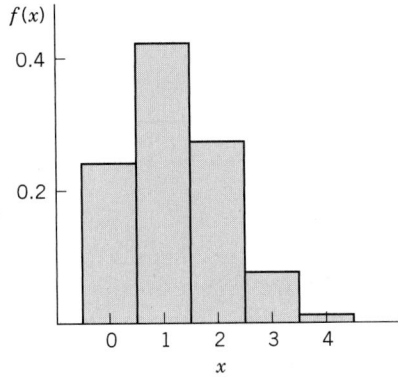

Figure 2   Probability histogram.

At this point, we digress briefly for an explanation of the role of probability distributions in statistical inference. To calculate the probabilities associated with the values of a random variable, we require a full knowledge of the uncertainties of the experimental outcomes. For instance, when X represents some numerical characteristic of a random sample from a population, we assume a known composition of the population in order that the distribution of X can be calculated numerically. In Example 5, the chances of observing the various values of X were calculated under the assumption that the proportion of all students who prefer Internet news was .3. Ordinarily, in practical applications, this population quantity would be unknown to us. Suppose the letter p stands for this unknown proportion of students who prefer Internet news. Statistical inference attempts to determine the values of p that are deemed plausible in light of the value of X actually observed in a sample. To fix ideas, suppose all four of the students sampled prefer Internet news. Based on this observation, is .3 a plausible value of p? Table 3 shows that if p were indeed .3, the chance of observing the extreme value $X = 4$ is only .0081. This very low probability casts doubt on the hypothesis that $p = .3$. This kind of statistical reasoning will be explored further in later chapters.

The probability distributions in Examples 4 and 5 were obtained by first assigning probabilities to the elementary outcomes using a process of logical deduction. When this cannot be done, one must turn to an empirical determination of the distribution. This involves repeating the experiment a large number of times and using the relative frequencies of the various values of X as approximations of the corresponding probabilities.

Example 6   A Probability Distribution Based on an Empirical Study

Let X denote the number of magazines to which a college senior subscribes. From a survey of 400 college seniors, suppose the frequency distribution of Table 4 was observed. Approximate the probability distribution of X.

TABLE 4   Frequency Distribution of the Number X of Magazine Subscriptions

Magazine Subscriptions (x)	Frequency	Relative Frequency[a]
0	61	.15
1	153	.38
2	106	.27
3	56	.14
4	24	.06
Total	400	1.00

[a]Rounded to second decimal.

SOLUTION   Viewing the relative frequencies as empirical estimates of the probabilities, we have essentially obtained an approximate determination of the probability distribution of X. The true probability distribution would emerge if a vast number (ideally, the entire population) of seniors were surveyed.

The reader should bear in mind an important distinction between a relative frequency distribution and the probability distribution. The former is a sample-based entity and is therefore susceptible to variation on different occasions of sampling. By contrast, the probability distribution is a stable entity that refers to the entire population. It is a theoretical construct that serves as a model for describing the variation in the population.

The probability distribution of X can be used to calculate the probabilities of events defined in terms of X. To illustrate this, consider the probability distribution of Table 5, which describes the number of homework assignments due next week for a randomly selected set of students taking at least 14 credits. What is the probability that X is equal to or larger than 2?

TABLE 5   A Probability Distribution for Number of Homework Assignments Due Next Week

Value x	Probability $f(x)$
0	.02
1	.23
2	.40
3	.25
4	.10

The event $[X \geq 2]$ is composed of $[X = 2]$, $[X = 3]$, and $[X = 4]$. Thus,

$$P[X \geq 2] = f(2) + f(3) + f(4)$$
$$= .40 + .25 + .10 = .75$$

Similarly, we also calculate

$$P[X \leq 2] = f(0) + f(1) + f(2)$$
$$= .02 + .23 + .40 = .65$$

Exercises

5.9   Listed to the right are the seven elementary outcomes of an experiment, their probabilities, and the value of a random variable X at each outcome. Obtain the probability distribution of X.

5.10   Two of the integers $\{1, 2, 6, 7, 9\}$ are chosen at random without replacement. Let X denote the sum of the two integers.

(a)   List all choices and the corresponding values of X.

(b)   List the distinct values of X.

(c)   Obtain the probability distribution of X.

Elementary Outcome	Probability	Value of X
e_1	.05	2
e_2	.30	0
e_3	.15	2
e_4	.10	0
e_5	.16	4
e_6	.11	0
e_7	.13	0

5.11 Let the random variable X represent the sum of the points in two tosses of a die.

(a) List the possible values of X.

(b) For each value of X, list the corresponding elementary outcomes.

(c) Obtain the probability distribution of X.

5.12 Examine if the following are legitimate probability distributions.

(a)

x	$f(x)$
−1	.2
2	.5
7	.2
9	.1

(b)

x	$f(x)$
1	.3
3	.5
4	.3
6	.2

(c)

x	$f(x)$
−2	.25
0	.50
2	.25
4	0

(d)

x	$f(x)$
0	.3
1	−.1
2	.8

5.13 For each case, list the values of x and $f(x)$ and examine if the specification represents a probability distribution. If it does not, state what properties are violated.

(a) $f(x) = \frac{1}{10}(x - 1)$ for $x = 2, 3, 4, 5$

(b) $f(x) = \frac{1}{3}(x - 2)$ for $x = 1, 2, 3, 4$

(c) $f(x) = \frac{1}{20}(2x + 4)$ for $x = -2, -1, 0, 1, 2$

(d) $f(x) = 3/2^x$ for $x = 2, 3, 4$

5.14 The probability distribution of X is given by the function

$$f(x) = \frac{1}{30}\binom{5}{x} \quad \text{for} \quad x = 1, 2, 3, 4$$

Find (a) $P[X = 3]$ (b) $P[X \text{ is even}]$.

5.15 Refer to Exercise 5.7. Suppose that for each purchase $P(B) = \frac{1}{2}$ and the decisions in dif-

ferent weeks are independent. Assign probabilities to the elementary outcomes and obtain the distribution of X.

5.16 Refer to Exercise 5.8. Assuming each choice is equally likely, determine the probability distribution of X.

5.17 From five cards bearing the numbers −3, −1, 0, 1, 3, one card will be drawn at random. Determine the probability distribution of the square of the number that appears on the selected card.

5.18 From the six marbles numbered as shown,

① ① ① ① ② ②

two marbles will be drawn at random without replacement. Let X denote the sum of the numbers on the selected marbles. List the possible values of X and determine the probability distribution.

5.19 Suppose, for a loaded die, the probabilities of the faces

1 2 3 4 5 6

are in the ratios $3:1:1:1:1:3$. Let X denote the number appearing on a single roll of the die.

(a) Determine the probability distribution of X.

(b) What is the probability of getting an even number?

5.20 A surprise quiz contains three multiple-choice questions: Question 1 has four suggested answers, Question 2 has three, and Question 3 has two. A completely unprepared student decides to choose the answers at random. Let X denote the number of questions the student answers correctly.

(a) List the possible values of X.

(b) Find the probability distribution of X.

(c) Find $P[\text{At least 1 correct}] = P[X \geq 1]$.

(d) Plot the probability histogram.

5.21 A probability distribution is partially given in the following table with the additional information that the even values of X are equally likely. Determine the missing entries in the table.

x	$f(x)$
1	.1
2	
3	0
4	
5	.3
6	

5.22 Consider the following setting of a random selection: A box contains 100 cards, of which 25 are numbered 1, 35 are numbered 2, 30 are numbered 3, 10 are numbered 4. One card will be drawn from the box and its number X observed. Give the probability distribution of X.

5.23 Two probability distributions are shown in the following tables. For each case, describe a specific setting of random selection (like the one given in Exercise 5.22) that yields the given probability distribution.

<table>
<tr><td colspan="2">(a)</td><td colspan="2">(b)</td></tr>
<tr><td>x</td><td>$f(x)$</td><td>x</td><td>$f(x)$</td></tr>
<tr><td>2</td><td>.32</td><td>-2</td><td>3/4</td></tr>
<tr><td>4</td><td>.44</td><td>0</td><td>4/14</td></tr>
<tr><td>6</td><td>.24</td><td>4</td><td>5/14</td></tr>
<tr><td></td><td></td><td>5</td><td>2/14</td></tr>
</table>

5.24 In a study of the life length of a species of mice, 120 newborn mice are observed. The numbers staying alive past the first, second, third, and fourth years are 106, 72, 25, and 0, respectively. Let X denote the life length (in discrete units of whole years) of this species of mice. Using these data, make an empirical determination of the probability distribution of X.

5.25 Use the probability distribution given here to calculate

(a) $P[X \le 3]$

(b) $P(X \ge 2]$

(c) $P[1 \le X \le 3]$

x	$f(x)$
0	.12
1	.25
2	.42
3	.11
4	.10

5.26 Of eight candidates seeking three positions at a counseling center, five have degrees in social science and three do not. If three candidates are selected at random, find the probability distribution of X, the number having social science degrees among the selected persons.

5.27 Based on recent records, the manager of a car painting center has determined the following probability distribution for the number of customers per day.

x	$f(x)$
0	.05
1	.20
2	.30
3	.25
4	.15
5	.05

(a) If the center has the capacity to serve two customers per day, what is the probability that one or more customers will be turned away on a given day?

(b) What is the probability that the center's capacity will not be fully utilized on a day?

(c) By how much must the capacity be increased so the probability of turning a customer away is no more than .10?

5.28   Suppose X denotes the number of telephone receivers in a single-family residential dwelling. From an examination of the phone subscription records of 381 residences in a city, the frequency distribution on the right is obtained.

(a) Based on these data, obtain an approximate determination of the probability distribution of X.

(b) Why is this regarded as an approximation?

(c) Plot the probability histogram.

No. of Receivers (x)	No. of Residences (Frequency)
0	2
1	82
2	161
3	89
4	47
Total	381

4. EXPECTATION (MEAN) AND STANDARD DEVIATION OF A PROBABILITY DISTRIBUTION

We will now introduce a numerical measure for the center of a probability distribution and another for its spread. In Chapter 2, we discussed the concepts of mean, as a measure of the center of a data set, and standard deviation, as a measure of spread. Because probability distributions are theoretical models in which the probabilities can be viewed as long-run relative frequencies, the sample measures of center and spread have their population counterparts.

To motivate their definitions, we first refer to the calculation of the mean of a data set. Suppose a die is tossed 20 times and the following data obtained.

$$4, \quad 3, \quad 4, \quad 2, \quad 5, \quad 1, \quad 6, \quad 6, \quad 5, \quad 2$$
$$2, \quad 6, \quad 5, \quad 4, \quad 6, \quad 2, \quad 1, \quad 6, \quad 2, \quad 4$$

The mean of these observations, called the sample mean, is calculated as

$$\bar{x} = \frac{\text{Sum of the observations}}{\text{Sample size}} = \frac{76}{20} = 3.8$$

Alternatively, we can first count the frequency of each point and use the relative frequencies to calculate the mean as

$$\bar{x} = 1\left(\frac{2}{20}\right) + 2\left(\frac{5}{20}\right) + 3\left(\frac{1}{20}\right) + 4\left(\frac{4}{20}\right) + 5\left(\frac{3}{20}\right) + 6\left(\frac{5}{20}\right) = 3.8$$

This second calculation illustrates the formula

$$\text{Sample mean } \bar{x} = \sum (\text{Value} \times \text{Relative frequency})$$

Rather than stopping with 20 tosses, if we imagine a very large number of tosses of a die, the relative frequencies will approach the probabilities, each of which is $\frac{1}{6}$ for a fair die. The mean of the (infinite) collection of tosses of a fair die should then be calculated as

$$1\left(\frac{1}{6}\right) + 2\left(\frac{1}{6}\right) + \cdots + 6\left(\frac{1}{6}\right) = \sum (\text{Value} \times \text{Probability}) = 3.5$$

Motivated by this example and the stability of long-run relative frequency, it is then natural to define the mean of a random variable X or its probability distribution as

$$\sum (\text{Value} \times \text{Probability}) \quad \text{or} \quad \sum x_i f(x_i)$$

where x_i's denote the distinct values of X. The mean of a probability distribution is also called the population mean for the variable X and is denoted by the Greek letter μ.

The mean of a random variable X is also called its expected value and, alternatively, denoted by $E(X)$. That is, the mean μ and expected value $E(X)$ are the same quantity and will be used interchangeably.

The **mean of X or population mean**

$$E(X) = \mu$$
$$= \sum (\text{Value} \times \text{Probability}) = \sum x_i f(x_i)$$

Here the sum extends over all the distinct values x_i of X.

Example 7 Calculating the Population Mean Number of Heads

With X denoting the number of heads in three tosses of a fair coin, calculate the mean of X.

SOLUTION The probability distribution of X was recorded in Table 1. From the calculations exhibited in Table 6 we find that the mean is 1.5.

The mean of a probability distribution has a physical interpretation. If a metal sheet is cut in the shape of the probability histogram, then μ represents the point on the base at which the sheet will balance. For instance, the mean $\mu = 1.5$ calculated in Example 7 is exactly at the center of mass for the distribution depicted in Figure 1. Because the amount of probability corresponds to the amount of mass in a bar, we interpret the balance point μ as the center of the probability distribution.

Like many concepts of probability, the idea of the mean or expectation originated from studies of gambling. When X refers to the financial gain in a game of chance, such as playing poker or participating in a state lottery, the name "expected gain" is more appealing than "mean gain." In the realm of statistics, both the names "mean" and "expected value" are widely used.

TABLE 6   Mean of the
Distribution
of Table 1

x	$f(x)$	$xf(x)$
0	$\dfrac{1}{8}$	0
1	$\dfrac{3}{8}$	$\dfrac{3}{8}$
2	$\dfrac{3}{8}$	$\dfrac{6}{8}$
3	$\dfrac{1}{8}$	$\dfrac{3}{8}$
Total	1	$\dfrac{12}{8} = 1.5 = \mu$

Example 8 Expected Value—Setting a Premium

A trip insurance policy pays $1000 to the customer in case of a loss due to theft or damage on a five-day trip. If the risk of such a loss is assessed to be 1 in 200, what is a fair premium for this policy?

SOLUTION   The probability that the company will be liable to pay $1000 to a customer is $\frac{1}{200}$ = .005. Therefore, the probability distribution of X, the payment per customer, is as follows.

Payment x	Probability $f(x)$
$0	.995
$1000	.005

We calculate

$$E(X) = 0 \times .995 + 1000 \times .005$$
$$= \$5.00$$

The company's expected cost per customer is $5.00 and, therefore, a premium equal to this amount is viewed as the fair premium. If this premium is charged and no other costs are involved, then the company will neither make a profit nor lose money in the long run. In practice, the premium is set at a higher price because it must include administrative costs and intended profit.

No Casino Game Has a Positive Expected Profit

© SuperStock, Inc.

Each year, thousands of visitors come to casinos to gamble. Although all count on being lucky and a few indeed return with a smiling face, most leave the casino with a light purse. But, what should a gambler's expectation be?

Consider a simple bet on the red of a roulette wheel that has 18 red, 18 black, and 2 green slots. This bet is at even money so a $10 wager on red has an expected profit of

$$E(\text{Profit}) = (10)\left(\frac{18}{38}\right) + (-10)\left(\frac{20}{38}\right) = -.526$$

The negative expected profit says we expect to lose an average of 52.6¢ on every $10 bet. Over a long series of bets, the relative frequency of winning will approach the probability $\frac{18}{38}$ and that of losing will approach $\frac{20}{38}$, so a player will lose a substantial amount of money.

Other bets against the house have a similar negative expected profit. How else could a casino stay in business?

The concept of expected value also leads to a numerical measure for the spread of a probability distribution—namely, the standard deviation. When we define the standard deviation of a probability distribution, the reasoning parallels that for the standard deviation discussed in Chapter 2.

Because the mean μ is the center of the distribution of X, we express variation of X in terms of the deviation $X - \mu$. We define the variance of X as the expected value of the squared deviation $(X - \mu)^2$. To calculate this expected value, we note that

$(X - \mu)^2$ Takes Value	With Probability
$(x_1 - \mu)^2$	$f(x_1)$
$(x_2 - \mu)^2$	$f(x_2)$
.	.
.	.
.	.
$(x_k - \mu)^2$	$f(x_k)$

The expected value of $(X - \mu)^2$ is obtained by multiplying each value $(x_i - \mu)^2$ by the probability $f(x_i)$ and then summing these products. This motivates the definition:

$$\text{Variance of } X = \sum (\text{Deviation})^2 \times (\text{Probability})$$
$$= \sum (x_i - \mu)^2 f(x_i)$$

The variance of X is abbreviated as $\text{Var}(X)$ and is also denoted by σ^2. The standard deviation of X is the positive square root of the variance and is denoted by $\text{sd}(X)$ or σ (a Greek lower-case sigma.)

The variance of X is also called the population variance and σ denotes the population standard deviation.

Variance and Standard Deviation of X

$$\sigma^2 = \text{Var}(X) = \sum (x_i - \mu)^2 f(x_i)$$
$$\sigma = \text{sd}(X) = +\sqrt{\text{Var}(X)}$$

Example 9 Calculating a Population Variance and Standard Deviation

Calculate the variance and the standard deviation of the distribution of X that appears in the left two columns of Table 7.

SOLUTION   We calculate the mean μ, the deviations $x - \mu$, $(x - \mu)^2$, and finally $(x - \mu)^2 f(x)$. The details are shown in Table 7.

TABLE 7 Calculation of Variance and Standard Deviation

x	$f(x)$	$xf(x)$	$(x - \mu)$	$(x - \mu)^2$	$(x - \mu)^2 f(x)$
0	.1	0	−2	4	.4
1	.2	.2	−1	1	.2
2	.4	.8	0	0	0
3	.2	.6	1	1	.2
4	.1	.4	2	4	.4
Total	1.0	2.0 = μ			1.2 = σ^2

$$\text{Var}(X) = \sigma^2 = 1.2$$
$$\text{sd}(X) = \sigma = \sqrt{1.2} = 1.095$$

An alternative formula for σ^2 often simplifies the numerical work (see Appendix A2.2).

Alternative Formula for Hand Calculation

$$\sigma^2 = \sum x_i^2 f(x_i) - \mu^2$$

Example 10   Alternative Calculation of Variance

We illustrate the alternative formula for σ^2 using the probability distribution in Example 9. See Table 8.

TABLE 8   Calculation of Variance by the Alternative Formula

x	$f(x)$	$xf(x)$	$x^2 f(x)$
0	.1	.0	.0
1	.2	.2	.2
2	.4	.8	1.6
3	.2	.6	1.8
4	.1	.4	1.6
Total	1.0	2.0 = μ	5.2 = $\sum x^2 f(x)$

$$\sigma^2 = 5.2 - (2.0)^2$$
$$= 1.2$$
$$\sigma = \sqrt{1.2} = 1.095$$

The standard deviation σ, rather than σ^2, is the appropriate measure of spread. Its unit is the same as that of X. For instance, if X refers to income in dollars, σ will have the unit dollar, whereas σ^2 has the rather artificial unit (dollar)^2.

Exercises

5.29 Given the following probability distribution:
(a) Construct the probability histogram.
(b) Find $E(X)$, σ^2, and σ.

x	f(x)
1	.4
2	.3
3	.2
4	.1

5.30 Find the mean and standard deviation of the following distribution.

x	f(x)
1	.3
2	.5
3	.1
4	.1

5.31 In bidding for a remodeling project, a carpenter determines that he will have a net profit of $5000 if he gets the contract and a net loss of $56 if his bid fails. If the probability of his getting the contract is .2, calculate his expected return.

5.32 A book club announces a sweepstakes in order to attract new subscribers. The prizes and the corresponding chances are listed here (typically, the prizes are listed in bold print in an advertisement flyer while the chances are entered in fine print or not mentioned at all).

Prize	Chance
$50,000	1 in one million
$ 5,000	1 in 250,000
$ 100	1 in 5,000
$ 20	1 in 500

Suppose you have just mailed in a sweepstakes ticket and X stands for your winnings.
(a) List the probability distribution of X. (*Caution:* What is not listed is the chance of winning nothing, but you can figure that out from the given information.)
(b) Calculate your expected winnings.

5.33 Calculate the mean and standard deviation for the probability distribution of Example 5.

5.34 An insurance policy pays $800 for the loss due to theft of a canoe. If the probability of a theft is assessed to be .05, find the expected payment. If the insurance company charges $50 for the policy, what is the expected profit per policy?

5.35   A construction company submits bids for two projects. Listed here are the profit and the probability of winning each project. Assume that the outcomes of the two bids are independent.

	Profit	Chance of Winning Bid
Project *A*	$ 75,000	.50
Project *B*	$120,000	.65

(a) List the possible outcomes (win/not win) for the two projects and find their probabilities.

(b) Let *X* denote the company's total profit out of the two contracts. Determine the probability distribution of *X*.

(c) If it costs the company $2000 for preparatory surveys and paperwork for the two bids, what is the expected net profit?

5.36   Refer to Exercise 5.35, but suppose that the projects are scheduled consecutively with *A* in the first year and *B* in the second year. The company's chance of winning project *A* is still .50. Instead of the assumption of independence, now assume that if the company wins project *A*, its chance of winning *B* becomes .80 due to a boost of its image, whereas its chance drops to .40 in case it fails to win *A*. Under this premise, do parts (a–c).

5.37   Upon examination of the claims records of 280 policy holders over a period of five years, an insurance company makes an empirical determination of the probability distribution of *X* = number of claims in five years.

(a) Calculate the expected value of *X*.

(b) Calculate the standard deviation of *X*.

x	f(x)
0	.315
1	.289
2	.201
3	.114
4	.063
5	.012
6	.006

5.38   Suppose the probability distribution of a random variable *X* is given by the function

$$f(x) = \frac{12}{25} \cdot \frac{1}{x} \qquad \text{for} \qquad x = 1, 2, 3, 4$$

Calculate the mean and standard deviation of this distribution.

5.39   The probability distribution of a random variable *X* is given by the function

$$f(x) = \frac{1}{84} \binom{5}{x} \binom{4}{3-x} \qquad \text{for} \qquad x = 0, 1, 2, 3$$

(a) Calculate the numerical probabilities and list the distribution.

(b) Calculate the mean and standard deviation of *X*.

*5.40   Given here are the probability distributions of two random variables *X* and *Y*.

x	f(x)	y	f(y)
1	.1	0	.2
2	.3	2	.4
3	.4	4	.3
4	.2	6	.1

(a) From the *X* distribution, determine the distribution of the random variable $8 - 2X$ and verify that it coincides with the *Y* distribution. (Hence, identify $Y = 8 - 2X$.)

(b) Calculate the mean and standard deviation of *X* (call these μ_X and σ_X, respectively).

(c) From the *Y* distribution, calculate the mean and standard deviation of *Y* (call these μ_Y and σ_Y, respectively).

(d) If $Y = a + bX$, then according to theory, we must have the relations $\mu_Y = a + b\mu_X$ and $\sigma_Y = |b|\sigma_X$. Verify these relations from your results in parts (b) and (c).

5.41   A salesman of small-business computer systems will contact four customers during a week. Each contact can result in either a sale, with probability .3, or no sale, with probability

.7. Assume that customer contacts are independent.

(a) List the elementary outcomes and assign probabilities.

(b) If X denotes the number of computer systems sold during the week, obtain the probability distribution of X.

(c) Calculate the expected value of X.

5.42   Refer to Exercise 5.41. Suppose the computer systems are priced at $2000, and let Y denote the salesman's total sales (in dollars) during a week.

(a) Give the probability distribution of Y.

(b) Calculate $E(Y)$ and see that it is the same as $2000 \times E(X)$.

5.43   Definition: The **median** of a distribution is the value m_0 of the random variable such that $P[X \leq m_0] \geq .5$ and $P[X \geq m_0] \geq .5$. In other words, the probability at or below m_0 is at least .5, and the probability at or above m_0 is at least .5. Find the median of the distribution given in Exercise 5.29.

5.44   Given the two probability distributions

x	$f(x)$	y	$f(y)$
1	.2	0	.1
2	.6	1	.2
3	.2	2	.4
		3	.2
		4	.1

(a) Construct probability histograms. Which distribution has a larger spread?

(b) Verify that both distributions have the same mean.

(c) Compare the two standard deviations.

5.   SUCCESSES AND FAILURES—BERNOULLI TRIALS

Often, an experiment can have only two possible outcomes. Example 5 concerned individual students who either preferred Internet or television news. Proportion .30 of the population preferred Internet news. Also, only two outcomes are possible for a single trial in the scenarios of Examples 1 and 2. In all these circumstances, a simple probability model can be developed for the chance variation in the outcomes. Moreover, the population proportion need not be known as in the previous examples. Instead, the probability distribution will involve this unknown population proportion as a parameter.

Sampling situations where the elements of a population have a dichotomy abound in virtually all walks of life. A few examples are:

Inspect a specified number of items coming off a production line and count the number of defectives.

Survey a sample of voters and observe how many favor a reduction of public spending on welfare.

Analyze the blood specimens of a number of rodents and count how many carry a particular viral infection.

Examine the case histories of a number of births and count how many involved delivery by Cesarean section.

Selecting a single element of the population is envisioned as a trial of the (sampling) experiment, so that each trial can result in one of two possible out-

Boy or girl?

A model for the potential sex of a newborn is the assignment of probability to each of the two outcomes. © Stephen Frisch/Stock Boston.

comes. Our ultimate goal is to develop a probability model for the number of outcomes in one category when repeated trials are performed.

An organization of the key terminologies, concerning the successive repetitions of an experiment, is now in order. We call each repetition by the simpler name—a **trial.** Furthermore, the two possible outcomes of a trial are now assigned the technical names success (S) and failure (F) just to emphasize the point that they are the only two possible results. These names bear no connotation of success or failure in real life. Customarily, the outcome of primary interest in a study is labeled success (even if it is a disastrous event). In a study of the rate of unemployment, the status of being unemployed may be attributed the statistical name success!

Further conditions on the repeated trials are necessary in order to arrive at our intended probability distribution. Repeated trials that obey these conditions are called Bernoulli trials after the Swiss mathematician Jacob Bernoulli.

Perhaps the simplest example of Bernoulli trials is the prototype model of tossing a coin, where the occurrences *head* and *tail* can be labeled S and F, respectively. For a fair coin, we assign probability $p = \frac{1}{2}$ to success and $q = \frac{1}{2}$ to failure.

Bernoulli Trials

1. Each trial yields one of two outcomes, technically called success (S) and failure (F).

2. For each trial, the probability of success $P(S)$ is the same and is denoted by $p = P(S)$. The probability of failure is then $P(F) = 1 - p$ for each trial and is denoted by q, so that $p + q = 1$.

3. Trials are independent. The probability of success in a trial remains unchanged given the outcomes of all the other trials.

Example 11 Sampling from a Population with Two Categories of Elements

Consider a lot (population) of items in which each item can be classified as either defective or nondefective. Suppose that a lot consists of 15 items, of which 5 are defective and 10 are nondefective.

Do the conditions for Bernoulli trials apply when sampling (1) with replacement and (2) without replacement?

SOLUTION

1. *Sampling with replacement.* An item is drawn at random (i.e., in a manner that all items in the lot are equally likely to be selected). The quality of the item is recorded and it is returned to the lot before the next drawing. The conditions for Bernoulli trials are satisfied. If the occurrence of a defective is labeled S, we have $P(S) = \frac{5}{15}$.

2. *Sampling without replacement.* In situation (2), suppose that 3 items are drawn one at a time but without replacement. Then the condition concerning the independence of trials is violated. For the first drawing, $P(S) = \frac{5}{15}$. If the first draw produces S, the lot then consists of 14 items, 4 of which are defective. Given this information about the result of the first draw, the conditional probability of obtaining an S on the second draw is then $\frac{4}{14} \neq \frac{5}{15}$, which establishes the lack of independence.

This violation of the condition of independence loses its thrust when the population is vast and only a small fraction of it is sampled. Consider sampling 3 items without replacement from a lot of 1500 items, 500 of which are defective. With S_1 denoting the occurrence of an S in the first draw and S_2 that in the second, we have

$$P(S_1) = \frac{500}{1500} = \frac{5}{15}$$

and

$$P(S_2 | S_1) = \frac{499}{1499}$$

For most practical purposes, the latter fraction can be approximated by $\frac{5}{15}$. Strictly speaking, there has been a violation of the independence of trials, but it is to such a negligible extent that the model of Bernoulli trials can be assumed as a good approximation.

Example 11 illustrates the important points:

> If elements are sampled from a dichotomous population at random and with replacement, the conditions for Bernoulli trials are satisfied.
>   When the sampling is made without replacement, the condition of the independence of trials is violated. However, if the population is large and only a small fraction of it (less than 10%, as a rule of thumb) is sampled, the effect of this violation is negligible and the model of the Bernoulli trials can be taken as a good approximation.

Example 12 further illustrates the kinds of approximations that are sometimes employed when using the model of the Bernoulli trials.

Example 12   Testing a New Antibiotic—Bernoulli Trials?

Suppose that a newly developed antibiotic is to be tried on 10 patients who have a certain disease and the possible outcomes in each case are cure (S) or no cure (F).
    Comment on the applicability of the Bernoulli trial model.

SOLUTION   Each patient has a distinct physical condition and genetic constitution that cannot be perfectly matched by any other patient. Therefore, strictly speaking, it may not be possible to regard the trials made on 10 different patients as 10 repetitions of an experiment under identical conditions, as the definition of Bernoulli trials demands. We must remember that the conditions of a probability model are abstractions that help to realistically simplify the complex mechanism governing the outcomes of an experiment. Identification with Bernoulli trials in such situations is to be viewed as an approximation of the real world, and its merit rests on how successfully the model explains chance variations in the outcomes.

Exercises

5.45 Is the model of Bernoulli trials plausible in each of the following situations? Discuss in what manner (if any) a serious violation of the assumptions can occur.

(a) Seven friends go to a blockbuster movie and each is asked whether the movie was excellent.

(b) A musical aptitude test is given to 10 students and the times to complete the test are recorded.

(c) Items coming off an assembly line are inspected and classified as defective or nondefective.

(d) Going house by house down the block and recording if the newspaper was delivered on time.

5.46 In each case, examine whether or not repetitions of the stated experiment conform to the model of Bernoulli trials. Where the model is appropriate, determine the numerical value of p or indicate how it can be determined.

(a) Roll a fair die and observe the number that shows up.

(b) Roll a fair die and observe whether or not the number 6 shows up.

(c) Roll two fair dice and observe the total of the points that show up.

(d) Roll two fair dice and observe whether or not both show the same number.

(e) Roll a loaded die and observe whether or not the number 6 shows up.

5.47 A jar contains 25 candies of which 8 are brown, 10 are yellow, and 7 are of other colors. Consider 4 successive draws of 1 candy at random from the jar and suppose the appearance of a yellow candy is the event of interest. For each of the following situations, state whether or not the model of Bernoulli trials is reasonable, and if so, determine the numerical value of p.

(a) After each draw, the selected candy is returned to the jar.

(b) After each draw, the selected candy is not returned to the jar.

(c) After each draw, the selected candy is returned to the jar and one new candy of the same color is added in the jar.

5.48 Refer to Exercise 5.47 and suppose instead that the mix consists of 2500 candies, of which 800 are brown, 1000 are yellow, and 700 are of other colors. Repeat parts (a–c) of Exercise 5.47 in this setting.

5.49 From four agricultural plots, two will be selected at random for a pesticide treatment. The other two plots will serve as controls. For each plot, denote by S the event that it is treated with the pesticide. Consider the assignment of treatment or control to a single plot as a trial.

(a) Is $P(S)$ the same for all trials? If so, what is the numerical value of $P(S)$?

(b) Are the trials independent? Why or why not?

5.50 Refer to Exercise 5.49. Now suppose for each plot a fair coin will be tossed. If a head shows up, the plot will be treated; otherwise, it will be a control. With this manner of treatment allocation, answer parts (a) and (b).

5.51 A market researcher intends to study the consumer preference between regular and decaffeinated coffee. Examine the plausibility of the model of Bernoulli trials in the following situations.

(a) One hundred consumers are randomly selected and each is asked to report the types of coffee (regular or decaffeinated) purchased in the five most recent occasions. If we consider each purchase as a trial, this inquiry deals with 500 trials.

(b) Five hundred consumers are randomly selected and each is asked about the most recent purchase of coffee. Here again the inquiry deals with 500 trials.

5.52 A backpacking party carries three emergency signal flares, each of which will light with a probability of .98. Assuming that the flares operate independently, find:

(a) The probability that at least one flare lights.

(b) The probability that exactly two flares light.

5.53 Consider Bernoulli trials with success probability $p = \frac{1}{3}$.

(a) Find the probability that four trials result in all failures.

(b) Given that the first four trials result in all failures, what is the conditional probability that the next four trials are all successes?

(c) Find the probability that the first success occurs in the fourth trial.

5.54 If in three Bernoulli trials $P[\text{All three are successes}] = .064$, what is the probability that all three are failures?

5.55 Consider four Bernoulli trials with success probability $p = .8$ in each trial. Find the probability that:

(a) All four trials result in successes.

(b) All are failures.

(c) There is at least one success.

5.56 A new driver who did not take driver's education has probability .7 of passing the driver's license exam. If tries are independent, find the probability that the driver (a) will not pass in two attempts, (b) will not pass in three attempts.

5.57 An animal either dies (D) or survives (S) in the course of a surgical experiment. The experiment is to be performed first with two animals. If both survive, no further trials are to be made. If exactly one animal survives, one more

animal is to undergo the experiment. Finally, if both animals die, two additional animals are to be tried.

(a) List the sample space.

(b) Assume that the trials are independent and the probability of survival in each trial is $\frac{1}{4}$. Assign probabilities to the elementary outcomes.

(c) Let X denote the number of survivors. Obtain the probability distribution of X by referring to part (b).

5.58 The accompanying table shows the percentages of residents in a large community when classified according to gender and presence of a particular allergy.

| | Allergy | |
	Present	Absent
Male	16	36
Female	9	39

Suppose that the selection of a person is considered a trial and the presence of the allergy is considered a success. For each case, identify the numerical value of p and find the required probability.

(a) Four persons are selected at random. What is the probability that none has the allergy?

(b) Four males are selected at random. What is the probability that none has the allergy?

(c) Two males and two females are selected at random. What is the probability that none has the allergy?

6. THE BINOMIAL DISTRIBUTION

This section deals with a basic distribution that models chance variation in repetitions of an experiment that has only two possible outcomes. The random variable X of interest is the frequency count of one of the categories. Previously, its distribution was calculated under the assumption that the population proportion is known. For instance, the probability distribution of Table 3, from Example 5, resulted from the specification that 30% of the population of students prefer news from the Internet. In a practical situation, however, the population proportion is usually an unknown quantity. When this is so, the probability distribution of X cannot be numerically determined. However, we will see that it is possible to construct a model for the probability distribution of X that contains the unknown population proportion as a parameter. The probability model serves as the major vehicle of drawing inferences about the population from observations of the random variable X.

A **probability model** is an assumed form of the probability distribution that describes the chance behavior for a random variable X.

Probabilities are expressed in terms of relevant population quantities, called the **parameters**.

Consider a **fixed number** n of Bernoulli trials with the success probability p in each trial. The number of successes obtained in n trials is a random variable that we denote by X. The probability distribution of this random variable X is called a binomial distribution.

The binomial distribution depends on the two quantities n and p. For instance, the distribution appearing in Table 1 is precisely the binomial distribution with $n = 3$ and $p = .5$, whereas that in Table 3 is the binomial distribution with $n = 4$ and $p = .3$.

The Binomial Distribution

Denote

$$n = \text{a fixed number of Bernoulli trials}$$
$$p = \text{the probability of success in each trial}$$
$$X = \text{the (random) number of successes in } n \text{ trials}$$

The random variable X is called a **binomial random variable**. Its distribution is called a **binomial distribution**.

A review of the developments in Example 5 will help motivate a formula for the general binomial distribution.

Example 13   Example 5 Revisited—An Example of the Binomial Distribution

The random variable X represents the number of students who prefer news from the Internet among a random sample of $n = 4$ students from a large university. Instead of the numerical value .3, we now denote the population proportion of students who prefer Internet news by the symbol p. Furthermore, we relabel the outcome "Internet" as a success (S) and "not Internet" as a failure (F). The elementary outcomes of sampling 4 students, the associated probabilities, and the value of X are listed as follows.

FFFF	SFFF	SSFF	SSSF	SSSS
	FSFF	SFSF	SSFS	
	FFSF	SFFS	SFSS	
	FFFS	FSSF	FSSS	
		FSFS		
		FFSS		

Value of X	0	1	2	3	4
Probability of each outcome	q^4	pq^3	p^2q^2	p^3q	p^4
Number of outcomes	1 $=\binom{4}{0}$	4 $=\binom{4}{1}$	6 $=\binom{4}{2}$	4 $=\binom{4}{3}$	1 $=\binom{4}{4}$

Because the population of students at a large university is vast, the trials can be treated as independent. Also, for an individual trial, $P(S) = p$ and $P(F) = q = 1 - p$. The event $[X = 0]$ has one outcome, FFFF, whose probability is

$$P[X = 0] = P(\text{FFFF}) = q \times q \times q \times q = q^4$$

To arrive at an expression for $P[X = 1]$, we consider the outcomes listed in the second column. The probability of SFFF is

$$P(\text{SFFF}) = p \times q \times q \times q = pq^3$$

and the same result holds for every outcome in this column. There are 4 outcomes so we obtain $P[X = 1] = 4pq^3$. The factor 4 is the number of outcomes with one S and three F's. Even without making a complete list of the outcomes, we can obtain this count. Every outcome has 4 places and the 1 place where S occurs can be selected from the total of 4 in $\binom{4}{1} = 4$ ways, while the remaining 3 places must be filled with an F. Continuing in the same line of reasoning, we see that the value $X = 2$ occurs with $\binom{4}{2} = 6$ outcomes, each of which has a probability of p^2q^2. Therefore $P[X = 2] = \binom{4}{2}p^2q^2$. After we work out the remaining terms, the binomial distribution with $n = 4$ trials can be presented as in Table 9.

TABLE 9   Binomial Distribution with $n = 4$ Trials

Value x	0	1	2	3	4
Probability $f(x)$	$\binom{4}{0}p^0q^4$	$\binom{4}{1}p^1q^3$	$\binom{4}{2}p^2q^2$	$\binom{4}{3}p^3q^1$	$\binom{4}{4}p^4q^0$

It would be instructive for the reader to verify that the numerical probabilities appearing in Table 3 are obtained by substituting $p = .3$ and $q = .7$ in the entries of Table 9.

Extending the reasoning of Example 13 to the case of a general number n of Bernoulli trials, we observe that there are $\binom{n}{x}$ outcomes that have exactly x successes and $n - x$ failures. The probability of every such outcome is $p^x q^{n-x}$. Therefore,

$$f(x) = P[X = x] = \binom{n}{x} p^x q^{n-x} \qquad \text{for} \qquad x = 0, 1, \ldots, n$$

is the formula for the binomial probability distribution with n trials.

The **binomial distribution** with n trials and success probability p is described by the function

$$f(x) = P[X = x] = \binom{n}{x} p^x (1 - p)^{n-x}$$

for the possible values $x = 0, 1, \ldots, n$.

Example 14   The Binomial Distribution and Genetics

According to the Mendelian theory of inherited characteristics, a cross fertilization of related species of red- and white-flowered plants produces a generation whose offspring contain 25% red-flowered plants. Suppose that a horticulturist wishes to cross 5 pairs of the cross-fertilized species. Of the resulting 5 offspring, what is the probability that:

(a)   There will be no red-flowered plants?

(b)   There will be 4 or more red-flowered plants?

SOLUTION   Because the trials are conducted on different parent plants, it is natural to assume that they are independent. Let the random variable X denote the number of red-flowered plants among the 5 offspring. If we identify the occurrence of a red as a success S, the Mendelian theory specifies that $P(S) = p = \frac{1}{4}$, and hence X has a binomial distribution with $n = 5$ and $p = .25$. The required probabilities are therefore

(a) $P[X = 0] = f(0) = (.75)^5 = .237$

(b) $P[X \geq 4] = f(4) + f(5) = \binom{5}{4}(.25)^4(.75)^1 + \binom{5}{5}(.25)^5(.75)^0$

$$= .015 + .001 = .016$$

To illustrate the manner in which the values of p influence the shape of the binomial distribution, the probability histograms for three binomial distributions with $n = 6$ and p values of .5, .3, and .7, respectively, are presented in Figure 3. When $p = .5$, the binomial distribution is symmetric with the highest probability occurring at the center (see Figure 3a).

For values of p smaller than .5, more probability is shifted toward the smaller values of x and the distribution has a longer tail to the right. Figure 3b, where the binomial histogram for $p = .3$ is plotted, illustrates this tendency. On the other hand, Figure 3c with $p = .7$ illustrates the opposite tendency: The value of p is higher than .5, more probability mass is shifted toward higher values of x, and the distribution has a longer tail to the left. Considering the histograms in Figures 3b and 3c, we note that the value of p in one histogram is the same as the value of q in the other. The probabilities in one histogram are exactly the same as those in the other, but their order is reversed. This illustrates a general property of the binomial distribution: When p and q are interchanged, the distribution of probabilities is reversed.

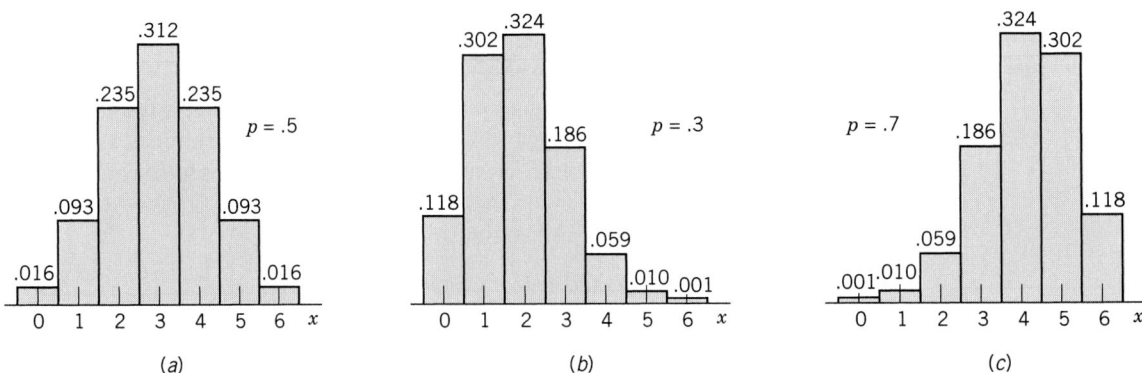

Figure 3   Binomial distributions for $n = 6$.

How to Use the Binomial Table (Appendix B, Table 2)

Although the binomial distribution is easily evaluated on a computer and some hand calculators, we provide a short table in Appendix B, Table 2. It covers selected sample sizes n ranging from 1 to 25 and several values of p. For a given pair (n, p), the table entry corresponding to each c represents the cumulative probability $P[X \le c] = \sum_{x=0}^{c} f(x)$, as is explained in the following scheme.

The Binomial Distribution

Value x	Probability $f(x)$
0	$f(0)$
1	$f(1)$
2	$f(2)$
.	.
.	.
.	.
n	$f(n)$
Total	1

Appendix B, Table 2 provides

c	Table Entry $\sum_{x=0}^{c} f(x) = P[X \le c]$
0	$f(0)$
1	$f(0) + f(1)$
2	$f(0) + f(1) + f(2)$
.	.
.	.
.	.
n	1.000

The probability of an individual value x can be obtained from this table by a subtraction of two consecutive entries. For example,

$$P[X = 2] = f(2) = \left(\begin{array}{c} \text{table entry at} \\ c = 2 \end{array}\right) - \left(\begin{array}{c} \text{table entry at} \\ c = 1 \end{array}\right)$$

Example 15   Binomial Distribution for the Number Cured

Suppose it is known that a new treatment is successful in curing a muscular pain in 50% of the cases. If it is tried on 15 patients, find the probability that:

(a)   At most 6 will be cured.

(b)   The number cured will be no fewer than 6 and no more than 10.

(c)   Twelve or more will be cured.

SOLUTION   Designating the cure of a patient by S and assuming that the results for individual patients are independent, we note that the binomial distribution with $n = 15$ and $p = .5$ is appropriate for $X =$ number of patients who are cured. To compute the required probabilities, we consult the binomial table for $n = 15$ and $p = .5$.

(a) $P[X \leq 6] = .304$, which is directly obtained by reading from the row $c = 6$.

(b) We are to calculate

$$P[6 \leq X \leq 10] = f(6) + f(7) + f(8) + f(9) + f(10)$$

$$= \sum_{x=6}^{10} f(x)$$

The table entry corresponding to $c = 10$ gives

$$P[X \leq 10] = \sum_{x=0}^{10} f(x) = .941$$

and the entry corresponding to $c = 5$ yields

$$P[X \leq 5] = \sum_{x=0}^{5} f(x) = .151$$

Because their difference represents the sum $\sum_{x=6}^{10} f(x)$, we obtain

$$P[6 \leq X \leq 10] = P[X \leq 10] - P[X \leq 5]$$
$$= .941 - .151$$
$$= .790$$

(c) To find $P[X \geq 12]$, we use the law of complement:

$$P[X \geq 12] = 1 - P[X \leq 11]$$
$$= 1 - .982$$
$$= .018$$

Note that $[X < 12]$ is the same event as $[X \leq 11]$.

(*An Aside:* Refer to our "muscular pain" example in Section 1 of Chapter 4. The mystery surrounding the numerical probability .018 is now resolved.)

The Mean and Standard Deviation of the Binomial Distribution

Although we already have a general formula that gives the binomial probabilities for any n and p, in later chapters we will need to know the mean and the standard deviation of the binomial distribution. The expression np for the mean is apparent from the following intuitive reasoning: If a fair coin is tossed 100 times, the expected number of heads is $100 \times \frac{1}{2} = 50$. Likewise, if the probability of an event is p, then in n trials the event is expected to happen np times. The formula for the standard deviation requires some mathematical derivation, which we omit.

The binomial distribution with n trials and success probability p has

$$\text{Mean} = np$$
$$\text{Variance} = npq \quad (\text{Recall: } q = 1 - p)$$
$$\text{sd} = \sqrt{npq}$$

Example 16   Calculating the Population Mean and Standard Deviation
of a Binomial Distribution

For the binomial distribution with $n = 3$ and $p = .5$, calculate the mean
and the standard deviation.

SOLUTION   Employing the formulas, we obtain

$$\text{Mean} = np = 3 \times .5 = 1.5$$
$$\text{sd} = \sqrt{npq} = \sqrt{3 \times .5 \times .5} = \sqrt{.75} = .866$$

The mean agrees with the results of Example 7. The reader may wish to check
the standard deviation by numerical calculations using the definition of σ.

Exercises

5.59   For each situation, state whether or not a binomial distribution holds for the random variable
X. Also, identify the numerical values of n and
p when a binomial distribution holds.

   (a)   A fair die is rolled 10 times, and X denotes the number of times 6 shows up.

   (b)   A fair die is rolled until 6 appears, and X
denotes the number of rolls.

   (c)   In a jar, there are ten marbles, of which
four are numbered 1, three are numbered
2, two are numbered 3, and one is numbered 4. Three marbles are drawn at random, one after another and with replacement, and X denotes the count of the
selected marbles that are numbered either 1 or 2.

   (d)   The same experiment as described in part
(c), but now X denotes the sum of the
numbers on the selected marbles.

5.60   Construct a tree diagram for three Bernoulli
trials. Attach probabilities in terms of p and q
to each outcome and then table the binomial
distribution for $n = 3$.

5.61   In each case, find the probability of x successes
in n Bernoulli trials with success probability p
for each trial.

   (a) $x = 2 \quad n = 3 \quad p = .35$

   (b) $x = 3 \quad n = 6 \quad p = .25$

   (c) $x = 2 \quad n = 6 \quad p = .75$

5.62   (a)   Plot the probability histograms for the binomial distributions for $n = 5$ and p
equal to .2, .5, and .8.

   (b)   Locate the means.

   (c)   Find $P[X \geq 4]$ for each of the three
cases.

5.63   If X has the binomial distribution with $n = 5$ and $p = .35$, calculate the probabilities $f(x) = P[X = x]$ for $x = 0, 1, \ldots, 5$ and find

(a) $P[X \leq 3]$

(b) $P[X \geq 3]$

(c) $P[X = 2 \text{ or } 4]$

5.64   Refer to Exercise 5.63. What is the most probable value of X (called the **mode** of a distribution)?

5.65   For the binomial distribution with $n = 4$ and $p = .45$, find the probability of:

(a) Three or more successes.

(b) At most three successes.

(c) Two or more failures.

5.66   For the binomial distribution with $n = 6$ and $p = \frac{2}{3}$, find the probability of:

(a) Fewer than 5 successes.

(b) More than 4 failures.

(c) No more than 5 and no less than 2 successes.

5.67   Suppose 15% of the trees in a forest have severe leaf damage from air pollution. If 5 trees are selected at random, find the probability that:

(a) Three of the selected trees have severe leaf damage.

(b) No more than two have severe leaf damage.

5.68   Rh-positive blood appears in 85% of the white population in the United States. If 8 people are sampled at random from that population, find the probability that:

(a) At least 6 of them have Rh-positive blood.

(b) At most 3 of them have Rh-negative blood, that is, an absence of Rh positive.

5.69   Using the binomial table, find the probability of:

(a) Four successes in 13 trials when $p = .3$.

(b) Eight failures in 13 trials when $p = .7$.

(c) Eight successes in 13 trials when $p = .3$. Explain why you get identical answers in parts (b) and (c).

5.70   Using the binomial table, find the probability of:

(a) Five or fewer successes in 8 trials when $p = .7$.

(b) No more than 11 and no less than 6 successes in 17 trials when $p = .6$.

(c) Exactly 40% successes in 20 trials when $p = .4$.

5.71   If $p = .7$, find the probability that

(a) More than 5 trials are needed in order to obtain 3 successes. (*Hint:* In other words, the event is: at most 2 successes in 5 trials.)

(b) More than 9 trials are needed in order to obtain 7 successes.

5.72   A survey report states that 70% of adult women visit their doctors for a physical examination at least once in two years. If 20 adult women are randomly selected, find the probability that

(a) Fewer than 14 of them have had a physical examination in the past two years.

(b) At least 17 of them have had a physical examination in the past two years.

5.73   Calculate the mean and standard deviation of the binomial distribution with

(a) $n = 19$ $p = .5$

(b) $n = 25$ $p = .2$

(c) $n = 25$ $p = .8$

5.74   (a) For the binomial distribution with $n = 3$ and $p = .6$, list the probability distribution $(x, f(x))$ in a table.

(b) From this table, calculate the mean and standard deviation by using the methods of Section 4.

(c) Check your results with the formulas mean $= np$, sd $= \sqrt{npq}$.

5.75   Suppose that 20% of the college seniors support an increase in federal funding for care of the elderly. If 20 college seniors are randomly selected, what is the probability that at most 3 of them support the increased funding?

5.76 Referring to Exercise 5.75, find:

(a) The expected number of college seniors, in a random sample of 20, supporting the increased funding.

(b) The probability that the number of sampled college seniors supporting the increased funding equals the expected number.

5.77 Suppose that, for a particular type of cancer, chemotherapy provides a 5-year survival rate of 80% if the disease could be detected at an early stage. Among 19 patients diagnosed to have this form of cancer at an early stage who are just starting the chemotherapy, find the probability that:

(a) Fourteen will survive beyond 5 years.

(b) Six will die within 5 years.

(c) The number of patients surviving beyond 5 years will be between 9 and 13 (both inclusive).

5.78 Referring to Exercise 5.77, find the expectation and standard deviation of the number of 5-year survivors.

5.79 According to a recent report of the American Medical Association, 7.85% of practicing physicians are in the specialty area of family practice. Assuming that the same rate prevails, find the mean and standard deviation of the number of physicians specializing in family practice out of a current random selection of 545 medical graduates.

5.80 According to the Mendelian theory of inheritance of genes, offspring of a dihybrid cross of peas could be any of the four types: round-yellow (*RY*), wrinkled-yellow (*WY*), round-green (*RG*) and wrinkled-green (*WG*), and their probabilities are in the ratio $9:3:3:1$.

(a) If X denotes the number of *RY* offspring from 130 such crosses, find the mean and standard deviation of X.

(b) If Y denotes the number of *WG* offspring from 85 such crosses, find the mean and standard deviation of Y.

5.81 The following table (see Exercise 5.58) shows the percentages of residents in a large community when classified according to gender and presence of a particular allergy. For each part below, find the mean and standard deviation of the specified random variable.

	Allergy	
	Present	Absent
Male	16	36
Female	9	39

(a) X stands for the number of persons having the allergy in a random sample of 40 persons.

(b) Y stands for the number of males having the allergy in a random sample of 40 males.

(c) Z stands for the number of females not having the allergy in a random sample of 40 females.

5.82 For a binomial distribution:

(a) If $n = 70$ and $p = .35$, find the mean and standard deviation.

(b) If mean $= 12$ and standard deviation $= 3$, find the numerical values of n and p.

The Following Exercise Requires a Computer

5.83 Many computer packages produce binomial probabilities. We illustrate the MINITAB commands for obtaining the binomial probabilities with $n = 5$ and $p = .33$. The probabilities $P[X = x]$ are obtained by first setting 0, 1, 2, 3, 4 in C*1* and then selecting:

Calc > **Probability distributions.**

Type 5 in **Number of trials**

and .33 in **Probability.**

Enter C*1* in **Input constant.** Click **OK.**

which produces the output

```
Probability Density Function

Binomial with n = 5 and p = 0.33

x     P(X=x)
0       0.135013
1       0.332493
2       0.327531
3       0.161321
4       0.039728
5       0.003914
```

To obtain the cumulative probabilities $P(X \le x)$, click **Cumulative probability** instead of **Probability.** The resulting output is

```
Cumulative Distribution Function

Binomial with n = 5 and p = 0.33

x     P(X<=x)
0       0.13501
1       0.46751
2       0.79504
3       0.95636
4       0.99609
5       1.00000
```

Using the computer, calculate

(a) $P[X \le 8]$ and $P[X = 8]$ when $p = .67$ and $n = 12$

(b) $P[10 \le X \le 15]$ when $p = .43$ and $n = 35$

Poor Linus. Chance did not even favor him with half correct. Reprinted by permission of United Features Syndicate, Inc. © 1968.

7.   THE BINOMIAL DISTRIBUTION IN CONTEXT

Requests for credit cards must be processed to determine if the applicant meets certain financial standards. In many instances, such as when the applicant already has a long-term good credit record, only a short review is required. Usually this consists of a credit score assigned on the basis of the answers to questions on the application and then a computerized check of credit records. Many other cases require a full review with manual checks of information to determine the credit worthiness of the applicant.

Each week, a large financial institution selects a sample of 20 incoming applications and counts the number requiring full review. From data collected over several weeks, it is observed that about 40% of the applications require full review. If we take this long-run relative frequency as the probability, what is an unusually large number of full reviews and what is an unusually small number?

Let X be the number in the sample that require a full review. From the binomial table, with $n = 20$ and $p = .4$, we get

$$P[X \leq 3] = .016$$
$$P[X \geq 13] = 1 - P[X \leq 12] = 1 - .979 = .021$$

Taken together, the probability of X being 3 or less or 13 or more is .037, so those values should occur less than four times in 100 samples. That is, they could be considered unusual. In Exercise 5.84, you will be asked to show that either including 4 or including 12 will lead to a combined probability greater than .05. That is, the large and small values should then occur more than 1 in 20 times. For many people, this would be too frequent to be considered rare or unusual.

For the count X, we expect $np = 20 \times .4 = 8$ applications in the sample to require a full review. The standard deviation of this count is $\sqrt{20(.4)(1 - .4)} = 2.191$. Alternatively, when n is moderate or large, we could describe as unusual two or more standard deviations from the mean. A value at least two standard deviations, or $2(2.191) = 4.382$, above the mean of 8 must be 13 or more. A value 2 or more standard deviations below the mean must be 3 or less. These values correspond exactly to the values above that which we called unusual. In other cases, the two standard deviations approach provides a reasonable and widely used approximation.

p Charts for Trend

A series of sample proportions should be visually inspected for trend. A graph called a p chart helps identify times when the population proportion has changed from its long-time stable value. Because many sample proportions will be graphed, it is customary to set control limits at 3 rather than 2 standard deviations.

When p_0 is the expected or long-run proportion, we obtain a lower control limit by dividing the lower limit for X, $np_0 - 3\sqrt{np_0(1 - p_0)}$, by the sample size. Doing the same with the upper bound, we obtain an upper control limit.

Lower control limit   Upper control limit

$$p_0 - 3\sqrt{\frac{p_0(1 - p_0)}{n}} \qquad p_0 + 3\sqrt{\frac{p_0(1 - p_0)}{n}}$$

In the context of credit applications that require a full review, $.4 = p_0$, so the control limits are

$$p_0 - 3\sqrt{\frac{p_0(1 - p_0)}{n}} = .4 - 3\sqrt{\frac{.4(1 - .4)}{20}} = .4 - 3(.110) = .07$$

$$p_0 + 3\sqrt{\frac{p_0(1 - p_0)}{n}} = .4 + 3\sqrt{\frac{.4(1 - .4)}{20}} = .4 + 3(.110) = .73$$

The **centerline** is drawn as a solid line at the expected or long-run proportion $p_0 = .4$, and the two control limits each at a distance of three standard deviations of the sample proportion from the centerline are also drawn as horizontal lines as in Figure 4. Sample proportions that fall outside of the control limits are considered unusual and should result in a search for a cause that may include a change in the mix of type of persons requesting credit cards.

The number of applications requiring full review out of the 20 in the sample were recorded for 19 weeks:

11 7 8 4 9 10 4 8 8 7 10 6 9 10 7 7 6 9 10

After converting to sample proportions by dividing by 20, we can graph the points in a p chart as in Figure 4. All the points are in control, and the financial institution does not appear to have reached the point where the mix of applicants includes more marginal cases.

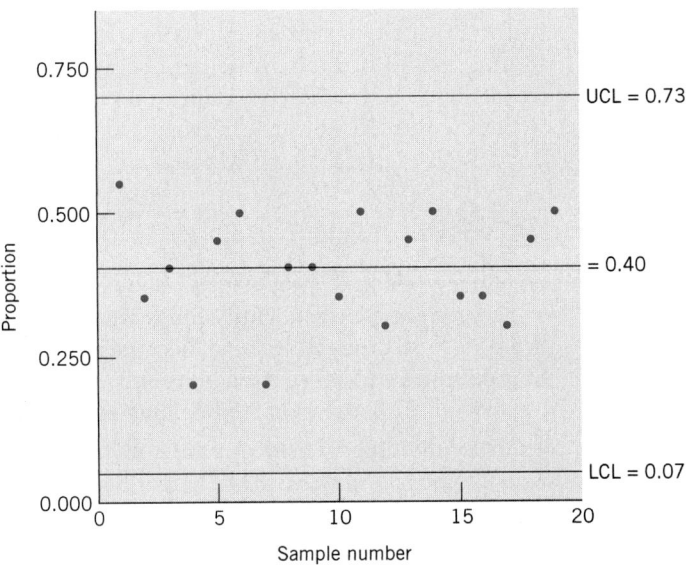

Figure 4   A p chart for the proportion of applications requiring a full review.

Exercises

5.84 Refer to the credit card application approval process on page 207 where unusual values are defined.

(a) Show that if 4 is included as an unusual value, then the probability $P[X \leq 4$ or $X \geq 13]$ is greater than .05.

(b) Show that if 12 is included as an unusual value, then the probability $P[X \leq 3$ or $X \geq 12]$ is greater than .05.

5.85 Refer to the credit card application approval process on page 206.

(a) Make a p chart using the centerline and control limits calculated for $p_0 = .4$.

(b) Suppose the next five weeks bring 12, 10, 15, 11, and 16 applications requiring full review. Graph the corresponding proportions on your p chart.

(c) Identify any weeks where the chart signals "out of control."

5.86 Several fast food restaurants advertise quarter-pound hamburgers. This could be interpreted as meaning half the hamburgers made have an uncooked weight of at least a quarter-pound and half have a weight that is less. An inspector checks 20 uncooked hamburgers at each restaurant.

(a) Make a p chart using the centerline and control limits calculated for $p_0 = .5$.

(b) Suppose that five restaurants have 8, 11, 7, 15, and 10 underweight hamburgers in samples of size 20. Graph the corresponding proportions on your p chart.

(c) Identify any restaurants where the chart signals "out of control."

5.87 Refer to Exercise 5.86.

(a) What are the unusual values for the number of underweight hamburgers in the sample if they correspond to proportions outside of the control limits of the p chart?

(b) Use the binomial table to find the probability of observing one of these unusual values.

5.88 Refer to Exercise 5.86. A syndicated newspaper story reported that inspectors found 22 of 24 hamburgers underweight at restaurant W and fined that restaurant. Draw new control limits on your chart, from Exercise 5.86, for one new sample of size 24. Plot the new proportion and determine if this point is "out of control."

USING STATISTICS WISELY

1. The assignment of a value to each possible outcome, which creates a random variable, should quantify an important feature of the outcomes.

2. Describe the chance behavior of a discrete random variable X by its probability distribution

$$f(x) = P[X = x] \qquad \text{for each possible value } x$$

3. Summarize a probability distribution, or the random variable, by its

$$\text{Mean:} \qquad \mu = \sum_{\text{all } x} x \cdot f(x)$$

$$\text{Variance:} \qquad \sigma^2 = \sum_{\text{all } x} (x - \mu)^2 \cdot f(x)$$

4. If the use of Bernoullie trials is reasonable, probabilities concerning the number of successes in n Bernoulli trials can be calculated using the formula for the binomial distribution

$$f(x) = \binom{n}{x} p^x (1 - p)^{n-x} \qquad \text{for } x = 0, 1, \ldots, n$$

having mean np and variance $np(1 - p)$.

5. Never use the formula $np(1 - p)$ for the variance of a count of successes in n trials without first checking that the conditions for Bernoulli trials, independent trials with the same probability of success for each trial, hold. If the conditions are satisfied, then the binomial distribution is appropriate.

KEY IDEAS AND FORMULAS

The outcomes of an experiment are quantified by assigning each of them a numerical value related to a characteristic of interest. The rule for assigning the numerical value is called a **random variable** X.

A random variable having a finite number of values, or a sequence of values like a count, is called **discrete**. If a random variable can take any value in an interval, it is called a **continuous** random variable.

The **probability distribution** of X, or simply **distribution**, describes the manner in which probability is distributed over the possible values of X. Specifically, it is a list or formula giving the pairs x and $f(x) = P[X = x]$.

A probability distribution serves as a model for explaining variation in a population.

A **probability histogram** graphically displays a discrete distribution using bars whose area equals the probability.

A probability distribution has a

$$\text{Mean} \qquad \mu = \sum (\text{Value} \times \text{Probability}) = \sum x f(x)$$

which is interpreted as the **population mean**. This quantity is also called the **expected value** $E(X)$. Although X is a variable, $E(X)$ is a constant.

The population variance is

$$\sigma^2 = E(X - \mu)^2 = \sum (x - \mu)^2 f(x)$$

The **standard deviation** σ is its square root. The standard deviation is a measure of the spread or variation of the population.

Bernoulli trials are defined by the characteristics: (1) two possible outcomes, **success** (S) or **failure** (F) for each trial; (2) a constant probability of success; and (3) independence of trials.

Sampling from a finite population without replacement violates the requirement of independence. If the population is large and the sample size small, the trials can be treated as independent for all practical purposes.

The number of successes X in a fixed number of Bernoulli trials is called a binomial random variable. Its probability distribution, called the binomial distribution, is given by

$$f(x) = \binom{n}{x} p^x q^{n-x} \qquad \text{for} \qquad x = 0, \ldots, n$$

where $n =$ number of trials, $p =$ probability of success in each trial, and $q = 1 - p$.

The binomial distribution has

$$\text{Mean} = np$$
$$\text{Standard deviation} = \sqrt{npq}$$

A probability model is an assumed model for the probability distribution of a random variable. Probabilities are expressed in terms of parameters which are features of the population. For example, the binomial distribution is a probability model and the proportion p is a parameter.

A p chart displays sample proportions to reveal trends or changes in the population proportion over time.

TECHNOLOGY

Calculating the binomial probabilities $P[X = x]$ and $P[X \leq x]$

MINITAB

Calculating the binomial probability $P[X = x] = \binom{n}{x} p^x (1 - p)^{n-x}$

The following commands illustrate the calculation of $P[X = 14]$ under the binomial distribution having $n = 21$ and $p = .57$:

Dialog box:

Calc > Probability distributions > Binomial.
Select **Probability.**
Type *21* in **Number of Trials** and *.57* in **Probability of success.**
Select **Input constant**, and enter *14*. Click **OK.**

Alternatively to get all of the binomial probabilities for $n = 21$, you can first

Enter 0, 1, 2, . . ., 21 in *C1.*

Calc > Probability distributions > Binomial.
Select **Probability.**
Type *21* in **Number of Trials** and *.57* in **Probability of success.**
Then, instead of clicking **Input constant**,
Enter *C1* in Input column and *C2* in Optional storage. Click OK.

Calculating the cumulative binomial probability

$$P[X \leq c] = \sum_{x=0}^{c} \binom{n}{x} p^x (1 - p)^{n-x}$$

Follow the same steps as in the calculation of the probability of single terms except:

> Select **Cumulative probability** instead of **Probability**.
> **Calc > Probability distributions > Binomial.**
> Select **Cumulative Probability.**
> Type *21* in **Number of trials** and *.57* in **Probability of success.**
> Select **Input constant** and enter *14*. Click **OK.**

EXCEL

Calculating the binomial probability $P[X = x] = \binom{n}{x} p^x (1 - p)^{n-x}$

We illustrate the calculation of $P[X = 14]$ under the binomial distribution having $n = 21$ and $p = .57$.

With the cursor in a blank cell, select the f_x icon, or select **Insert** and then **Function.** Choose **Statistical** and then **BINOMDIST.** Click **OK.**
Type *14* in **Number_s,** *21* in **Trials,** *.57* in **Probability_s,** and *False* in **Cumulative.** Click **OK.**

Calculating the cumulative binomial probability

$$P[X \leq c] = \sum_{x=0}^{c} \binom{n}{x} p^x (1 - p)^{n-x}$$

Follow the same steps as in the calculation of $P[X = x]$ except:

Type *True* in **Cumulative.**

TI-84/-83 PLUS

Calculating the binomial probability $P[X = x] = \binom{n}{x} p^x (1 - p)^{n-x}$

The following commands illustrate the calculation of $P[X = 14]$ under the binomial distribution having $n = 21$ and $p = .57$:

> Press **2nd VARS** to read the *probability distribution* menu.
> Select **0:binompdf (** and press **ENTER.**
> With **binompdf(** on the home screen, type *10, .57, 4)* to give:

$$\text{binompdf(} 10, .57, 4)$$

> Press **ENTER.**

Calculating the cumulative binomial probability

$$P[X \le c] = \sum_{x=0}^{c} \binom{n}{x} p^x (1 - p)^{n-x}$$

Follow the same steps as in the calculation of $P[X = x]$ except replace the second step by

Select **A:binompdf(** and press **ENTER.**

8.   REVIEW EXERCISES

5.89 Let X denote the difference (no. of heads − no. of tails) in three tosses of a coin.

(a) List the possible values of X.

(b) List the elementary outcomes associated with each value of X.

5.90 Suppose there are two boxes. Box 1 contains 20 articles, of which 5 are defective, and Box 2 contains 30 articles, of which 6 are defective. One article is randomly selected from each box, and the selections from the two boxes are independent. Let X denote the total number of defective articles obtained.

(a) List the possible values of X and identify the elementary outcomes associated with each value.

(b) Determine the probability distribution of X.

*5.91 Refer to Exercise 5.90 and now suppose that the sampling is done in two stages: First, a box is selected at random and then, from the selected box, two articles are drawn at random and without replacement. Let Y denote the number of defective articles in the sample.

(a) List the elementary outcomes concerning the possible selections of the box and the possible compositions of the sample (use a tree diagram). Find their probabilities. (*Hint:* Use conditional probability and the multiplication rule.)

(b) Determine the probability distribution of Y.

5.92 Refer to Exercise 5.90, but now suppose that the contents of the two boxes are pooled together into a single larger box. Then two arti-

cles are drawn at random and without replacement. Let W denote the number of defective articles in the sample. Obtain the probability distribution of W.

*5.93 In a tennis championship, player A competes against player B in consecutive sets, and the game continues until one player wins three sets. Assume that, for each set, $P(\text{A wins}) = .4$, $P(\text{B wins}) = .6$, and the outcomes of different sets are independent. Let X stand for the number of sets played.

(a) List the possible values of X and identify the elementary outcomes associated with each value.

(b) Obtain the probability distribution of X.

5.94 The probability distribution of a random variable X is given by the formula

$$f(x) = \frac{32}{31} \left(\frac{1}{2^x} \right) \qquad \text{for} \qquad x = 1, 2, 3, 4, 5$$

(a) Calculate the numerical value of $f(x)$ for each x and make a table of the probability distribution.

(b) Plot the probability histogram.

5.95 In an assortment of 12 lightbulbs, there are 4 with broken filaments. A customer takes 3 bulbs from the assortment without inspecting the filaments. Find the probability distribution of the number X of defective bulbs that the customer may get.

5.96 For the following probability distribution:

(a) Calculate μ.

(b) Calculate σ^2 and σ.

(c) Plot the probability histogram and locate μ.

x	$f(x)$
0	.3
1	.4
2	.3

5.97 For the following probability distribution:
(a) Calculate $E(X)$.
(b) Calculate $sd(X)$.
(c) Draw the probability histogram and locate the mean.

x	$f(x)$
2	.1
3	.3
4	.3
5	.2
6	.1

5.98 Refer to Exercise 5.97.
(a) List the x values that lie in the interval $\mu - \sigma$ to $\mu + \sigma$ and calculate $P[\mu - \sigma \leq X \leq \mu + \sigma]$.
(b) List the x values that lie in the interval $\mu - 2\sigma$ to $\mu + 2\sigma$ and calculate $P[\mu - 2\sigma \leq X \leq \mu + 2\sigma]$.

5.99 A student buys a lottery ticket for $1. For every 1000 tickets sold, two bicycles are to be given away in a drawing.
(a) What is the probability that the student will win a bicycle?
(b) If each bicycle is worth $200, determine the student's expected gain.

5.100 In the finals of a tennis match, the winner will get $60,000 and the loser $15,000. Find the expected winnings of player B if (a) the two finalists are evenly matched and (b) player B has probability .8 of winning.

5.101 The number of overnight emergency calls X to the answering service of a heating and air conditioning firm have the probabilities .05, .1, .15, .35, .20, and .15 for 0, 1, 2, 3, 4, and 5 calls, respectively.
(a) Find the probability of fewer than 3 calls.
(b) Determine $E(X)$ and $sd(X)$.

5.102 Suppose the number of parking tickets X issued during a police officer's shift has the probability distribution

x	0	1	2	3	4
$f(x)$	0.13	0.14	0.43	0.20	0.10

(a) Find the mean and standard deviation of the number of parking tickets issued.
(b) Let $A = [X \leq 2]$ and $B = [X \geq 1]$. Find $P(A|B) = P(X \leq 2|X \geq 1)$.
(c) Suppose the numbers of tickets issued on different days are independent. What is the probability that, over the next five days, no parking tickets will be issued on exactly one of the days?

5.103 A botany student is asked to match the popular names of three house plants with their obscure botanical names. Suppose the student never heard of these names and is trying to match by sheer guess. Let X denote the number of correct matches.
(a) Obtain the probability distribution of X.
(b) What is the expected number of matches?

5.104 The number of days, X, that it takes the post office to deliver a letter between City A and City B has the probability distribution

x	$f(x)$
3	.5
4	.3
5	.2

Find:
(a) The expected number of days.
(b) The standard deviation of the number of days.

5.105 A roulette wheel has 38 slots, of which 18 are red, 18 black, and 2 green. A gambler will play three times, each time betting $5 on red. The gambler gets $10 if red occurs and loses the bet otherwise. Let X denote the net gain of the gambler in 3 plays (for instance, if he loses all three times, then $X = -15$).

(a) Obtain the probability distribution of X.

(b) Calculate the expected value of X.

(c) Will the expected net gain be different if the gambler alternates his bets between red and black? Why or why not?

5.106 Suppose that X can take the values 0, 1, 2, 3, and 4, and the probability distribution of X is incompletely specified by the function

$$f(x) = \frac{1}{4}\left(\frac{3}{4}\right)^x \quad \text{for} \quad x = 0, 1, 2, 3$$

Find (a) $f(4)$ (b) $P[X \geq 2]$ (c) $E(X)$ and (d) sd(X).

5.107 Let $\overline{X}$ = average number of dots resulting from two tosses of a fair die. For instance, if the faces 4 and 5 show, the corresponding value of $\overline{X}$ is $(4 + 5)/2 = 4.5$. Obtain the probability distribution of $\overline{X}$.

5.108 Refer to Exercise 5.107. On the same graph, plot the probability histograms of

X_1 = No. of points on the first toss of the die

$\overline{X}$ = Average number of points in two tosses of the die

5.109 *The cumulative probabilities for a distribution.* A probability distribution can also be described by a function that gives the accumulated probability at or below each value of X. Specifically,

$$F(c) = P[X \leq c] = \sum_{x \leq c} f(x)$$

Cumulative distribution function at c = Sum of probabilities of all values $x \leq c$

For the probability distribution given here, we calculate

x	$f(x)$	$F(x)$
1	.07	.07
2	.12	.19
3	.25	
4	.28	
5	.18	
6	.10	

$$F(1) = P[X \leq 1] = f(1) = .07$$
$$F(2) = P[X \leq 2] = f(1) + f(2) = .19$$

(a) Complete the $F(x)$ column in this table.

(b) Now cover the $f(x)$ column with a strip of paper. From the $F(x)$ values, reconstruct the probability function $f(x)$. [*Hint:* $f(x) = F(x) - F(x - 1)$.]

*5.110 *Runs.* In a row of six plants two are infected with a leaf disease and four are healthy. If we restrict attention to the portion of the sample space for exactly two infected plants, the model of randomness (or lack of contagion) assumes that any two positions, for the infected plants in the row are as likely as any other.

(a) Using the symbols I for infected and H for healthy, list all possible occurrences of two I's and four H's in a row of 6.

(*Note:* There are $\binom{6}{2} = 15$ elementary outcomes.)

(b) A random variable of interest is the number of runs X that is defined as the number of unbroken sequences of letters of the same kind. For example, the arrangement IHHHIH has four runs, IIHHHH has two. Find the value of X associated with each outcome you listed in part (a).

(c) Obtain the probability distribution of X under the model of randomness.

*5.111 Refer to part (c) of Exercise 5.110. Calculate the mean and standard deviation of X.

5.112 Let the random variable Y denote the proportion of times a head occurs in three tosses of a coin, that is, $Y =$ (No. of heads in 3 tosses)/3.

(a) Obtain the probability distribution Y.

(b) Draw the probability histogram.

(c) Calculate the $E(Y)$ and sd(Y).

5.113 Is the model of Bernoulli trials plausible in each of the following situations? Identify any serious violations of the conditions.

(a) A dentist records if each tooth in the lower jaw has a cavity or has none.

(b) Persons applying for a driver's license will

be recorded as writing left- or right-handed.

(c) For each person taking a seat at a lunch counter, observe the time it takes to be served.

(d) Each day of the first week in April is recorded as being either clear or cloudy.

(e) Cars selected at random will or will not pass state safety inspection.

5.114 Give an example (different from those appearing in Exercise 5.113) of repeated trials with two possible outcomes where:

(a) The model of Bernoulli trials is reasonable.

(b) The condition of independence is violated.

(c) The condition of equal $P(S)$ is violated.

5.115 If the probability of having a male child is .5, find the probability that the third child is the first son.

5.116 If the probability of getting caught copying someone else's exam is .1, find the probability of not getting caught in three attempts. Assume independence.

5.117 A stoplight on the way to class is red 70% of the time. What is the probability of hitting a red light:

(a) 2 days in a row?

(b) 3 days in a row?

(c) 2 out of 3 days?

5.118 A basketball team scores 45% of the times it gets the ball. Find the probability that the first basket occurs on its third possession. (Assume independence.)

5.119 If in three Bernoulli trials the probability that the first two trials are both failures is 4/49, what is the probability that the first two are successes and the third is a failure?

5.120 The proportion of people having the blood type O in a large southern city is .4. For two randomly selected donors:

(a) Find the probability of at least one type O.

(b) Find the expected number of type O.

(c) Repeat parts (a) and (b) if there are three donors.

5.121 A viral infection is spread by contact with an infected person. Let the probability that a healthy person gets the infection in one contact be $p = .4$.

(a) An infected person has contact with six healthy persons. Specify the distribution of $X = $ No. of persons who contract the infection.

(b) Find $P[X \leq 3]$, $P[X = 0]$, and $E(X)$.

5.122 The probability that a voter will believe a rumor about a politician is .3. If 20 voters are told individually:

(a) Find the probability that none of the 20 believes the rumor.

(b) Find the probability that seven or more believe the rumor.

(c) Determine the mean and standard deviation of the number who believe the rumor.

5.123 National safety statistics suggest that about 33% of the persons treated in an emergency room because of moped accidents are under 16 years of age. Suppose you count the number of persons under 16 among the next 14 moped accident victims to come to the emergency room.

(a) Find the mean of X.

(b) Find the standard deviation of X.

(c) Find the probability that the first injured person is under 16 years old and the second is at least 16 years old.

5.124 In rolling a fair die 4 times, find the probability that 6 appears no more than twice.

5.125 A school newspaper claims that 70% of the students support its view on a campus issue. A random sample of 20 students is taken, and 10 students agree with the newspaper. Find $P[10$ or less agree] if 70% support the view and comment on the plausibility of the claim.

5.126 For each situation, state if a binomial distribution is reasonable for the random variable X. Justify your answer.

(a) A multiple-choice examination consists of 10 problems, each of which has 5 suggested answers. A student marks answers by pure guesses (i.e., one answer is chosen at random out of the 5), and X denotes the number of marked answers that are wrong.

(b) A multiple-choice examination has two parts: Part 1 has 8 problems, each with 5 suggested answers, and Part 2 has 10 problems, each with 4 suggested answers. A student marks answers by pure guesses, and X denotes the total number of problems that the student correctly answers.

(c) Twenty married couples are interviewed about exercise, and X denotes the number of persons (out of the 40 people interviewed) who are joggers.

5.127 For the binomial distribution with $n = 14$ and $p = .4$, determine
(a) $P[3 \leq X \leq 9]$
(b) $P[3 < X \leq 9]$
(c) $P[3 < X < 9]$
(d) $E(X)$ (e) $sd(X)$

5.128 Using the binomial table:
(a) List the probability distribution for $n = 5$ and $p = .4$.
(b) Plot the probability histogram.
(c) Calculate $E(X)$ and $Var(X)$ from the entries in the list from part (a).
(d) Calculate $E(X) = np$ and $Var(X) = npq$ and compare your answer with part (c).

*5.129 For a binomial distribution with $p = .15$, find the smallest number n such that 1 success is more probable than no successes in n trials.

5.130 Using the binomial table, find:
(a) The probability of 3 successes in 8 trials when $p = .3$.
(b) The probability of 7 failures in 16 trials when $p = .7$.
(c) The probability of 3 or fewer successes in 8 trials when $p = .4$.
(d) The probability of more than 12 successes in 16 trials when $p = .6$.

(e) The probability of between 8 and 13 successes (both inclusive) in 16 trials with $p = .6$.

5.131 Using the binomial table, find the probability of:
(a) Three or less successes for $p = .1, .2, .3, .4, .5$ when $n = 12$.
(b) Three or less successes for $p = .1, .2, .3, .4, .5$ when $n = 18$.

5.132 A sociologist feels that only half of the high school seniors capable of graduating from college go to college. Of 17 high school seniors who have the ability to graduate from college, find the probability that 12 or more will go to college if the sociologist is correct. Assume that the seniors will make their decisions independently. Also find the expected number.

5.133 Only 30% of the people in a large city feel that its mass transit system is adequate. If 20 persons are selected at random, find the probability that 10 or more will feel that the system is adequate. Find the probability that exactly 10 will feel that the system is adequate.

5.134 Jones claims to have extrasensory perception (ESP). In order to test the claim, a psychologist shows Jones five cards that carry different pictures. Then Jones is blindfolded and the psychologist selects one card and asks Jones to identify the picture. This process is repeated 16 times. Suppose, in reality, that Jones has no ESP but responds by sheer guesses.
(a) What is the probability that the identifications are correct at most 3 times?
(b) What is the probability that the identifications are wrong at least 10 times?
(c) Find the expected value and standard deviation of the number of correct identifications.

5.135 A financial planner will get a $500 bonus if she recruits a new client in each of the 3 months of the next quarter. Otherwise there will be no bonus. Suppose that, for each single month, the probability is .4 that she will recruit a new client. Let X be the amount of bonus money she will receive.

(a) Find the probability distribution of X assuming the results for different months are independent.

(b) Find the expected value of X.

*5.136 *Geometric distribution.* Instead of performing a fixed number of Bernoulli trials, an experimenter performs trials until the first success occurs. The number of successes is now fixed at 1, but the number of trials Y is now random. It can assume any of the values 1, 2, 3, and so on with no upper limit.

(a) Show that
$$f(y) = q^{y-1}p \qquad \text{for} \qquad y = 1, 2, \ldots$$

(b) Find the probability of 3 or fewer trials when $p = .5$.

*5.137 *Poisson distribution for rare events.* The Poisson distribution is often appropriate when the probability of an event (success) is small. It has served as a probability model for the number of plankton in a liter of water, calls per hour to an answering service, and earthquakes in a year. The Poisson distribution also approximates the binomial when the expected value np is small but n is large. The Poisson distribution with mean m has the form

$$f(x) = e^{-m}\frac{m^x}{x!} \qquad \text{for} \qquad x = 0, 1, 2, \ldots$$

where e is the exponential number or 2.718 (rounded) and $x!$ is the number $x(x - 1)(x - 2)\cdots 1$ with $0! = 1$. Given $m = 3$ and $e^{-3} = .05$, find: (a) $P[X = 0]$, (b) $P[X = 1]$.

5.138 An inspector will sample bags of potato chips to see if they fall short of the weight, 14 ounces, printed on the bag. Samples of 20 bags will be selected and the number with weight less than 14 ounces will be recorded.

(a) Make a p chart using the centerline and control limits calculated for $p_0 = .5$.

(b) Suppose that samples from ten different days have

11 8 14 10 13 12 7 14 10 13

underweight bags. Graph the corresponding proportions on your p chart.

(c) Identify any days where the chart signals "out of control."

6

The Normal Distribution

Bell-shaped Distribution of Heights of Red Pine Seedlings

Trees are a renewable resource that is continually studied to both monitor the current status and improve this valuable natural resource. One researcher followed the growth of red pine seedlings. The heights (mm) of 1456 three-year-old seedlings are summarized in the histogram. This histogram suggests a distribution with a single peak and which falls off in a symmetric manner. The histogram of the heights of adult males, or of adult females, has a similar pattern. A bell-shaped distribution is likely to be appropriate for the size of many things in nature.

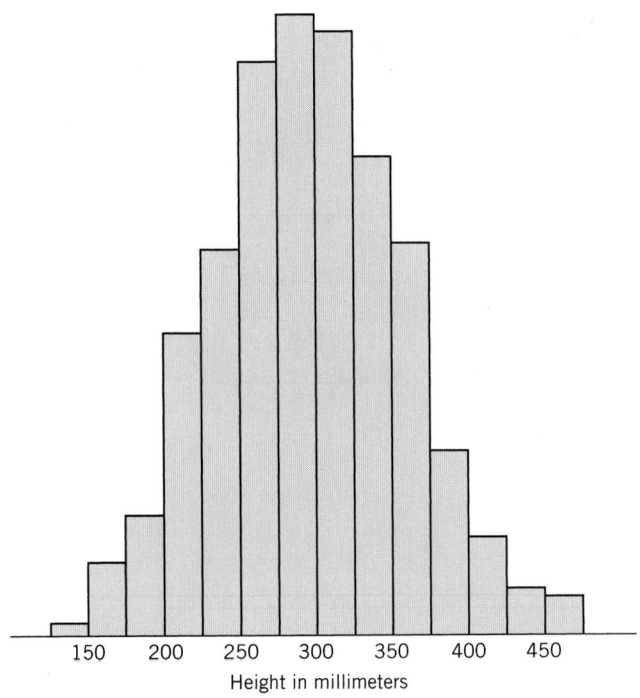

Height in millimeters

1. PROBABILITY MODEL FOR A CONTINUOUS RANDOM VARIABLE

Up to this point, we have limited our discussion to probability distributions of discrete random variables. Recall that a discrete random variable takes on only some isolated values, usually integers representing a count. We now turn our attention to the probability distribution of a continuous random variable—one that can ideally assume any value in an interval. Variables measured on an underlying continuous scale, such as weight, strength, life length, and temperature, have this feature.

Just as probability is conceived as the long-run relative frequency, the idea of a continuous probability distribution draws from the relative frequency histogram for a large number of measurements. The reader may wish to review Section 3.3 of Chapter 2 where grouping of data in class intervals and construction of a relative frequency histogram were discussed. We have remarked that with an increasing number of observations in a data set, histograms can be constructed with class intervals having smaller widths. We will now pursue this point in order to motivate the idea of a continuous probability distribution. To focus the discussion let us consider that the weight X of a newborn baby is the continuous random variable of our interest. How do we conceptualize the probability distribution of X? Initially, suppose that the birth weights of 100 babies are recorded, the data grouped in class intervals of 1 pound, and the relative frequency histogram in Figure 1a is obtained. Recall that a relative frequency histogram has the following properties:

1. The total area under the histogram is 1.
2. For two points a and b such that each is a boundary point of some class, the relative frequency of measurements in the interval a to b is the area under the histogram above this interval.

For example, Figure 1a shows that the interval 7.5 to 9.5 pounds contains a proportion .28 + .25 = .53 of the 100 measurements.

Next, we suppose that the number of measurements is increased to 5000 and they are grouped in class intervals of .25 pound. The resulting relative frequency histogram appears in Figure 1b. This is a refinement of the histogram in Figure 1a in that it is constructed from a larger set of observations and exhibits relative frequencies for finer class intervals. (Narrowing the class interval without increasing the number of observations would obscure the overall shape of the distribution.) The refined histogram in Figure 1b again has the properties 1 and 2 stated above.

By proceeding in this manner, even further refinements of relative frequency histograms can be imagined with larger numbers of observations and smaller class intervals. In pursuing this conceptual argument, we ignore the difficulty that the accuracy of the measuring device is limited. In the course of refining the histograms, the jumps between consecutive rectangles tend to dampen out, and the top of the histogram approximates the shape of a smooth curve, as illustrated in Figure 1c. Because probability is interpreted as long-run relative

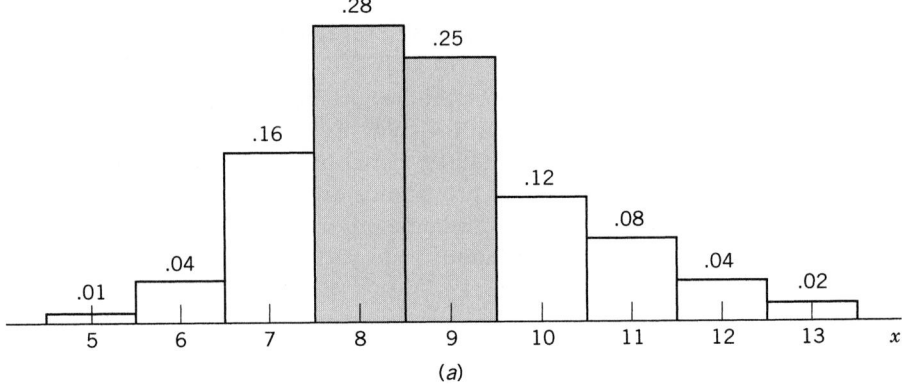

(a)

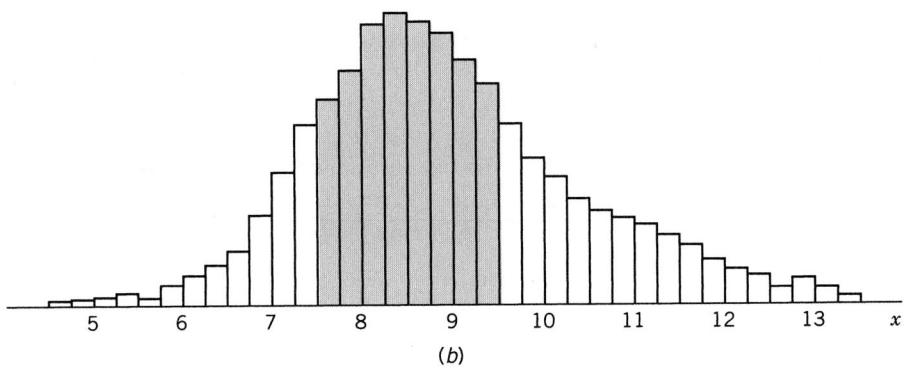

(b)

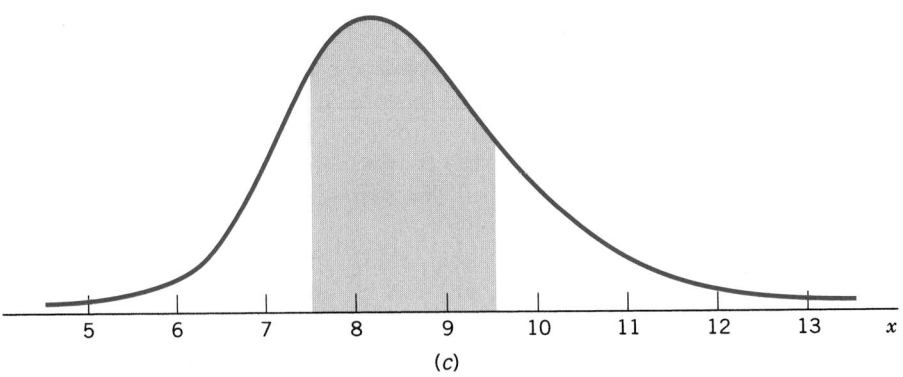

(c)

Figure 1   Probability density curve viewed as a limiting form of relative frequency histograms.

frequency, the curve obtained as the limiting form of the relative frequency histograms represents the manner in which the total probability 1 is distributed over the interval of possible values of the random variable X. This curve is called the probability density curve of the continuous random variable X. The mathematical function $f(x)$ whose graph produces this curve is called the probability density function of the continuous random variable X.

The properties 1 and 2 that we stated earlier for a relative frequency histogram are shared by a probability density curve that is, after all, conceived as a limiting smoothed form of a histogram. Also, since a histogram can never protrude below the x axis, we have the further fact that $f(x)$ is nonnegative for all x.

The probability density function $f(x)$ describes the distribution of probability for a continuous random variable. It has the properties:

1. The total area under the probability density curve is 1.
2. $P[a \leq X \leq b]$ = area under the probability density curve between a and b.
3. $f(x) \geq 0$ for all x.

Unlike the description of a discrete probability distribution, the probability density $f(x)$ for a continuous random variable does not represent the probability that the random variable will exactly equal the value x, or the event $[X = x]$. Instead, a probability density function relates the probability of an interval $[a, b]$ to the area under the curve in a strip over this interval. A single point x, being an interval with a width of 0, supports 0 area, so $P[X = x] = 0$.

With a continuous random variable, the probability that $X = x$ is **always** 0. It is only meaningful to speak about the probability that X lies in an interval.

The deduction that the probability at every single point is zero needs some clarification. In the birth-weight example, the statement $P[X = 8.5$ pounds] $= 0$ probably seems shocking. Does this statement mean that no child can have a birth weight of 8.5 pounds? To resolve this paradox, we need to recognize that the accuracy of every measuring device is limited, so that here the number 8.5 is actually indistinguishable from all numbers in an interval surrounding it, say, [8.495, 8.505]. Thus, the question really concerns the probability of an interval surrounding 8.5, and the area under the curve is no longer 0.

When determining the probability of an interval a to b, we need not be concerned if either or both endpoints are included in the interval. Since the probabilities of $X = a$ and $X = b$ are both equal to 0,

$$P[a \leq X \leq b] = P[a < X \leq b] = P[a \leq X < b] = P[a < X < b]$$

In contrast, these probabilities may not be equal for a discrete distribution.

Fortunately, for important distributions, areas have been extensively tabulated. In most tables, the entire area to the left of each point is tabulated. To obtain the probabilities of other intervals, we must apply the following rules.

$$P[a < X < b] = \text{(Area to left of } b) - \text{(Area to the left of } a)$$

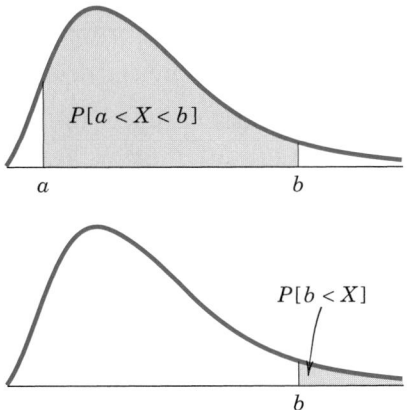

$$P[b < X] = 1 - \text{(Area to left of } b)$$

SPECIFICATION OF A PROBABILITY MODEL

A probability model for a continuous random variable is specified by giving the mathematical form of the probability density function. If a fairly large number of observations of a continuous random variable are available, we may try to approximate the top of the staircase silhouette of the relative frequency histogram by a mathematical curve.

In the absence of a large data set, we may tentatively assume a reasonable model that may have been suggested by data from a similar source. Of course, any model obtained in this way must be closely scrutinized to verify that it conforms to the data at hand. Section 6 addresses this issue.

FEATURES OF A CONTINUOUS DISTRIBUTION

As is true for relative frequency histograms, the probability density curves of continuous random variables could possess a wide variety of shapes. A few of these are illustrated in Figure 2. Many statisticians use the term skewed for a long tail in one direction.

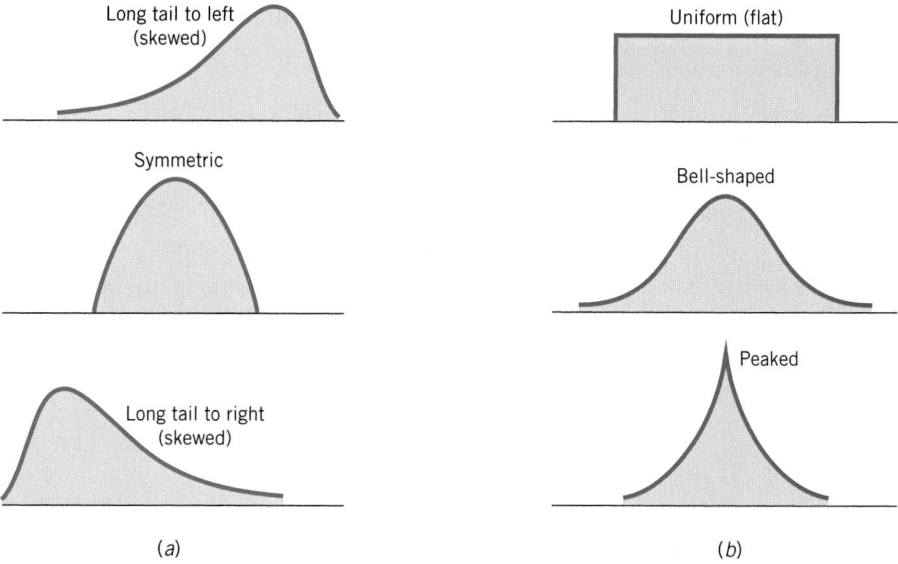

Figure 2   Different shapes of probability density curves. (*a*) Symmetry and deviations from symmetry. (*b*) Different peakedness.

A continuous random variable X also has a mean, or expected value $E(X)$, as well as a variance and a standard deviation. Their interpretations are the same as in the case of discrete random variables, but their formal definitions involve integral calculus and are therefore not pursued here. However, it is instructive to see in Figure 3 that the mean $\mu = E(X)$ marks the balance point of the probability mass. The median, another measure of center, is the value of X that divides the area under the curve into halves.

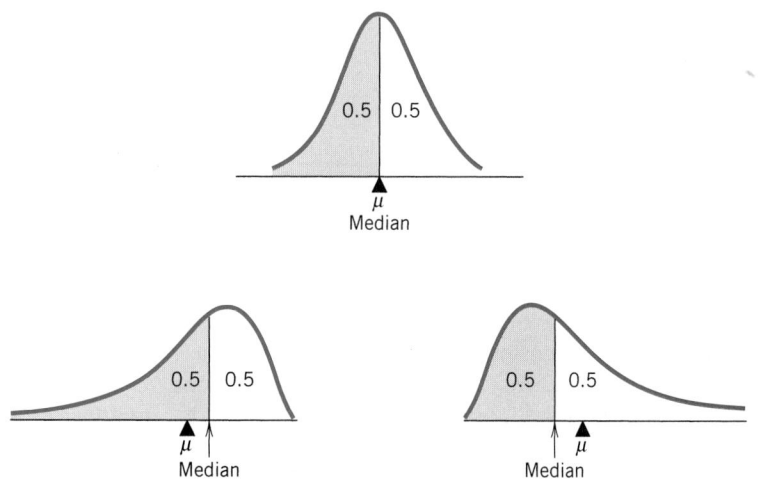

Figure 3   Mean as the balance point and median as the point of equal division of the probability mass.

Besides the median, we can also define the quartiles and other percentiles of a probability distribution.

The population $100p$-th percentile is an x value that supports area p to its left and $1 - p$ to its right.

$$\text{Lower (first) quartile} = \text{25th percentile}$$
$$\text{Second quartile (or median)} = \text{50th percentile}$$
$$\text{Upper (third) quartile} = \text{75th percentile}$$

The quartiles for two distributions are shown in Figure 4.

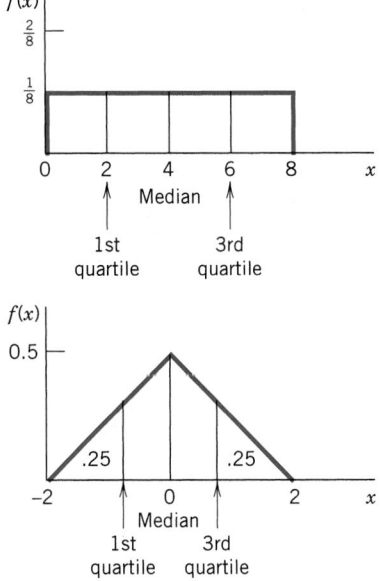

Figure 4   Quartiles of two continuous distributions.

Statisticians often find it convenient to convert random variables to a dimensionless scale. Suppose X, a real estate salesperson's commission for a month, has mean \$4000 and standard deviation \$500. Subtracting the mean produces the deviation $X - 4000$ measured in dollars. Then, dividing by the standard deviation, expressed in dollars, yields the dimensionless variable $Z = (X - 4000)/500$. Moreover, the standardized variable Z can be shown to have mean 0 and standard deviation 1. (See Appendix A2.1 for details.)

The standardized variable

$$Z = \frac{X - \mu}{\sigma} = \frac{\text{Variable} - \text{Mean}}{\text{Standard deviation}}$$

has mean 0 and sd 1.

Exercises

6.1   Which of the functions sketched below could be a probability density function for a continuous random variable? Why or why not?

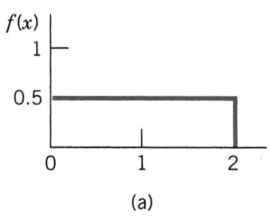

(a)

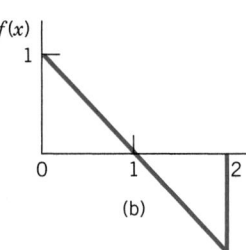

(b)

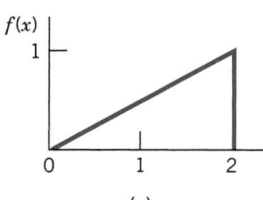

(c)

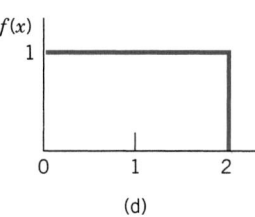

(d)

6.2   Determine the following probabilities from the curve $f(x)$ diagrammed in Exercise 6.1(a).
   (a) $P[\,0 < X < .5\,]$
   (b) $P[\,.5 < X < 1\,]$
   (c) $P[\,1.5 < X < 2\,]$
   (d) $P[\,X = 1\,]$

6.3   For the curve $f(x)$ graphed in Exercise 6.1(c), which of the two intervals $[\,0 < X < .5\,]$ or $[\,1.5 < X < 2\,]$ is assigned a higher probability?

6.4   Determine the median and the quartiles for the probability distribution depicted in Exercise 6.1(a).

6.5   Determine the median and the quartiles for the curve depicted in Exercise 6.1(c).

6.6   Determine the 15th percentile of the curve in Exercise 6.1(a).

6.7   If a student is more likely to be late than on time for the 1:20 PM history class:
   (a)   Determine if the median of the student's arrival time distribution is earlier than, equal to, or later than 1:20 PM.
   (b)   On the basis of the given information, can you determine if the mean of the student's arrival time distribution is earlier than, equal to, or later than 1:20 PM? Comment.

6.8   Which of the distributions in Figure 3 are compatible with the following statements?
   (a)   The distribution of starting salaries for computer programmers has a mean of $45,000, but half of the newly employed programmers make less than $40,000 annually.
   (b)   In spite of recent large increases in salary, half of the professional football players still make less than the average salary.

6.9   Find the standardized variable Z if X has
   (a)   Mean 15 and standard deviation 4.
   (b)   Mean 61 and standard deviation 9.
   (c)   Mean 161 and variance 25.

6.10   Males 20 to 29 years old have a mean height of 69 inches with a standard deviation of 2.8 inches. Females 20 to 29 years old have a mean height of 64 inches with a standard deviation of 2.6 inches. (Based on *Statistical Abstract of the U.S. 2002*, Table 189.)

(a) Find the standardized variable for the heights of males.

(b) Find the standardized variable for the heights of females.

(c) For a 67-inch-tall person, find the value of the standardized variable for males.

(d) For a 67-inch-tall person, find the value of the standardized variable for females.

Compare your answer with part (c) and comment.

6.11   Find the standardized variable Z if X has

(a) Mean 8 and standard deviation 2.

(b) Mean 350 and standard deviation 8.

(c) Mean 888 and variance 81.

2.   THE NORMAL DISTRIBUTION—ITS GENERAL FEATURES

The normal distribution, which may already be familiar to some readers as the curve with the bell shape, is sometimes associated with the names of Pierre Laplace and Carl Gauss, who figured prominently in its historical development. Gauss derived the normal distribution mathematically as the probability distribution of the error of measurements, which he called the "normal law of errors." Subsequently, astronomers, physicists, and, somewhat later, data collectors in a wide variety of fields found that their histograms exhibited the common feature of first rising gradually in height to a maximum and then decreasing in a symmetric manner. Although the normal curve is not unique in exhibiting this form, it has been found to provide a reasonable approximation in a great many situations. Unfortunately, at one time during the early stages of the development of statistics, it had many overzealous admirers. Apparently, they felt that all real-life data must conform to the bell-shaped normal curve, or otherwise, the process of data collection should be suspect. It is in this context that the distribution became known as the normal distribution. However, scrutiny of data has often revealed inadequacies of the normal distribution. In fact, the universality of the normal distribution is only a myth, and examples of quite nonnormal distributions abound in virtually every field of study. Still, the normal distribution plays a central role in statistics, and inference procedures derived from it have wide applicability and form the backbone of current methods of statistical analysis.

Does the $\bar{x}$ ranch have average beef?

A **normal distribution** has a bell-shaped density[1] as shown in Figure 5. It has

$$\text{Mean} = \mu$$
$$\text{Standard deviation} = \sigma$$

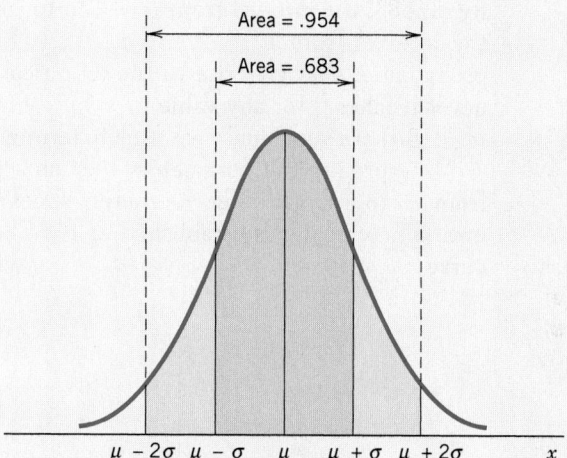

Figure 5   Normal distribution.

The probability of the interval extending

One sd on each side of the mean: $P[\mu - \sigma < X < \mu + \sigma] = .683$

Two sd on each side of the mean: $P[\mu - 2\sigma < X < \mu + 2\sigma] = .954$

Three sd on each side of the mean: $P[\mu - 3\sigma < X < \mu + 3\sigma] = .997$

Notation

The normal distribution with a mean of μ and a standard deviation of σ is denoted by $N(\mu, \sigma)$.

---

[1]The formula, which need not concern us, is

$$f(x) = \frac{1}{\sqrt{2\pi}\sigma} e^{-\frac{1}{2}\left(\frac{x-\mu}{\sigma}\right)^2} \quad \text{for} \quad -\infty < x < \infty$$

where π is the area of a circle having unit radius, or approximately 3.1416, and e is approximately 2.7183.

Although we are speaking of the importance of the normal distribution, our remarks really apply to a whole class of distributions having bell-shaped densities. There is a normal distribution for each value of its mean μ and its standard deviation σ.

A few details of the normal curve merit special attention. The curve is symmetric about its mean μ, which locates the peak of the bell (see Figure 5). The interval running one standard deviation in each direction from μ has a probability of .683, the interval from $\mu - 2\sigma$ to $\mu + 2\sigma$ has a probability of .954, and the interval from $\mu - 3\sigma$ to $\mu + 3\sigma$ has a probability of .997. It is these probabilities that give rise to the empirical rule stated in Chapter 2. The curve never reaches 0 for any value of x, but because the tail areas outside ($\mu - 3\sigma$, $\mu + 3\sigma$) are very small, we usually terminate the graph at these points.

Interpreting the parameters, we can see in Figure 6 that a change of mean from μ_1 to a larger value μ_2 merely slides the bell-shaped curve along the axis until a new center is established at μ_2. There is no change in the shape of the curve.

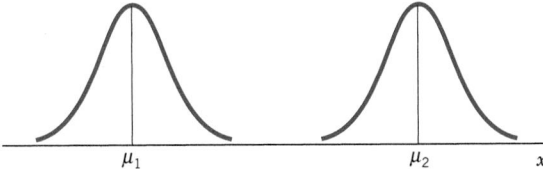

Figure 6   Two normal distributions with different means but the same standard deviation.

A different value for the standard deviation results in a different maximum height of the curve and changes the amount of the area in any fixed interval about μ (see Figure 7). The position of the center does not change if only σ is changed.

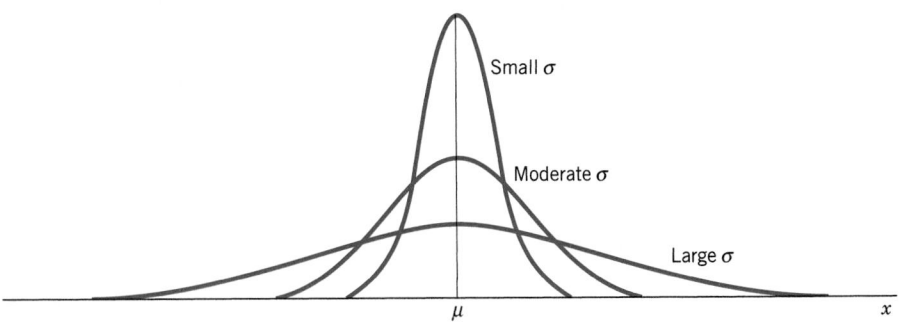

Figure 7   Decreasing σ increases the maximum height and the concentration of probability about μ.

3. THE STANDARD NORMAL DISTRIBUTION

The particular normal distribution that has a mean of 0 and a standard deviation of 1 is called the standard normal distribution. It is customary to denote the standard normal variable by Z. The standard normal curve is illustrated in Figure 8.

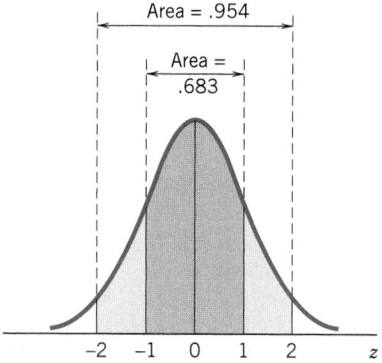

Figure 8   The standard normal curve.

The standard normal distribution has a bell-shaped density with

$$\text{Mean } \mu = 0$$
$$\text{Standard deviation } \sigma = 1$$

The standard normal distribution is denoted by $N(0, 1)$.

USE OF THE STANDARD NORMAL TABLE (APPENDIX B, TABLE 3)

The standard normal table in the appendix gives the area to the left of a specified value of z as

$$P[Z \leq z] = \text{Area under curve to the left of } z$$

For the probability of an interval $[a, b]$,

$$P[a \leq Z \leq b] = [\text{Area to left of } b] - [\text{Area to left of } a]$$

The following properties can be observed from the symmetry of the standard normal curve about 0 as exhibited in Figure 9.

1. $P[Z \leq 0] = .5$
2. $P[Z \leq -z] = 1 - P[Z \leq z] = P[Z \geq z]$

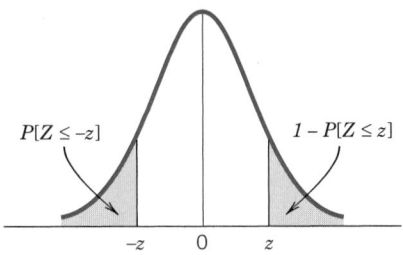

Figure 9   Equal normal tail probabilities.

Example 1   Determining Standard Normal Probabilities for Tail Events

Find $P[Z \leq 1.37]$ and $P[Z > 1.37]$.

SOLUTION   From the normal table, we see that the probability or area to the left of 1.37 is .9147. (See Table 1.) Consequently, $P[Z \leq 1.37] = .9147$. Moreover, because $[Z > 1.37]$ is the complement of $[Z \leq 1.37]$,

$$P[Z > 1.37] = 1 - P[Z \leq 1.37] = 1 - .9147 = .0853$$

as we can see in Figure 10. An alternative method is to use symmetry to show that $P[Z > 1.37] = P[Z < -1.37]$, which can be obtained directly from the normal table.

TABLE 1   How to Read from
Appendix B, Table 3 for
$z = 1.37 = 1.3 + .07$

z	.00	$\cdots$	.07	$\cdots$
.0			$\vdots$	
.			$\vdots$	
.			$\vdots$	
.			$\downarrow$	
1.3	$----------\rightarrow$		.9147	
.				
.				
.				

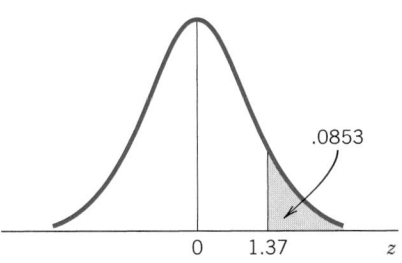

Figure 10   An upper tail normal probability.

Example 2   Determining the Standard Normal Probability of an Interval

Calculate $P[-.155 < Z < 1.60]$.

SOLUTION   From Appendix B, Table 3, we see that

$$P[Z \leq 1.60] = \text{Area to left of } 1.60 = .9452$$

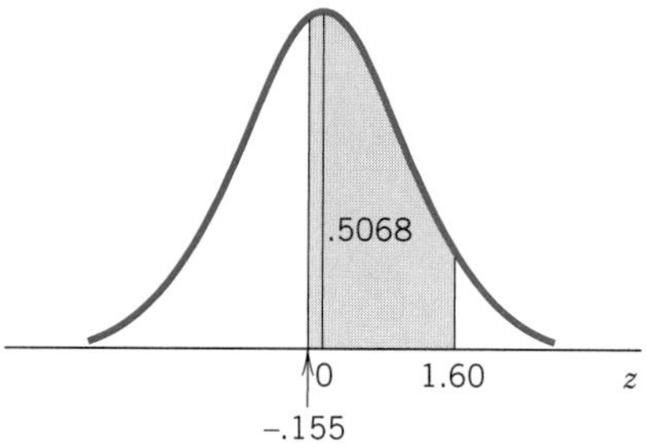

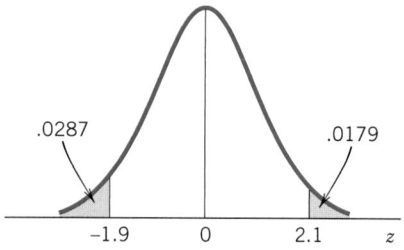

Figure 11   Normal probability
of an interval.

We interpolate[2] between the entries for $-.15$ and $-.16$ to obtain

$$P[Z \leq -.155] = \text{Area to left of } -.155 = .4384$$

Therefore,

$$P[-.155 < Z < 1.60] = .9452 - .4384 = .5068$$

which is the shaded area in Figure 11.

Example 3   Determining the Standard Normal Probability Outside of an Interval

Find $P[Z < -1.9 \text{ or } Z > 2.1]$.

SOLUTION   The two events $[Z < -1.9]$ and $[Z > 2.1]$ are incompatible, so we add their probabilities:

$$P[Z < -1.9 \text{ or } Z > 2.1] = P[Z < -1.9] + P[Z > 2.1]$$

As indicated in Figure 12,

Figure 12   Normal probabilities
for Example 3.

---

[2]Since $z = -.155$ is halfway between $-.15$ and $-.16$, the interpolated value is halfway between the table entries .4404 and .4364. The result is .4384.

$P[Z > 2.1]$ is the area to the right of 2.1, which is $1 - $ [Area to left of 2.1] $= 1 - .9821 = .0179$. The normal table gives $P[Z < -1.9] = .0287$ directly. Adding these two quantities, we get

$$P[Z < -1.9 \text{ or } Z > 2.1] = .0287 + .0179 = .0466$$

Example 4   Determining an Upper Percentile
of the Standard Normal Distribution

Locate the value of z that satisfies $P[Z > z] = .025$.

SOLUTION   If we use the property that the total area is 1, the area to the left of z must be $1 - .0250 = .9750$. The marginal value with the tabular entry .9750 is $z = 1.96$ (diagrammed in Figure 13).

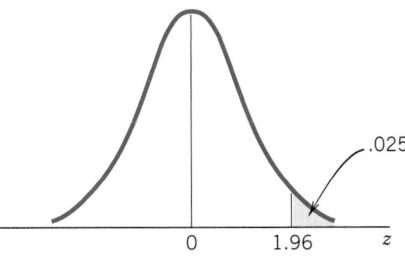

z	.00	$\cdots$	.06	$\cdots$
.0				
$\cdot$				
$\cdot$				
$\cdot$				
1.9	$\leftarrow\!\!-\!\!-\!\!-\!\!-\!\!-\!\!-\!\!-\!\!-$		.9750	
$\cdot$				
$\cdot$				
$\cdot$				

Figure 13 $P[Z > 1.96] = .025$.

Example 5   Determining z for Given Equal Tail Areas

Obtain the value of z for which $P[-z \leq Z \leq z] = .90$.

SOLUTION   We observe from the symmetry of the curve that

$$P[Z < -z] = P[Z > z] = .05$$

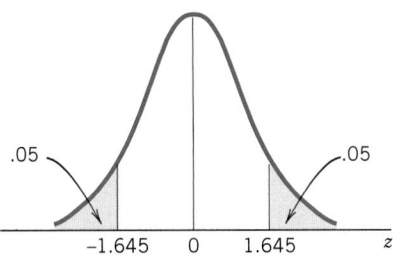

Figure 14 $P[Z < -1.645 \text{ or } Z > 1.645] = .10$.

From the normal table, we see that $z = 1.65$ gives $P[Z < -1.65] = .0495$ and $z = 1.64$ gives $P[Z < -1.64] = .0505$. Because .05 is halfway between these two probabilities, we interpolate between the two z values to obtain $z = 1.645$ (see Figure 14).

Suggestion: The preceding examples illustrate the usefulness of a sketch to depict an area under the standard normal curve. A correct diagram shows how to combine the left-side areas given in the normal table.

Exercises

6.12 Find the area under the standard normal curve to the left of

(a) $z = 1.19$ (b) $z = .18$

(c) $z = -1.81$ (d) $z = -2.3$

6.13 Find the area under the standard normal curve to the left of

(a) $z = .83$ (b) $z = 1.03$

(c) $z = -1.03$ (d) $z = -1.35$

6.14 Find the area under the standard normal curve to the right of

(a) $z = 1.19$

(b) $z = .64$

(c) $z = -1.14$

(d) $z = -1.525$ (interpolate)

6.15 Find the area under the standard normal curve to the right of

(a) $z = .83$

(b) $z = 2.83$

(c) $z = -1.23$

(d) $z = 1.635$ (interpolate)

6.16 Find the area under the standard normal curve over the interval

(a) $z = -.85$ to $z = .85$

(b) $z = -1.06$ to $z = 1.06$

(c) $z = .32$ to $z = 2.65$

(d) $z = -.745$ to $z = 1.244$ (interpolate)

6.17 Find the area under the standard normal curve over the interval

(a) $z = -.44$ to $z = .44$

(b) $z = -1.33$ to $z = 1.33$

(c) $z = .40$ to $z = 2.03$

(d) $z = 1.405$ to $z = 2.306$ (interpolate)

6.18 Identify the z values in the following diagrams of the standard normal distribution (interpolate, as needed).

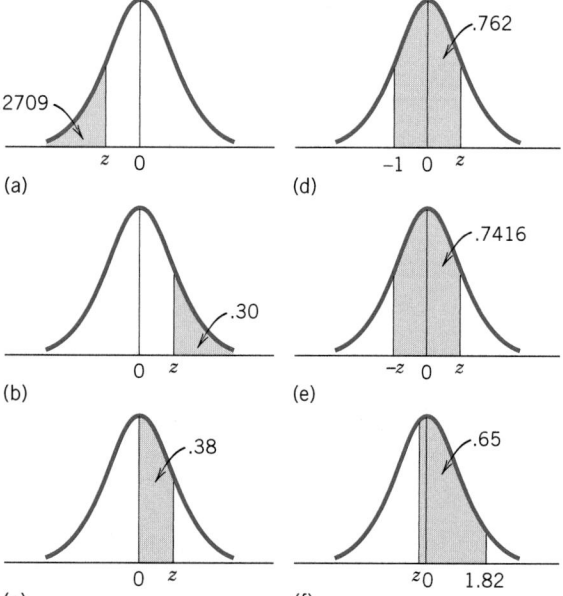

6.19 Identify the z values in the following diagrams of the standard normal distribution (interpolate, as needed).

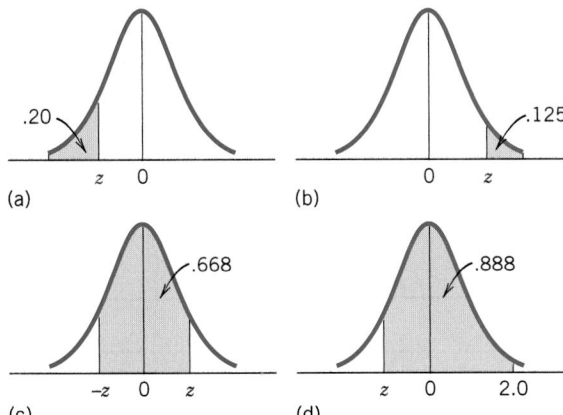

(a)

(b)

(c)

(d)

6.20 For a standard normal random variable Z, find
  (a) $P[Z < .52]$
  (b) $P[Z < -.52]$
  (c) $P[Z > 1.79]$
  (d) $P[Z > -1.79]$
  (e) $P[-1.3 < Z < 2.1]$
  (f) $P[.06 < Z < .8]$

  (g) $P[-1.62 < Z < -.54]$
  (h) $P[|Z| < 1.64]$

6.21 Find the z value in each of the following cases.
  (a) $P[Z < z] = .1762$
  (b) $P[Z > z] = .10$
  (c) $P[-z < Z < z] = .954$
  (d) $P[-.6 < Z < z] = .50$

6.22 Find the quartiles of the standard normal distribution.

6.23 Find
  (a) $P[Z < .33]$.
  (b) The 33rd percentile of the standard normal distribution.
  (c) $P[Z < .70]$.
  (d) The 70th percentile of the standard normal distribution.

6.24 Find
  (a) $P[Z < .46]$.
  (b) The 46th percentile of the standard normal distribution.
  (c) $P[Z < .85]$.
  (d) The 85th percentile of the standard normal distribution.

4. PROBABILITY CALCULATIONS WITH NORMAL DISTRIBUTIONS

Fortunately, no new tables are required for probability calculations regarding the general normal distribution. Any normal distribution can be set in correspondence to the standard normal by the following relation.

> If X is distributed as $N(\mu, \sigma)$, then the standardized variable
> $$Z = \frac{X - \mu}{\sigma}$$
> has the standard normal distribution.

This property of the normal distribution allows us to cast a probability problem concerning X into one concerning Z. To find the probability that X lies in a given interval, convert the interval to the z scale and then calculate the probability by using the standard normal table (Appendix B, Table 3).

Example 6   Converting a Normal Probability to a Standard Normal Probability

Given that X has the normal distribution $N(60, 4)$, find $P[55 \leq X \leq 63]$.

SOLUTION   Here, the standardized variable is $Z = \dfrac{X - 60}{4}$. The distribution of X is shown in Figure 15, where the distribution of Z and the z scale are also displayed below the x scale. In particular,

$$x = 55 \text{ gives } z = \frac{55 - 60}{4} = -1.25$$

$$x = 63 \text{ gives } z = \frac{63 - 60}{4} = .75$$

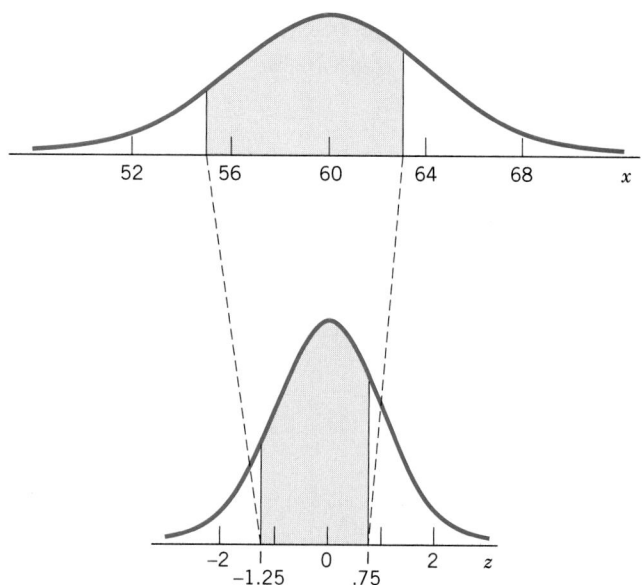

Figure 15   Converting to the z scale.

Therefore,

$$P[55 \leq X \leq 63] = P[-1.25 \leq Z \leq .75]$$

Using the normal table, we find $P[Z \leq .75] = .7734$ and $P[Z \leq -1.25] = .1056$, so the required probability is $.7734 - .1056 = .6678$.

The working steps employed in Example 6 can be formalized into the rule:

If X is distributed as $N(\mu, \sigma)$, then

$$P[a \leq X \leq b] = P\left[\frac{a - \mu}{\sigma} \leq Z \leq \frac{b - \mu}{\sigma}\right]$$

where Z has the standard normal distribution.

Example 7   Probabilities Concerning Calories in a Lunch Salad

The number of calories in a salad on the lunch menu is normally distributed with mean $=$ 200 and sd $=$ 5. Find the probability that the salad you select will contain:

   (a) More than 208 calories.
   (b) Between 190 and 200 calories.

SOLUTION   Letting X denote the number of calories in the salad, we have the standardized variable

$$Z = \frac{X - 200}{5}$$

   (a) The z value corresponding to $x = 208$ is

$$z = \frac{208 - 200}{5} = 1.6$$

Therefore,

$$P[X > 208] = P[Z > 1.6]$$
$$= 1 - P[Z \leq 1.6]$$
$$= 1 - .9452 = .0548$$

   (b) The z values corresponding to $x = 190$ and $x = 200$ are

$$\frac{190 - 200}{5} = -2.0 \quad \text{and} \quad \frac{200 - 200}{5} = 0$$

respectively. We calculate

$$P[190 \leq X \leq 200] = P[-2.0 \leq Z \leq 0]$$
$$= .5 - .0228 = .4772$$

Example 8   Determining a Percentile of a Normal Population

The raw scores in a national aptitude test are normally distributed with mean $=$ 506 and sd $=$ 81.

   (a) What proportion of the candidates scored below 574?
   (b) Find the 70th percentile of the scores.

SOLUTION   If we denote the raw score by X, the standardized score

$$Z = \frac{X - 506}{81}$$

is distributed as $N(0, 1)$.

(a) The z score corresponding to 574 is

$$z = \frac{574 - 506}{81} = .8395$$

So

$$P[X < 574] = P[Z < .8395] = .799$$

Thus, 79.9%, or about 80%, of the candidates scored below 574. In other words, the score 574 nearly locates the 80th percentile.

(b) We first find the 70th percentile in the z scale and then convert it to the x scale. From the standard normal table, we find

$$P[Z \le .524] = .70$$

The standardized score $z = .524$ corresponds to

$$x = 506 + 81(.524)$$
$$= 548.44$$

Therefore, the 70th percentile score is about 548.4.

Exercises

6.25 If X is normally distributed with $\mu = 60$ and $\sigma = 4$, find
(a) $P[X < 55]$
(b) $P[X \le 67]$
(c) $P[X > 66]$
(d) $P[X > 51]$
(e) $P[53 \le X \le 69]$
(f) $P[61 < X < 64]$

6.26 If X is normally distributed with a mean of 20 and a standard deviation of 5, find
(a) $P[X < 18]$
(b) $P[X \le 39]$
(c) $P[X > 37]$
(d) $P[X > 11]$
(e) $P[22 \le X \le 31]$
(f) $P[22 < X < 31]$

6.27 If X has a normal distribution with $\mu = 130$ and $\sigma = 5$, find b such that
(a) $P[X < b] = .975$
(b) $P[X > b] = .025$
(c) $P[X < b] = .305$

6.28 If X is normally distributed with a mean of 24 and a standard deviation of 3, find b such that
(a) $P[X < b] = .7995$
(b) $P[X > b] = .001$
(c) $P[X < b] = .063$

6.29 Scores on a certain nationwide college entrance examination follow a normal distribution with a mean of 500 and a standard deviation of 100. Find the probability that a student will score:
(a) Over 650.
(b) Less than 250.
(c) Between 325 and 675.

6.30 Refer to Exercise 6.29.
(a) If a school only admits students who score over 680, what proportion of the student pool would be eligible for admission?
(b) What limit would you set that makes 50% of the students eligible?
(c) What should be the limit if only the top 15% are to be eligible?

6.31 According to the children's growth chart that doctors use as a reference, the heights of two-year-old boys are nearly normally distributed with a mean of 34.5 inches and a standard deviation of 1.3 inches. If a two-year-old boy is selected at random, what is the probability that he will be between 32.5 and 36.5 inches tall?

6.32 The time it takes a symphony orchestra to play Beethoven's *Ninth Symphony* has a normal distribution with a mean of 64.3 minutes and a standard deviation of 1.15 minutes. The next time it is played, what is the probability that it will take between 62.5 and 67.7 minutes?

6.33 The weights of apples served at a restaurant are normally distributed with a mean of 5 ounces and standard deviation of 1.2 ounces. What is the probability that the next person served will be given an apple that weighs less than 4 ounces?

6.34 The diameter of hail hitting the ground during a storm is normally distributed with a mean of .5 inch and a standard deviation of .1 inch. What is the probability that:

(a) A hailstone picked up at random will have a diameter greater than .71 inch?

(b) Two hailstones picked up in a row will have diameters greater than .6 inch? (Assume independence of the two diameters.)

(c) By the end of the storm, what proportion of the hailstones would have had diameters greater than .71 inch?

6.35 According to current U.S. Census Bureau data, the heights of 20- to 29-year-old women can be well approximated by a normal distribution with mean 64.1 inches and standard deviation 3.4 inches.

(a) What is the probability that the height of a randomly selected woman 20 to 29 years old exceeds 70 inches?

(b) What is the probability that the height of a randomly selected woman 20 to 29 years old is less than or equal to 60 inches?

6.36 Suppose the weights of packages of lettuce coming off a packaging line have a normal distribution with mean 8.1 ounces and standard deviation .1 ounce.

(a) If every package is labeled 8 ounces, what proportion of the packages weigh less than the labeled amount.

(b) If only 2.5% of the packages exceed weight w, what is the value of w?

6.37 The time for an emergency medical squad to arrive at the sports center at the edge of town is distributed as a normal variable with $\mu = 17$ minutes and $\sigma = 3$ minutes.

(a) Determine the probability that the time to arrive is:

(i) More than 22 minutes.

(ii) Between 13 and 21 minutes.

(iii) Between 15.5 and 18.5 minutes.

(b) Which arrival period of duration 1 minute is assigned the highest probability by the normal distribution?

6.38 The force required to puncture a cardboard mailing tube with a sharp object is normally distributed with mean 32 pounds and standard deviation 4 pounds. What is the probability that a tube will puncture if it is struck by

(a) A 25-pound blow with the object?

(b) A 35-pound blow with the object?

5. THE NORMAL APPROXIMATION TO THE BINOMIAL

The binomial distribution, introduced in Chapter 5, pertains to the number of successes X in n independent trials of an experiment. When the success probability p is not too near 0 or 1 and the number of trials is large, the normal distribution serves as a good approximation to the binomial probabilities. Bypassing the mathematical proof, we concentrate on illustrating the manner in which this approximation works.

Figure 16 presents the binomial distribution for the number of trials n being 5, 12, and 25 when $p = .4$. Notice how the distribution begins to assume the dis-

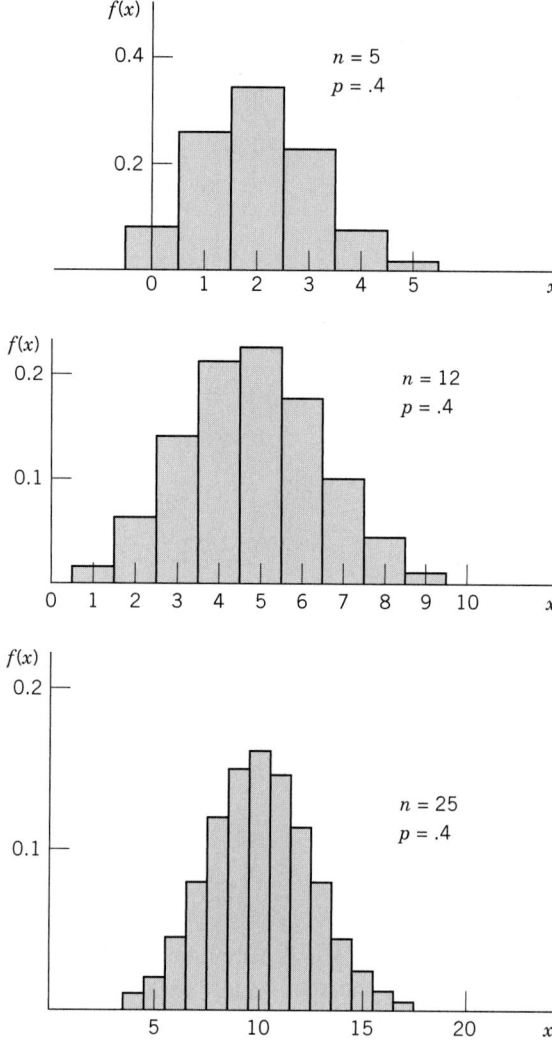

Figure 16   The binomial distributions for $p = .4$
and $n = 5, 12, 25$.

tinctive bell shape for increasing n. Even though the binomial distributions with
$p = .4$ are not symmetric, the lack of symmetry becomes negligible for large n.
But how do we approximate the binomial probability

$$P[X = x] = \binom{n}{x} p^x (1 - p)^{n-x}$$

by a normal probability? The normal probability assigned to a single value x is
zero. However, as shown in Figure 17, the probability assigned to the interval
$x - \frac{1}{2}$ to $x + \frac{1}{2}$ is the appropriate comparison. The addition and subtraction of
$\frac{1}{2}$ is called the **continuity correction**.

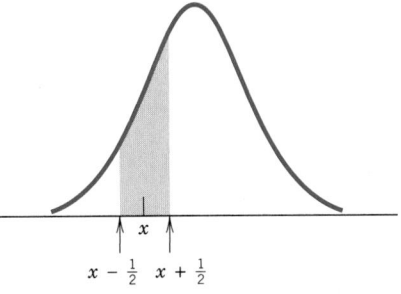

Figure 17   Idea of continuity correction.

For $n = 15$ and $p = .4$, the binomial distribution assigns

$$P[X = 7] = .787 - .610 = .177$$

Recall from Chapter 5 that the binomial distribution has

$$\text{Mean} = np = 15(.4) = 6$$
$$\text{sd} = \sqrt{np(1 - p)} = \sqrt{15(.4)(.6)} = 1.897$$

To obtain an approximation, we select the normal distribution with the same mean, $\mu = 6$, and same $\sigma = 1.897$. The normal approximation is then the probability assigned to the interval $7 - \frac{1}{2}$ to $7 + \frac{1}{2}$.

$$P[6.5 < X < 7.5] = P\left[\frac{6.5 - 6}{1.897} < \frac{X - 6}{1.897} < \frac{7.5 - 6}{1.897}\right]$$
$$\approx P[.264 < Z < .791] = .7855 - .6041 = .1814$$

Of course $n = 15$ is small, so the approximation .1814 differs somewhat from the exact value .177. However, the accuracy of the approximation increases with the number of trials n.

The normal approximation to the binomial applies when n is large and the success probability p is not too close to 0 or 1. The binomial probability of $[a \leq X \leq b]$ is approximated by the normal probability of $[a - \frac{1}{2} \leq X \leq b + \frac{1}{2}]$.

The Normal Approximation to the Binomial

When np and $n(1 - p)$ are both large,[3] say, greater than 15, the binomial distribution is well approximated by the normal distribution having mean $= np$ and sd $= \sqrt{np(1 - p)}$. That is,

$$Z = \frac{X - np}{\sqrt{np(1 - p)}} \qquad \text{is approximately } N(0, 1)$$

---

[3]To be consistent with the rule in Chapter 13 some suggest using the normal approximation when np is greater than 5. The exact calculation of binomial probabilities using statistical software is always preferable.

Example 9   Normal Approximation to the Binomial

Let X have a binomial distribution with $p = .4$ and $n = 150$. Approximate the probability that:

      (a) X is between 52 and 71 both inclusive.

      (b) X is greater than 67.

SOLUTION

      (a)   We calculate the mean and standard deviation of X.

$$\text{Mean} = np = 150(.4) = 60$$

$$\text{sd} = \sqrt{np(1 - p)} = \sqrt{150(.4)(.6)} = \sqrt{36} = 6$$

The standardized variable is

$$Z = \frac{X - 60}{6}$$

The event $[52 \leq X \leq 71]$ includes both endpoints. The appropriate continuity correction is to *subtract* $\frac{1}{2}$ from the lower end and *add* $\frac{1}{2}$ to the upper end. We then approximate

$$P[51.5 \leq X \leq 71.5] = P\left[\frac{51.5 - 60}{6} \leq \frac{X - 60}{6} \leq \frac{71.5 - 60}{6}\right]$$

$$\approx P[-1.417 \leq Z \leq 1.917]$$

From the normal table, we interpolate

$$P[-1.417 \leq Z \leq 1.917] = .9724 - .0782 = .8942$$

and approximate $P[52 \leq X \leq 71]$ by the normal probability .8942.

      (b)   For $[X > 67]$, we reason that 67 is not included so that $[X \geq 67 + .5]$ or $[X \geq 67.5]$ is the event of interest:

$$P[X \geq 67.5] = P\left[\frac{X - 60}{6} \geq \frac{67.5 - 60}{6}\right]$$

$$\approx P[Z \geq 1.25] = 1 - .8944$$

The normal approximation to the binomial gives $P[X > 67] \approx .1056$.

Example 10   A Normal Probability Approximation for a Survey

A recent study reported that 54% of the adults in the United States drink at least one cup of coffee a day. If this is still the current rate, what is the probability that in a random sample of 1000 adults the number that drink at least one cup of coffee a day will be (a) less than 519 and (b) 556 or more?

SOLUTION   Let X be the number of adults in the sample of 1000 adults who drink at least one cup of coffee a day. Under the assumption that the proportion remains at .54, the distribution of X is well modeled by the binomial distribution with $n = 1000$ and $p = .54$. Since n is large and

$$np = 540, \qquad \sqrt{np(1 - p)} = \sqrt{284.4} = 15.76$$

the binomial distribution of X is approximately $N(540, 15.76)$.

(a) Because X is a count, the event $[X < 519]$ is the same as $[X \leq 518]$. Using the continuity correction, we have

$$P[X \leq 519] \approx P\left[Z \leq \frac{518.5 - 540}{15.76}\right]$$
$$= P[Z \leq -1.364]$$
$$= .0863$$

(b)

$$P[X \geq 556] \approx P\left[Z \geq \frac{555.5 - 540}{15.76}\right]$$
$$= P[Z \geq .9835]$$
$$= 1 - .8373 = .1627$$

Remark: If the object is to calculate binomial probabilities, today the best practice is to evaluate them directly using an established statistical computing package. The numerical details need not concern us. However, the fact that

$$\frac{X - np}{\sqrt{np(1 - p)}} \qquad \text{is approximately normal}$$

when np and $n(1 - p)$ are both large remains important. We will use it in later chapters when discussing inferences about proportions. Because the continuity correction will not be crucial, we will drop it for the sake of simplicity. Beyond this chapter, we will employ the normal approximation but *without* the continuity correction.

Exercises

6.39 Let the number of successes X have a binomial distribution with $n = 25$ and $p = .6$.

(a) Find the exact probabilities of each of the following:

$$X = 17 \qquad 11 \leq X \leq 18 \qquad 11 < X < 18$$

(b) Apply the normal approximation to each situation in part (a).

6.40 Let the number of successes X have a binomial distribution with $n = 25$ and $p = .6$.

(a) Find the exact probability of each of the following:

$$X = 14 \qquad 13 \leq X \leq 19 \qquad 13 < X < 19$$

(b) Apply the normal approximation to each situation in part (a).

6.41 Let the number of successes X have a binomial distribution with $p = .25$ and $n = 300$. Approximate the probability of (a) $X = 80$ (b) $X \leq 65$ and (c) $68 \leq X \leq 89$.

6.42   Let the number of successes X have a binomial distribution with $n = 200$ and $p = .65$. Use the normal distribution to approximate the probability of

(a) $X = 130$

(b) $X \leq 150$

(c) $137 \leq X \leq 152$

6.43   State whether or not the normal approximation to the binomial is appropriate in each of the following situations.

(a) $n = 90, p = .24$

(b) $n = 100, p = .03$

(c) $n = 120, p = .98$

(d) $n = 61, p = .4$

6.44   State whether or not the normal approximation to the binomial is appropriate in each of the following situations.

(a) $n = 500, p = .23$

(b) $n = 10, p = .4$

(c) $n = 300, p = .02$

(d) $n = 150, p = .97$

(e) $n = 100, p = .71$

6.45   Copy Figure 16 and add the standard score scale $z = (x - np)/\sqrt{np(1 - p)}$ underneath the x-axis for $n = 5, 12, 25$. Notice how the distributions center on zero and most of the probability lies between $z = -2$ and $z = 2$.

6.46   The median age of residents of the United States is 35.6 years. If a survey of 200 residents is taken, approximate the probability that at least 110 will be under 35.6 years of age.

6.47   The unemployment rate in a city is 7.9%. A sample of 300 persons is selected from the labor force. Approximate the probability that

(a)   Less than 18 unemployed persons are in the sample

(b)   More than 30 unemployed persons are in the sample

6.48   A survey reports that 96% of the people think that violence has increased in the past five years. Out of a random sample of 50 persons, 48 expressed the opinion that citizens have become more violent in the past five years. Does the normal approximation seem appropriate for $X =$ the number of persons who expressed the opinion that citizens have become more violent in the past five years? Explain.

6.49   Of the customers visiting the stereo section of a large electronics store, only 20% make a purchase. If 80 customers visit the stereo section tomorrow, find the probability that more than 20 will make a purchase.

6.50   The weekly amount spent by a company for travel has approximately a normal distribution with mean \$850 and standard deviation \$40. What is the probability that the actual expenses will exceed \$870 in 20 or more weeks during the next year?

6.51   With reference to Exercise 6.50, calculate the probability that the actual expenses would exceed \$880 for between 10 and 16 weeks, inclusive during the next year.

6.52   In a large midwestern university, 30% of the students live in apartments. If 200 students are randomly selected, find the probability that the number of them living in apartments will be between 55 and 70 inclusive.

6.53   According to a study of mobility, 33% of U.S. residents in the age group 20 to 24 years moved to different housing in 2002 from where they lived in 2001. (Based on *Statistical Abstract of the U.S. 2003*, Table 34.) If the same percentage holds today, give the approximate probability that in a random sample of 100 residents 20 to 24 years old, there will be 39 or more persons who have moved in the past year.

6.54   Suppose that 20% of the trees in a forest are infested with a certain type of parasite.

(a)   What is the probability that, in a random sample of 300 trees, the number of trees having the parasite will be between 49 and 71 inclusive?

*(b)   After sampling 300 trees, suppose that 72 trees are found to have the parasite. Does this provide strong evidence that the population proportion is higher than 20%? Base your answer on $P[X \geq 72]$ when 20% are infested.

*6.   CHECKING THE PLAUSIBILITY OF A NORMAL MODEL

Does a normal distribution serve as a reasonable model for the population that produced the sample? One reason for our interest in this question is that many commonly used statistical procedures require the population to be nearly normal. If a normal distribution is tentatively assumed to be a plausible model, the investigator must still check this assumption once the sample data are obtained.

Although they involve subjective judgment, graphical procedures prove most helpful in detecting serious departures from normality. Histograms can be inspected for lack of symmetry. The thickness of the tails can be checked for conformance with the normal by comparing the proportions of observations in the intervals $(\bar{x} - s, \bar{x} + s)$, $(\bar{x} - 2s, \bar{x} + 2s)$, and $(\bar{x} - 3s, \bar{x} + 3s)$ with those suggested by the empirical guidelines for the bell-shaped (normal) distribution.

A more effective way to check the plausibility of a normal model is to construct a special graph, called a **normal-scores plot**, of the sample data. In order to describe this method, we will first explain the meaning of normal scores, indicate how the plot is constructed, and then explain how to interpret the plot. For an easy explanation of the ideas, we work with a small sample size. In practical applications, at least 15 or 20 observations are needed to detect a meaningful pattern in the plot.

The term **normal scores** refers to an idealized sample from the standard normal distribution—namely, the z values that divide the standard normal distribution into equal-probability intervals. For purposes of discussion, suppose the sample size is $n = 4$. Figure 18 shows the standard normal distribution where four points are located on the z axis so the distribution is divided into five segments of equal probability $\frac{1}{5} = .2$. These four points, denoted by m_1, m_2, m_3, and m_4, are precisely the normal scores for a sample of size $n = 4$. Using Appendix B, Table 3, we find that

$$m_1 = -.84$$
$$m_2 = -.25$$
$$m_3 = .25$$
$$m_4 = .84$$

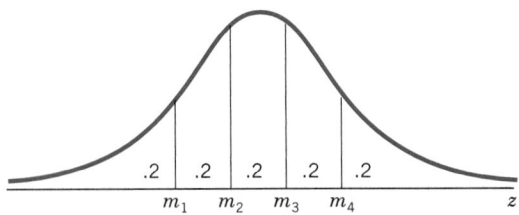

Figure 18   The $N(0, 1)$ distribution and the normal scores for $n = 4$.

A normal-scores plot allows us to visually assess how well a sample mimics the idealized normal sample. To construct a normal-scores plot:

1. Order the sample data from smallest to largest.
2. Obtain the normal scores.
3. Pair the ith largest observation with the ith largest normal score and plot the pairs in a graph.

Example 11   Making a Normal-Scores Plot for Sample Size 4

Suppose a random sample of size 4 has produced the observations 68, 82, 44, and 75. Construct a normal-scores plot.

SOLUTION   The ordered observations and the normal scores are shown in Table 2, and the normal-scores plot of the data is given in Figure 19.

TABLE 2   Normal Scores

Normal Scores	Ordered Sample
$m_1 = -.84$	44
$m_2 = -.25$	68
$m_3 = .25$	75
$m_4 = .84$	82

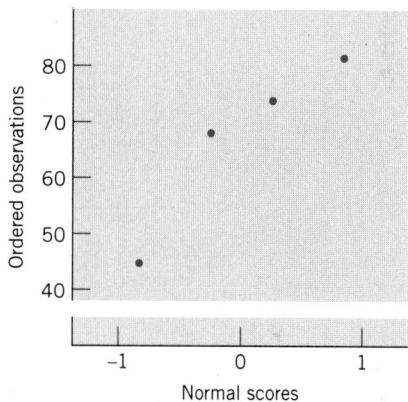

Figure 19   Normal-scores plot of Table 2 data.

INTERPRETATION OF THE PLOT

How does the normal-scores plot of a data set help in checking normality? To explain the main idea, we continue our discussion with the data of Example 11. Let μ and σ denote the mean and standard deviation of the population from which the sample was obtained. The normal scores that are the idealized z observations can then be converted to the x scale by the usual relation $x = \mu + \sigma z$. The actual x observations and the corresponding idealized observations are given in Table 3. If the population were indeed normal, we would expect the two columns of Table 3 to be close. In other words, a plot of the observed x values versus the normal scores would produce a straight-line pattern, where the intercept of the line would indicate the value of μ and the slope of the line would indicate σ.

TABLE 3   Idealized Sample

Observed x Values	Idealized x Values
44	$\mu + \sigma\, m_1$
68	$\mu + \sigma\, m_2$
75	$\mu + \sigma\, m_3$
82	$\mu + \sigma\, m_4$

> A straight line pattern in a normal-scores plot supports the plausibility of a normal model. A curve appearance indicates a departure from normality.

The normal-scores plot of a data set is easily obtained using the computer. The use of the MINITAB package is illustrated here with the data set of Exercise 2.126, Chapter 2, concerning the strength measurements of southern pine. (See Exercise 6.80 for the commands.)

MINITAB actually uses one of the many slight variants of the normal scores described here, but the plots are very similar if more than 20 observations are plotted. Notice that the plot in Figure 20 conforms quite well to the straight-line pattern expected for normal observations. The largest five observations, however, give some evidence of being too large with respect to the others.

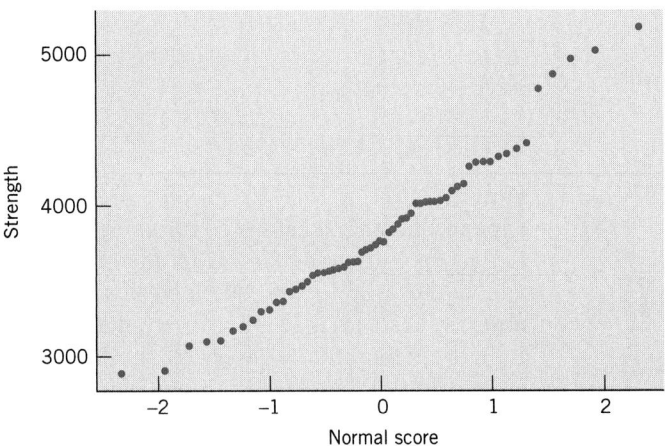

Figure 20   A normal-scores plot of the lumber strength data from Exercise 2.126. Nearly a straight line.

*7.   TRANSFORMING OBSERVATIONS TO ATTAIN NEAR NORMALITY

A valid application of many powerful techniques of statistical inference, especially those suited to small or moderate samples, requires that the population distribution be reasonably close to normal. When the sample measurements appear to have been taken from a population that departs drastically from normality, an appropriate conversion to a new variable may bring the distribution close to normal. Efficient techniques can then be safely applied to the converted data, whereas their application to the original data would have been questionable. Inferential methods requiring the assumption of normality are discussed in later chapters. The goal of our discussion here is to show how a transformation can improve the approximation to a normal distribution.

There is no rule for determining the best transformation in a given situation. For any data set that does not have a symmetric histogram, we consider a variety of transformations.

Some Useful Transformations

Make large values larger: Make large values smaller:

$$x^3, \quad x^2 \qquad\qquad \sqrt{x}, \quad \sqrt[4]{x}, \quad \log_e x, \quad \frac{1}{x}$$

You may recall that $\log_e x$ is the natural logarithm. Fortunately, computers easily calculate and order the transformed values, so that several transformations in a list can be quickly tested. Note, however, that the observations must be positive if we intend to use $\sqrt{x}$, $\sqrt[4]{x}$, and $\log_e x$.

The selection of a good transformation is largely a matter of trial and error. If the data set contains a few numbers that appear to be detached far to the right, $\sqrt{x}$, $\sqrt[4]{x}$, and $\log_e x$, or negative powers that would pull these stragglers closer to the other data points should be considered.

Example 12   A Transformation to Improve Normality

A forester records the volume of timber, measured in cords, for 49 plots selected in a large forest. The data are given in Table 4 and the corresponding histogram appears in Figure 21a. The histogram exhibits a long tail to the right, so it is reasonable to consider the transformations $\sqrt{x}$, $\sqrt[4]{x}$, $\log_e x$, and $1/x$. Transform the data to near normality.

SOLUTION   The most satisfactory result, obtained with

$$\text{Transformed data} \;=\; \sqrt[4]{\text{Volume}}$$

is illustrated in Table 5 and Figure 21b. The latter histogram more nearly resembles a symmetric bell-shaped pattern expected for normal populations.

TABLE 4   Volume of Timber in Cords

39.3	14.8	6.3	.9	6.5
3.5	8.3	10.0	1.3	7.1
6.0	17.1	16.8	.7	7.9
2.7	26.2	24.3	17.7	3.2
7.4	6.6	5.2	8.3	5.9
3.5	8.3	44.8	8.3	13.4
19.4	19.0	14.1	1.9	12.0
19.7	10.3	3.4	16.7	4.3
1.0	7.6	28.3	26.2	31.7
8.7	18.9	3.4	10.0	

Courtesy of Professor Alan Ek.

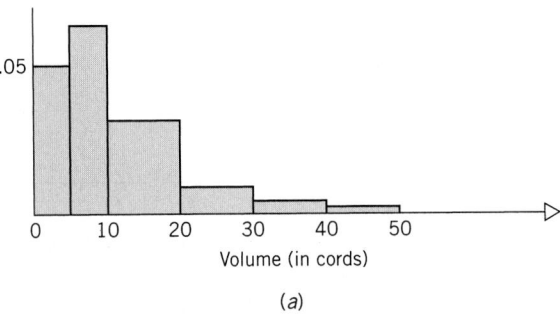

(a)

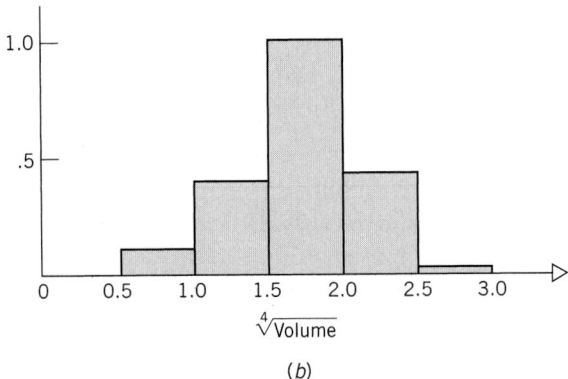

(b)

Figure 21   An illustration of the transformation technique. (a) Histogram of timber volume. (b) Histogram of $\sqrt[4]{\text{Volume}}$.

TABLE 5 The Transformed Data $\sqrt[4]{\text{Volume}}$

2.50	1.96	1.58	.97	1.60
1.37	1.70	1.78	1.07	1.63
1.57	2.03	2.02	.91	1.68
1.28	2.26	2.22	2.05	1.34
1.64	1.60	1.51	1.70	1.56
1.37	1.70	2.59	1.70	1.91
2.10	2.09	1.94	1.17	1.86
2.11	1.79	1.36	2.02	1.44
1.00	1.66	2.31	2.26	2.37
1.72	2.09	1.36	1.78	

HOW MUCH TIMBER IS IN THIS FOREST?
The volume of timber available for making lumber can only be estimated by sampling the number of trees in randomly selected plots within the forest. The distribution of tree size must also be taken into account. © Planet Earth Pictures/FPG International/Getty Images

USING STATISTICS WISELY

1. A sketch of the bell-shaped normal curve and the area of interest can prevent blunders when determining probabilities and percentiles.

2. Never apply the normal approximation to the binomial, treating

$$Z = \frac{X - np}{\sqrt{np(1 - p)}}$$

as standard normal, when the expected number of successes (or failures) is too small. That is, when either

$$np \quad \text{or} \quad n(1 - p) \quad \text{is 15 or less}$$

3. Do not just assume that data come from a normal distribution. When there are at least 20 to 25 observations, it is good practice to construct a normal-scores plot to check this assumption.

KEY IDEAS AND FORMULAS

The probability distribution for a continuous random variable X is specified by a probability density curve. The function that specifies this curve is called a **probability density function**. It can be symmetric about the mean of X or it can be skewed, meaning that it has a long tail to either the left or the right.

The probability that X lies in an interval from a to b is determined by the area under the probability density curve between a and b. The total area under the curve is 1, and the curve is never negative.

The population **100 p-th percentile** is an x value that has probability p to its left and probability $1 - p$ to its right.

When X has mean μ and standard deviation σ, the **standardized variable**

$$Z = \frac{X - \mu}{\sigma}$$

has mean 0 and standard deviation 1.

The **normal distribution** has a symmetric bell-shaped curve centered at the mean. The intervals extending one, two, and three standard deviations around the mean contain the probabilities .683, .954, and .997, respectively.

If X is normally distributed with mean μ and standard deviation σ, then

$$Z = \frac{X - \mu}{\sigma}$$

has the **standard normal distribution**.

When the number of trials n is large and the success probability p is not too near 0 or 1, the binomial distribution is well approximated by a normal distribution with mean np and sd $= \sqrt{np(1 - p)}$. Specifically, the probabilities for a binomial variable X can be approximately calculated by treating

$$Z = \frac{X - np}{\sqrt{np(1 - p)}}$$

as standard normal. For a moderate number of trials n, the approximation is improved by appropriately adjusting by $\frac{1}{2}$ called a **continuity correction**.

The **normal scores** are an ideal sample from a standard normal distribution. Plotting each ordered observation versus the corresponding normal score creates a **normal-scores plot**, which provides a visual check for possible departures from a normal distribution.

Transformation of the measurement scale often helps to convert a long-tailed distribution to one that resembles a normal distribution.

TECHNOLOGY

Probability and Percentiles for the Standard Normal and General Normal Distribution

MINITAB

MINITAB uses the same steps for calculations with the standard normal and cases of other means and standard deviations. We illustrate with the calculation of $P[X \le 8]$ when X is normal with mean 5 and standard deviation 12.5.

> **Calc > Probability Distributions > Normal.**
> Select **Cumulative Probability.** Type *5* in **Mean.**
> Type *12.5* in **Standard deviation.**
> Select **Input Constant** and type 8. Click **OK.**

The default settings Mean 0 and Standard deviation 1 simplify the steps for obtaining standard normal probabilities.

The inverse problem of finding b so that $P[X \le b] = a$, where a is a specified probability, is illustrated with finding b so that $P[X \le b] = .9700$ when X is normal with mean 5 and standard deviation 12.5.

> **Calc > Probability Distributions > Normal.**
> Select **Inverse Probability.** Type *5* in **Mean.**
> Type *12.5* in **Standard deviation.**
> Select **Input Constant** and type *.9700.* Click **OK.**

EXCEL

EXCEL uses the function *NORMSDIST* for standard normal probabilities and *NORMDIST* for a general normal distribution. We illustrate with the calculation of $P[X \leq 8]$ when X is normal with mean 5 and standard deviation 12.5.

Select the f_x icon, or select **Insert** and then **Function.**
Choose **Statistical** and then **NORMDIST.** Click **OK.**
Type *8* in **X,** *5* in **Mean,** *12.5* in **Standard_dev** and *False* in **Cumulative.**
Click **OK.**

The inverse problem of finding b so that $P[X \leq b] = a$, where a is a specified probability, is illustrated with finding b so that $P[X \leq b] = .9700$ when X is normal with mean 5 and standard deviation 12.5.

Select the f_x icon, or select **Insert** and then **Function.**
Choose **Statistical,** and then **NORMINV.** Click **OK.**
Type *.9700* in **Probability,** *5* in **Mean,** *12.5* in **Standard_dev.** Click **OK.**

To solve the standard normal inverse problem replace **NORMINV** by **NORM-SINV.**

TI-84/-83 PLUS

We illustrate with the calculation of $P[X \leq 8]$ when X is normal with mean 5 and standard deviation 12.5.

In the Home screen, press **2nd VARS**

From the *DISTR* menu, select **2: Normalcdf(.**

Type entries to obtain **Normalcdf(—1E99, 8, 5, 12.5),**
Then press **ENTER.**

The inverse problem of finding b so that $P[X \leq b] = a$, where a is a specified probability, is illustrated with finding b so that $P[X \leq b] = .9700$ when X is normal with mean 5 and standard deviation 12.5.

In the Home screen, press **2nd VARS.**
From the *DISTR* mean, select **3: InvNorm(.**
Type entries to obtain **InvNorm(.9700, 5, 12.5).**
Then press **ENTER.**

8.   REVIEW EXERCISES

6.55   Determine (a) the median and (b) the quartiles for the distribution shown in the following illustration.

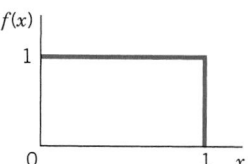

6.56   For X having the density in Exercise 6.55, find (a) $P[X > .8]$ (b) $P[.5 \le X \le .8]$ and (c) $P[.5 < X < .8]$.

6.57   In the context of the height of red pine seedlings presented at the front of the chapter, describe the reasoning that leads from a histogram to the concept of a probability density curve. (Think of successive histograms based on 100 heights, 500 heights, 1456 heights, and then an unlimited number.)

6.58   For a standard normal random variable Z, find
(a) $P[Z < 1.36]$
(b) $P[Z > 1.225]$
(c) $P[.63 < Z < 1.98]$
(d) $P[-1.37 < Z < 1.055]$

6.59   For the standard normal distribution, find the value z such that
(a)   Area to its left is .0838
(b)   Area to its left is .047
(c)   Area to its right is .2611
(d)   Area to its right is .12

6.60   Find the 20th, 40th, 60th, and 80th percentiles of the standard normal distribution.

6.61   If Z is a standard normal random variable, what is the probability that
(a)   Z exceeds .62?
(b)   Z lies in the interval $(-1.40, 1.40)$?
(c) $|Z|$ exceeds 3.0?
(d) $|Z|$ is less than 2.0?

6.62   The distribution of raw scores in a college qualification test has mean 580 and standard deviation 80.
(a)   If a student's raw score is 696, what is the corresponding standardized score?
(b)   If the standardized score is $-.8$, what is the raw score?
(c)   Find the interval of standardized scores corresponding to the raw scores of 380 to 560.
(d)   Find the interval of the raw scores corresponding to the standardized scores of -1.4 to 1.4.

6.63   If X is normally distributed with $\mu = 100$ and $\sigma = 8$, find
(a) $P[X < 107]$
(b) $P[X < 97]$
(c) $P[X > 110]$
(d) $P[X > 90]$
(e) $P[95 < X < 106]$
(f) $P[103 < X < 114]$
(g) $P[88 < X < 100]$
(h) $P[60 < X < 108]$

6.64   If X has a normal distribution with $\mu = 100$ and $\sigma = 5$, find b such that
(a) $P[X < b] = .6700$
(b) $P[X > b] = .0110$
(c) $P[|X - 100| < b] = .966$

6.65   Suppose that a student's verbal score X from next year's Graduate Record Exam can be considered an observation from a normal population having mean 499 and standard deviation 120. Find
(a) $P[X > 600]$
(b)   90th percentile of the distribution
(c)   Probability that the student scores below 400

6.66   The lifting capacities of a class of industrial workers are normally distributed with mean 65 pounds and standard deviation 8 pounds. What proportion of these workers can lift an 80-pound load?

6.67 The bonding strength of a drop of plastic glue is normally distributed with mean 100 pounds and standard deviation 8 pounds. A broken plastic strip is repaired with a drop of this glue and then subjected to a test load of 98 pounds. What is the probability that the bonding will fail?

6.68 *Grading on a curve.* The scores on an examination are normally distributed with mean μ = 70 and standard deviation σ = 8. Suppose that the instructor decides to assign letter grades according to the following scheme (left endpoint included).

Scores	Grade
Less than 58	F
58 to 66	D
66 to 74	C
74 to 82	B
82 and above	A

Find the percentage of students in each grade category.

6.69 Suppose the duration of trouble-free operation of a new vacuum cleaner is normally distributed with mean 530 days and standard deviation 100 days.

(a) What is the probability that the vacuum cleaner will work for at least two years without trouble?

(b) The company wishes to set the warranty period so that no more than 10% of the vacuum cleaners would need repair services while under warranty. How long a warranty period must be set?

6.70 Suppose the amount of a popular sport drink in bottles leaving the filling machine has a normal distribution with mean 101.5 milliliters (ml) and standard deviation 1.6 ml.

(a) If the bottles are labeled 100 ml, what proportion of the bottles contain less than the labeled amount.

(b) If only 5% of the bottles have contents that exceed a specified amount v, what is the value of v?

6.71 Suppose the amount of sun block lotion in plastic bottles leaving a filling machine has a normal distribution. The bottles are labeled 300 milliliters (ml) but the actual mean is 302 ml and the standard deviation is 2 ml.

(a) What is the probability that an individual bottle will contain less than 299 ml?

(b) If only 5% of the bottles have contents that exceed a specified amount v, what is the value of v?

*6.72 *A property of the normal distribution.* Suppose the random variable X is normally distributed with mean μ and standard deviation σ. If Y is a linear function of X—that is, $Y = a + bX$, where a and b are constants—then Y is also normally distributed with

$$\text{Mean} = a + b\mu$$
$$\text{sd} = |b|\sigma$$

For instance, if X is distributed as $N(25, 2)$ and $Y = 7 - 3X$, then the distribution of Y is normal with Mean $= 7 - 3(25) = -68$ and sd $= |-3| \times 2 = 6$.

(a) At the "low" setting of a water heater, the temperature X of water is normally distributed with Mean $= 102°F$ and sd $= 4°F$. If Y refers to the temperature measurement in the centigrade scale, that is, $Y = \frac{5}{9}(X - 32)$, what is the distribution of Y?

(b) Referring to part (a), find the probability of $[35 \le Y \le 42]$.

Remark: The relation between a general normal and the standard normal is only a special case of this property. Specifically, the standardized variable Z is the linear function.

$$Z = \frac{X - \mu}{\sigma} = -\frac{\mu}{\sigma} + \frac{1}{\sigma}X$$

where Z has

$$\text{Mean} = -\frac{\mu}{\sigma} + \frac{1}{\sigma}\mu = 0$$

$$\text{sd} = \frac{1}{\sigma}\sigma = 1$$

6.73 Let X denote the number of successes in n Bernoulli trials with a success probability of p.

(a) Find the exact probabilities of each of the following:

(i) $X \leq 6$ when $n = 25, p = .4$

(ii) $10 \leq X \leq 18$ when $n = 20, p = .7$

(iii) $X \geq 12$ when $n = 16, p = .5$

(b) Use a normal approximation for each situation in part (a).

6.74 It is known from past experience that 7% of the tax bills are paid late. If 20,000 tax bills are sent out, find the probability that:

(a) Less than 1350 are paid late.

(b) 1480 or more are paid late.

6.75 A particular program, say, program A, previously drew 30% of the television audience. To determine if a recent rescheduling of the programs on a competing channel has adversely affected the audience of program A, a random sample of 400 viewers is to be asked whether or not they currently watch this program.

(a) If the percentage of viewers watching program A has not changed, what is the probability that fewer than 105 out of a sample of 400 will be found to watch the program?

(b) If the number of viewers of the program is actually found to be less than 105, will this strongly support the suspicion that the population percentage has dropped? Use your calculation from part (a).

6.76 The number of successes X has a binomial distribution. State whether or not the normal approximation is appropriate in each of the following situations: (a) $n = 400, p = .23$ (b) $n = 20, p = .03$ (c) $n = 90, p = .98$.

6.77 Because 10% of the reservation holders are "no-shows," a U.S. airline sells 400 tickets for a flight that can accommodate 370 passengers.

(a) Find the probability that one or more reservation holders will not be accommodated on the flight.

(b) Find the probability of fewer than 350 passengers on the flight.

6.78 On a Saturday afternoon, 147 customers will be observed during check-out and the number paying by card, credit or debit, will be recorded. Records from the store suggest that 43% of customers pay by card. Approximate the probability that:

(a) More than 60 will pay by card.

(b) Between 60 and 70 will pay by card.

6.79 In all of William Shakespeare's works, he used $884,647^4$ different words. Of these, 14,376 appeared only once. In 1985 a 429-word poem was discovered that may have been written by Shakespeare. To keep the probability calculations simple, assume that the choices between a new word and one from the list of 884,647 are independent for each of the 429 words. Approximate the probability that a new word will not be on the list, by the relative frequency of words used once.

(a) Find the expected number of new words in the poem.

(b) Use the normal approximation to the binomial to determine the probability of finding 12 or more new words in the poem. Use the continuity correction.

(c) Use the normal approximation to the binomial to determine the probability of finding 2 or fewer new words in the poem. Use the continuity correction.

(d) Use the normal approximation to the binomial to determine the probability of finding more than 2 but less than 12 new words in the poem. On the basis of your answer, decide if 9 = actual number of new words not in the list is consistent with Shakespeare having written the poem or if it contradicts this claim.

The Following Exercises Require a Computer

6.80 *Normal-scores plot.* Use a computer program to make a normal-scores plot for the volume of timber data in Table 4. Comment on the departure from normality displayed by the normal-scores plot.

[4]See R. Thisted and B. Efron (1987). "Did Shakespeare write a newly-discovered poem?" *Biometrika*, **74**, 445–455.

We illustrate a normal-scores plot using MINITAB. With the data set in column 1, the MINITAB commands

---

Calc > Calculator.

Type C2 in **Store.** Type *NSCOR(C1)* in **Expression.** Click **OK.**

Graph > Scatterplot. Select *Simple.* Type *C1* under **Y** variables and *C2* under **X** variables. Click **OK.**

---

will create a normal-scores plot for the observations in *C1.* (MINITAB uses a variant of the normal scores, m_i, that we defined.

*6.81 Use MINITAB or another package program to make a normal-scores plot of the malt extract data in Table D.8 of the Data Bank.

*6.82 Use MINITAB or another package program to make a normal-scores plot of the computer anxiety scores in Table D.4 of the Data Bank.

*6.83 ***Transformations and normal-scores plots.*** The MINITAB computer language makes it possible to easily transform data. With the data already set in column 1, the commands

---

Dialog box:

Calc > Calculator. Type C2 in **Store.** Type *LOGE(C1)* in **Expression.** Click **OK.**

Calc > Calculator. Type C3 in **Store.** Type *SQRT(C1)* in **Expression.** Click **OK.**

Calc > Calculator. Type C4 in **Store.** Type *SQRT(C2)* in **Expression.** Click **OK.**

---

will place the natural logarithm ln X in C2, $\sqrt{x}$ in C3, and $x^{1/4}$ in C4. Normal-scores plots can then be constructed as in Exercise 6.80.

Take the square root of the timber data in Table 4 and then create a normal-scores plot.

7

Variation in Repeated Samples— Sampling Distributions

Bowling Averages

A bowler records individual game scores and the average for a three-game series.

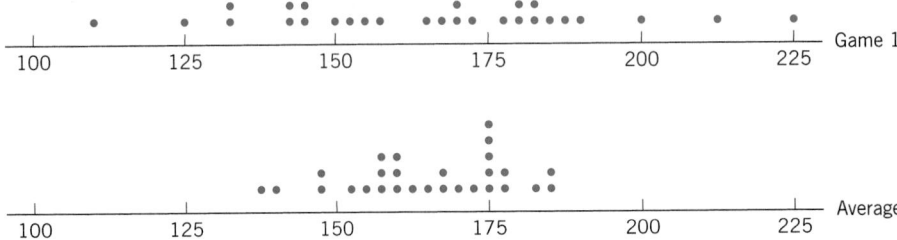

Bowlers are well aware that their three-game averages are less variable than their single-game scores. Sample means always have less variability than individual observations. © Bob Daemmrich/Stock Boston/Picture Quest.

1. INTRODUCTION

At the heart of statistics lie the ideas of inference. They enable the investigator to argue from the particular observations in a sample to the general case. These generalizations are founded on an understanding of the manner in which variation in the population is transmitted, by sampling, to variation in statistics like the sample mean. This key concept is the subject of this chapter.

Typically, we are interested in learning about some numerical feature of the population, such as the proportion possessing a stated characteristic, the mean and standard deviation of the population, or some other numerical measure of center or variability.

> A numerical feature of a population is called a **parameter**.

The true value of a population parameter is an unknown constant. It can be correctly determined only by a complete study of the population. The concepts of statistical inference come into play whenever this is impossible or not practically feasible.

If we only have access to a sample from the population, our inferences about a parameter must then rest on an appropriate sample-based quantity. Whereas a parameter refers to some numerical characteristic of the population, a sample-based quantity is called a **statistic**.

> A **statistic** is a numerical valued function of the sample observations.

For example, the sample mean

$$\overline{X} = \frac{X_1 + \cdots + X_n}{n}$$

is a statistic because its numerical value can be computed once the sample data, consisting of the values of $X_1, \ldots, X_n$, are available. Likewise, the sample median and the sample standard deviation are also sample-based quantities so each is a statistic.

A sample-based quantity (statistic) must serve as our source of information about the value of a parameter. Three points are crucial:

1. Because a sample is only a part of the population, the numerical value of a statistic cannot be expected to give us the exact value of the parameter.

2. The observed value of a statistic depends on the particular sample that happens to be selected.

3. There will be some variability in the values of a statistic over different occasions of sampling.

A brief example will help illustrate these important points. Suppose an urban planner wishes to study the average commuting distance of workers from their home to their principal place of work. Here the statistical population consists of the commuting distances of all the workers in the city. The mean of this finite but vast and unrecorded set of numbers is called the population mean, which we denote by μ. We want to learn about the parameter μ by collecting data from a sample of workers. Suppose 80 workers are randomly selected and the (sample) mean of their commuting distances is found to be $\bar{x} = 8.3$ miles. Evidently, the population mean μ cannot be claimed to be exactly 8.3 miles. If one were to observe another random sample of 80 workers, would the sample mean again be 8.3 miles? Obviously, we do not expect the two results to be identical. Because the commuting distances do vary in the population of workers, the sample mean would also vary on different occasions of sampling. In practice, we observe only one sample and correspondingly a single value of the sample mean such as $\bar{x} = 8.3$. However, it is the idea of the variability of the $\bar{x}$ values in repeated sampling that contains the clue to determining how precisely we can hope to determine μ from the information on $\bar{x}$.

2. THE SAMPLING DISTRIBUTION OF A STATISTIC

The fact that the value of the sample mean, or any other statistic, will vary as the sampling process is repeated is a key concept. Because any statistic, the sample mean in particular, varies from sample to sample, it is a random variable and has its own probability distribution. The variability of the statistic, in repeated sampling, is described by this probability distribution.

> The probability distribution of a statistic is called its sampling distribution.

The qualifier "sampling" indicates that the distribution is conceived in the context of repeated sampling from a population. We often drop the qualifier and simply say the **distribution of a statistic.**

Although in any given situation, we are limited to one sample and the corresponding single value for a statistic, over repeated samples from a population the statistic varies and has a sampling distribution. The sampling distribution of a statistic is determined from the distribution $f(x)$ that governs the population, and it also depends on the sample size n. Let us see how the distribution of $\overline{X}$ can be determined in a simple situation where the sample size is 2 and the population consists of 3 units.

Example 1   Illustration of a Sampling Distribution

A population consists of three housing units, where the value of X, the number of rooms for rent in each unit, is shown in the illustration.

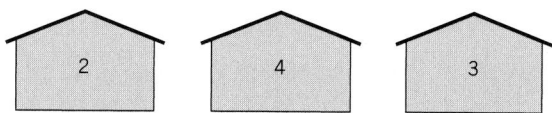

Consider drawing a random sample of size 2 with replacement. That is, we select a unit at random, put it back, and then select another unit at random. Denote by X_1 and X_2 the observation of X obtained in the first and second drawing, respectively. Find the sampling distribution of $\overline{X} = (X_1 + X_2)/2$.

SOLUTION   The population distribution of X is given in Table 1, which simply formalizes the fact that each of the X values 2, 3, and 4 occurs in $\frac{1}{3}$ of the population of the housing units.

TABLE 1   The Population Distribution

x	$f(x)$
2	$\dfrac{1}{3}$
3	$\dfrac{1}{3}$
4	$\dfrac{1}{3}$

Because each unit is equally likely to be selected, the observation X_1 from the first drawing has the same distribution as given in Table 1. Since the sampling is with replacement, the second observation X_2 also has this same distribution.

The possible samples (x_1, x_2) of size 2 and the corresponding values of $\overline{X}$ are

(x_1, x_2)	(2, 2)	(2, 3)	(2, 4)	(3, 2)	(3, 3)	(3, 4)	(4, 2)	(4, 3)	(4, 4)
$\overline{x} = \dfrac{x_1 + x_2}{2}$	2	2.5	3	2.5	3	3.5	3	3.5	4

The nine possible samples are equally likely so, for instance, $P[\overline{X} = 2.5] = \frac{2}{9}$. Continuing in this manner, we obtain the distribution of $\overline{X}$, which is given in Table 2.

This sampling distribution pertains to repeated selection of random samples of size 2 with replacement. It tells us that if the random sampling is repeated a large number of times, then in about $\frac{1}{9}$, or 11%, of the cases, the sample mean would be 2, and in $\frac{2}{9}$, or 22%, of the cases, it would be 2.5, and so on.

TABLE 2   The Probability Distribution of $\overline{X} = (X_1 + X_2)/2$

Value of $\overline{X}$	Probability
2	$\dfrac{1}{9}$
2.5	$\dfrac{2}{9}$
3	$\dfrac{3}{9}$
3.5	$\dfrac{2}{9}$
4	$\dfrac{1}{9}$

Figure 1 shows the probability histograms of the distributions in Tables 1 and 2.

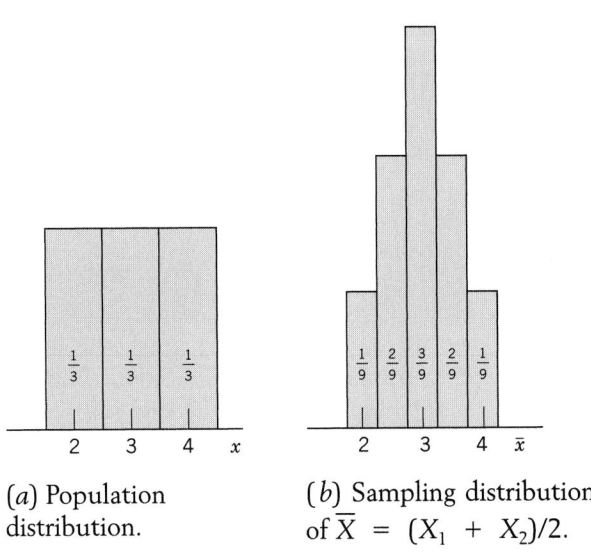

(a) Population distribution.

(b) Sampling distribution of $\overline{X} = (X_1 + X_2)/2$.

Figure 1   Idea of a sampling distribution.

In the context of Example 1, suppose instead the population consists of 300 housing units, of which 100 units have 2 rooms, 100 units have 3 rooms, and

100 units have 4 rooms for rent. When we sample two units from this large population, it would make little difference whether or not we replace the unit after the first selection. Each observation would still have the same probability distribution—namely, $P[X = 2] = P[X = 3] = P[X = 4] = \frac{1}{3}$, which characterizes the population.

When the population is very large and the sample size relatively small, it is inconsequential whether or not a unit is replaced before the next unit is selected. Under these conditions, too, we refer to the observations as a random sample. What are the key conditions required for a sample to be random?

> The observations $X_1, X_2, \ldots, X_n$ are a random sample of size n from the population distribution if they result from independent selections and each observation has the same distribution as the population.

More concisely, under the independence and same distribution conditions, we refer to the observations as a random sample.

Because of variation in the population, the random sample will vary and so will $\overline{X}$, the sample median, or any other statistic.

Example 2   The Sample Mean and Median Each Have a Sampling Distribution

A large population is described by the probability distribution

x	$f(x)$
0	.2
3	.3
12	.5

Let X_1, X_2, X_3 be a random sample of size 3 from this distribution.

(a)   List all the possible samples and determine their probabilities.

(b)   Determine the sampling distribution of the sample mean.

(c)   Determine the sampling distribution of the sample median.

SOLUTION   (a)   Because we have a random sample, each of the three observations X_1, X_2, X_3 has the same distribution as the population and they are independent. So, the sample 0, 3, 0 has probability $(.2) \times (.3) \times (.2) = 0.12$. The calculations for all $3 \times 3 \times 3 = 27$ possible samples are given in Table 3.

(b)   The probabilities of all samples giving the same value $\bar{x}$ are added to obtain the sampling distribution on the second page of Table 3.

(c)   The calculations and sampling distribution of the median are also given on the second page of Table 3.

TABLE 3   Sampling Distributions

Population Distribution

x	$f(x)$
0	.2
3	.3
12	.5

Population mean: $E(X) = 0(.2) + 3(.3) + 12(.5) = 6.9 = \mu$

Population variance: $Var(X) = 0^2(.2) + 3^2(.3) + 12^2(.5) - 6.9^2$
$$= 27.09 = \sigma^2$$

	Possible Samples x_1 x_2 x_3	Sample Mean $\bar{x}$	Sample Median m	Probability
1	0 0 0	0	0	(.2)(.2)(.2) = .008
2	0 0 3	1	0	(.2)(.2)(.3) = .012
3	0 0 12	4	0	(.2)(.2)(.5) = .020
4	0 3 0	1	0	(.2)(.3)(.2) = .012
5	0 3 3	2	3	(.2)(.3)(.3) = .018
6	0 3 12	5	3	(.2)(.3)(.5) = .030
7	0 12 0	4	0	(.2)(.5)(.2) = .020
8	0 12 3	5	3	(.2)(.5)(.3) = .030
9	0 12 12	8	12	(.2)(.5)(.5) = .050
10	3 0 0	1	0	(.3)(.2)(.2) = .012
11	3 0 3	2	3	(.3)(.2)(.3) = .018
12	3 0 12	5	3	(.3)(.2)(.5) = .030
13	3 3 0	2	3	(.3)(.3)(.2) = .018
14	3 3 3	3	3	(.3)(.3)(.3) = .027
15	3 3 12	6	3	(.3)(.3)(.5) = .045
16	3 12 0	5	3	(.3)(.5)(.2) = .030
17	3 12 3	6	3	(.3)(.5)(.3) = .045
18	3 12 12	9	12	(.3)(.5)(.5) = .075
19	12 0 0	4	0	(.5)(.2)(.2) = .020
20	12 0 3	5	3	(.5)(.2)(.3) = .030
21	12 0 12	8	12	(.5)(.2)(.5) = .050
22	12 3 0	5	3	(.5)(.3)(.2) = .030
23	12 3 3	6	3	(.5)(.3)(.3) = .045
24	12 3 12	9	12	(.5)(.3)(.5) = .075
25	12 12 0	8	12	(.5)(.5)(.2) = .050
26	12 12 3	9	12	(.5)(.5)(.3) = .075
27	12 12 12	12	12	(.5)(.5)(.5) = .125
				Total = 1.000

(Continued)

TABLE 3 (*Cont.*)   Sampling Distribution of $\overline{X}$

$\overline{x}$	$f(\overline{x})$
0	.008
1	.036 = .012 + .012 + .012
2	.054 = .018 + .018 + .018
3	.027
4	.060 = .020 + .020 + .020
5	.180 = .030 + .030 + .030
	+ .030 + .030 + .030
6	.135 = .045 + .045 + .045
8	.150 = .050 + .050 + .050
9	.225 = .075 + .075 + .075
12	.125 = .125

$$
\begin{aligned}
E(\overline{X}) = \sum \overline{x} f(\overline{x}) &= 0(.008) + 1(.036) + 2(.054) + 3(.027) \\
&\quad + 4(.060) + 5(.180) + 6(.135) \\
&\quad + 8(.150) + 9(.225) + 12(.125) \\
&= 6.9 \text{ same as } E(X), \text{ pop. mean}
\end{aligned}
$$

$$
\begin{aligned}
\operatorname{Var}(\overline{X}) = \sum \overline{x}^2 f(\overline{x}) - \mu^2 &= 0^2(.008) + 1^2(.036) + 2^2(.054) + 3^2(.027) \\
&\quad + 4^2(.060) + 5^2(.180) + 6^2(.135) \\
&\quad + 8^2(.150) + 9^2(.225) + 12^2(.125) - (6.9)^2 \\
&= 9.03 = \frac{27.09}{3} = \frac{\sigma^2}{3}
\end{aligned}
$$

$\operatorname{Var}(\overline{X})$ is one-third of the population variance.

Sampling Distribution of the Median m

m	$f(m)$
0	.104 = .008 + .012 + .020 + .012
	+ .020 + .012 + .020
3	.396 = .018 + .030 + .030 + .018 + .030
	+ .018 + .027 + .045 + .030 + .045
	+ .030 + .030 + .045
12	.500 = .050 + .075 + .050 + .075
	+ .050 + .075 + .125

Mean of the distribution of sample median
$$
= 0(.104) + 3(.396) + 12(.500) = 7.188 \neq 6.9 = \mu
$$
Different from the mean of the population distribution

Variance of the distribution of sample median
$$
= 0^2(.104) + 3^2(.396) + 12^2(.500) - (7.188)^2 = 23.897
$$
[not one-third of the population variance 27.09]

To illustrate the idea of a sampling distribution, we considered simple populations with only three possible values and small sample sizes $n = 2$ and $n = 3$. The calculation gets more tedious and extensive when a population has many values of X and n is large. However, the procedure remains the same. Once the population and sample size are specified:

1. List all possible samples of size n.
2. Calculate the value of the statistic for each sample.
3. List the distinct values of the statistic obtained in step 2. Calculate the corresponding probabilities by identifying all the samples that yield the same value of the statistic.

We leave the more complicated cases to statisticians who can sometimes use additional mathematical methods to derive exact sampling distributions.

Instead of a precise determination, one can turn to the computer in order to approximate a sampling distribution. The idea is to program the computer to actually draw a random sample and calculate the statistic. This procedure is then repeated a large number of times and a relative frequency histogram constructed from the values of the statistic. The resulting histogram will be an approximation to the sampling distribution. This approximation will be used in Example 4.

Exercises

7.1 Identify each of the following as either a parameter or a statistic.

(a) Sample standard deviation.

(b) Sample interquartile range.

(c) Population 10th percentile.

(d) Sample first quartile.

(e) Population median.

7.2 Identify the parameter, statistic, and population when they appear in each of the following statements.

(a) During the first half of 2003, twenty different movies received the distinction of generating the most box office revenue for a weekend.

(b) A survey of 400 minority persons living in Chicago revealed that 41 were out of work.

(c) Out of a sample of 100 dog owners who applied for dog licenses in northern Wisconsin, 18 had a Labrador retriever.

7.3 Data obtained from asking the wrong questions at the wrong time or in the wrong place can lead to misleading summary statistics. Explain why the following collection procedures are likely to produce useless data.

(a) To evaluate the number of students who are employed at least part time, the investigator interviews students who are taking an evening class.

(b) To study the pattern of spending of persons earning the minimum wage, a survey is taken during the first three weeks of December.

7.4 Explain why the following collection procedures are likely to produce data that fail to yield the desired information.

(a) To evaluate public opinion about a new global trade agreement, an interviewer asks persons, "Do you feel that this unfair trade agreement should be canceled?"

(b) To determine how eighth-grade girls feel about having boys in the classroom, a random sample from a private girls' school is polled.

7.5 From the set of numbers $\{3, 5, 7\}$, a random sample of size 2 will be selected with replacement.

(a) List all possible samples and evaluate $\bar{x}$ for each.

(b) Determine the sampling distribution of $\overline{X}$.

7.6 A random sample of size 2 will be selected, with replacement, from the set of numbers $\{0, 2, 4\}$.

(a) List all possible samples and evaluate $\bar{x}$ and s^2 for each.

(b) Determine the sampling distribution of $\overline{X}$.

(c) Determine the sampling distribution of S^2.

7.7 A bride-to-be asks a prospective wedding photographer to show a sample of her work. She provides ten pictures. Should the bride-to-be consider this a random sample of the quality of pictures she will get? Comment.

7.8 To determine the time a cashier spends on a customer in the express lane, the manager decides to record the time to check-out for the customer who is being served at 10 past the hour, 20 past the hour, and so on. Will measurements collected in this manner be a random sample of the times a cashier spends on a customer?

7.9 *Using a physical device to generate random samples.* Using a die, generate a sample and evaluate the statistic. Then repeat many times and obtain an estimate of the sampling distribution. In particular, investigate the sampling distribution of the median for a sample of size 3 from the population distribution.

x	$f(x)$
1	2/6
2	3/6
4	1/6

(a) Roll the die. Assign $X = 1$ if 1 or 2 dots show. Complete the assignment of values so that X has the population distribution when the die is fair.

(b) Roll the die two more times and obtain the median of the three observed values of X.

(c) Repeat to obtain a total of 25 samples of size 3. Calculate the relative frequencies, among the 75 values, of 1, 2, and 4. Compare with the population probabilities and explain why they should be close.

(d) Obtain the 25 values of the sample median and create a frequency table. Explain how the distribution in this table approximates the actual sampling distribution. It is easy to see how this approach extends to any sample size.

7.10 Referring to Exercise 7.9, use a die to generate samples of size 3. Investigate the sampling distribution of the number of times a value 1 occurs in a sample of size 3.

(a) Roll the die and assign the value for X. Repeat until you obtain a total of 25 samples of size 3. Calculate the relative frequencies, among the 75 values, of 1, 2, and 4. Compare with the population probabilities and explain why they should be close.

(b) Let a random variable Y equal 1 if at least one value of X in the sample is 1. Set Y equal to 0 otherwise. Then the relative frequency of $[Y = 1]$ is an estimate of $P[Y = 1]$, the probability of at least one value 1 in the sample of size 3. Give your estimate.

---

3.   DISTRIBUTION OF THE SAMPLE MEAN AND THE CENTRAL LIMIT THEOREM

Statistical inference about the population mean is of prime practical importance. Inferences about this parameter are based on the sample mean

$$\overline{X} = \frac{X_1 + X_2 + \cdots + X_n}{n}$$

and its sampling distribution. Consequently, we now explore the basic properties of the sampling distribution of $\overline{X}$ and explain the role of the normal distribution as a useful approximation.

In particular, we want to relate the sampling distribution of $\overline{X}$ to the population from which the random sample was selected. We denote the parameters
of the population by

$$\text{Population mean} \ = \ \mu$$
$$\text{Population standard deviation} \ = \ \sigma$$

The sampling distribution of $\overline{X}$ also has a mean $E(\overline{X})$ and a standard deviation sd$(\overline{X})$. These can be expressed in terms of the population mean μ and
standard deviation σ. (The interested reader can consult Appendix A.4 for
details.)

Mean and Standard Deviation of $\overline{X}$

The distribution of the sample mean, based on a random sample of size n,
has

$$E(\overline{X}) \ = \ \mu \qquad (\ = \ \text{Population mean})$$

$$\text{Var}(\overline{X}) \ = \ \frac{\sigma^2}{n} \qquad \left(= \ \frac{\text{Population variance}}{\text{Sample size}} \right)$$

$$\text{sd}(\overline{X}) \ = \ \frac{\sigma}{\sqrt{n}} \qquad \left(= \ \frac{\text{Population standard deviation}}{\sqrt{\text{Sample size}}} \right)$$

The first result shows that the distribution of $\overline{X}$ is centered at the population mean μ in the sense that expectation serves as a measure of center of a distribution. The last result states that the standard deviation of $\overline{X}$ equals the population standard deviation divided by the square root of the sample size. That is,
the variability of the sample mean is governed by the two factors: the population variability σ and the sample size n. Large variability in the population in-
duces large variability in $\overline{X}$, thus making the sample information about μ less
dependable. However, this can be countered by choosing n large. For instance,
with $n \ = \ 100$, the standard deviation of $\overline{X}$ is $\sigma/\sqrt{100} \ = \ \sigma/10$, a tenth of the
population standard deviation. With increasing sample size, the standard devia-
tion $\sigma/\sqrt{n}$ decreases and the distribution of $\overline{X}$ tends to become more concentrated around the population mean μ.

Example 3   **The Mean and Variance of the Sampling Distribution of $\overline{X}$**

Calculate the mean and standard deviation for the population distribution given in Table 1 and for the distribution of $\overline{X}$ given in Table 2. Verify the relations $E(\overline{X}) = \mu$ and $sd(\overline{X}) = \sigma/\sqrt{n}$.

SOLUTION   The calculations are performed in Table 4.

TABLE 4   Mean and Variance of $\overline{X} = (X_1 + X_2)/2$

	Population Distribution				Distribution of $\overline{X} = (X_1 + X_2)/2$			
x	$f(x)$	$xf(x)$	$x^2f(x)$		$\overline{x}$	$f(\overline{x})$	$\overline{x}f(\overline{x})$	$\overline{x}^2f(\overline{x})$
2	$\frac{1}{3}$	$\frac{2}{3}$	$\frac{4}{3}$		2	$\frac{1}{9}$	$\frac{2}{9}$	$\frac{4}{9}$
3	$\frac{1}{3}$	$\frac{3}{3}$	$\frac{9}{3}$		2.5	$\frac{2}{9}$	$\frac{5}{9}$	$\frac{12.5}{9}$
4	$\frac{1}{3}$	$\frac{4}{3}$	$\frac{16}{3}$		3	$\frac{3}{9}$	$\frac{9}{9}$	$\frac{27}{9}$
Total	1	3	$\frac{29}{3}$		3.5	$\frac{2}{9}$	$\frac{7}{9}$	$\frac{24.5}{9}$
					4	$\frac{1}{9}$	$\frac{4}{9}$	$\frac{16}{9}$
					Total	1	3	$\frac{84}{9}$

$$\mu = 3$$

$$\sigma^2 = \frac{29}{3} - (3)^2 = \frac{2}{3}$$

$$E(\overline{X}) = 3 = \mu$$

$$\text{Var}(\overline{X}) = \frac{84}{9} - (3)^2 = \frac{1}{3}$$

By direct calculation, $sd(\overline{X}) = 1/\sqrt{3}$. This is confirmed by the relation

$$sd(\overline{X}) = \frac{\sigma}{\sqrt{n}} = \sqrt{\frac{2}{3}}\bigg/\sqrt{2} = \frac{1}{\sqrt{3}}$$

We now state two important results concerning the shape of the sampling distribution of $\overline{X}$. The first result gives the exact form of the distribution of $\overline{X}$ when the population distribution is normal:

> ### $\overline{X}$ Is Normal When Sampling from a Normal Population
>
> In random sampling from a **normal** population with mean μ and standard deviation σ, the sample mean $\overline{X}$ has the normal distribution with mean μ and standard deviation $\sigma/\sqrt{n}$.

Eample 4   Determining Probabilities Concerning $\overline{X}$—Normal Populations

The weight of a pepperoni and cheese pizza from a local provider is a random variable whose distribution is normal with mean 16 ounces and standard deviation 1 ounce. You intend to purchase four pepperoni and cheese pizzas. What is the probability that:

(a) The average weight of the four pizzas will be greater than 17.1 ounces?

(b) The total weight of the four pizzas will not exceed 61.0 ounces?

SOLUTION   Because the population is normal, the distribution of the sample mean $\overline{X} = (X_1 + X_2 + X_3 + X_4)/4$ is exactly normal with mean 16 ounces and standard deviation $1/\sqrt{4} = .5$ ounce.

(a)   Since $\overline{X}$ is $N(16, .5)$

$$P[\overline{X} > 17.1] = P\left[\frac{\overline{X} - 16}{.5} > \frac{17.1 - 16}{.5}\right]$$
$$= P[Z > 2.20] = 1 - .9861 = .0139$$

Only rarely, just over one time in a hundred purchases of four pizzas, would the average weight exceed 17.1 ounces.

(b)   The event that the total weight $X_1 + X_2 + X_3 + X_4 = 4\overline{X}$ does not exceed 61.0 ounces is the same event that the average weight $\overline{X}$ is less than or equal to $61.0/4 = 15.25$. Consequently

$$P[X_1 + X_2 + X_3 + X_4 \leq 61.0] = P[\overline{X} \leq 15.25]$$
$$= P\left[\frac{\overline{X} - 16}{.5} \leq \frac{15.25 - 16}{.5}\right]$$
$$= P[Z \leq -1.50] = .0668$$

Only about seven times in one hundred purchases would the total weight be less than 61.0 ounces.

When sampling from a nonnormal population, the distribution of $\overline{X}$ depends on the particular form of the population distribution that prevails. A surprising result, known as the central limit theorem, states that when the sample size n is large, the distribution $\overline{X}$ is approximately normal, regardless of the shape of the population distribution. In practice, the normal approximation is usually adequate when n is greater than 30.

Central Limit Theorem

Whatever the population, the distribution of $\overline{X}$ is approximately normal when n is large.

In random sampling from an arbitrary population with mean μ and standard deviation σ, when n is large, the distribution of $\overline{X}$ is approximately normal with mean μ and standard deviation $\sigma/\sqrt{n}$. Consequently,

$$Z = \frac{\overline{X} - \mu}{\sigma/\sqrt{n}} \quad \text{is approximately } N(0, 1)$$

Whether the population distribution is continuous, discrete, symmetric, or asymmetric, the central limit theorem asserts that as long as the population variance is finite, the distribution of the sample mean $\overline{X}$ is nearly normal if the sample size is large. In this sense, the normal distribution plays a central role in the development of statistical procedures. Although a proof of the theorem requires higher mathematics, we can empirically demonstrate how this result works.

Example 5   Demonstrating the Central Limit Theorem

Consider a population having a discrete uniform distribution that places a probability of .1 on each of the integers 0, 1, . . . , 9. This may be an appropriate model for the distribution of the last digit in telephone numbers or the first overflow digit in computer calculations. The line diagram of this distribution appears in Figure 2. The population has $\mu = 4.5$ and $\sigma = 2.872$.

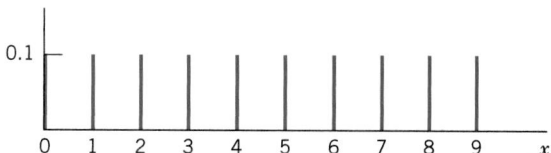

Figure 2   Uniform distribution on the integers 0, 1, . . . , 9.

By means of a computer, 100 random samples of size 5 were generated from this distribution, and $\overline{x}$ was computed for each sample. The results of this repeated random sampling are presented in Table 5. The relative frequency histogram in Figure 3 is constructed from the 100 observed values of $\overline{x}$. Although the population distribution (Figure 2) is far from normal, the top of the histogram of the $\overline{x}$ values (Figure 3) has the appearance of a bell-shaped curve, even for the small sample size of 5. For larger sample sizes, the normal distribution would give an even closer approximation.

TABLE 5 Samples of Size 5 from a Discrete Uniform Distribution

Sample Number	Observations	Sum	Mean $\bar{x}$	Sample Number	Observations	Sum	Mean $\bar{x}$
1	4, 7, 9, 0, 6	26	5.2	51	4, 7, 3, 8, 8	30	6.0
2	7, 3, 7, 7, 4	28	5.6	52	2, 0, 3, 3, 2	10	2.0
3	0, 4, 6, 9, 2	21	4.2	53	4, 4, 2, 6, 3	19	3.8
4	7, 6, 1, 9, 1	24	4.8	54	1, 6, 4, 0, 6	17	3.4
5	9, 0, 2, 9, 4	24	4.8	55	2, 4, 5, 8, 9	28	5.6
6	9, 4, 9, 4, 2	28	5.6	56	1, 5, 5, 4, 0	15	3.0
7	7, 4, 2, 1, 6	20	4.0	57	3, 7, 5, 4, 3	22	4.4
8	4, 4, 7, 7, 9	31	6.2	58	3, 7, 0, 7, 6	23	4.6
9	8, 7, 6, 0, 5	26	5.2	59	4, 8, 9, 5, 9	35	7.0
10	7, 9, 1, 0, 6	23	4.6	60	6, 7, 8, 2, 9	32	6.4
11	1, 3, 6, 5, 7	22	4.4	61	7, 3, 6, 3, 6	25	5.0
12	3, 7, 5, 3, 2	20	4.0	62	7, 4, 6, 0, 1	18	3.6
13	5, 6, 6, 5, 0	22	4.4	63	7, 9, 9, 7, 5	37	7.4
14	9, 9, 6, 4, 1	29	5.8	64	8, 0, 6, 2, 7	23	4.6
15	0, 0, 9, 5, 7	21	4.2	65	6, 5, 3, 6, 2	22	4.4
16	4, 9, 1, 1, 6	21	4.2	66	5, 0, 5, 2, 9	21	4.2
17	9, 4, 1, 1, 4	19	3.8	67	2, 9, 4, 9, 1	25	5.0
18	6, 4, 2, 7, 3	22	4.4	68	9, 5, 2, 2, 6	24	4.8
19	9, 4, 4, 1, 8	26	5.2	69	0, 1, 4, 4, 4	13	2.6
20	8, 4, 6, 8, 3	29	5.8	70	5, 4, 0, 5, 2	16	3.2
21	5, 2, 2, 6, 1	16	3.2	71	1, 1, 4, 2, 0	8	1.6
22	2, 2, 9, 1, 0	14	2.8	72	9, 5, 4, 5, 9	32	6.4
23	1, 4, 5, 8, 8	26	5.2	73	7, 1, 6, 6, 9	29	5.8
24	8, 1, 6, 3, 7	25	5.0	74	3, 5, 0, 0, 5	13	2.6
25	1, 2, 0, 9, 6	18	3.6	75	3, 7, 7, 3, 5	25	5.0
26	8, 5, 3, 0, 0	16	3.2	76	7, 4, 7, 6, 2	26	5.2
27	9, 5, 8, 5, 0	27	5.4	77	8, 1, 0, 9, 1	19	3.8
28	8, 9, 1, 1, 8	27	5.4	78	6, 4, 7, 9, 3	29	5.8
29	8, 0, 7, 4, 0	19	3.8	79	7, 7, 6, 9, 7	36	7.2
30	6, 5, 5, 3, 0	19	3.8	80	9, 4, 2, 9, 9	33	6.6
31	4, 6, 4, 2, 1	17	3.4	81	3, 3, 3, 3, 3	15	3.0
32	7, 8, 3, 6, 5	29	5.8	82	8, 7, 7, 0, 3	25	5.0
33	4, 2, 8, 5, 2	21	4.2	83	5, 3, 2, 1, 1	12	2.4
34	7, 1, 9, 0, 9	26	5.2	84	0, 4, 5, 2, 6	17	3.4
35	5, 8, 4, 1, 4	22	4.4	85	3, 7, 5, 4, 1	20	4.0
36	6, 4, 4, 5, 1	20	4.0	86	7, 4, 5, 9, 8	33	6.6
37	4, 2, 1, 1, 6	14	2.8	87	3, 2, 9, 0, 5	19	3.8
38	4, 7, 5, 5, 7	28	5.6	88	4, 6, 6, 3, 3	22	4.4
39	9, 0, 5, 9, 2	25	5.0	89	1, 0, 9, 3, 7	20	4.0
40	3, 1, 5, 4, 5	18	3.6	90	2, 9, 6, 8, 5	30	6.0
41	9, 8, 6, 3, 2	28	5.6	91	4, 8, 0, 7, 6	25	5.0
42	9, 4, 2, 2, 8	25	5.0	92	5, 6, 7, 6, 3	27	5.4
43	8, 4, 7, 2, 2	23	4.6	93	3, 6, 2, 5, 6	22	4.4
44	0, 7, 3, 4, 9	23	4.6	94	0, 1, 1, 8, 4	14	2.8
45	0, 2, 7, 5, 2	16	3.2	95	3, 6, 6, 4, 5	24	4.8
46	7, 1, 9, 9, 9	35	7.0	96	9, 2, 9, 8, 6	34	6.8
47	4, 0, 5, 9, 4	22	4.4	97	2, 0, 0, 6, 8	16	3.2
48	5, 8, 6, 3, 3	25	5.0	98	0, 4, 5, 0, 5	14	2.8
49	4, 5, 0, 5, 3	17	3.4	99	0, 3, 7, 3, 9	22	4.4
50	7, 7, 2, 0, 1	17	3.4	100	2, 5, 0, 0, 7	14	2.8

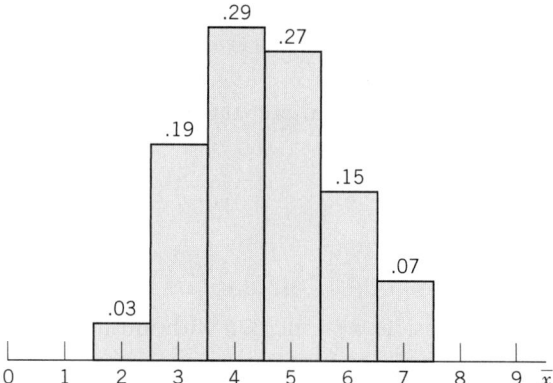

Figure 3   Relative frequency histogram
of the $\bar{x}$ values recorded in Table 5.

Calculating from the 100 simulated $\bar{x}$ values in Table 5, we find the sample mean and standard deviation to be 4.54 and 1.215, respectively. These are in close agreement with the theoretical values for the mean and standard deviation of $\bar{X}$: $\mu = 4.5$ and $\sigma/\sqrt{n} = 2.872/\sqrt{5} = 1.284$.

It might be interesting for the reader to collect similar samples by reading the last digits of numbers from a telephone directory and then to construct a histogram of the $\bar{x}$ values.

Another graphic example of the central limit theorem appears in Figure 4, where the population distribution represented by the solid curve is a continuous

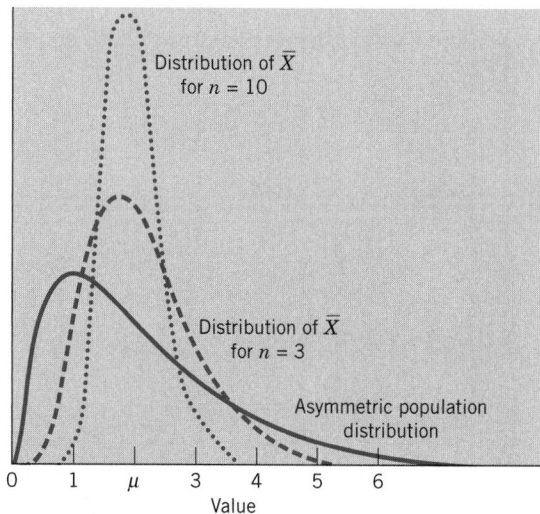

Figure 4   Distributions of $\bar{X}$ for $n = 3$ and $n = 10$ in sampling from an asymmetric population.

asymmetric distribution with $\mu = 2$ and $\sigma = 1.41$. The distributions of the sample mean $\overline{X}$ for sample sizes $n = 3$ and $n = 10$ are plotted as dashed and dotted curves on the graph. These indicate that with increasing n, the distributions become more concentrated around μ and look more like the normal distribution.

Example 6   Probability Calculations for $\overline{X}$—Large Sample Size

Consider a population with mean 82 and standard deviation 12.

(a)   If a random sample of size 64 is selected, what is the probability that the sample mean will lie between 80.8 and 83.2?

(b)   With a random sample of size 100, what is the probability that the sample mean will lie between 80.8 and 83.2?

SOLUTION

(a)   We have $\mu = 82$ and $\sigma = 12$. Since $n = 64$ is large, the central limit theorem tells us that the distribution of $\overline{X}$ is approximately normal with

$$\text{Mean} = \mu = 82$$

$$\text{Standard deviation} = \frac{\sigma}{\sqrt{n}} = \frac{12}{\sqrt{64}} = 1.5$$

To calculate $P[80.8 < \overline{X} < 83.2]$, we convert to the standardized variable

$$Z = \frac{\overline{X} - \mu}{\sigma/\sqrt{n}} = \frac{\overline{X} - 82}{1.5}$$

The z values corresponding to 80.8 and 83.2 are

$$\frac{80.8 - 82}{1.5} = -.8 \quad \text{and} \quad \frac{83.2 - 82}{1.5} = .8$$

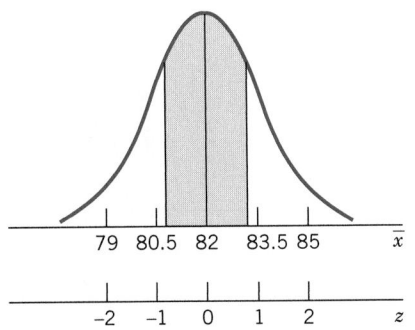

Consequently,

$$P[80.8 < \overline{X} < 83.2] = P[-.8 < Z < .8]$$
$$= .7881 - .2119 \quad \text{(using the normal table)}$$
$$= .5762$$

(b)   We now have $n = 100$, so $\sigma/\sqrt{n} = 12/\sqrt{100} = 1.2$, and

$$Z = \frac{\overline{X} - 82}{1.2}$$

Therefore,

$$P[80.8 < \overline{X} < 83.2] = P\left[\frac{80.8 - 82}{1.2} < Z < \frac{83.2 - 82}{1.2}\right]$$
$$= P[-1.0 < Z < 1.0]$$
$$= .8413 - .1587$$
$$= .6826$$

Note that the interval (80.8, 83.2) is centered at $\mu = 82$. The probability that $\overline{X}$ will lie in this interval is larger for $n = 100$ than for $n = 64$.

Example 7   Probability Calculations for $\overline{X}$, Mean Gripping Strength

Suppose that the population distribution of the gripping strengths of industrial workers is known to have a mean of 110 and standard deviation of 10. For a random sample of 75 workers, what is the probability that the sample mean gripping strength will be:

(a)   Between 109 and 112?

(b)   Greater than 111?

SOLUTION   Here the population mean and the standard deviation are $\mu = 110$ and $\sigma = 10$, respectively. The sample size $n = 75$ is large, so the central limit theorem ensures that the distribution of $\overline{X}$ is approximately normal with

$$\text{Mean} = 110$$

$$\text{Standard deviation} = \frac{\sigma}{\sqrt{n}} = \frac{10}{\sqrt{75}} = 1.155$$

(a)   To find $P[109 < \overline{X} < 112]$, we convert to the standardized variable

$$Z = \frac{\overline{X} - 110}{1.155}$$

and calculate the z values

$$\frac{109 - 110}{1.155} = -.866 \qquad \frac{112 - 110}{1.155} = 1.732$$

The required probability is

$$P[109 < \overline{X} < 112] = P[-.866 < Z < 1.732]$$
$$= .958 - .193$$
$$= .765$$

(b)

$$P[\overline{X} > 111] = P\left[Z > \frac{111 - 110}{1.155}\right]$$
$$= P[Z > .866]$$
$$= 1 - P[Z \leq .866]$$
$$= 1 - .807$$
$$= .193$$

A natural question that arises is how large should n be for the normal approximation to be used for the distribution of $\overline{X}$? The nature of the approximation depends on the extent to which the population distribution deviates from a normal form. If the population distribution is normal, then $\overline{X}$ is exactly normally distributed for all n, small or large. As the population distribution increasingly departs from normality, larger values of n are required for a good approximation. Ordinarily, $n > 30$ provides a satisfactory approximation.

Exercises

7.11   A population has mean 79 and standard deviation 9. Calculate $E(\overline{X})$ and $sd(\overline{X})$ for a random sample of size (a) 4 and (b) 25.

7.12   A population has mean 30 and standard deviation 16. Calculate the expected value and standard deviation of $\overline{X}$ for a random sample of size (a) 9 and (b) 16.

7.13   A population has standard deviation 20. What is the standard deviation of $\overline{X}$ for a random sample of size (a) $n = 25$ (b) $n = 100$ and (c) $n = 400$?

7.14   A population has standard deviation 53. What is the standard deviation of $\overline{X}$ for a random sample of size (a) 36 and (b) 144?

7.15   Using the sampling distribution determined for $\overline{X} = (X_1 + X_2)/2$ in Exercise 7.5, verify that $E(\overline{X}) = \mu$ and $sd(\overline{X}) = \sigma/\sqrt{2}$.

7.16   Using the sampling distribution determined for $\overline{X} = (X_1 + X_2)/2$ in Exercise 7.6, verify that $E(\overline{X}) = \mu$ and $sd(\overline{X}) = \sigma/\sqrt{2}$.

7.17   A population has distribution

Value	Probability
1	.2
2	.6
3	.2

Let X_1 and X_2 be independent and each have the same distribution as the population.

(a)   Determine the missing elements in the table for the sampling distribution of $\overline{X} = (X_1 + X_2)/2$.

$\overline{x}$	Probability
1.0	
1.5	
2.0	.44
2.5	.24
3.0	

(b) Find the expected value of $\overline{X}$.

(c) If the sample size is increased to 36, give the mean and variance of $\overline{X}$.

7.18 A normal population has $\mu = 24$ and $\sigma = 5$. For sample size $n = 4$, determine (a) mean of $\overline{X}$ (b) standard deviation of $\overline{X}$ and (c) distribution of $\overline{X}$.

7.19 A normal population has mean 37 and standard deviation 6. For a random sample of size $n = 6$, determine:

(a) Mean of $\overline{X}$.

(b) Standard deviation of $\overline{X}$.

(c) Distribution of $\overline{X}$.

7.20 A population has distribution

Value	Probability
0	.2
2	.1
4	.7

Let X_1 and X_2 be independent and each have the same distribution as the population.

(a) Determine the missing elements in the table for the sampling distribution of $\overline{X} = (X_1 + X_2)/2$.

$\bar{x}$	Probability
0	
1	
2	.29
3	
4	.49

(b) Find the expected value of $\overline{X}$.

(c) If the sample size is increased to 25, give the mean and variance of $\overline{X}$.

7.21 Suppose the weights of packages of lettuce coming off a packaging line have a normal distribution with mean 8.1 ounces and standard deviation .1 ounce.

(a) If every package is labeled 8 ounces, what proportion of the packages weigh less than the labeled amount?

(b) If two packages are randomly selected, specify the mean, standard deviation, and distribution of the average weight.

(c) If two packages are randomly selected, what is the probability that the average weight is less than 8 ounces?

7.22 Suppose the amount of a popular sport drink in bottles leaving the filling machine has a normal distribution with mean 101.5 milliliters (ml) and standard deviation 1.6 ml.

(a) If the bottles are labeled 100 ml, what proportion of the bottles contain less than the labeled amount.

(b) If four bottles are randomly selected, find the mean and standard deviation of the average content.

(c) What is the probability that the average content is less than 100 ml?

7.23 The distribution of personal income of full-time retail clerks working in a large eastern city has $\mu = \$31,000$ and $\sigma = \$5000$.

(a) What is the approximate distribution for $\overline{X}$ based on a random sample of 100 persons?

(b) Evaluate $P[\overline{X} > 31,500]$.

7.24 A random sample of size 100 is taken from a population having a mean of 23 and a standard deviation of 4. The shape of the population distribution is unknown.

(a) What can you say about the probability distribution of the sample mean $\overline{X}$?

(b) Find the probability that $\overline{X}$ will exceed 23.6.

7.25 The lengths of the trout fry in a pond at the fish hatchery are approximately normally distributed with mean 3.4 inches and standard deviation .8 inch. Three dozen fry will be netted and their lengths measured.

(a) What is the probability that the sample mean length of the 36 netted trout fry will be less than 3.2 inches?

(b) Why might the fish in the net not represent a random sample of trout fry in the pond?

7.26 The heights of male students at a university have a nearly normal distribution with mean 70 inches and standard deviation 2.8 inches. If 5 male students are randomly selected to make up an intramural basketball team, what is the probability that the heights of the team will average over 72.0 inches?

7.27 According to the growth chart that doctors use as a reference, the heights of two-year-old boys are normally distributed with mean 34.5 inches and standard deviation 1.3 inches. For a random sample of 6 two-year-old boys, find the probability that the sample mean will be between 34.1 and 35.2 inches.

7.28 The weight of an almond is normally distributed with mean .05 ounce and standard deviation .015 ounce. Find the probability that a package of 100 almonds will weigh between 4.8 and 5.3 ounces. That is, find the probability that $\overline{X}$ will be between .048 and .053 ounce.

*7.29 Refer to Table 5.

(a) Calculate the sample median for each sample.

(b) Construct a frequency table and make a histogram.

(c) Compare the histogram for the median with that given in Figure 3 for the sample mean. Does your comparison suggest that the sampling distribution of the mean or median has the smaller variance?

7.30 The number of days, X, that it takes the post office to deliver a letter between City A and City B has the probability distribution

x	3	4	5
$f(x)$	.5	.3	.2

(a) Find the expected number days and the standard deviation of the number of days.

(b) A company in City A sends a letter to a company in City B with a return receipt request that is to be mailed immediately upon receiving the letter. Find the probability distribution of total number of days from the time the letter is mailed until the return receipt arrives back at the company in City A. Assume the two delivery times are independent.

(c) A single letter will be sent from City A on each of 100 different days. What is the approximate probability that more than 25 of the letters will take 5 days to reach City B?

7.31 The number of complaints per day, X, received by a cable TV distributor has the probability distribution

x	0	1	2	3
$f(x)$	.4	.3	.1	.2

(a) Find the expected number of complaints per day.

(b) Find the standard deviation of the number of complaints.

(c) What is the probability distribution of total number of complaints received in two days? Assume the numbers of complaints on different days are independent.

(d) What is the approximate probability that the distributor will receive more than 125 complaints in 90 days?

4.   STATISTICS IN CONTEXT

Troy, a Canadian importer of cut flowers, must effectively deal with uncertainty every day that he is in business. For instance, he must order enough of each kind of flower to supply his regular wholesale customers and yet not have too many left each day. Fresh flowers are no longer fresh on the day after arrival.

Troy purchases his fresh flowers from growers in the United States, Mexico, and Central and South America. Because most of the growers purchase their growing stock and chemicals from the United States, all of the selling prices are quoted in U.S. dollars. On a typical day, he purchases tens of thousands of cut flowers. Troy knows their price in U.S. dollars, but this is not his ultimate cost. Because of a fluctuating exchange rate, he does not know his ultimate cost at the time of purchase.

As with most businesses, Troy takes about a month to pay his bills. He must pay in Canadian dollars, so fluctuations in the Canadian/U.S. exchange rate from the time of purchase to the time the invoice is paid are a major source of uncertainty. Can this uncertainty be quantified and modeled?

The Canadian dollar to U.S. dollar exchange rate equals the number of Canadian dollars which must be paid for each U.S. dollar. Data from several years will provide the basis for a model. As given in Table 6, the exchange rate was 1.1603 in December 1990 and 1.1560 in January 1991 (reading across each row). It was 1.4761 in March 2003.

If the exchange rate is 1.20, one dollar and twenty cents Canadian is required to pay for each single U.S. dollar. An invoice for 1000 U.S. dollars would cost Troy 1200 Canadian dollars while it would cost 1210 dollars if the exchange rate were 1.21. It is the change in the exchange rate from time of purchase to payment that creates uncertainty.

TABLE 6 Monthly Canadian to U.S. Dollar Exchange Rates 12/1990–3/2003

1.1603	1.1560	1.1549	1.1572	1.1535	1.1499	1.1439	1.1493
1.1452	1.1370	1.1279	1.1302	1.1467	1.1571	1.1825	1.1928
1.1874	1.1991	1.1960	1.1924	1.1907	1.2225	1.2453	1.2674
1.2725	1.2779	1.2602	1.2471	1.2621	1.2698	1.2789	1.2820
1.3080	1.3215	1.3263	1.3174	1.3308	1.3173	1.3424	1.3644
1.3830	1.3808	1.3836	1.3826	1.3783	1.3540	1.3503	1.3647
1.3893	1.4132	1.4005	1.4077	1.3762	1.3609	1.3775	1.3612
1.3552	1.3509	1.3458	1.3534	1.3693	1.3669	1.3752	1.3656
1.3592	1.3693	1.3658	1.3697	1.3722	1.3694	1.3508	1.3381
1.3622	1.3494	1.3556	1.3725	1.3942	1.3804	1.3843	1.3775
1.3905	1.3872	1.3869	1.4128	1.4271	1.4409	1.4334	1.4166
1.4298	1.4452	1.4655	1.4869	1.5346	1.5218	1.5452	1.5404
1.5433	1.5194	1.4977	1.5176	1.4881	1.4611	1.4695	1.4890
1.4932	1.4771	1.4776	1.4674	1.4722	1.4486	1.4512	1.4608
1.4689	1.4957	1.4770	1.4778	1.4828	1.4864	1.5125	1.5426
1.5219	1.5032	1.5216	1.5587	1.5578	1.5411	1.5245	1.5308
1.5399	1.5679	1.5717	1.5922	1.5788	1.5997	1.5964	1.5877
1.5815	1.5502	1.5318	1.5456	1.5694	1.5761	1.5780	1.5715
1.5592	1.5414	1.5121	1.4761				

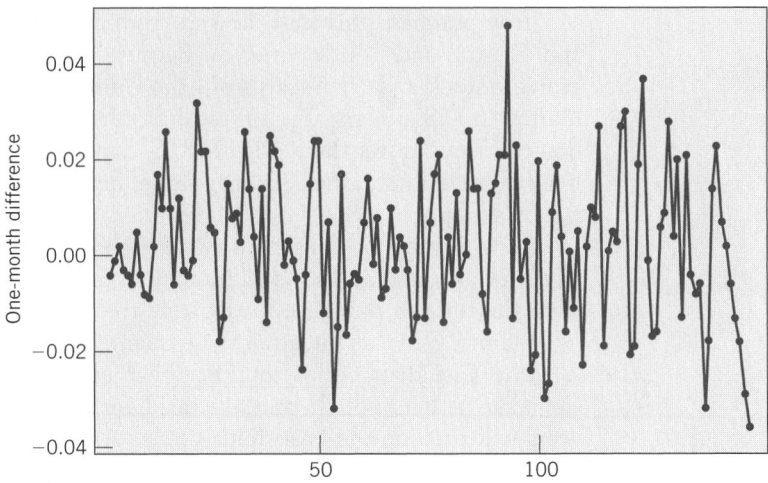

Figure 5   Time plot of one-month differences in the Canadian / U. S. exchange rate.

Although the exchange rate changes every day, we consider the monthly rates. The value of the difference

Current month exchange rate − Previous month exchange rate

would describe the change in cost, per dollar invoiced, resulting from the one-month delay between purchasing and paying for a shipment of cut flowers. If the rate goes down, Troy makes money. If the rate goes up, he loses money. Figure 5 gives a time plot of these differences for the period 1/1991 – 3/2003. The differences appear to be stable over time.

Figure 6 gives a histogram and a normal-scores plot. It seems reasonable to model the monthly change in exchange rate as a random variable having a normal distribution. The $n = 147$ monthly differences have mean $\bar{x} = .0022$ and standard deviation $s = .01588$.

According to the methods developed in the next chapter, 0 is a plausible value for the mean. Our approximating normal distribution has $\mu = 0$ and $\sigma = .0156$. We have successfully modeled the uncertainty in the exchange rate over a one-month period. If Troy paid all of his invoices in exactly one month, this then is the variability that he would face.

Over a three-month period, Troy would pay three times and the total uncertainty would be the sum of three independent mean 0 normal random variables. The variance of the sum is $3\sigma^2 = 3(.0156)^2 = .000730$, so the standard deviation is

$$\sqrt{3(.0156)^2} = .0270$$

Although the variance is 3 times as large as that of a single difference, the standard deviation does not increase that fast. The standard deviation is $\sqrt{3}(.0156)$.

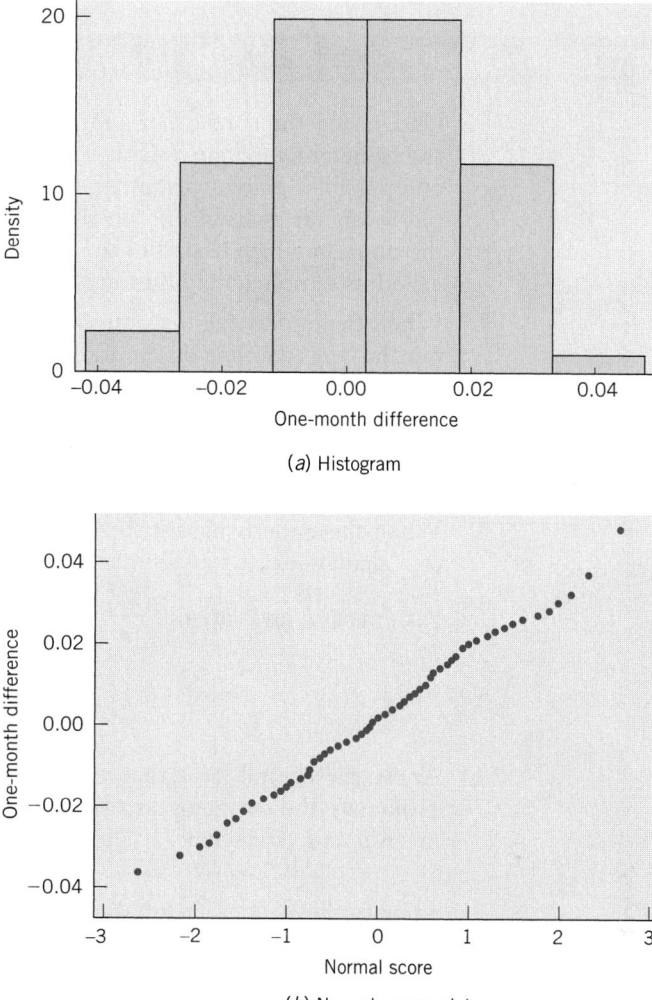

Figure 6   (*a*) Histogram and (*b*) normal-scores plot of differences.

Exercises

7.32   Refer to the model for monthly differences in the exchange rate. Find the mean and standard deviation for the sampling distribution of the sample mean of three monthly differences. Relate these quantities to the population mean and standard deviation for the sum of three differences given in the Statistics in Context section.

7.33   Refer to the Statistics in Context section concerning the flower importer.

(a)   Suppose it takes the importer two months to pay his invoices. Make a histogram of the differences two months apart. (The two-month differences may not be independent but the histogram is the correct summary of uncertainty for some two-month period in the future.)

(b)   Compare your histogram in part (a) with that for the one-month changes. Which is more variable?

USING STATISTICS WISELY

1. Understand the concept of a sampling distribution. Each observation is the value of a random variable so a sample of n observations varies from one possible sample to another. Consequently, a statistic such as a sample mean varies from one possible sample to another. The probability distribution which describes the chance behavior of the sample mean is called its sampling distribution.

2. When the underlying distribution has mean μ and variance σ^2, remember that the sampling distribution of $\overline{X}$ has

$$\text{Mean of } \overline{X} \;=\; \mu \;=\; \text{Population mean}$$

$$\text{Variance of } \overline{X} \;=\; \mu^2 \;=\; \frac{\text{Population variance}}{n}$$

3. When the underlying distribution is normal with mean μ and variance σ^2, calculate exact probabilities for $\overline{X}$ using the normal distribution with mean μ and variance $\dfrac{\sigma^2}{n}$.

$$P[\overline{X} \leq b] \;=\; P\left[Z \leq \frac{b - \mu}{\sigma/\sqrt{n}}\right]$$

4. Apply the central limit theorem, when the sample size is large, to approximate the sampling distribution of $\overline{X}$ by a normal distribution with mean μ and variance σ^2/n. The probability $P[\overline{X} \leq b]$ is approximately equal to the standard normal probability $P\left[Z \leq \dfrac{b - \mu}{\sigma/\sqrt{n}}\right]$.

5. Do not confuse the population distribution, which describes the variation for a single random variable, with the sampling distribution of a statistic.

6. When the population distribution is noticeably nonnormal, do not try to conclude that the sampling distribution of $\overline{X}$ is normal unless the sample size is at least moderately large, 30 or more.

KEY IDEAS

The observations $X_1, X_2, \ldots, X_n$ are a random sample of size n from the population distribution if they result from independent selections and each observation has the same distribution as the population. Under these conditions, we refer to the observations as a random sample.

A parameter is a numerical characteristic of the population. It is a constant, although its value is typically unknown to us. The object of a statistical analysis of sample data is to learn about the parameter.

A numerical characteristic of a sample is called a statistic. The value of a statistic varies in repeated sampling.

When random sampling from a population, a statistic is a random variable. The probability distribution of a statistic is called its sampling distribution.

The sampling distribution of $\overline{X}$ has mean μ and standard deviation $\sigma/\sqrt{n}$, where μ = population mean, σ = population standard deviation, and n = sample size.

With increasing n, the distribution of $\overline{X}$ is more concentrated around μ.

If the population distribution is normal, $N(\mu, \sigma)$, the distribution of $\overline{X}$ is $N(\mu, \sigma/\sqrt{n})$.

Regardless of the shape of the population distribution, the distribution of $\overline{X}$ is approximately $N(\mu, \sigma/\sqrt{n})$, provided that n is large. This result is called the central limit theorem.

5.   REVIEW EXERCISES

7.34 A population consists of the four numbers {0, 2, 4, 6}. Consider drawing a random sample of size 2 with replacement.

(a)   List all possible samples and evaluate $\bar{x}$ for each.

(b)   Determine the sampling distribution of $\overline{X}$.

(c)   Write down the population distribution and calculate its mean μ and standard deviation σ.

(d)   Calculate the mean and standard deviation of the sampling distribution of $\overline{X}$ obtained in part (b), and verify that these agree with μ and $\sigma/\sqrt{2}$, respectively.

7.35 Refer to Exercise 7.34 and, instead of $\overline{X}$, consider the statistic

Sample range R = Largest observation − Smallest observation

For instance, if the sample observations are (2, 6), the range is 6 − 2 = 4.

(a)   Calculate the sample range for all possible samples.

(b)   Determine the sampling distribution of R.

7.36 What sample size is required in order that the standard deviation of $\overline{X}$ be:

(a) $\frac{1}{4}$ of the population standard deviation?

(b) $\frac{1}{7}$ of the population standard deviation?

(c)   12% of the population standard deviation?

7.37 A population has distribution

Value	Probability
0	.3
1	.4
2	.3

Let X_1 and X_2 be independent and each have the same distribution as the population.

(a)   Determine the missing elements in the table for the sampling distribution of $\overline{X}$ = $(X_1 + X_2)/2$.

$\bar{x}$	Probability
0.0	
0.5	.24
1.0	.34
1.5	
2.0	

(b)   Find the expected value of $\overline{X}$.

(c)   If the sample size is increased to 81, give the mean and variance of $\overline{X}$.

7.38 A population has distribution

Value	Probability
1	.6
3	.3
5	.1

Let X_1 and X_2 be independent and each have the same distribution as the population.

(a) Determine the missing elements in the table for the sampling distribution of $\overline{X} = (X_1 + X_2)/2$.

$\bar{x}$	Probability
1	
2	.36
3	.21
4	
5	

(b) Find the expected value of $\overline{X}$.

(c) If the sample size is increased to 25, give the mean and variance of $\overline{X}$.

7.39 Suppose a population distribution is normal with mean 60 and standard deviation 10. For a random sample of size $n = 9$

(a) What are the mean and standard deviation of $\overline{X}$?

(b) What is the distribution of $\overline{X}$? Is this distribution exact or approximate?

(c) Find the probability that $\overline{X}$ lies between 56 and 64.

7.40 The weights of pears in an orchard are normally distributed with mean .32 pound and standard deviation .08 pound.

(a) If one pear is selected at random, what is the probability that its weight will be between .28 and .34 pound?

(b) If $\overline{X}$ denotes the average weight of a random sample of four pears, what is the probability that $\overline{X}$ will be between .28 and .34 pound?

7.41 Suppose that the size of pebbles in a river bed is normally distributed with mean 12.1 mm and standard deviation 3.2 mm. A random sample of 9 pebbles will be measured. Let $\overline{X}$ denote the average size of the sampled pebbles.

(a) What is the distribution of $\overline{X}$?

(b) What is the probability that $\overline{X}$ is smaller than 10?

(c) What percentage of the pebbles in the river bed are of size smaller than 10?

7.42 A random sample of size 150 is taken from a population that has mean 37 and standard deviation 7. The population distribution is not normal.

(a) Is it reasonable to assume a normal distribution for the sample mean $\overline{X}$? Why or why not?

(b) Find the probability that $\overline{X}$ lies between 36 and 38.

(c) Find the probability that $\overline{X}$ exceeds 38.5.

7.43 The distribution for the time it takes a student to complete the fall class registration has a mean of 54 minutes and a standard deviation of 8 minutes. For a random sample of 81 students:

(a) Determine the mean and standard deviation of $\overline{X}$.

(b) What can you say about the distribution of $\overline{X}$?

7.44 Refer to Exercise 7.43. Evaluate (a) $P[\overline{X} > 55]$ (b) $P[52.3 < \overline{X} < 55]$ and (c) $P[\overline{X} < 55]$.

7.45 The mean and standard deviation of the strength of a packaging material are 55 and 7 pounds, respectively. If 45 specimens of this material are tested

(a) What is the probability that the sample mean strength $\overline{X}$ will be between 54 and 56 pounds?

(b) Find the interval centered at 55, where $\overline{X}$ will lie with probability .95.

7.46 Consider a random sample of size $n = 49$ from a population that has a standard deviation of $\sigma = 14$.

(a) Find the probability that the sample mean $\overline{X}$ will lie within 2 units of the population mean—that is, $P[-2 \leq \overline{X} - \mu \leq 2]$.

(b) Find the number k so that $P[-k \leq \overline{X} - \mu \leq k] = .90$.

(c) What is the probability that $\overline{X}$ will differ from μ by more than 4 units?

7.47 The daily number of kayaks sold, X, at a water sports store has the probability distribution

x	0	1	2
$f(x)$	.5	.3	.2

(a) Find the expected number of kayaks sold in a day.

(b) Find the standard deviation of the number of kayaks sold in a day.

(c) Find the probability distribution of the total number of kayaks sold in the next two days. Suppose that the number of sales on different days are independent.

(d) Over the next 64 days, what is the approximate probability that at least 53 kayaks will be sold?

(e) How many kayaks should the store order to have approximate probability .95 of meeting the total demand in the next 64 days?

7.48 Suppose packages of cream cheese coming from an automated processor have weights that are normally distributed. For one day's production run, the mean is 8.2 ounces and the standard deviation is 0.1 ounce.

(a) If the packages of cream cheese are labeled 8 ounces, what proportion of the packages weigh less than the labeled amount?

(b) If only 5% of the packages exceed a specified weight w, what is the value of w?

(c) Suppose two packages are selected at random from the day's production. What is the probability that the average weight of the two packages is less than 8.3 ounces?

(d) Suppose 5 packages are selected at random from the day's production. What is the probability that at most one package weighs at least 8.3 ounces?

7.49 Suppose the amount of sun block lotion in plastic bottles leaving a filling machine has a normal distribution. The bottles are labeled 300 milliliters (ml) but the actual mean is 302 ml and the standard deviation is 2 ml.

(a) What is the probability that an individual bottle will contain less than 299 ml?

(b) If only 5% of the bottles have contents that exceed a specified amount v, what is the value of v?

(c) Two bottles can be purchased together in a twin-pack. What is the probability that the mean content of bottles in a twin-pack is less than 299 ml? Assume the contents of the two bottles are independent.

(d) If you purchase two twin-packs of the lotion, what is the probability that only one of the twin-packs has a mean bottle content less than 299 ml?

CLASS PROJECTS

1. (a) Count the number of occupants X including the driver in each of 20 passing cars. Calculate the mean $\bar{x}$ of your sample.

   (b) Repeat part (a) 10 times.

   (c) Collect the data sets of the individual car counts x from the entire class and construct a relative frequency histogram.

   (d) Collect the $\bar{x}$ values from the entire class (10 from each student) and construct a relative frequency histogram for $\bar{x}$, choosing appropriate class intervals.

   (e) Plot the two relative frequency histograms and comment on the closeness of their shapes to the normal distribution.

2. (a) Collect a sample of size 7 and compute $\bar{x}$ and the sample median.

   (b) Repeat part (a) 30 times.

   (c) Plot dot diagrams for the values of the two statistics in part (a). These plots reflect the individual sampling distributions.

   (d) Compare the amount of variation in $\overline{X}$ and the median.

   In this exercise, you might record weekly soft-drink consumptions, sentence lengths, or hours of sleep for different students.

COMPUTER PROJECT

1. Conduct a simulation experiment on the computer to verify the central limit theorem. Generate $n = 6$ observations from the continuous distribution that is uniform on 0 to 1. Calculate $\overline{X}$. Repeat 150 times. Make a histogram of the $\overline{X}$ values and a normal-scores plot. Does the distribution of $\overline{X}$ appear to be normal for $n = 6$? You may wish to repeat with $n = 20$.

   If MINITAB is available, you could use the commands

Calc > Random Data > Uniform.
Type *150* after **Generate** and *C1 – C6* in **Store in Column(s)** Click **OK.**
Calc > Row Statistics. Click **Mean** and type *C1 – C6* in **Input Variables.**
Type C7 in **Store.** Click **OK.**

The 150 means in C7 can then be summarized using the dialog box sequence described in Chapter 2.

Stat > Basic Statistics > Graphical summary.
Type C7 in **Variables.** Click **OK.**

8

Drawing Inferences from Large Samples

Building Strong Evidence
from Diverse Individual Cases

One of the major contributions of statistics to modern thinking is the understanding that information on single, highly variable observations can be combined in great numbers to obtain very precise information about a population.

Although each individual is satisfied or not satisfied with his or her job, a sample survey can obtain accurate information about the population proportion that are satisfied.

© Ron Chapple/FPG International/Getty Images.

73% are satisfied with their present job.
56% are satisfied with their standard of living.

1.   INTRODUCTION

Inferences are generalizations about a population that are made on the basis of a sample collected from the population. For instance, a researcher interested in the growth of pine trees plants 40 seedlings. The heights of these 40 plants would be a sample that is hopefully representative of the population consisting of all current and future seedlings that could be planted.

More specifically, we begin by modeling the population by a probability distribution which has a numerical feature of interest called a **parameter**. A random sample from the population distribution will provide information about the parameter.

The problem of statistical inference arises when we wish to make generalizations about a population when only a sample will be available. Once a sample is observed, its main features can be determined by the methods of descriptive summary discussed in Chapters 2 and 3. However, more often than not, our principal concern is with not just the particular data set, but what can be said about the population based on the information extracted from analyzing the sample data. We call these generalizations **statistical inferences** or just **inferences**.

Consider a study on the effectiveness of a diet program in which 30 participants report their weight loss. We have on hand a sample of 30 measurements of weight loss. But is the goal of the study confined to this particular group of 30 persons? No, it is not. We need to evaluate the effectiveness of the diet program for the population of potential users. The sample measurements must, of course, provide the basis for any conclusions.

> **Statistical inference** deals with drawing conclusions about population parameters from an analysis of the sample data.

Any inference about a population parameter will involve some uncertainty because it is based on a sample rather than the entire population. To be meaningful, a statistical inference must include a specification of the uncertainty that is determined using the ideas of probability and the sampling distribution of the statistic.

The purpose of an investigation, spelled out in a clear statement of purpose as described in Chapter 1, can reveal the nature of the inference required to answer important questions.

The two most important types of inferences are (1) **estimation of parameter(s)** and (2) **testing of statistical hypotheses**. The true value of a parameter is an unknown constant that can be correctly ascertained only by an exhaustive study of the population, if indeed that were possible. Our objective may be to obtain a guess or an estimate of the unknown true value along with a determination of its accuracy. This type of inference is called **estimation of parameters**. An alternative objective may be to examine whether the sample data support or contradict the investigator's conjecture about the true value of the parameter. This latter type of inference is called **testing of statistical hypotheses**.

Example 1   Types of Inference: Point Estimation, Interval Estimation, and Testing Hypotheses

To study the growth of pine trees at an early stage, a nursery worker records 40 measurements of the heights of one-year-old red pine seedlings. This set of measurements appears in Table 1.

TABLE 1   Heights of One-Year-Old Red Pine Seedlings Measured in Centimeters

2.6	1.9	1.8	1.6	1.4	2.2	1.2	1.6
1.6	1.5	1.4	1.6	2.3	1.5	1.1	1.6
2.0	1.5	1.7	1.5	1.6	2.1	2.8	1.0
1.2	1.2	1.8	1.7	0.8	1.5	2.0	2.2
1.5	1.6	2.2	2.1	3.1	1.7	1.7	1.2

Courtesy of Professor Alan Ek.

Employing the ideas of Chapter 2, we can calculate a descriptive summary for this set of measurements.

Sample mean $\bar{x}$ = 1.715   Sample standard deviation s = .475
Sample median = 1.6   First quartile = 1.5   Third quartile = 2.0

However, the target of our investigation is not just the particular set of measurements recorded, but also concerns the vast (infinite) population of the heights of all possible one-year-old pine seedlings. The population distribution of the heights is unknown to us and so are the parameters such as the population mean μ and the population standard deviation σ. If we take the view that the 40 observations represent a random sample from the population distribution of heights, one goal of this study may be to "learn about μ." More specifically, depending on the purpose of the study, we may wish to do one, two, or all three of the following:

1. Estimate a single value for the unknown μ (**point estimation**).
2. Determine an interval of plausible values for μ (**interval estimation**).
3. Decide whether or not the mean height μ is 1.9 centimeters, which was previously found to be the mean height of a different stock of pine seedlings (**testing statistical hypotheses**).

Example 2   Inferences about an Unknown Proportion

A market researcher wishes to determine what proportion of new-car buyers in her city are satisfied with their new car one year after the purchase. She feels, correctly, that this assessment could be made quickly and effectively by sampling a small fraction of the new-car buyers. The persons selected will be asked if they are satisfied as opposed to not satisfied. Suppose that a sample of 200 randomly selected new-car purchasers are interviewed and 168 say they are satisfied.

A descriptive summary of this finding is provided by

$$\text{Sample proportion satisfied} = \hat{p} = \frac{168}{200} = .84$$

Here the target of our investigation is the proportion of new-car purchasers who are satisfied, p, in the entire collection of new-car buyers in the city. The value of p is unknown. The sample proportion $\hat{p} = .84$ sheds some light on p, but it is subject to some error since it draws only on a part of the population. The investigator would like to evaluate its margin of error and provide an interval of plausible values of p.

She may wish to test the hypothesis that proportion satisfied, p, for her city is not lower than the value given by a nationwide vehicle satisfaction study.

2.   POINT ESTIMATION OF A POPULATION MEAN

The object of point estimation is to calculate, from the sample data, a single number that is likely to be close to the unknown value of the parameter. The available information is assumed to be in the form of a random sample X_1, X_2, ... , X_n of size n taken from the population. We wish to formulate a statistic such that its value computed from the sample data would reflect the value of the population parameter as closely as possible.

> A statistic intended for estimating a parameter is called a **point estimator**, or simply an **estimator**. The standard deviation of an estimator is called its **standard error**: S.E.

When we estimate a population mean from a random sample, perhaps the most intuitive estimator is the sample mean,

$$\overline{X} = \frac{X_1 + X_2 + \cdots + X_n}{n}$$

For instance, to estimate the mean height of the population of pine seedlings in Example 1, we would naturally compute the mean of the sample measurements. Employing the estimator $\overline{X}$, with the data of Table 1, we get the result $\overline{x} = 1.715$ centimeters, which we call a **point estimate**, or simply an **estimate of μ**.

Without an assessment of accuracy, a single number quoted as an estimate may not serve a very useful purpose. We must indicate the extent of variability in the distribution of the estimator. The standard deviation, alternatively called the **standard error** of the estimator, provides information about its variability.

In order to study the properties of the sample mean $\overline{X}$ as an estimator of the population mean μ, let us review the results from Chapter 7.

1. $E(\overline{X}) = \mu$.

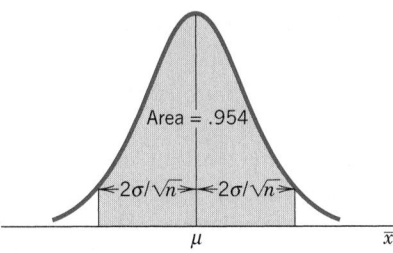

Figure 1   Approximate normal distribution of $\overline{X}$.

2. $\text{sd}(\overline{X}) = \dfrac{\sigma}{\sqrt{n}}$ so S.E.$(\overline{X}) = \dfrac{\sigma}{\sqrt{n}}.$

3. With large n, $\overline{X}$ is nearly normally distributed with mean μ and standard deviation $\sigma/\sqrt{n}$.

The first two results show that the distribution of $\overline{X}$ is centered around μ and its standard error is $\sigma/\sqrt{n}$, where σ is the population standard deviation and n the sample size.

To understand how closely $\overline{X}$ is expected to estimate μ, we now examine the third result, which is depicted in Figure 1. Recall that, in a normal distribution, the interval running two standard deviations on either side of the mean contains probability .954. Thus, prior to sampling, the probability is .954 that the estimator $\overline{X}$ will be at most a distance $2\sigma/\sqrt{n}$ from the true parameter value μ. This probability statement can be rephrased by saying that **when we are estimating μ by $\overline{X}$, the 95.4%** error margin is $2\sigma/\sqrt{n}$.

Use of the probability .954, which corresponds to the multiplier 2 of the standard error, is by no means universal. The following notation will facilitate our writing of an expression for the $100(1 - \alpha)\%$ error margin, where $1 - \alpha$ denotes the desired high probability such as .95 or .90.

Notation

$$z_{\alpha/2} = \text{Upper } \alpha/2 \text{ point of standard normal distribution}$$

That is, the area to the right of $z_{\alpha/2}$ is $\alpha/2$, and the area between $-z_{\alpha/2}$ and $z_{\alpha/2}$ is $1 - \alpha$ (see Figure 2).

A few values of $z_{\alpha/2}$ obtained from the normal table appear in Table 2 for easy reference. To illustrate the notation, suppose we want to determine the 90% error margin. We then set $1 - \alpha = .90$ so $\alpha/2 = .05$ and, from Table 2, we have $z_{.05} = 1.645$. Therefore, when estimating μ by $\overline{X}$, the 90% error margin is $1.645\,\sigma/\sqrt{n}$.

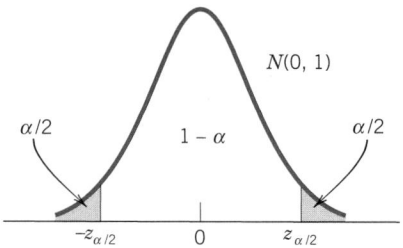

Figure 2   The notation $z_{\alpha/2}$.

TABLE 2   Values of $z_{\alpha/2}$

$1 - \alpha$	.80	.85	.90	.95	.99
$z_{\alpha/2}$	1.28	1.44	1.645	1.96	2.58

A minor difficulty remains in computing the standard error of $\overline{X}$. The expression involves the unknown population standard deviation σ, but we can estimate σ by the sample standard deviation.

$$S = \sqrt{\frac{\sum_{i=1}^{n} (X_i - \overline{X})^2}{n - 1}}$$

When n is large, the effect of estimating the standard error $\sigma/\sqrt{n}$ by $S/\sqrt{n}$ can be neglected. We now summarize.

Point Estimation of the Mean

Parameter: Population mean μ.
  Data: $X_1, \ldots, X_n$ (a random sample of size n)
Estimator: $\overline{X}$ (sample mean)

$$\text{S.E.}(\overline{X}) = \frac{\sigma}{\sqrt{n}} \qquad \text{Estimated S.E.}(\overline{X}) = \frac{S}{\sqrt{n}}$$

For large n, the $100(1 - \alpha)\%$ error margin is $z_{\alpha/2}\,\sigma/\sqrt{n}$.
(If σ is unknown, use S in place of σ.)

Example 3   Point Estimation of the Mean Height of Seedlings

From the data of Example 1, consisting of 40 measurements of the heights of one-year-old red pine seedlings, give a point estimate of the population mean height and state a 95% error margin.

SOLUTION   The sample mean and the standard deviation computed from the 40 measurements in Table 1 are

$$\bar{x} = \frac{\sum x_i}{40} = 1.715$$

$$s = \sqrt{\frac{\sum (x_i - \bar{x})^2}{39}} = \sqrt{.2254} = .475$$

To calculate the 95% error margin, we set $1 - \alpha = .95$ so that $\alpha/2 = .025$ and $z_{\alpha/2} = 1.96$. Therefore, the 95% error margin is

$$\frac{1.96\ s}{\sqrt{n}} = \frac{1.96 \times .475}{\sqrt{40}} = .15 \text{ centimeter}$$

Our estimate of the population mean height is 1.715 centimeters. We do not expect the population mean to be exactly this value and we attach an error of plus and minus .15.

Caution: (a) Standard error should not be interpreted as the "typical" error in a problem of estimation as the word "standard" may suggest. For instance, when S.E.$(\overline{X}) = .3$, we should not think that the error $(\overline{X} - \mu)$ is likely to be .3, but rather, prior to observing the data, the probability is approximately .954 that the error will be within $\pm 2(\text{S.E.}) = \pm .6$.

(b) An estimate and its variability are often reported in either of the forms: estimate $\pm$S.E. or estimate $\pm 2(\text{S.E.})$. In reporting a numerical result such as 53.4 ± 4.6, we must specify whether 4.6 represents S.E., $2(\text{S.E.})$, or some other multiple of the standard error.

DETERMINING THE SAMPLE SIZE

During the planning stage of an investigation, it is important to address the question of sample size. Because sampling is costly and time-consuming, the investigator needs to know beforehand the sample size required to give the desired precision.

In order to determine how large a sample is needed for estimating a population mean, we must specify

$$d = \text{Desired error margin}$$

and

$$1 - \alpha = \text{Probability associated with error margin}$$

Referring to the expression for a $100(1 - \alpha)\%$ error margin, we then equate:

$$z_{\alpha/2} \frac{\sigma}{\sqrt{n}} = d$$

This gives an equation in which n is unknown. Solving for n, we obtain

$$n = \left[\frac{z_{\alpha/2}\,\sigma}{d}\right]^2$$

which determines the required sample size. Of course, the solution is rounded to the next higher integer, because a sample size cannot be fractional.

This determination of sample size is valid provided $n > 30$, so that the normal approximation to $\overline{X}$ is satisfactory.

To be $100(1 - \alpha)\%$ sure that the error of estimation $|\overline{X} - \mu|$ does not exceed d, the **required sample size** is

$$n = \left[\frac{z_{\alpha/2}\,\sigma}{d}\right]^2$$

If σ is completely unknown, a small-scale preliminary sampling is necessary to obtain an estimate of σ to be used in the formula to compute n.

Example 4   Determining a Sample Size for Collecting Water Samples

A limnologist wishes to estimate the mean phosphate content per unit volume of lake water. It is known from studies in previous years that the standard deviation has a fairly stable value of $\sigma = 4$. How many water samples must the limnologist analyze to be 90% certain that the error of estimation does not exceed 0.8 milligrams?

SOLUTION   Here $\sigma = 4$ and $1 - \alpha = .90$, so $\alpha/2 = .05$. The upper .05 point of the $N(0, 1)$ distribution is $z_{.05} = 1.645$. The tolerable error is $d = .8$. Computing

$$n = \left[\frac{1.645 \times 4}{.8}\right]^2 = 67.65$$

we determine that the required sample size is $n = 68$.

Exercises

8.1   For estimating a population mean with the sample mean $\overline{X}$, find (i) the standard error of $\overline{X}$ and (ii) the $100(1 - \alpha)\%$ error margin in each case.

(a) $n = 152$ $\sigma = 22$ $1 - \alpha = .95$

(b) $n = 85$ $\sigma = 8.2$ $1 - \alpha = .99$

(c) $n = 295$ $\sigma = 56$ $1 - \alpha = .92$

8.2   Determine the point estimate of the population mean μ and its $100(1 - \alpha)\%$ margin of error in each case.

(a) $n = 120$ $\bar{x} = 81.2$
     $s = 5.96$ $1 - \alpha = .975$

(b) $n = 240$ $\bar{x} = 925$
     $s = 87$ $1 - \alpha = .88$

(c) $n = 1200$ $\bar{x} = .628$
     $s = .095$ $1 - \alpha = .90$

8.3   Consider the problem of estimating a population mean μ based on a random sample of size n from the population. Compute a point estimate of μ and the estimated standard error in each of the following cases.

(a) $n = 70$ $\sum x_i = 752$ $\sum (x_i - \bar{x})^2 = 235$
(b) $n = 90$ $\sum x_i = 2653$ $\sum (x_i - \bar{x})^2 = 546$
(c) $n = 160$ $\sum x_i = 3985$ $\sum (x_i - \bar{x})^2 = 745$

8.4   Determine a 95.4% error margin for the estimation of μ in each of the three cases in Exercise 8.3.

8.5   A credit company randomly selected 50 contested items and recorded the dollar amount being contested. These contested items had sample mean $\bar{x} = 75.43$ dollars and $s = 24.73$ dollars. Construct a point estimate for the population mean contested amount, μ, and give its 90% error margin.

8.6   A manager at a power company monitored the employee time required to process high-efficiency lamp bulb rebates. A random sample of 40 applications gave a sample mean time of 3.8 minutes and a standard deviation of 1.2 minutes. Construct a point estimate for the population mean time to process, μ, and give its 90% error margin.

8.7   When estimating μ from a large sample, suppose that one has found the 95% error margin of $\overline{X}$ to be 4.2. From this information, determine:

(a) The estimated S.E. of $\overline{X}$.

(b) The 90% error margin.

8.8   For each case, determine the sample size n that is required for estimating the population mean. The population standard deviation σ and the desired error margin are specified.

(a) $\sigma = 4.8$ 95% error margin $= .75$
(b) $\sigma = 135$ 80% error margin $= 4.5$
(c) $\sigma = .082$ 98% error margin $= .025$

8.9   When estimating the mean of a population, how large a sample is required in order that the 95% error margin be:

(a) $\frac{1}{8}$ of the population standard deviation?

(b) 15% of the population standard deviation?

8.10   Assume that the standard deviation of the number of violent incidents in one hour of children's shows on television is 3.2. An investigator would like to be 99% sure that the true mean number of violent incidents per hour is estimated within 1.4 incidents. For how many randomly selected hours does she need to count the number of violent incidents?

8.11   Referring to Exercise 8.5, suppose that the survey of 50 contested items was, in fact, a pilot study intended to give an idea of the population standard deviation. Assuming $\sigma = \$25$, determine the sample size that is needed for estimating the population mean amount contested with a 98% error margin of $5.00.

8.12   Assume that the standard deviation of the heights of five-year-old boys is 3.5 inches. How many five-year-old boys need to be sampled if we want to be 90% sure that the population mean height is estimated within .5 inch?

8.13   Let the abbreviation PSLT stand for the percent of the gross family income that goes into paying state and local taxes. Suppose one wants to estimate the mean PSLT for the population of all families in New York City with gross incomes in the range $35,000 to $40,000. If $\sigma = 2.5$, how many such families should be surveyed if one wants to be 90% sure of being able to estimate the true mean PSLT within .5?

8.14   To estimate μ with a 90% error margin of 2.9 units, one has determined that the required sample size is 108. What then is the required sample size if one wants the 95% error margin to be 1.8 units?

3.   CONFIDENCE INTERVAL FOR A POPULATION MEAN

For point estimation, a single number lies in the forefront even though a standard error is attached. Instead, it is often more desirable to produce an interval of values that is likely to contain the true value of the parameter.

Ideally, we would like to be able to collect a sample and then use it to calculate an interval that would definitely contain the true value of the parameter. This goal, however, is not achievable because of sample-to-sample variation. Instead, we insist that before sampling the proposed interval will contain the true value with a specified high probability. This probability, called the level of confidence, is typically taken as .90, .95, or .99.

To develop this concept, we first confine our attention to the construction of a confidence interval for a population mean μ, assuming that the population is normal and the standard deviation σ is *known*. This restriction helps to simplify the initial presentation of the concept of a confidence interval. Later on, we will treat the more realistic case where σ is also unknown.

A probability statement about $\overline{X}$ based on the normal distribution provides the cornerstone for the development of a confidence interval. From Chapter 7, recall that when the population is normal, the distribution of $\overline{X}$ is also normal. It has mean μ and standard deviation $\sigma/\sqrt{n}$. Here μ is unknown, but $\sigma/\sqrt{n}$ is a known number because the sample size n is known and we have assumed that σ is known.

The normal table shows that the probability is .95 that a normal random variable will lie within 1.96 standard deviations from its mean. For $\overline{X}$, we then have

$$P\left[\mu - 1.96\frac{\sigma}{\sqrt{n}} < \overline{X} < \mu + 1.96\frac{\sigma}{\sqrt{n}}\right] = .95$$

as shown in Figure 3.

Now, the relation

$$\mu - 1.96\frac{\sigma}{\sqrt{n}} < \overline{X} \qquad \text{is the same as} \qquad \mu < \overline{X} + 1.96\frac{\sigma}{\sqrt{n}}$$

and

$$\overline{X} < \mu + 1.96\frac{\sigma}{\sqrt{n}} \qquad \text{is the same as} \qquad \overline{X} - 1.96\frac{\sigma}{\sqrt{n}} < \mu$$

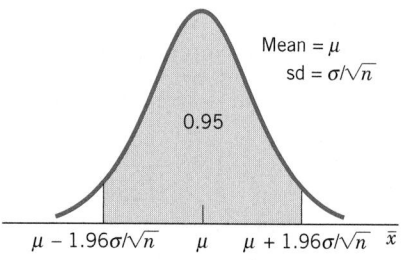

Figure 3   Normal distribution of $\overline{X}$.

as we can see by transposing $1.96\sigma/\sqrt{n}$ from one side of an inequality to the other. Therefore, the event

$$\left[\mu - 1.96\frac{\sigma}{\sqrt{n}} < \overline{X} < \mu + 1.96\frac{\sigma}{\sqrt{n}}\right]$$

is equivalent to

$$\left[\overline{X} - 1.96\frac{\sigma}{\sqrt{n}} < \mu < \overline{X} + 1.96\frac{\sigma}{\sqrt{n}}\right]$$

In essence, both events state that the difference $\overline{X} - \mu$ lies between $-1.96\sigma/\sqrt{n}$ and $1.96\sigma/\sqrt{n}$. Thus, the probability statement

$$P\left[\mu - 1.96\frac{\sigma}{\sqrt{n}} < \overline{X} < \mu + 1.96\frac{\sigma}{\sqrt{n}}\right] = .95$$

can also be expressed as

$$P\left[\overline{X} - 1.96\frac{\sigma}{\sqrt{n}} < \mu < \overline{X} + 1.96\frac{\sigma}{\sqrt{n}}\right] = .95$$

This second form tells us that, before we sample, the random interval from $\overline{X} - 1.96\sigma/\sqrt{n}$ to $\overline{X} + 1.96\sigma/\sqrt{n}$ will include the unknown parameter μ with a probability of .95. Because σ is assumed to be known, both the upper and lower endpoints can be computed as soon as the sample data are available. Guided by the above reasonings, we say that the interval

$$\left(\overline{X} - 1.96\frac{\sigma}{\sqrt{n}}, \quad \overline{X} + 1.96\frac{\sigma}{\sqrt{n}}\right)$$

or its realization $(\overline{x} - 1.96\sigma/\sqrt{n}, \overline{x} + 1.96\sigma/\sqrt{n})$ is a 95% confidence interval for μ when the population is normal and σ known.

Example 5   Calculating a Confidence Interval—Normal Population σ Known

The daily carbon monoxide (CO) emission from a large production plant will be measured on 25 randomly selected weekdays. The production process is always being modified and the current mean value of daily CO emissions μ is unknown. Data collected over several years confirm that, for each year, the distribution of CO emission is normal with a standard deviation of .8 ton.

Suppose the sample mean is found to be $\overline{x} = 2.7$ tons. Construct a 95% confidence interval for the current daily mean emission μ.

SOLUTION   The population is normal, and the observed value $\overline{x} = 2.7$.

$$\left(2.7 - 1.96\frac{.8}{\sqrt{25}}, \quad 42.7 + 1.96\frac{.8}{\sqrt{25}}\right) = (2.39, 3.01)$$

is a 95% confidence interval for μ. Since μ is unknown, we cannot determine whether or not μ lies in this interval.

Referring to the confidence interval obtained in Example 5, we must **not** speak of the probability of the fixed interval (2.39, 3.01) covering the true mean μ. The particular interval (2.39, 3.01) either does or does not cover μ, and we will never know which is the case.

We need not always tie our discussion of confidence intervals to the choice of a 95% level of confidence. An investigator may wish to specify a different high probability. We denote this probability by $1 - \alpha$ and speak of a $100(1 - \alpha)\%$ confidence interval. The only change is to replace 1.96 with $z_{\alpha/2}$, where $z_{\alpha/2}$ denotes the upper $\alpha/2$ point of the standard normal distribution (i.e., the area to the right of $z_{\alpha/2}$ is $\alpha/2$, as shown in Figure 2).

In summary, when the population is normal and σ is known, a $100(1 - \alpha)\%$ confidence interval for μ is given by

$$\left(\overline{X} - z_{\alpha/2} \frac{\sigma}{\sqrt{n}}, \quad \overline{X} + z_{\alpha/2} \frac{\sigma}{\sqrt{n}} \right)$$

INTERPRETATION OF CONFIDENCE INTERVALS

To better understand the meaning of a confidence statement, we use the computer to perform repeated samplings from a normal distribution with $\mu = 100$ and $\sigma = 10$. Ten samples of size 7 are selected, and a 95% confidence interval $\overline{x} \pm 1.96 \times 10/\sqrt{7}$ is computed from each. For the first sample, $\overline{x} = 104.3$ and the interval is 104.3 ± 7.4, or 96.9 to 111.7. This and the other intervals are illustrated in Figure 4, where each vertical line segment represents one confi-

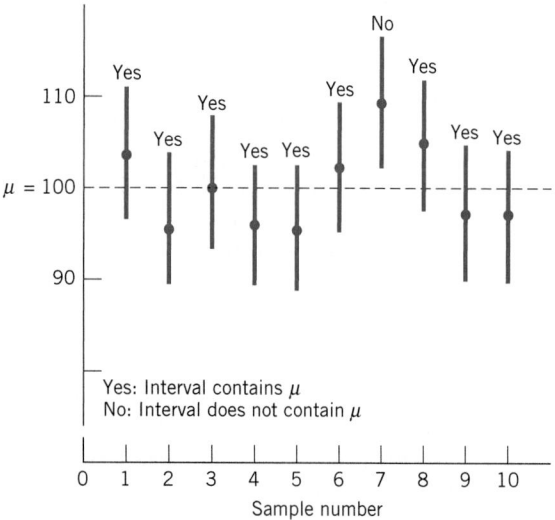

Figure 4   Interpretation of the confidence interval for μ.

dence interval. The midpoint of a line is the observed value of $\overline{X}$ for that particular sample. Also note that all the intervals are of the same length $2 \times 1.96\sigma/\sqrt{n} = 14.8$. Of the 10 intervals shown, 9 cover the true value of μ. This is not surprising, because the specified probability .95 represents the long-run relative frequency of these intervals covering the true $\mu = 100$.

Because confidence interval statements are the most useful way to communicate information obtained from a sample, certain aspects of their formulation merit special emphasis. Stated in terms of a 95% confidence interval for μ, these are:

1.   Before we sample, a confidence interval $(\overline{X} - 1.96\sigma/\sqrt{n}, \overline{X} + 1.96\sigma/\sqrt{n})$ is a random interval that attempts to cover the true value of the parameter μ.

2.   The probability

$$P\left[\overline{X} - 1.96\,\frac{\sigma}{\sqrt{n}} < \mu < \overline{X} + 1.96\,\frac{\sigma}{\sqrt{n}}\right] = .95$$

interpreted as the long-run relative frequency over many repetitions of sampling asserts that about 95% of the intervals will cover μ.

3.   Once $\overline{x}$ is calculated from an observed sample, the interval

$$\left(\overline{x} - 1.96\,\frac{\sigma}{\sqrt{n}}, \quad \overline{x} + 1.96\,\frac{\sigma}{\sqrt{n}}\right)$$

which is a realization of the random interval, is presented as a 95% confidence interval for μ. A numerical interval having been determined, it is no longer sensible to speak about the probability of its covering a fixed quantity μ.

4.   In any application we never know if the 95% confidence interval covers the unknown mean μ. Relying on the long-run relative frequency of coverage in property 2, we adopt the terminology confidence once the interval is calculated.

At this point, one might protest, "I have only one sample and I am not really interested in repeated sampling." But if the confidence estimation techniques presented in this text are mastered and followed each time a problem of interval estimation arises, then over a lifetime approximately 95% of the intervals will cover the true parameter. Of course, this is contingent on the validity of the assumptions underlying the techniques—independent normal observations here.

LARGE SAMPLE CONFIDENCE INTERVALS FOR μ

Having established the basic concepts underlying confidence interval statements, we now turn to the more realistic situation for which the population standard deviation σ is unknown. We require the sample size n to be large in or-

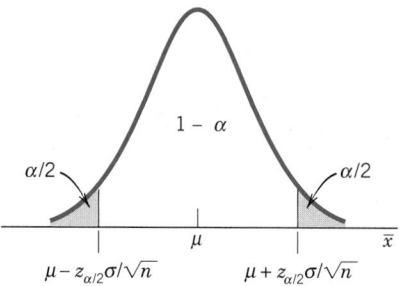

Figure 5   Normal distribution of $\overline{X}$.

der to dispense with the assumption of a normal population. The central limit theorem then tells us that $\overline{X}$ is nearly normal whatever the form of the population. Referring to the normal distribution of $\overline{X}$ in Figure 5 and the discussion accompanying Figure 2, we again have the probability statement

$$P\left[\overline{X} - z_{\alpha/2}\frac{\sigma}{\sqrt{n}} < \mu < \overline{X} + z_{\alpha/2}\frac{\sigma}{\sqrt{n}}\right] = 1 - \alpha$$

(Strictly speaking, this probability is approximately $1 - \alpha$ for a nonnormal population.) Even though the interval

$$\left(\overline{X} - z_{\alpha/2}\frac{\sigma}{\sqrt{n}}, \quad \overline{X} + z_{\alpha/2}\frac{\sigma}{\sqrt{n}}\right)$$

will include μ with the probability $1 - \alpha$, it does not serve as a confidence interval because it involves the unknown quantity σ. However, because n is large, replacing $\sigma/\sqrt{n}$ with its estimator $S/\sqrt{n}$ does not appreciably affect the probability statement. Summarizing, we find that the large sample confidence interval for μ has the form

Estimate $\pm$ (z Value)(Estimated standard error)

Large Sample Confidence Interval for μ

When n is large, a $100(1 - \alpha)\%$ confidence interval for μ is given by

$$\left(\overline{X} - z_{\alpha/2}\frac{S}{\sqrt{n}}, \quad \overline{X} + z_{\alpha/2}\frac{S}{\sqrt{n}}\right)$$

where S is the sample standard deviation.

Example 6   A Confidence Interval for Mean Weekly Work Time

Many students work part time during the school year. To determine the hours a week a typical student employee works, a sample of 68 working students was selected at a large midwestern university. The responses to the question

"How many hours did you work last week?" had a mean of 14.3 hours and a standard deviation of 5.7 hours.

Compute (a) 90% and (b) 80% confidence intervals for the mean number of hours worked per week.

SOLUTION   The sample size $n = 68$ is large, so a normal approximation for the distribution of the sample mean $\overline{X}$ is appropriate. From the sample data, we know that

$$\overline{x} = 14.3 \text{ hours} \quad \text{and} \quad s = 5.7 \text{ hours}$$

(a) With $1 - \alpha = .90$, we have $\alpha/2 = .05$, and $z_{.05} = 1.645$,

$$1.645 \frac{s}{\sqrt{n}} = \frac{1.645 \times 5.7}{\sqrt{68}} = 1.14$$

The 90% confidence interval for the population mean of number of hours worked μ becomes

$$\left(\overline{x} - 1.645 \frac{s}{\sqrt{n}}, \ \overline{x} + 1.645 \frac{s}{\sqrt{n}} \right) = (14.6 - 1.14, 14.6 + 1.14)$$

or approximately $(13.5, 15.7)$ hours per week. This means that we can be 90% confident that the mean hours per week μ is in the interval 13.5 to 15.7 hours. We have this confidence because 90% of the random samples of 68 working students would produce intervals $\overline{x} \pm 1.645 \, s/\sqrt{n}$ that contain μ.

(b) With $1 - \alpha = .80$, we have $\alpha/2 = .10$, and $z_{.10} = 1.28$, so

$$1.28 \frac{s}{\sqrt{n}} = \frac{1.28 \times 5.7}{\sqrt{68}} = .88$$

The 80% confidence interval for μ becomes

$$(14.6 - .88, 14.6 + .88) \quad \text{or} \quad (13.7, 15.5)$$

Comparing the two confidence intervals, we note that the 80% confidence interval is shorter than the 90% interval. A shorter interval seems to give a more precise location for μ but suffers from a lower long-run frequency of being correct.

CONFIDENCE INTERVAL FOR A PARAMETER

The concept of a confidence interval applies to any parameter, not just the mean. It requires that a lower limit L and an upper limit U be computed from the sample data. Then the random interval from L to U must have the specified probability of covering the true value of the parameter. The large sample $100(1 - \alpha)\%$ confidence interval for μ has

$$L = \overline{X} - z_{\alpha/2} \frac{S}{\sqrt{n}} \qquad U = \overline{X} + z_{\alpha/2} \frac{S}{\sqrt{n}}$$

> **Definition of a Confidence Interval for a Parameter**
>
> An interval (L, U) is a $100(1 - \alpha)\%$ confidence interval for a parameter if
> $$P[L < \text{Parameter} < U] = 1 - \alpha$$
> and the endpoints L and U are computable from the sample.

Example 7   A Confidence Interval for the Mean Time to Complete a Test

Madison recruits for the fire department need to complete a timed test that simulates working conditions. It includes placing a ladder against a building, pulling out a section of fire hose, dragging a weighted object, and crawling in a simulated attic environment. The times, in seconds, for recruits to complete the test for Madison firefighter are

425	389	380	421	438	331	368	417	403	416	385	315
427	417	386	386	378	300	321	286	269	225	268	317
287	256	334	342	269	226	291	280	221	283	302	308
296	266	238	286	317	276	254	278	247	336	296	259
270	302	281	228	317	312	327	288	395	240	264	246
294	254	222	285	254	264	277	266	228	347	322	232
365	356	261	293	354	236	285	303	275	403	268	250
279	400	370	399	438	287	363	350	278	278	234	266
319	276	291	352	313	262	289	273	317	328	292	279
289	312	334	294	297	304	240	303	255	305	252	286
297	353	350	276	333	285	317	296	276	247	339	328
267	305	291	269	386	264	299	261	284	302	342	304
336	291	294	323	320	289	339	292	373	410	257	406
374	268										

Obtain a 95% confidence interval for the mean time of recruits who complete the test.

SOLUTION   A computer calculation gives

```
SAMPLE SIZE    158
MEAN           307.77
STD DEV        51.852
```

Since $1 - \alpha = .95$, $\alpha/2 = .025$, and $z_{.025} = 1.96$, the large sample 95% confidence interval for μ becomes

$$\left(\bar{x} - z_{\alpha/2} \frac{s}{\sqrt{n}}, \quad \bar{x} + z_{\alpha/2} \frac{s}{\sqrt{n}} \right)$$

$$= \left(307.77 - 1.96 \frac{51.852}{\sqrt{158}}, \quad 307.77 + 1.96 \frac{51.852}{\sqrt{158}} \right)$$

or

$$(299.69, 315.86) \text{ seconds}$$

When the sample size is large, the sample also contains information on the shape of the distribution that can be elicited by graphical displays. Figure 6 gives the stem-and-leaf display, with the data rounded to two places, accompanied by the boxplot. The confidence interval pertains to the mean of a population with a long right-hand tail.

```
STEM-AND-LEAF of TIME N   158
LEAF UNIT = 10
      2 | 2222223333
      2 | 44444555555555
      2 | 666666666666666667777777777777
      2 | 8888888888888888899999999999999999
      3 | 000000000001111111111
      3 | 222222233333333
      3 | 444555555
      3 | 6667777
      3 | 88888899
      3 | 00001111
      4 | 22233
```

Figure 6   A stem-and-leaf display and boxplot give more information about the form of the population.

Exercises

8.15   Determine a 90% confidence interval for μ if $n = 63, \bar{x} = 81.3$, and $s = 5.8$.

8.16   Determine a 99% confidence interval for μ if $n = 90, \bar{x} = .920$, and $s = .057$.

8.17   Each day of the year, a large sample of cellular phone calls is selected and a 95% confidence interval is calculated for the mean length of all cellular phone calls on that day. Of these 365 confidence intervals, one for each day of the year, approximately how many will cover their respective population means? Explain your reasoning.

8.18   A forester measures 100 needles off a pine tree and finds $\bar{x} = 3.1$ centimeters and $s = 0.7$ centimeter. She reports that a 95% confidence interval for the mean needle length is

$$3.1 - 1.96 \frac{0.7}{\sqrt{100}} \quad \text{to} \quad 3.1 + 1.96 \frac{0.7}{\sqrt{100}}$$

$$\text{or} \quad (2.96, 3.24)$$

(a)   Is the statement correct?

(b)   Does the interval (2.96, 3.24) cover the true mean? Explain.

8.19   In a study on the nutritional qualities of fast foods, the amount of fat was measured for a random sample of 35 hamburgers of a particular restaurant chain. The sample mean and standard deviation were found to be 30.2 and 3.8 grams, respectively. Use these data to construct a 95% confidence interval for the mean fat content in hamburgers served in these restaurants.

8.20   In the same study described in Exercise 8.19, the sodium content was also measured for the sampled hamburgers, and the sample mean and standard deviation were 658 and 47 milligrams, respectively. Determine a 98% confidence interval for the true mean sodium content.

8.21   An entomologist sprayed 120 adult Melon flies with a specific low concentration of malathion and observed their survival times. The mean and standard deviation were found to be 18.3 and 5.2 days, respectively. Use these data to construct a 99% confidence interval for the true mean survival time.

8.22   From a random sample of 70 high school seniors in a large school district, the mean and standard deviation of the verbal scores in the Scholastic Aptitude Test (SAT) are found to be 424 and 45, respectively. Based on this sample, construct a 98% confidence interval for the mean verbal score in the SAT for the population of all seniors in this school district.

8.23   Refer to Exercise 8.22. The sample mean and standard deviation of the math scores in the SAT are found to be 493 and 72, respectively. Determine a 95% confidence interval for the mean math score of all seniors in the school district.

8.24   Referring to Example 7, where the 158 times to complete the firefighter test have mean 307.77 and standard deviation 51.852, obtain a 99% confidence interval for the mean time of all possible recruits who would complete the test.

8.25   Based on a survey of 140 employed persons in a city, the mean and standard deviation of the commuting distances between home and the principal place of business are found to be 8.6 and 4.3 miles, respectively. Determine a 90% confidence interval for the mean commuting distance for the population of all employed persons in the city.

8.26   A manager at a power company monitored the employee time required to process high-efficiency lamp bulb rebates. A random sample of 40 applications gave a sample mean time of 3.8 minutes and a standard deviation of 1.2 minutes. Construct a 90% confidence interval for the mean time to process μ.

8.27   A credit company randomly selected 50 contested items and recorded the dollar amount being contested. These contested items had a sample mean $\bar{x} = 75.43$ dollars and $s = 24.73$ dollars. Construct a 95% confidence interval for the mean amount contested, μ.

8.28   In a study to determine whether a certain stimulant produces hyperactivity, 55 mice were injected with 10 micrograms of the stimulant. Afterward, each mouse is given a hyperactivity rating score. The mean score was $\bar{x} = 14.9$ and $s = 2.8$. Give a 95% confidence interval for the population mean score μ.

8.29   Refer to the Statistics in Context section of Chapter 7 concerning monthly changes in the Canadian to U.S. exchange rate. A computer calculation gives $\bar{x} = .0021$ and $s = .0159$ for the $n = 147$ monthly changes. Find a 95% confidence interval for the mean monthly change.

8.30   Determine a 90% confidence interval for the mean amount contested in Exercise 8.27.

8.31   Refer to the 40 height measurements given in Table 1 and their summary statistics reported in Example 1. Calculate a 99% confidence interval for the population mean height.

8.32   Radiation measurements on a sample of 65 microwave ovens produced $\bar{x} = .11$ and $s = .06$. Determine a 95% confidence interval for the mean radiation.

8.33   Refer to the data on the growth of female salmon in the marine environment in Table D.7 of the Data Bank. A computer calculation gives a 95% confidence interval.

```
One-Sample Z: Fmarine
```

The assumed sigma = 41.05

Variable	N	Mean	StDev	95.0% CI
Fmarine	40	429.15	41.05	(416.43, 441.87)

(a)   Does the 95% confidence interval cover the true mean growth of all female salmon in that marine environment?

(b)   Why are you 95% confident that the interval (416.43, 441.87) covers the true mean?

8.34   Refer to the data on the girth, in centimeters, of grizzly bears in Table D.8 of the Data Bank. A computer calculation gives

```
One-Sample Z: Girth
```

The assumed sigma = 21.79

Variable	N	Mean	StDev	95.0% CI
Girth	61	93.39	21.79	(87.93, 98.86)

(a) Does the 95% confidence interval cover the true mean girth of all grizzly bears in the area of the study?

(b) Why are you 95% confident that the interval (87.93, 98.86) covers the true mean?

8.35   The amount of PCBs (polychlorinated biphenyls) was measured in 40 samples of soil that were treated with contaminated sludge. The following summary statistics were obtained.

$$\bar{x} = 3.56 \qquad s = .5 \text{ ppm}$$

(a) Obtain a 95% confidence interval for the population mean μ, amount of PCBs in the soil.

Answer parts (b), (c), and (d) Yes, No, or Cannot tell. Explain your answer.

(b) Does the sample mean PCB content lie in your interval obtained in part (a)?

(c) Does the population mean PCB content lie in your interval obtained in part (a)?

(d) It is likely that 95% of the data lie in your interval obtained in part (a)?

8.36   A national fast food chain, with thousands of franchise locations, needed to audit the books at each location. They first selected a sample of 50 locations and performed the audit. They determined that a 95% confidence interval for the mean time to complete an audit is

$$(28.4 \text{ hours}, 52.7 \text{ hours})$$

Answer the following questions "Yes," "No," or "Cannot tell" and justify your answer.

(a) Does the population mean lie in the interval (28.4, 52.7)?

(b) Does the sample mean lie in the interval (28.4, 52.7)?

(c) For a future sample of 50 franchise locations, will the sample mean fall in the interval (28.4, 52.7)?

(d) Does 95% of the sample data lie in the interval (28.4, 52.7)?

4.   TESTING HYPOTHESES ABOUT A POPULATION MEAN

Broadly speaking, the goal of testing statistical hypotheses is to determine if a claim or conjecture about some feature of the population, a parameter, is strongly supported by the information obtained from the sample data. Here we illustrate the testing of hypotheses concerning a population mean μ. The available data will be assumed to be a random sample of size n from the population of interest. Further, the sample size n will be large ($n > 30$ for a rule of thumb).

The formulation of a hypotheses testing problem and then the steps for solving it require a number of definitions and concepts. We will introduce these key statistical concepts

Null hypothesis and the alternative hypothesis

Type I and Type II errors

Level of significance

Rejection region

P–value

in the context of a specific problem to help integrate them with intuitive reasoning.

> PROBLEM: *Can an upgrade reduce the mean transaction time at automated teller machines?* At peak periods, customers are subject to unreasonably long waits before receiving cash. To help alleviate this difficulty, the bank wants to reduce the time it takes a customer to complete a transaction. From extensive records, it is found that the transaction times have a distribution with mean 270 and standard deviation 24 seconds. The teller machine vendor suggests that a new software and hardware upgrade will reduce the mean time for a customer to complete a transaction. For experimental verification, a random sample of 38 transaction times will be taken at a machine with the upgrade and the sample mean $\overline{X}$ calculated. How should the result be used toward a statistical validation of the claim that the true (population) mean transaction time is less than 270 seconds?

Whenever we seek to establish a claim or conjecture on the basis of strong support from sample data, the problem is called one of hypothesis testing.

FORMULATING THE HYPOTHESES

In the language of statistics, the claim or the research hypothesis that we wish to establish is called the alternative hypothesis H_1. The opposite statement, one that nullifies the research hypothesis, is called the null hypothesis H_0. The word "null" in this context means that the assertion we are seeking to establish is actually void.

Formulation of H_0 and H_1

When our goal is to establish an assertion with substantive support obtained from the sample, the negation of the assertion is taken to be the null hypothesis H_0 and the assertion itself is taken to be the alternative hypothesis H_1.

Our initial question, "Is there strong evidence in support of the claim?" now translates to "Is there strong evidence for rejecting H_0?" The first version typically appears in the statement of a practical problem, whereas the second version is ingrained in the conduct of a statistical test. It is crucial to understand the correspondence between the two formulations of a question.

Before claiming that a statement is established statistically, adequate evidence from data must be produced to support it. A close analogy can be made to a court trial where the jury clings to the null hypothesis of "not guilty" unless there is convincing evidence of guilt. The intent of the hearings is to establish the assertion that the accused is guilty, rather than to prove that he or she is innocent.

	Court Trial	Testing Statistical Hypothesis
Requires strong evidence to establish:	Guilt.	Conjecture (research hypothesis).
Null hypothesis (H_0):	Not guilty.	Conjecture is false.
Alternative hypothesis (H_1):	Guilty.	Conjecture is true.
Attitude:	Uphold "not guilty" unless there is a strong evidence of guilt.	Retain the null hypothesis unless it makes the sample data very unlikely to happen.

False rejection of H_0 is a more serious error than failing to reject H_0 when H_1 is true.

Once H_0 and H_1 are formulated, our goal is to analyze the sample data in order to choose between them.

A decision rule, or a test of the null hypothesis, specifies a course of action by stating what sample information is to be used and how it is to be used in making a decision. Bear in mind that we are to make one of the following two decisions:

Decisions

Either

      Reject H_0 and conclude that H_1 is substantiated

or

      Retain H_0 and conclude that H_1 fails to be substantiated

---

Rejection of H_0 amounts to saying that H_1 is substantiated, whereas nonrejection or retention of H_0 means that H_1 fails to be substantiated. A key point is that a decision to reject H_0 must be based on strong evidence. Otherwise, the claim H_1 could not be established beyond a reasonable doubt.

In our problem of evaluating the upgraded teller machine, let μ be the population mean transaction time. Because μ is claimed to be lower than 270 seconds, we formulate the alternative hypothesis as $H_1 : \mu < 270$. According to the description of the problem, the researcher does not care to distinguish between the situations that $\mu = 270$ and $\mu > 270$ for the claim is false in either case. For this reason, it is customary to write the null hypothesis simply as a statement of no difference. Accordingly, we formulate the

Testing Problem

Test $H_0 : \mu = 270$ versus $H_1 : \mu < 270$

TEST CRITERION AND REJECTION REGION

Naturally, the sample mean $\overline{X}$, calculated from the measurements of $n = 38$ randomly selected transaction times, ought to be the basis for rejecting H_0 or not. The question now is: For what sort of values of $\overline{X}$ should we reject H_0? Because the claim states that μ is low (a left-sided alternative), only low values of $\overline{X}$ can contradict H_0 in favor of H_1. Therefore, a reasonable decision rule should be of the form

Reject H_0 if $\overline{X} \leq c$

Retain H_0 if $\overline{X} > c$

This decision rule is conveniently expressed as $R : \overline{X} \leq c$, where R stands for the rejection of H_0. Also, in this context, the set of outcomes $[\overline{X} \leq c]$ is called the rejection region or critical region, and the cutoff point c is called the critical value.

The cutoff point c must be specified in order to fully describe a decision rule. To this end, we consider the case when H_0 holds, that is, $\mu = 270$. Rejection of H_0 would then be a wrong decision, amounting to a false acceptance of the claim — a serious error. For an adequate protection against this kind of error, we must ensure that $P[\overline{X} \leq c]$ is very small when $\mu = 270$. For example, suppose that we wish to hold a low probability of $\alpha = .05$ for a wrong rejection of H_0. Then our task is to find the c that makes

$$P[\overline{X} \leq c] = .05 \qquad \text{when} \qquad \mu = 270$$

We know that, for large n, the distribution of $\overline{X}$ is approximately normal with mean μ and standard deviation $\sigma / \sqrt{n}$, whatever the form of the underlying population. Here $n = 38$ is large, and we initially assume that σ is known. Specifically, we assume that $\sigma = 24$ seconds, the same standard deviation as with the original money machines. Then, when $\mu = 270$, the distribution of $\overline{X}$ is $N(270, 24/\sqrt{38})$ so

$$Z = \frac{\overline{X} - 270}{24/\sqrt{38}}$$

has the $N(0, 1)$ distribution.

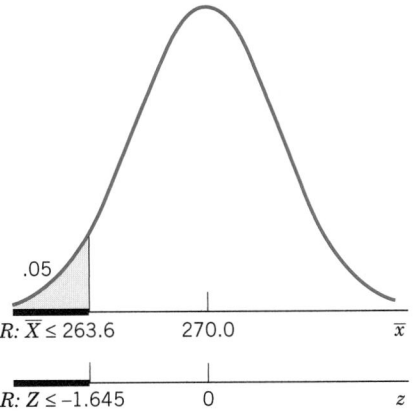

.05

$R: \overline{X} \le 263.6$ 270.0 $\overline{x}$

$R: Z \le -1.645$ 0 z

Figure 7   Rejection region with the cutoff $c = 263.6$.

Because $P[Z \le -1.645] = .05$, the cutoff c on the $\overline{x}$ scale must be 1.645 standard deviations below $\mu_0 = 270$ or

$$c = 270 - 1.645 \left(\frac{24}{\sqrt{38}} \right) = 270 - 6.40 = 263.60$$

Our decision rule is now completely specified by the rejection region (see Figure 7)

$$R: \overline{X} \le 263.6$$

that has $\alpha = .05$ as the probability of wrongly rejecting H_0.

Instead of locating the rejection region on the scale of $\overline{X}$, we can cast the decision criterion on the standardized scale as well:

$$Z = \frac{\overline{X} - \mu_0}{\sigma/\sqrt{n}} = \frac{\overline{X} - 270}{24/\sqrt{38}}$$

and set the rejection region as $R: Z \le -1.645$ (see Figure 7). This form is more convenient because the cutoff -1.645 is directly read off the normal table, whereas the determination of c involves additional numerical work.

The random variable $\overline{X}$ whose value serves to determine the action is called the **test statistic**.

A **test of the null hypothesis** is a course of action specifying the set of values of a test statistic $\overline{X}$, for which H_0 is to be rejected.

This set is called the **rejection region** of the test.

A test is completely specified by a test statistic and the rejection region.

TWO TYPES OF ERROR AND THEIR PROBABILITIES

Up to this point we only considered the probability of rejecting H_0 when, in fact, H_0 is true and illustrated how a decision rule is determined by setting this probability equal to .05. The following table shows all the consequences that might arise from the use of a decision rule.

Decision Based on Sample	Unknown True Situation	
	H_0 True $\mu = 270$	H_1 True $\mu < 270$
Reject H_0	Wrong rejection of H_0 (Type I error)	Correct decision
Retain H_0	Correct decision	Wrong retention of H_0 (Type II error)

In particular, when our sample-based decision is to reject H_0, either we have a correct decision (if H_1 is true) or we commit a Type I error (if H_0 is true). On the other hand, a decision to retain H_0 either constitutes a correct decision (if H_0 is true) or leads to a Type II error. To summarize:

Two Types of Error

Type I error: Rejection of H_0 when H_0 is true

Type II error: Nonrejection of H_0 when H_1 is true

α = Probability of making a Type I error
(also called the level of significance)

β = Probability of making a Type II error

In our problem of evaluating the upgraded teller machine, the rejection region is of the form $R : \overline{X} \leq c$; so that,

$$\alpha = P[\overline{X} \leq c] \quad \text{when} \quad \mu = 270 \quad (H_0 \ true)$$
$$\beta = P[\overline{X} > c] \quad \text{when} \quad \mu < 270 \quad (H_1 \ true)$$

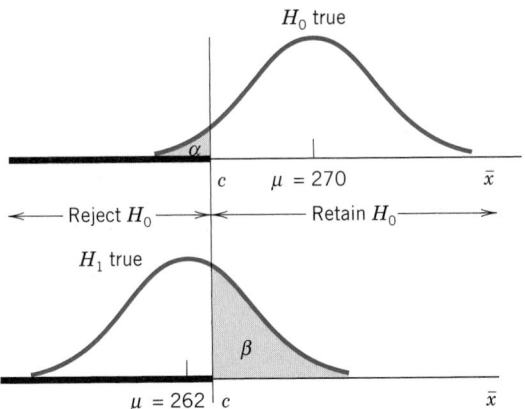

Figure 8   The error probabilities α and β.

Of course, the probability β depends on the numerical value of μ that prevails under H_1. Figure 8 shows the Type I error probability α as the shaded area under the normal curve that has $\mu = 270$ and the Type II error probability β as the shaded area under the normal curve that has $\mu = 262$, a case of H_1 being true.

From Figure 8, it is apparent that no choice of the cutoff c can minimize both the error probabilities α and β. If c is moved to the left, α gets smaller but β gets larger, and if c is moved to the right, just the opposite effects take place. In view of this dilemma and the fact that a wrong rejection of H_0 is the more serious error, we hold α at a predetermined low level such as .10, .05, or .01 when choosing a rejection region. We will not pursue the evaluation of β, but we do note that if the β turns out to be uncomfortably large, the sample size must be increased.

PERFORMING A TEST

When determining the rejection region for this example, we assumed that $\sigma = 24$ seconds, the same standard deviation as with the original money machines. Then, when $\mu = 270$, the distribution of $\overline{X}$ is $N(270, 24/\sqrt{38})$ and the rejection region $R: \overline{X} \leq 263.6$ was arrived at by fixing $\alpha = .05$ and referring

$$Z = \frac{\overline{X} - 270}{24/\sqrt{38}}$$

to the standard normal distribution.

In practice, we are usually not sure about the assumption that $\sigma = 24$, the standard deviation of the transaction times using the upgraded machine, is the same as with the original teller machine. But that does not cause any problem as long as the sample size is large. When n is large ($n > 30$), the normal approximation for $\overline{X}$ remains valid even if σ is estimated by the sample standard deviation S. Therefore, for testing $H_0: \mu = \mu_0$ versus $H_1: \mu < \mu_0$ with level of significance α, we employ the test statistic

$$Z = \frac{\overline{X} - \mu_0}{S/\sqrt{n}}$$

and set the rejection region $R: Z \leq -z_\alpha$. This test is commonly called a large sample normal test or a Z test.

Example 8   A Test of Hypotheses to Establish That $\mu < 270$

Referring to the automated teller machine transaction times, suppose that, from the measurements of a random sample of 38 transaction times, the sample mean and standard deviation are found to be 261 and 22 seconds, respectively. Test the null hypothesis $H_0: \mu = 270$ versus $H_1: \mu < 270$ using a 2.5% level of significance and state whether or not the claim $\mu < 270$ is substantiated.

SOLUTION   Because $n = 38$ and the null hypothesis specifies that μ has the value $\mu_0 = 270$, we employ the test statistic

$$Z = \frac{\overline{X} - 270}{S/\sqrt{38}}$$

The rejection region should consist of small values of Z because H_1 is left-sided. For a 2.5% level of significance, we take $\alpha = .025$, and since $z_{.025} = 1.96$, the rejection region is (see Figure 9) $R: Z \leq -1.96$.

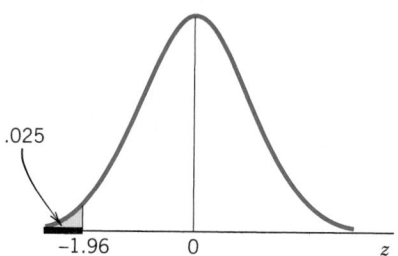

.025

−1.96   0 z

Figure 9   Rejection region for Z.

With the observed values $\overline{x} = 261$ and $s = 22$, we calculate the test statistic

$$z = \frac{261 - 270}{22/\sqrt{38}} = -2.52$$

Because this observed z is in R, the null hypothesis is rejected at the level of significance $\alpha = .025$. We conclude that the claim of a reduction in the mean transaction time is strongly supported by the data.

P–VALUE: HOW STRONG IS A REJECTION OF H_0?

Our test in Example 8 was based on the fixed level of significance $\alpha = .025$, and we rejected H_0 because the observed $z = -2.52$ fell in the rejection region $R: Z \leq -1.96$. A strong evidence against H_0 emerged due to the fact that

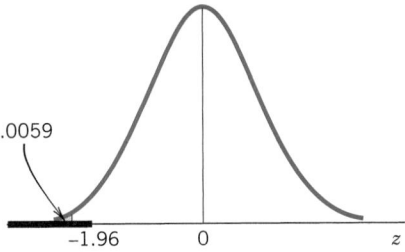

Figure 10   *P*–value with left-sided rejection region.

a small α was used. The natural question at this point is: How small an α could we use and still arrive at the conclusion of rejecting H_0? To answer this question, we consider the observed $z = -2.52$ itself as the cutoff point (critical value) and calculate the rejection probability

$$P[Z \leq -2.52] = .0059$$

The smallest possible α that would permit rejection of H_0, on the basis of the observed $z = -2.52$, is therefore .0059 (see Figure 10). It is called the significance probability or *P*–value of the observed z. This very small *P*–value, .0059, signifies a strong rejection of H_0 or that the result is highly statistically significant.

The *P*–value is the probability, calculated under H_0, that the test statistic takes a value equal to or more extreme than the value actually observed. The *P*–value serves as a measure of the strength of evidence against H_0. A small *P*–value means that the null hypothesis is strongly rejected or the result is highly statistically significant.

Our illustrations of the basic concepts of hypothesis tests thus far focused on a problem where the alternative hypothesis is of the form $H_1: \mu < \mu_0$, called a left-sided alternative. If the alternative hypothesis in a problem states that the true μ is larger than its null hypothesis value of μ_0, we formulate the right-sided alternative $H_1: \mu > \mu_0$ and use a right-sided rejection region $R: Z \geq z_\alpha$.

We illustrate the right-sided case in an example after summarizing the main steps that are involved in the conduct of a statistical test.

The Steps for Testing Hypotheses

1. Formulate the null hypothesis H_0 and the alternative hypothesis H_1.
2. Test criterion: State the test statistic and the form of the rejection region.
3. With a specified α, determine the rejection region.
4. Calculate the test statistic from the data.
5. Draw a conclusion: State whether or not H_0 is rejected at the specified α and interpret the conclusion in the context of the problem. Also, it is a good statistical practice to calculate the *P*–value and strengthen the conclusion.

Example 9   Evaluating a Weight Loss Diet—Calculation of a P–Value

A brochure inviting subscriptions for a new diet program states that the participants are expected to lose over 22 pounds in five weeks. Suppose that, from the data of the five-week weight losses of 56 participants, the sample mean and standard deviation are found to be 23.5 and 10.2 pounds, respectively. Could the statement in the brochure be substantiated on the basis of these findings? Test with $\alpha = .05$. Also calculate the P–value and interpret the result.

SOLUTION   Let μ denote the population mean weight loss from five weeks of participation in the program. Because our aim is to substantiate the assertion that $\mu > 22$ pounds, we formulate the hypotheses

$$H_0 : \mu = 22 \quad \text{versus} \quad H_1 : \mu > 22$$

The sample size is $n = 56$. Denoting the sample mean weight loss of the 56 participants by $\overline{X}$ and the standard deviation by S, our test statistic is

$$Z = \frac{\overline{X} - \mu_0}{S/\sqrt{n}} = \frac{\overline{X} - 22}{S/\sqrt{56}}$$

Because H_1 is right-sided, the rejection region should be of the form $R : Z \geq c$. Because $z_{.05} = 1.645$, the test with level of significance .05 has the rejection region (see Figure 11) $R : Z \geq 1.645$.

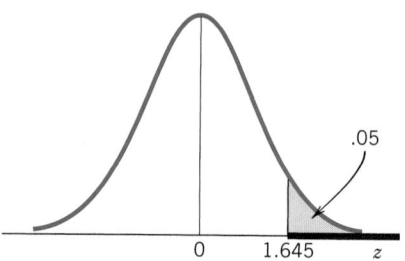

.05

0   1.645 z

Figure 11   Right-sided rejection region with $\alpha = .05$.

With the observed values $\overline{x} = 23.5$ and $s = 10.2$, we calculate

$$z = \frac{23.5 - 22}{10.2/\sqrt{56}} = 1.10$$

Because 1.10 is not in R, we do not reject the null hypothesis. We conclude that, with level of significance $\alpha = .05$, the stated claim that $\mu > 22$ is not substantiated.

Because our observed z is 1.10 and larger values are more extreme, the significance probability of this result is

$$P\text{–value} = P[Z \geq 1.10] = .1357 \quad \text{(from the normal table)}$$

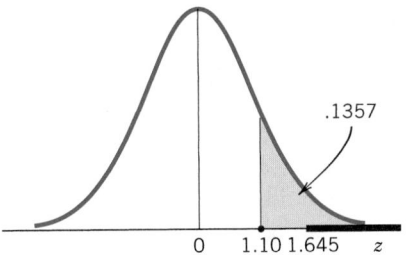

Figure 12   *P*−value with right-sided rejection region.

That is, .1357 is the smallest α at which H_0 could be rejected (Figure 12). This is not ordinarily considered a negligible chance so we conclude that the data do not provide a strong basis for rejection of H_0.

The preceding hypotheses are called one-sided hypotheses, because the values of the parameter μ under the alternative hypothesis lie on one side of those under the null hypothesis. The corresponding tests are called one-sided tests or one-tailed tests. By contrast, we can have a problem of testing the null hypothesis

$$H_0: \mu = \mu_0$$

versus the two-sided alternative or two-sided hypothesis

$$H_1: \mu \neq \mu_0$$

Here H_0 is to be rejected if $\overline{X}$ is too far away from μ_0 in either direction, that is, if Z is too small or too large. For a level α test, we divide the rejection probability α equally between the two tails and construct the rejection region

$$R: Z \leq -z_{\alpha/2} \quad \text{or} \quad Z \geq z_{\alpha/2}$$

which can be expressed in the more compact notation

$$R: |Z| \geq z_{\alpha/2}$$

Example 10   Testing Hypotheses about the Mean Height of Seedlings

Consider the data of Example 1 concerning the height measurements of 40 pine seedlings. Do these data indicate that the population mean height is different from 1.9 centimeters?

SOLUTION   We are seeking evidence in support of $\mu \neq 1.9$ so the hypotheses should be formulated as

$$H_0: \mu = 1.9 \quad \text{versus} \quad H_1: \mu \neq 1.9$$

The sample size $n = 40$ being large, we will employ the test statistic

$$Z = \frac{\overline{X} - \mu_0}{S/\sqrt{n}} = \frac{\overline{X} - 1.9}{S/\sqrt{40}}$$

The two-sided form of H_1 dictates that the rejection region must also be two-sided.

Let us choose $\alpha = .05$, then $\alpha/2 = .025$ and $z_{.025} = 1.96$. Consequently, for $\alpha = .05$, the rejection region is (see Figure 13)

$$R : |Z| \geq 1.96$$

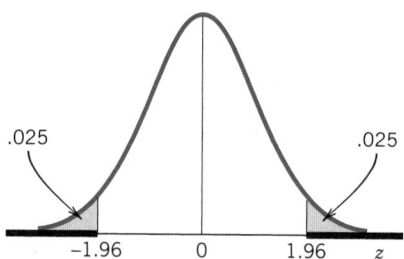

.025   .025

−1.96   0   1.96 z

Figure 13   Two-sided rejection region with $\alpha = .05$.

From Example 1, $\bar{x} = 1.715$ and $s = .475$, so the observed value of the test statistic is

$$z = \frac{\bar{x} - \mu_0}{s/\sqrt{n}} = \frac{1.715 - 1.9}{.475/\sqrt{40}} = -2.46$$

Because $|z| = 2.46$ is larger than 1.96, we reject H_0 at $\alpha = .05$.

In fact, the large value $|z| = 2.46$ seems to indicate a much stronger rejection of H_0 than that arising from the choice of $\alpha = .05$. How small an α can we set and still reject H_0? This is precisely the idea underlying the significance probability or the P-value. We calculate (see Figure 14)

$$P\text{–value} = P[|Z| \geq 2.46]$$
$$= P[Z \leq -2.46] + P[Z \geq 2.46]$$
$$= 2 \times .0069 = .0138$$

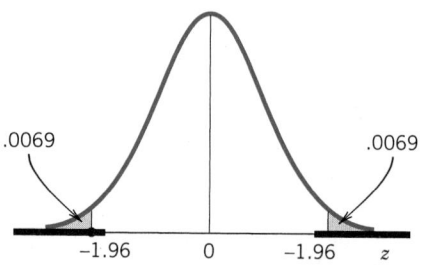

.0069   .0069

−1.96   0   −1.96 z

Figure 14 P–value with two-sided rejection region.

With α as small as .0138, H_0 is still rejected. This very small P–value gives strong support for H_1.

In summary:

Large Sample Tests for μ

When the sample size is large, a Z test concerning μ is based on the normal test statistic

$$Z = \frac{\overline{X} - \mu_0}{S/\sqrt{n}}$$

The rejection region is one- or two-sided depending on the alternative hypothesis. Specifically,

$$H_1: \mu > \mu_0 \quad \text{requires} \quad R: Z \geq z_\alpha$$
$$H_1: \mu < \mu_0 \quad\quad\quad\quad R: Z \leq -z_\alpha$$
$$H_1: \mu \neq \mu_0 \quad\quad\quad\quad R: |Z| \geq z_{\alpha/2}$$

Because the central limit theorem prevails for large n, no assumption is required as to the shape of the population distribution.

Exercises

8.37 Stated here are some claims or research hypotheses that are to be substantiated by sample data. In each case, identify the null hypothesis H_0 and the alternative hypothesis H_1 in terms of the population mean μ.

(a) The mean time a health insurance company takes to pay claims is less than 14 working days.

(b) The average person watching a movie at a local multiplex theater spends over $2.50 on refreshments.

(c) The mean hospital bill for a birth in the city is less than $3000.

(d) The mean time between purchases of a brand of mouthwash by loyal customers is different from 60 days.

8.38 From an analysis of the sample data, suppose that the decision has been to reject the null hypothesis. In the context of each part (a–d) of Exercise 8.37, answer the following questions:

In what circumstance is it a correct decision?

When is it a wrong decision, and what type of error is then made?

8.39 From an analysis of the sample data, suppose that the decision has been made to retain the null hypothesis. In the context of each part (a–d) of Exercise 8.37, answer the following questions.

In what circumstance is it a correct decision?

When is it a wrong decision, and what type of error is then made?

8.40 For each situation (a–d) in Exercise 8.37, state which of the following three forms of the rejection region is appropriate when σ is known.

$$R: \overline{X} \leq c \quad \text{(left-sided)}$$
$$R: \overline{X} \geq c \quad \text{(right-sided)}$$
$$R: |\overline{X} - \mu_0| \geq c \quad \text{(two-sided)}$$

8.41 Each part (a–d) of this problem gives the population standard deviation σ, the statement of a claim about μ, the sample size n, and the desired level of significance α. Formulate (i) the hypotheses, (ii) the test statistic Z, and (iii) the rejection region. (The answers to part (a) are provided.)

(a) $\sigma = 2$ claim: $\mu > 30$, $n = 55$, $\alpha = .05$

[Answers: (i) $H_0:\mu = 30$, $H_1:\mu > 30$

(ii) $Z = \dfrac{\overline{X} - 30}{2/\sqrt{55}}$ (iii) $R: Z \geq 1.645$]

(b) $\sigma = .085$ claim: $\mu < .15$ $n = 125$ $\alpha = .025$

(c) $\sigma = 8.6$ claim: $\mu \neq 80$ $n = 38$ $\alpha = .01$

(d) $\sigma = 1.23$ claim: $\mu \neq 0$ $n = 40$ $\alpha = .06$

8.42 Suppose that the observed values of the sample mean in the contexts of parts (a–d) of Exercise 8.41 are given as follows. Calculate the test statistic Z and state the conclusion with the specified α.

(a) $\overline{x} = 30.54$ (b) $\overline{x} = .136$

(c) $\overline{x} = 77.35$ (d) $\overline{x} = -.59$

8.43 Suppose you are to verify the claim that $\mu > 30$ on the basis of a random sample of size 70 and you know that $\sigma = 5.6$.

(a) If you set the rejection region to be $R:\overline{X} \geq 31.31$, what is the level of significance of your test?

(b) Find the numerical value of c so that the test $R:\overline{X} \geq c$ has a 5% level of significance.

8.44 A sample of 40 sales receipts from a grocery store has $\overline{x} = \$141$ and $s = \$30.2$. Use these values to perform a test of $H_0:\mu = 150$ against $H_1:\mu < 150$ with $\alpha = .05$.

8.45 Use the values in Exercise 8.44 to test

$$H_0:\mu = 150 \quad \text{versus} \quad H_1:\mu \neq 150$$

with $\alpha = .05$.

8.46 An investigator at a large midwestern university wants to determine the typical weekly amount of time students work on part-time jobs. More particularly, he wants to test the null hypothesis that the mean time is 15 hours versus a two-sided alternative. A sample of 39 students who hold part-time jobs is summarized by the computer output

Descriptive Statistics: hours

Variable	N	Mean	Median	StDev
Hours	39	16.69	15.00	7.61

(a) Perform the hypothesis test at the 1% level of significance.

(b) Calculate the significance probability and interpret the result.

8.47 Refer to the data on the growth of female salmon growth in the marine environment in Table D.7 of the Data Bank. A computer calculation for a test of $H_0 : \mu = 411$ versus $H_1 : \mu \neq 411$ is given below.

Test of mu = 411 vs mu not = 411
The assumed sigma = 41.05

Variable	N	Mean	StDev
Fmarine	40	429.15	41.05

Variable	Z	P
Fmarine	2.80	0.005

(a) What is the conclusion if you test with $\alpha = .01$?

(b) What mistake could you have made in part (a)?

(c) Before you collected the data, what was the probability of making the mistake in part (a)?

(d) Give a long-run relative frequency interpretation of the probability in part (c).

(e) Give the P–value of your test with $\alpha = .01$.

8.48 Refer to the data on the girth, in centimeters, of grizzly bears in Table D.8 of the Data Bank. A computer calculation for a test of $H_0 : \mu = 100$ centimeters versus $H_1 : \mu \neq 100$ gives

```
One-Sample Z: Girth

Test of mu = 100 vs mu not = 100
The assumed sigma = 21.79

Variable    N    Mean    StDev
Girth       61   93.39   21.79

Variable    Z       P
Girth       -2.37   0.018
```

(a) What is the conclusion if you test with $\alpha = .02$?

(b) What mistake could you have made in part (a)?

(c) Before you collected the data, what was the probability of making the mistake in part (a)?

(d) Give a long-run relative frequency interpretation of the probability in part (c).

(e) Give the P–value of your test with $\alpha = .01$.

8.49 A manager at a power company monitored the employee time required to process high-efficiency lamp bulb rebates. A random sample of 40 applications gave a sample mean time of 3.8 minutes and a standard deviation of 1.2 minutes. Is the claim that $\mu > 3.5$ minutes substantiated by these data? Test with $\alpha = .10$.

8.50 A credit company randomly selected 50 contested items and recorded the dollar amount being contested. These contested items had sample mean $\bar{x} = 75.43$ dollars and $s = 24.73$ dollars. Is the claim "μ differs from 85 dollars" substantiated by these data? Test with $\alpha = .01$.

8.51 Calculating from a random sample of 36 observations, one obtains the results $\bar{x} = 80.4$ and $s = 16.2$. In the context of each of the following hypothesis testing problems, determine the P–value of these results and state whether or not it signifies a strong rejection of H_0:

(a) To test $H_0: \mu = 74$ versus $H_1: \mu > 74$

(b) To test $H_0: \mu = 85$ versus $H_1: \mu < 85$

(c) To test $H_0: \mu = 76$ versus $H_1: \mu \neq 76$

8.52 In a given situation, suppose H_0 was not rejected at $\alpha = .02$. Answer the following questions as "yes," "no," or "can't tell" as the case may be.

(a) Would H_0 also be retained at $\alpha = .01$?

(b) Would H_0 also be retained at $\alpha = .05$?

(c) Is the P–value smaller than .02?

8.53 A company wishing to improve its customer service collected hold times from 75 randomly selected incoming calls to its hot line that were put on hold. These calls had sample mean hold time $\bar{x} = 3.4$ minutes and $s = 2.4$ minutes. Is the claim that $\mu > 3.0$ minutes substantiated by these data? Test with $\alpha = .05$.

8.54 A company's mixed nuts are sold in cans and the label says that 25% of the contents is cashews. Suspecting that this might be an overstatement, an inspector takes a random sample of 35 cans and measures the percent weight of cashews [i.e., 100(weight of cashews/weight of all nuts)] in each can. The mean and standard deviation of these measurements are found to be 23.5 and 3.1, respectively. Do these results constitute strong evidence in support of the inspector's belief? (Answer by calculating and interpreting the P–value).

8.55 Biological oxygen demand (BOD) is an index of pollution that is monitored in the treated effluent of paper mills on a regular basis. From 43 determinations of BOD (in pounds per day) at a particular paper mill during the spring and summer months of 1992, the mean and standard deviation were found to be 3246 and 757, respectively. The company had set the target that the mean BOD should be 3000 pounds per day. Do the sample data indicate that the actual amount of BOD is significantly off the target? (Use $\alpha = .05$.)

8.56 Refer to Exercise 8.55. Along with the determinations of BOD, the discharge of suspended solids (SS) was also monitored at the same site. The mean and standard deviation of the 43 determinations of SS were found to be 5710 and 1720 pounds per day, respectively. Do these results strongly support the company's claim that the true mean SS is lower than 6000 pounds per day? (Answer by calculating and interpreting the P–value.)

5.   INFERENCES ABOUT A POPULATION PROPORTION

The reasoning leading to estimation of a mean also applies to the problem of estimation of a population proportion. Example 2 considers sampling $n = 500$ persons to infer about the proportion of the population that is unemployed. When n elements are randomly sampled from the population, the data will consist of the count X of the number of sampled elements possessing the characteristic. Common sense suggests the sample proportion

$$\hat{p} = \frac{X}{n}$$

as an estimator of p. The hat notation reminds us that $\hat{p}$ is a statistic.

When the sample size n is only a small fraction of the population size, the sample count X has the binomial distribution with mean np and standard deviation $\sqrt{npq}$, where $q = 1 - p$. Recall from Chapter 6 that, when n is large, the binomial variable X is well approximated by a normal with mean np and standard deviation $\sqrt{npq}$. That is,

$$Z = \frac{X - np}{\sqrt{npq}}$$

is approximately standard normal. This statement can be converted into a statement about proportions by dividing the numerator and the denominator by n. In particular,

$$Z = \frac{(X - np)/n}{(\sqrt{npq})/n} = \frac{\hat{p} - p}{\sqrt{pq/n}}$$

This last form, illustrated in Figure 15, is crucial to all inferences about a population proportion p. It shows that $\hat{p}$ is approximately normally distributed with mean p and standard deviation $\sqrt{pq/n}$.

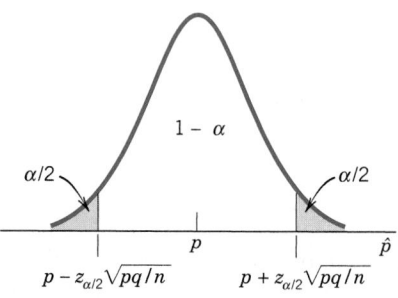

Figure 15   Approximate normal distribution of $\hat{p}$.

POINT ESTIMATION OF p

Intuitively, the sample proportion $\hat{p}$ is a reasonable estimator of the population proportion p. When the count X has a binomial distribution,

$$E(X) = np \qquad sd(X) = \sqrt{npq}$$

Since $\hat{p} = X/n$, the properties of expectation give

$$E(\hat{p}) = p$$
$$sd(\hat{p}) = \sqrt{pq/n}$$

In other words, the sampling distribution of $\hat{p}$ has a mean equal to the population proportion. The second result shows that the standard error of the estimator $\hat{p}$ is

$$S.E.(\hat{p}) = \sqrt{\frac{pq}{n}}$$

The estimated standard error can be obtained by substituting the sample estimate $\hat{p}$ for p and $\hat{q} = 1 - \hat{p}$ for q in the formula, or

$$\text{Estimated S.E.}(\hat{p}) = \sqrt{\frac{\hat{p}\hat{q}}{n}}$$

When n is large, prior to sampling, the probability is approximately .954 that the error of estimation $|\hat{p} - p|$ will be less than $2 \times$ (estimated S.E.).

Point Estimation of a Population Proportion

Parameter: Population proportion p

Data: $X =$ Number having the characteristic in a random sample of size n

Estimator: $\hat{p} = \dfrac{X}{n}$

$$S.E.(\hat{p}) = \sqrt{\frac{pq}{n}} \qquad \text{and} \qquad \text{estimated S.E.}(\hat{p}) = \sqrt{\frac{\hat{p}\hat{q}}{n}}$$

For large n, an approximate $100(1 - \alpha)\%$ error margin is $z_{\alpha/2}\sqrt{\hat{p}\hat{q}/n}$.

Example 11 Estimating the Proportion of Purchasers

A large mail-order club that offers monthly specials wishes to try out a new item. A trial mailing is sent to a random sample of 250 members selected from the list of over 9000 subscribers. Based on this sample mailing, 70 of the members decide to purchase the item. Give a point estimate of the proportion of club members that could be expected to purchase the item and attach a 95.4% error margin.

SOLUTION   The number in the sample represents only a small fraction of the total membership, so the count can be treated as if it were a binomial variable.

Here $n = 250$ and $X = 70$, so the estimate of the population proportion is

$$\hat{p} = \frac{70}{250} = .28$$

$$\text{Estimated S.E.}(\hat{p}) = \sqrt{\frac{\hat{p}\hat{q}}{n}} = \sqrt{\frac{.28 \times .72}{250}} = .028$$

$$95.4\% \text{ error margin} = 2 \times .028 = .056$$

Therefore, the estimated proportion is $\hat{p} = .28$, with a 95.4% error margin of .06 (rounded to two decimals).

Producers of breakfast cereals continually experiment with new products. Each promising new cereal must be market-tested on a sample of potential purchasers. An added twist here is that youngsters are a major component of the market. In order to elicit accurate information from young people, one firm developed a smiling face scale.

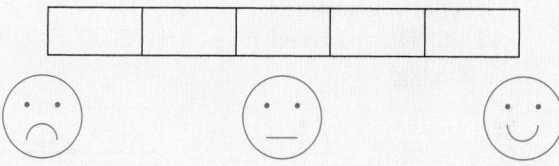

After tasting a new product, respondents are asked to check one box to rate the taste. A good product should have most of the youngsters responding in the top two boxes. Grouping these into a single top category and the lower three boxes into a lower category, we are in the situation of estimating the proportion of the market population that would rate taste in the top category.

Out of a sample of 42 youngsters, 30 rated a new cereal in the top category.

CONFIDENCE INTERVAL FOR p

A large sample confidence interval for a population proportion can be obtained from the approximate normality of the sample proportion $\hat{p}$. Since $\hat{p}$ is nearly normal with mean p and standard deviation $\sqrt{pq/n}$, the random interval

$\hat{p} \pm z_{\alpha/2}\sqrt{pq/n}$ is a candidate. However, the standard deviation involves the unknown parameter p, so we use the estimated standard deviation $\sqrt{\hat{p}\hat{q}/n}$ to set the endpoints of the confidence interval. Notice again that the common form of the confidence interval is

$$\text{Estimate} \pm (z \text{ value})(\text{estimated standard error})$$

Large Sample Confidence Interval for p

For large n, a $100(1 - \alpha)\%$ confidence interval for p is given by

$$\left(\hat{p} - z_{\alpha/2}\sqrt{\frac{\hat{p}\hat{q}}{n}}, \quad \hat{p} + z_{\alpha/2}\sqrt{\frac{\hat{p}\hat{q}}{n}} \right)$$

Example 12   A Confidence Interval for the Proportion Satisfied

Consider the data in Example 2 where 168 out of a random sample of 200 new car purchasers were satisfied with their car after one year. Compute a 95% confidence interval for the population proportion of satisfied new car purchasers.

SOLUTION   The sample size $n = 200$ is large so a normal approximation to the distribution of $\hat{p}$ is justified. Since $1 - \alpha = .95$, we have $\alpha/2 = .025$ and $z_{.025} = 1.96$. The observed $\hat{p} = 168/200 = .84$, and $\hat{q} = 1 - .84 = .16$. We calculate

$$z_{.025}\sqrt{\frac{\hat{p}\hat{q}}{n}} = 1.96\sqrt{\frac{.84 \times .16}{200}} = 1.96 \times .0259 = .051$$

Therefore, a 95% confidence interval for the population proportion of satisfied new-car buyers is $.840 \pm .051$, or $(.789, .891)$.

Because our procedure will produce true statements 95% of the time, we can be 95% confident that the proportion of satisfied new-car buyers is between .789 and .891.

DETERMINING THE SAMPLE SIZE

Note that, prior to sampling, the numerical estimate $\hat{p}$ of p is not available. Therefore, for a $100(1 - \alpha)\%$ error margin for the estimation of p, we use the expression $z_{\alpha/2}\sqrt{pq/n}$. The required sample size is obtained by equating $z_{\alpha/2}\sqrt{pq/n} = d$, where d is the specified error margin. We then obtain

$$n = pq\left[\frac{z_{\alpha/2}}{d}\right]^2$$

If the value of p is known to be roughly in the neighborhood of a value p^*, then n can be determined from

$$n = p^*(1 - p^*)\left[\frac{z_{\alpha/2}}{d}\right]^2$$

Without prior knowledge of p, pq can be replaced by its maximum possible value $\frac{1}{4}$ and n determined from the relation

$$n = \frac{1}{4}\left[\frac{z_{\alpha/2}}{d}\right]^2$$

Example 13   Selecting a Sample Size for Estimating a Proportion

A public health survey is to be designed to estimate the proportion p of a population having defective vision. How many persons should be examined if the public health commissioner wishes to be 98% certain that the error of estimation is below .05 when:

(a)   There is no knowledge about the value of p?

(b) p is known to be about .3?

SOLUTION   The tolerable error is $d = .05$. Also $1 - \alpha = .98$, so $\alpha/2 = .01$. From the normal table, we know that $z_{.01} = 2.33$.

(a)   Since p is unknown, the conservative bound on n yields

$$\frac{1}{4}\left[\frac{2.33}{.05}\right]^2 = 543$$

A sample of size 543 would suffice.

(b)   If $p^* = .3$, the required sample size is

$$n = (.3 \times .7)\left[\frac{2.33}{.05}\right]^2 = 456$$

LARGE SAMPLE TESTS ABOUT p

We consider testing $H_0: p = p_0$ versus $H_1: p \neq p_0$. With a large number of trials n, the sample proportion

$$\hat{p} = \frac{X}{n}$$

is approximately normally distributed. Under the null hypothesis, p has the specified value p_0 and the distribution of $\hat{p}$ is approximately $N(p_0, \sqrt{p_0 q_0/n})$. Consequently, the standardized statistic

$$Z = \frac{\hat{p} - p_0}{\sqrt{p_0 q_0/n}}$$

has the $N(0, 1)$ distribution. Since the alternative hypothesis is two-sided, the rejection region of a level α test is given by

$$R:|Z| \geq z_{\alpha/2}$$

For one-sided alternatives, we use a one-tailed test in exactly the same way we discussed in Section 4 in connection with tests about μ.

Example 14   Testing for a Change in the Proportion below the Poverty Level

A five-year-old census recorded that 20% of the families in a large community lived below the poverty level. To determine if this percentage has changed, a random sample of 400 families is studied and 70 are found to be living below the poverty level. Does this finding indicate that the current percentage of families earning incomes below the poverty level has changed from what it was five years ago?

SOLUTION   Let p denote the current population proportion of families living below the poverty level. Because we are seeking evidence to determine whether p is *different* from .20, we wish to test

$$H_0:p = .20 \qquad \text{versus} \qquad H_1:p \neq .20$$

The sample size $n = 400$ being large, the Z test is appropriate. The test statistic is

$$Z = \frac{\hat{p} - .2}{\sqrt{.2 \times .8/400}}$$

If we set $\alpha = .05$, the rejection region is $R:|Z| \geq 1.96$. From the sample data the computed value of Z is

$$z = \frac{(70/400) - .2}{\sqrt{.2 \times .8/400}} = \frac{.175 - .2}{.020} = -1.25$$

Because $|z| = 1.25$ is smaller than 1.96, the null hypothesis is not rejected at $\alpha = .05$. We conclude that the data do not provide strong evidence that a change in the percentage of families living below the poverty level has occurred.

The significance probability of the observed value of Z is

$$\begin{aligned} P\text{–value} &= P[|Z| \geq 1.25] \\ &= P[Z \leq -1.25] + P[Z \geq 1.25] \\ &= 2 \times .1056 = .2112 \end{aligned}$$

We would have to inflate α to more than .21 in order to reject the null hypothesis. Thus, the evidence against H_0 is really weak.

Exercises

8.57 To estimate a population proportion p, suppose that n units are randomly sampled and x number of the sampled units are found to have the characteristic of interest. For each case, provide a point estimate of p and determine its 95% error margin.

(a) $n = 50$, $x = 28$

(b) $n = 410$, $x = 75$

(c) $n = 2500$, $x = 2001$

8.58 For each case in Exercise 8.57, determine the 98% error margin of the estimate.

8.59 In a psychological experiment, individuals are permitted to react to a stimulus in one of two ways, say, A or B. The experimenter wishes to estimate the proportion p of persons exhibiting reaction A. How many persons should be included in the experiment to be 90% confident that the error of estimation is within .03 if the experimenter:

(a) Knows that p is about .3?

(b) Has no idea about the value of p?

8.60 A national safety council wishes to estimate the proportion of automobile accidents that involve pedestrians. How large a sample of accident records must be examined to be 98% certain that the estimate does not differ from the true proportion by more than .03? (The council believes that the true proportion is below .25.)

8.61 A sample of 78 university students revealed that 49 carried their books and notes in a backpack.

(a) Estimate the population proportion of students who carry their books and notes in a backpack.

(b) Obtain the estimated S.E.

(c) Obtain a 95% confidence interval for the proportion of students who use backpacks.

8.62 Each year, an insurance company reviews its claim experience in order to set future rates.

Regarding their damage-only automobile insurance policies, at least one claim was made on 2073 of the 12,299 policies in effect for the year. Treating these data as a random sample for the population of all possible damage-only policies that could be issued, find a 95% confidence interval for the population proportion of at least one claim.

8.63 An automobile club which pays for emergency road services (ERS) requested by its members wishes to estimate the proportions of the different types of ERS requests. Upon examining a sample of 2927 ERS calls, it finds that 1499 calls related to starting problems, 849 calls involved serious mechanical failures requiring towing, 498 calls involved flat tires or lockouts, and 81 calls were for other reasons.

(a) Estimate the true proportion of ERS calls that involved serious mechanical problems requiring towing and determine its 95% margin of error.

(b) Calculate a 98% confidence interval for the true proportion of ERS calls that related to starting problems.

8.64 In a survey of 700 year 2005 high school graduates, it was found that 50.3% had enrolled in colleges, 11.1% in vocational institutions, while the other 38.6% did not pursue any further studies. Based on these findings:

(a) Calculate a 95% confidence interval for the college enrollment rate of year 2005 high school graduates.

(b) Determine a 90% confidence interval for the true percent of the high school graduates enrolling in vocational institutions.

8.65 Identify the null and the alternative hypotheses in the following situations.

(a) A university official believes that the proportion of students that currently hold part-time jobs has increased from the value .26 that prevailed four years ago.

(b) A cable company claims that, because of improved procedures, the proportion of its cable subscribers that have complaints against the cable company is now less than .13.

(c) Referring to part (b), suppose a consumer advocate feels the proportion of cable subscribers that have complaints against the cable company this year is greater than .13. She will conduct a survey to challenge the cable company's claim.

(d) An inspector wants to establish that 2 × 4 lumber at a mill does not meet a specification that requires at most 5% break under a standard load.

8.66 Given here are the descriptive statements of some claims that one intends to establish on the basis of data. In each case, identify the null and the alternative hypotheses in terms of a population proportion p.

(a) Of smokers who eventually quit smoking, less than 40% are able to do so in just one attempt.

(b) On a particular freeway, over 25% of the cars that use a lane restricted exclusively to multipassenger cars use the lane illegally.

(c) At a particular clinic, less than 20% of the patients wait over half an hour to see the doctor.

8.67 Each part of this problem specifies a claim about a population proportion, the sample size n, and the desired level of significance α. Formulate (i) the hypotheses, (ii) the test statistic, and (iii) the rejection region. (The answers to part (a) are provided for illustration.)

(a) Claim: $p < .32$ $n = 120$ $\alpha = .05$

[Answers:

(i) $H_0: p = .32$ $H_1: p < .32$

(ii) $Z = \dfrac{\hat{p} - .32}{\sqrt{.32 \times .68/120}} = \dfrac{\hat{p} - .32}{.0426}$

(iii) $R: Z \leq -1.645$]

(b) Claim: $p > .75$ $n = 228$ $\alpha = .02$

(c) Claim: $p \neq .60$ $n = 77$ $\alpha = .02$

(d) Claim: $p < .56$ $n = .86$ $\alpha = .10$

8.68 Given here are the observed sample proportions $\hat{p}$ in the contexts of parts (a–d) of Exercise 8.67. Calculate the test statistic and draw a conclusion of the test at the specified level of significance.

(a) $\hat{p} = .233$

(b) $\hat{p} = .818$

(c) $\hat{p} = .709$

(d) $\hat{p} = .387$

8.69 Assuming that n is large, write the test statistic and determine the rejection region in each case.

(a) $H_0: p = .4$ versus $H_1: p < .4$, $\alpha = .10$

(b) $H_0: p = .6$ versus $H_1: p \neq .6$, $\alpha = .10$

8.70 An educator wishes to test $H_0: p = .3$ against $H_1: p > .3$, where $p = $ proportion of college football players who graduate in four years.

(a) State the test statistic and the rejection region for a large sample test having $\alpha = .05$.

(b) If 19 out of a random sample of 48 players graduated in four years, what does the test conclude? Calculate the P–value and interpret the result.

8.71 A concerned group of citizens wants to show that less than half the voters support the President's handling of a recent crisis. Let $p = $ proportion of voters who support the handling of the crisis.

(a) Determine H_0 and H_1.

(b) If a random sample of 500 voters gives 228 in support, what does the test conclude? Use $\alpha = .05$. Also evaluate the P–value.

8.72 Refer to Exercise 8.63. Perform a test of hypotheses to determine whether the proportion of ERS calls involving flat tires or lockouts was significantly smaller than .19, the true proportion for previous years. (Use a 5% level of significance.)

8.73 Refer to Exercise 8.64. Perform a test of hypotheses to determine whether the college enrollment rate in the population of year 2005 high school graduates was significantly higher than 48%. (Calculate the P–value and interpret the result.)

8.74   An independent bank concerned about its customer base decided to conduct a survey of bank customers. Out of 505 customers who returned the survey form, 258 rated the overall bank services as excellent.

(a)   Test, at level $\alpha = .10$, the null hypothesis that the proportion of customers who would rate the overall bank services as excellent is .46 versus a two-sided alternative.

(b)   Calculate the P–value and comment on the strength of evidence.

8.75   An independent bank concerned about its customer base decided to conduct a survey of bank customers. Out of 505 customers who returned the survey form, 258 rated the overall bank services as excellent.

(a)   Find a 90% confidence interval for the proportion of customers who would rate the overall bank services as excellent.

(b)   The bank has 8200 customers. Convert your confidence interval for the proportion in part (a) into a 90% confidence interval for the total number of customers who would rate the overall bank services as excellent.

8.76   From telephone interviews with 980 adults, it was found that 78% of those persons supported tougher legislation for antipollution measures. Does this poll substantiate the conjecture that more than 75% of the adult population are in favor of tougher legislation for antipollution measures? (Answer by calculating the P–value.)

8.77   Refer to the box with the smiling face scale for rating cereals. Using the data that 30 out of 42 youngsters in a sample rated a cereal in the top category, find an approximate 95% confidence interval for the corresponding population proportion.

USING STATISTICS WISELY

1.   Calculate the estimated standard error $s/\sqrt{n}$ to accompany the point estimate $\bar{x}$ of a population mean.

2.   Understand the interpretation of a $100(1 - \alpha)\%$ confidence interval. When the sample size is large, before the data are collected

$$\left(\overline{X} - z_{\alpha/2} \frac{S}{\sqrt{n}}, \quad \overline{X} + z_{\alpha/2} \frac{S}{\sqrt{n}} \right)$$

is a random interval that will cover the fixed unknown mean μ with probability $1 - \alpha$. The long-run frequency interpretation of the probability $1 - \alpha$ says that, in many repeated applications of this method, about proportion $1 - \alpha$ of the times the interval will cover the respective population mean.

3.   When conducting a test of hypothesis, formulate the assertion that the experiment seeks to confirm as the alternative hypothesis.

4.   When the sample size is large, base a test of the null hypothesis $H_0 : \mu = \mu_0$ on the test statistic

$$\frac{\overline{X} - \mu_0}{S/\sqrt{n}}$$

which has, approximately a standard normal distribution. The rejection region is one-sided or two-sided corresponding to the alternative hypothesis.

5. Understand the interpretation of a level α test. If the null hypothesis is true, before the data are collected, the probability is α that the experiment will produce observations that lead to the rejection of the null hypothesis. Consequently, after many independent experiments, the proportion that lead to rejection of the null hypothesis will be nearly α.

6. To obtain a precise estimate of a proportion usually requires a sample size of a few hundred.

7. Remember that the statistical procedures presented in this chapter will not be valid if the large sample is not randomly selected but collected from convenient units.

KEY IDEAS AND FORMULAS

Statistical concepts and methods provide the framework that allows us to learn about the population from a sample of observations. The process begins by modeling the population as a probability distribution, which has some numerical feature of interest called a parameter. Then, given a sample from this population distribution called the data, we make a generalization or statistical inference about the parameter.

Two basic forms of inference are (1) estimation of a population parameter and (2) testing statistical hypotheses.

A parameter can be estimated in two ways: by quoting (1) a single numerical value (point estimation) or (2) an interval of plausible values (interval estimation).

The statistic whose value gives a point estimate is called an estimator. The standard deviation of a point estimator is also called its standard error.

To be meaningful, a point estimate must be accompanied by an evaluation of its error margin.

A $100(1 - \alpha)\%$ confidence interval is an interval that, before sampling, will cover the true value of the parameter with probability $1 - \alpha$. The interval must be computable from the sample data.

If random samples are repeatedly drawn from a population and a $100(1 - \alpha)\%$ confidence interval is calculated from each, then about $100(1 - \alpha)\%$ of those intervals will include the true value of the parameter. We never know what happens in a single application. Our confidence draws from the success rate of $100(1 - \alpha)\%$ over many applications.

A statistical hypothesis is a statement about a population parameter.

A statement or claim, which is to be established with a strong support from the sample data, is formulated as the alternative hypothesis (H_1). The null hypothesis (H_0) says that the claim is void.

A test of the null hypothesis is a decision rule that tells us when to reject H_0 and when not to reject H_0. A test is specified by a test statistic and a rejection region (critical region).

A wrong decision may occur in one of the two ways:

A false rejection of H_0 (Type I error)

Failure to reject H_0 when H_1 is true (Type II error)

Errors cannot always be prevented when making a decision based on a sample. It is their probabilities that we attempt to keep small.

A Type I error is considered to be more serious. The maximum Type I error probability of a test is called its level of significance and is denoted by α.

The significance probability or P-value of an observed test statistic is the smallest α for which this observation leads to the rejection of H_0.

Main steps in testing statistical hypotheses

1. Formulate the null hypotheses H_0 and the alternative hypothesis H_1.
2. Test criterion: State the test statistic and the form of the rejection region.
3. With a specified α, determine the rejection region.
4. Calculate the test statistic from the data.
5. Draw a conclusion: State whether or not H_0 is rejected at the specified α and interpret the conclusion in the context of the problem. Also, it is a good statistical practice to calculate the P-value and strengthen the conclusion.

The Type II error probability is denoted by β.

Inferences about a Population Mean When n Is Large

When n is large, we need not be concerned about the shape of the population distribution. The central limit theorem tells us that the sample mean $\overline{X}$ is nearly normally distributed with mean μ and standard deviation $\sigma/\sqrt{n}$. Moreover, $\sigma/\sqrt{n}$ can be estimated by $S/\sqrt{n}$.

Parameter of interest is

$$\mu = \text{Population mean}$$

Inferences are based on

$$\overline{X} = \text{Sample mean}$$

1. A point estimator of μ is the sample mean $\overline{X}$.

$$\text{Estimated standard error} = \frac{S}{\sqrt{n}}$$

$$\text{Approximate } 100(1 - \alpha)\% \text{ error margin} = z_{\alpha/2}\frac{S}{\sqrt{n}}$$

2. A $100(1 - \alpha)\%$ confidence interval for μ is

$$\left(\overline{X} - z_{\alpha/2}\frac{S}{\sqrt{n}}, \quad \overline{X} + z_{\alpha/2}\frac{S}{\sqrt{n}} \right)$$

3. The test of the null hypothesis concerning μ, called the large sample normal test or Z test, uses the test statistic

$$Z = \frac{\overline{X} - \mu_0}{S/\sqrt{n}}$$

where μ_0 is the value of μ that marks the boundary between H_0 and H_1. Given a level of significance α,

Reject $H_0: \mu = \mu_0$ in favor of $H_1: \mu > \mu_0$ if $Z \geq z_\alpha$

Reject $H_0: \mu = \mu_0$ in favor of $H_1: \mu < \mu_0$ if $Z \leq -z_\alpha$

Reject $H_0: \mu = \mu_0$ in favor of $H_1: \mu \neq \mu_0$ if $|Z| \geq z_{\alpha/2}$

The first two alternative hypotheses, $H_1: \mu > \mu_0$ and $H_1: \mu < \mu_0$ are one-sided hypothesis and the third is a two-sided hypothesis. The rejection regions correspond to the alternative hypothesis so tests in the first two cases are called one-sided tests or one-tailed tests. Those in the third case are two-tailed tests.

Inference about a Population Proportion When n Is Large

Parameter of interest:

p = Population proportion of individuals possessing stated characteristic

Inferences are based on $\hat{p} = \dfrac{x}{n}$, the sample proportion.

1. A point estimator of p is $\hat{p}$.

   Estimated standard error = $\sqrt{\dfrac{\hat{p}\,\hat{q}}{n}}$ where $\hat{q} = 1 - \hat{p}$

   $100(1 - \alpha)\%$ error margin = $z_{\alpha/2}\sqrt{\dfrac{\hat{p}\,\hat{q}}{n}}$

2. A $100(1 - \alpha)\%$ confidence interval for p is

$$\left(\hat{p} - z_{\alpha/2}\sqrt{\dfrac{\hat{p}\,\hat{q}}{n}}, \quad \hat{p} + z_{\alpha/2}\sqrt{\dfrac{\hat{p}\,\hat{q}}{n}} \right)$$

3. To test hypotheses about p, the test statistic is

$$Z = \dfrac{\hat{p} - p_0}{\sqrt{p_0 q_0 / n}}$$

where p_0 is the value of p that marks the boundary between H_0 and H_1. The rejection region is right-sided, left-sided, or two-sided according to $H_1: p > p_0$, $H_1: p < p_0$, or $H_1: p \neq p_0$, respectively.

TECHNOLOGY

Large Sample Confidence Intervals and Tests Concerning a Mean

The software programs use a known value for the population standard deviation σ. If this is not given in your application, you need to obtain the sample standard deviation using the technology described in Chapter 2.

MINITAB

Confidence Intervals for μ

We illustrate the calculation of a 99% confidence interval for μ when we have determined that the sample standard deviation is 8.2 (or the known population $\sigma = 8.2$).

> **Data:** *C1*
>
> **Stat > Basic Statistics > 1-Sample Z.**
> Type *C1* in **Samples in columns** and *8.2* in **Standard deviation.**
> Click **Options** and type *99* in **Confidence level.**
> Click **OK.** Click **OK.**

Tests of Hypotheses Concerning μ

We illustrate the calculation of an $\alpha = .01$ level test of $H_0 : \mu = 32$ versus a one-sided alternative $H_1 : \mu > 32$ when we have determined that the sample standard deviation is 8.2 (or the known population $\sigma = 8.2$).

> **Data:** *C1*
>
> **Stat > Basic Statistics > 1-Sample Z.**
> Type *C1* in **Samples in Columns.**
> Type *8.2* in **Sigma.** Following **Test mean,** type *32,* the value of the mean under the null hypothesis.
> Click **Options** and type *99* in **Confidence level.**
> In the **Alternative** cell select **greater than,** the direction of the alternative hypothesis. Click **OK.** Click **OK.**

If the sample size and mean are available, instead of the second step, type these values in the corresponding cells.

EXCEL

Confidence Intervals for μ

We illustrate the calculation of a 99% confidence interval for μ when we have determined that the sample standard deviation is 8.2 (or the known population $\sigma = 8.2$).

> Select **Insert and then Function.**
> Choose **Statistical** and then **CONFIDENCE.**
> Enter *1 − .99* or *.01* for **Alpha,** *8.2* for **Standard_dev,** and *100* for **size.**

> (Add and subtract this value to and from $\bar{x}$ to obtain the confidence interval.)

Tests of Hypotheses Concerning μ

We illustrate the calculation of a test of $H_0 : \mu = 32$ versus a one-sided alternative $H_1 : \mu > 32$ when we have determined that the sample standard deviation is 8.2 (or the known population $\sigma = 8.2$). Start with the data entered in column A.

> Select **Insert** and then **Function.** Choose **Statistical** and then **ZTEST.**
> Highlight the data in column A for **Array.** Enter *32* in **X** and *8.2* in **Sigma.**
> (Leave **Sigma** blank and the sample standard deviation will be used.)
> Click **OK.**

The program returns the one-sided P–value for right-sided alternatives. For two-sided alternatives, you need to double the P–value if $\bar{x}$ is above $\mu_0 = 32$.

TI-84/-83 PLUS

Confidence Intervals for μ

We illustrate the calculation of a 99% confidence interval for μ when we have determined that the sample standard deviation is 8.2 (or the known population $\sigma = 8.2$). Start with the data entered in **L₁.**

> Press **STAT** and select **TESTS** and then **7: Zinterval.**
> Select **Data** with *List* set to **L₁** and **FREQ** to 1.
> Following σ: enter *8.2.*
> Enter *.99* following **C-Level:** Select **Calculate.**
> Then press **ENTER.**

If instead the sample size and sample mean are available, the second step is Select **Stats** (instead of **Data**) and enter the sample size and mean for n and $\bar{x}$.

Tests of Hypotheses Concerning μ

We illustrate the calculation of an $\alpha = .01$ level test of $H_0 : \mu = 32$ versus a one-sided alternative $H_1 : \mu > 32$ when we have determined that the sample standard deviation is 8.2 (or the known population $\sigma = 8.2$). Start with the data entered in column **L₁.**

> Press **STAT** and select **TESTS** and then **1: Z - Test.**
> Select **Data** with **List** set to **L₁** and **Freq** to 1.
> Following σ: enter *8.2.* Enter *32* for μ_0.

Select the direction of the alternate hypothesis. Select **Calculate.**
Press **ENTER.**

The calculator will return the P–value.

If instead the sample size and sample mean are available, the second step is

Select **Stats** (instead of **Data**) and type in the sample size and mean.

6. REVIEW EXERCISES

8.78   For estimating a population mean with the sample mean $\overline{X}$, calculate (i) the standard error of $\overline{X}$ and (ii) $100(1 - \alpha)\%$ error margin in each case.

(a) $n = 42 \quad \sigma = .825 \quad 1 - \alpha = .98$

(b) $n = 110 \quad \sigma = 98 \quad 1 - \alpha = .80$

(c) $n = 880 \quad \sigma = 55 \quad 1 - \alpha = .90$

8.79   Consider the problem of estimating μ based on a random sample of size n. Compute a point estimate of μ and the estimated standard error when

(a) $n = 80 \quad \Sigma x_i = 752$
$\Sigma (x_i - \overline{x})^2 = 345$

(b) $n = 169 \quad \Sigma x_i = 1290$
$\Sigma (x_i - \overline{x})^2 = 842$

8.80   The time it takes for a taxi to drive from the office to the airport was recorded on 40 occasions. It was found that $\overline{x} = 47$ minutes and $s = 5$ minutes. Give

(a)   An estimate of μ = population mean time to drive.

(b)   An approximate 95.4% error margin.

8.81   By what factor should the sample size be increased to reduce the standard error of $\overline{X}$ to

(a) $\frac{1}{2}$ its original value?

(b) $\frac{1}{4}$ its original value?

8.82   A food service manager wants to be 95% certain that the error in the estimate of the mean

number of sandwiches dispensed over the lunch hour is 10 or less. What sample size should be selected if a preliminary sample suggests

(a) $\sigma = 40$?

(b) $\sigma = 80$?

8.83   A zoologist wishes to estimate the mean blood sugar level of a species of animal when injected with a specified dosage of adrenaline. A sample of 55 animals of a common breed are injected with adrenaline, and their blood sugar measurements are recorded in units of milligrams per 100 milliliters of blood. The mean and standard deviation of these measurements are found to be 126.9 and 10.5, respectively.

(a)   Give a point estimate of the population mean and find a 95.4% error margin.

(b)   Determine a 90% confidence interval for the population mean.

8.84   On a Thursday, students were asked to report the number of hours sleep they received the previous night. The results are summarized in the computer output

Descriptive Statistics: sleep(hrs)

Variable	N	Mean	Median	StDev	SE Mean
Sleep	59	7.144	7.000	1.207	0.157

Find a 98% confidence interval for the mean number of hours sleep.

8.85 After feeding a special diet to 80 mice, the scientist measures their weight in grams and obtains $\bar{x} = 35$ grams and $s = 4$ grams. He states that a 90% confidence interval for μ is given by

$$\left(35 - 1.645\,\frac{4}{\sqrt{80}}, \quad 35 + 1.645\,\frac{4}{\sqrt{80}}\right)$$

or (34.26, 35.74)

(a) Was the confidence interval calculated correctly? If not, provide the correct result.

(b) Does the interval (34.26, 35.74) cover the true mean? Explain your answer.

*8.86 Computing from a large sample, one finds the 95% confidence interval for μ to be (6.8, 14.2). Based on this information alone, determine the 90% confidence interval for μ.

[*Hint:* A $100(1 - \alpha)\%$ confidence interval for μ is centered at $\bar{x}$ and has half-width $= z_{\alpha/2}\,s/\sqrt{n}$.]

8.87 In each case, identify the null hypothesis (H_0) and the alternative hypothesis (H_1) using the appropriate symbol for the parameter of interest.

(a) A consumer group plans to test-drive several cars of a new model in order to document that its average highway mileage is less than 50 miles per gallon.

(b) Confirm the claim that the mean number of pages per transmission sent by a campus fax station is more than 3.4.

(c) A chiropractic method will be tried on a number of persons suffering from persistent backache in order to demonstrate the claim that its success rate is higher than 50%.

(d) The setting of an automatic dispenser needs adjustment when the mean fill differs from the intended amount of 16 ounces. Several fills will be accurately measured in order to decide if there is a need for resetting.

(e) The content of fat in a gourmet chocolate ice cream is more than the amount, 4%, that is printed on the label.

8.88 Suppose you are to verify the claim that $\mu < -6.5$ with a sample of size 40 and you know $\sigma = 2.2$.

(a) If you set the rejection region to be $R : \bar{X} \le -7.1$, what is the level of significance of your test?

(b) Find the numerical value of c so that the test $R : \bar{X} \le c$ has $\alpha = .10$.

8.89 Suppose you are to verify the claim that $\mu \ne 32$ with a sample of size 100 and you know $\sigma = 10.6$.

(a) If you set the rejection region to be $R : |\bar{X} - 32| \ge 2.47$, what is the numerical value of α?

(b) If you want the test $R : |\bar{X} - 32| \ge c$ to have $\alpha = .05$, what must the numerical value of c be?

8.90 In a given situation, suppose H_0 was rejected at $\alpha = .05$. Answer the following questions as "yes," "no," or "can't tell" as the case may be.

(a) Would H_0 also be rejected at $\alpha = .03$?

(b) Would H_0 also be rejected at $\alpha = .10$?

(c) Is the P-value larger than .05?

8.91 Refer to the data on the amount of reflected light from urban areas in Table D.3b of the Data Bank. A computer calculation for a test of $H_0 : \mu = 84$ versus $H_1 : \mu \ne 84$ has the output

```
Test of mu = 84 vs mu not = 84
The assumed sigma = 4.979

Variable      N    Mean    StDev
Lighturb     40  82.075   4.979

Variable             Z       P
Lighturb         -2.45   0.014
```

(a) What is the conclusion of the test when $\alpha = .03$.

(b) Use the value for Z to test the null hypothesis $H_0 : \mu = 84$ versus the one-sided alternative $H_0 : \mu < 84$ at the $\alpha = .01$ level of significance.

8.92 A company wishing to improve its customer service collected hold times from 75 randomly selected incoming calls to its hot line that were put on hold. These calls had sample mean hold time $\bar{x} = 3.4$ minutes and $s = 2.3$ minutes.

Is the claim that $\mu > 3.0$ minutes substantiated by these data? Test with $\alpha = .05$.

8.93   The daily number of kayaks sold, X, at a water sports store has the probability distribution

x	0	1	2
$f(x)$	.5	.3	.2

(a)   Find the expected number of kayaks sold in a day.

(b)   Find the standard deviation of the number of kayaks sold in a day.

(c)   Suppose data from the next 64 different days give $\bar{x} = .84$ and standard deviation $s = .40$ number of kayaks sold. Can we conclude that the mean number of kayaks sold is greater than it used to be? Test with $\alpha = .05$.

8.94   In a large-scale, cost-of-living survey undertaken last January, weekly grocery expenses for families with one or two children were found to have a mean of $148 and a standard deviation of $25. To investigate the current situation, a random sample of families with one or two children is to be chosen and their last week's grocery expenses are to be recorded.

(a)   How large a sample should be taken if one wants to be 95% sure that the error of estimation of the population mean grocery expenses per week for families with one or two children does not exceed $2? (Use the previous s as an estimate of the current σ.)

(b)   A random sample of 100 families is actually chosen, and from the data of their last week's grocery bills, the mean and the standard deviation are found to be $155 and $22, respectively. Construct a 98% confidence interval for the current mean grocery expense per week for the population of families with one or two children.

8.95   A random sample of 2000 persons from the labor force of a large city are interviewed, and 175 of them are found to be unemployed.

(a)   Estimate the rate of unemployment based on the data.

(b)   Establish a 95% error margin for your estimate.

8.96   Referring to Exercise 8.95, compute a 98% confidence interval for the rate of unemployment.

8.97   Let $p =$ proportion of adults in a city who required a lawyer in the past year.

(a)   Determine the rejection region for an $\alpha = .05$ level test of

$$H_0: p = .25 \text{ against } H_1: p > .25$$

(b)   If 65 persons in a random sample of 190 required lawyer services, what does the test conclude?

8.98   Suppose that on the basis of a random sample of size 170, you are to verify the claim that a population proportion is larger than .40.

(a)   If you set the rejection region to be $R: \hat{p} \geq .453$, what is the level of significance of your test?

(b)   Determine the numerical value of c so that the test $R: \hat{p} \geq c$ has $\alpha = .025$.

8.99   Suppose that on the basis of a random sample of size 250, you are to verify the claim that a population proportion is different from .30.

(a)   If you set the rejection region to be $R: |\hat{p} - .30| \geq .06$, what is the level of significance of your test?

(b)   Determine the numerical value of c so the test $R: |\hat{p} - .30| \geq c$ has $\alpha = .10$.

8.100   A survey of 120 high school students shows that 39 of them work over 20 hours per week during the school year. Does it demonstrate the conjecture that more than 25% of the high school students work over 20 hours per week during the school year? (Answer by calculating the P–value.)

8.101   A marketing manager wishes to determine if lemon-scented and almond-scented dishwashing liquids are equally liked by consumers. Out of 250 consumers interviewed, 145 expressed their preference for the lemon-scented and the remaining 105 preferred the almond-scented.

(a) Do these data provide strong evidence that there is a difference in popularity between the two scented liquids? (Test with $\alpha = .05$.)

(b) Construct a 95% confidence interval for the population proportion of consumers who prefer the almond-scented liquid.

8.102 Each year, an insurance company reviews its claim experience in order to set future rates. Regarding their damage-only automobile insurance policies, at least one claim was made on 2073 of the 12,299 policies in effect for the year. Treat these data as a random sample for the population of all possible damage-only policies that could be issued.

(a) Test, at level $\alpha = .05$, the null hypothesis that the probability of at least one claim is 0.16 versus a two-sided alternative.

(b) Calculate the P–value and comment on the strength of evidence.

8.103 A genetic model suggests that 80% of the plants grown from a cross between two given strains of seeds will be of the dwarf variety. After breeding 200 of these plants, 136 were observed to be of the dwarf variety.

(a) Does this observation strongly contradict the genetic model?

(b) Construct a 95% confidence interval for the true proportion of dwarf plants obtained from the given cross.

8.104 One wishes to estimate the proportion of car owners who purchase more than $1,000,000 of liability coverage in their automobile insurance policies.

(a) How large a sample should be chosen to estimate the proportion with a 95% error margin of .008? (Use $p^* = .15$.)

(b) A random sample of 400 car owners is taken, and 56 of them are found to have chosen this extent of coverage. Construct a 95% confidence interval for the population proportion.

*8.105 *Finding the power of a test.* Consider the problem of testing $H_0 : \mu = 10$ versus $H_1 : \mu > 10$ with $n = 64$, $\sigma = 2$ (known), and $\alpha = .025$. The rejection region of this test is given by

$$R: \frac{\overline{X} - 10}{2/\sqrt{64}} \geq 1.96 \quad \text{or}$$

$$R: \overline{X} \geq 10 + 1.96 \frac{2}{\sqrt{64}} = 10.49$$

Suppose we wish to calculate the power of this test at the alternative $\mu_1 = 11$. Power = the probability of rejecting the null hypothesis when the alternative is true. Since our test rejects the null hypothesis when $\overline{X} \geq 10.49$, its power at $\mu_1 = 11$ is the probability

$$P[\overline{X} \geq 10.49 \text{ when the true mean } \mu_1 = 11]$$

If the population mean is 11, we know that $\overline{X}$ has the normal distribution with mean 11 and sd $= \sigma/\sqrt{n} = 2/\sqrt{64} = .25$. The standardized variable is

$$Z = \frac{\overline{X} - 11}{.25}$$

and we calculate

$$\text{Power} = P[\overline{X} \geq 10.49 \text{ when } \mu_1 = 11]$$
$$= P\left[Z \geq \frac{10.49 - 11}{.25}\right]$$
$$= P[Z \geq -2.04] = .9793$$
$$\text{(using normal table)}$$

Following the above steps, calculate the power of this test at the alternative:

(a) $\mu_1 = 10.5$

(b) $\mu_1 = 10.8$

8.106 Refer to the data on the computer attitude score (CAS) in Table D.4 of the Data Bank. A computer summary of a level $\alpha = .05$ test of $H_0: \mu = 2.6$ versus a two-sided alternative and a 95% confidence interval is given below.

```
One-Sample Z: CAS

Test of mu = 2.6 vs mu not = 2.6
The assumed sigma = 0.484
```

Variable	N	Mean	StDev
CAS	35	2.8157	0.4840

Variable	95.0% CI	Z	P
CAS	(2.6554, 2.9761)	2.64	0.008

(a) Will the 99% confidence interval for mean CAS be smaller or larger than the one in the printout? Verify your answer by determining the 99% confidence interval.

(b) Use the value for Z to test the null hypothesis $H_0 : \mu = 2.6$ versus the one-sided alternative $H_1 : \mu > 2.6$ at the $\alpha = .05$ level of significance.

8.107 Refer to the data on percent malt extract in Table D.8 of the Data Bank. A computer summary of a level $\alpha = .05$ test of $H_0 : \mu = 77$ versus a two-sided alternative and a 95% confidence interval is given below.

```
One-Sample Z: malt extract(%)


Test of mu = 77 vs mu not = 77
The assumed sigma = 1.101

Variable        N    Mean   StDev
malt extract    40   77.458  1.101

Variable          95.0% CI       Z      P
malt extract  ( 77.116, 77.799)  2.63   0.009
```

(a) Will the 98% confidence interval for mean malt extract be smaller or larger than the one in the printout? Verify your answer by determining the 98% confidence interval.

(b) Use the value for Z to test the null hypothesis $H_0: \mu = 77.0$ versus the one-sided alternative $H_1: \mu > 77.0$ at the $\alpha = .05$ level of significance.

The Following Exercises Require a Computer

8.108 Refer to the earthquake data given in Exercise 2.20 in Chapter 2 and file 2.20.DAT. Use MINITAB or some other package program to:

(a) Determine a large sample 95% confidence interval for mean earthquake size.

(b) Determine the P–value for the two-sided Z test of $H_0 : \mu = 6.2$ versus $H_1 : \mu \neq 6.2$.

8.109 Refer to the data on the heights of red pine seedlings in Example 1. Use MINITAB (or some other package program) to:

(a) Find a 97% percent confidence interval for the mean height.

(b) Test $H_0: \mu = 1.9$ versus $H_1 : \mu \neq 1.9$ centimeters with $\alpha = .03$.

8.110 Referring to speedy lizard data in Exercise 2.19, obtain a 95% confidence interval for the mean speed of that genus.

8.111 Refer to the male salmon data given in Table D.7 of the Data Bank. Use MINITAB or some other package program to find a 90% large sample confidence interval for the mean freshwater growth.

8.112 Refer to the physical fitness data given in Table D.5 of the Data Bank. Use MINITAB or some other package program to:

(a) Find a 97% large sample confidence interval for the pretest number of situps.

(b) Construct a histogram to determine if the underlying distribution is symmetric or has a long tail to one side.

8.113 Refer to the sleep data given in Table D.10 of the Data Bank. Use MINITAB or some other package program to:

(a) Find a 95% large sample confidence interval for the mean number of breathing pauses per hour (BPH).

(b) Construct a histogram to determine if the underlying distribution is symmetric or has a long tail to one side.

8.114 Refer to the grizzly bear given in Table D.8 of the Data Bank. Use MINITAB or some other package program to:

(a) Find a 95% large sample confidence interval for the mean weight in pounds of all bears living in that area.

(b) Construct a histogram to determine if the underlying distribution is symmetric or has a long tail to one side.

9

Small Sample Inferences for Normal Populations

Collecting a Sample of Lengths of Anacondas

Jesus Rivas, a herpetologist, is currently doing the definitive research on green anacondas. These snakes, some of the largest in the world, can grow to 25 feet in length. They have been known to swallow live goats and even people. Jesus Rivas and fellow researchers walk barefoot in shallow water in the Llanos grasslands shared by Venezuela and Colombia during the dry season. When they feel a snake with their feet, they grab and hold it with the help of another person. After muzzling the snake with a sock and tape, they place a string along an imaginary centerline from head to tail. The measured length of string is the recorded length of the anaconda.

Females are typically larger than males. The lengths (feet) of 21 females are

10.2	11.4	13.6	17.1	16.5	11.8	15.6
11.8	11.3	11.9	9.6	14.4	13.2	13.5
12.4	12.1	11.6	8.6	13.0	16.3	14.4

If the captured snakes can be treated as a random sample, a 95% confidence interval for the mean length of a female anaconda in that area is given in the computer output

```
One-Sample T: Length(ft)

Variable       N    Mean  StDev      95.0% CI
Length(ft)    21  12.871  2.262  ( 11.842, 13.901)
```

Both capturing and measuring make data collection difficult.
© Gary Braasch/The Image Bank/Getty Images

1. INTRODUCTION

In Chapter 8, we discussed inferences about a population mean when a large sample is available. Those methods are deeply rooted in the central limit theorem, which guarantees that the distribution of $\overline{X}$ is approximately normal. By the versatility of the central limit theorem, we did not need to know the specific form of the population distribution.

Many investigations, especially those involving costly experiments, require statistical inferences to be drawn from small samples ($n \leq 30$, as a rule of thumb). Since the sample mean $\overline{X}$ will still be used for inferences about μ, we must address the question, "What is the sampling distribution of $\overline{X}$ when n is not large?" Unlike the large sample situation, here we do not have an unqualified answer. In fact, when n is small, the distribution of $\overline{X}$ does depend to a considerable extent on the form of the population distribution. With the central limit theorem no longer applicable, more information concerning the population is required for the development of statistical procedures. In other words, the appropriate methods of inference depend on the restrictions met by the population distribution.

In this chapter, we describe how to set confidence intervals and test hypotheses **when it is reasonable to assume that the population distribution is normal.**

We begin with inferences about the mean μ of a normal population. Guided by the development in Chapter 8, it is again natural to focus on the ratio

$$\frac{\overline{X} - \mu}{S/\sqrt{n}}$$

when σ is also unknown. The sampling distribution of this ratio, called Student's *t* distribution, is introduced next.

2. STUDENT'S *t* DISTRIBUTION

When $\overline{X}$ is based on a random sample of size n from a normal $N(\mu, \sigma)$ population, we know that $\overline{X}$ is exactly distributed as $N(\mu, \sigma/\sqrt{n})$. Consequently, the standardized variable

$$Z = \frac{\overline{X} - \mu}{\sigma/\sqrt{n}}$$

has the standard normal distribution.

Because σ is typically unknown, an intuitive approach is to estimate σ by the sample standard deviation S. Just as we did in the large sample situation, we consider the ratio

$$T = \frac{\overline{X} - \mu}{S/\sqrt{n}}$$

Its probability density function is still symmetric about zero. Although estimating σ with S does not appreciably alter the distribution in large samples, it does make a substantial difference if the sample is small. The new notation T is required in order to distinguish it from the standard normal variable Z. In fact, this ratio is no longer standardized. Replacing σ by the sample quantity S introduces more variability in the ratio, making its standard deviation larger than 1.

The distribution of the ratio T is known in statistical literature as "Student's t distribution." This distribution was first studied by a British chemist W. S. Gosset, who published his work in 1908 under the pseudonym "Student." The brewery for which he worked apparently did not want the competition to know that it was using statistical techniques to better understand and improve its fermentation process.

Student's t Distribution

If $X_1, \ldots, X_n$ is a random sample from a normal population $N(\mu, \sigma)$ and

$$\overline{X} = \frac{1}{n} \sum X_i \quad \text{and} \quad S^2 = \frac{\sum (X_i - \overline{X})^2}{n - 1}$$

then the distribution of

$$T = \frac{\overline{X} - \mu}{S/\sqrt{n}}$$

is called Student's t distribution with $n - 1$ degrees of freedom

---

The qualification "with $n - 1$ degrees of freedom" is necessary, because with each different sample size or value of $n - 1$, there is a different t distribution. The choice $n - 1$ coincides with the divisor for the estimator S^2 that is based on $n - 1$ degrees of freedom.

The t distributions are all symmetric around 0 but have tails that are more spread out than the $N(0, 1)$ distribution. However, with increasing degrees of freedom, the t distributions tend to look more like the $N(0, 1)$ distribution. This agrees with our previous remark that for large n the ratio

$$\frac{\overline{X} - \mu}{S/\sqrt{n}}$$

is approximately standard normal. The density curves for t with 2 and 5 degrees of freedom are plotted in Figure 1 along with the $N(0, 1)$ curve.

Appendix B, Table 4, gives the upper α points t_α for some selected values of α and the degrees of freedom (abbreviated d.f.).

The curve is symmetric about zero so the lower α point is simply $-t_\alpha$. The entries in the last row marked "d.f. = infinity" in Appendix B, Table 4, are exactly the percentage points of the $N(0, 1)$ distribution.

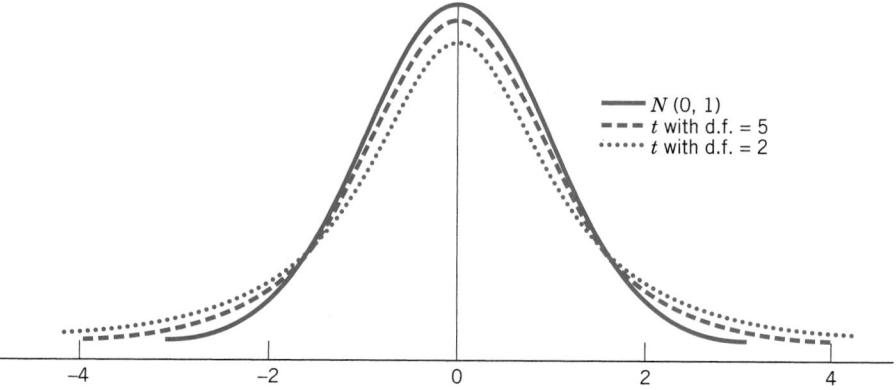

Figure 1   Comparison of $N(0, 1)$ and t density curves.

Example 1   Obtaining Percentage Points of t Distributions

Using Appendix B, Table 4, determine the upper .10 point of the t distribution with 5 degrees of freedom. Also find the lower .10 point.

SOLUTION   With d.f. = 5, the upper .10 point of the t distribution is found from Appendix B, Table 4, to be $t_{.10} = 1.476$. Since the curve is symmetric about 0, the lower .10 point is simply $-t_{.10} = -1.476$. See Figure 2.

Percentage Points
of t distributions

d.f. \ α	...	.10
.	.	.
.	.	.
.	.	.
5	...	1.476

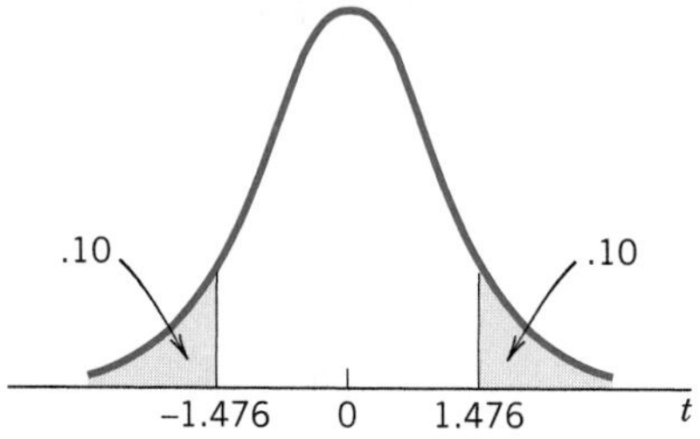

Figure 2   The upper and lower .10 points of the t distribution with d.f. = 5.

Example 2   Determining a Central Interval Having Probability .90
For the t distribution with d.f. $= 9$, find the number b such that
$P[-b < T < b] = .90$.

SOLUTION   In order for the probability in the interval $(-b, b)$ to be .90, we must have a probability of .05 to the right of b and, correspondingly, a probability of .05 to the left of $-b$ (see Figure 3). Thus, the number b is the upper $\alpha = .05$ point of the t distribution. Reading Appendix B, Table 4, at $\alpha = .05$ and d.f. $= 9$, we find $t_{.05} = 1.833$, so $b = 1.833$.

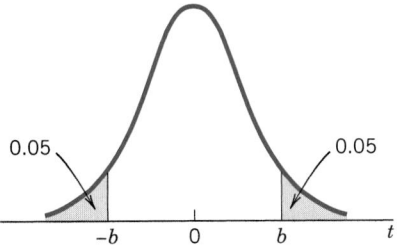

Figure 3   Finding the 5th percentile.

Exercises

9.1   Using the table for the t distributions, find:
   (a)   The upper .05 point when d.f. $= 6$.
   (b)   The lower .025 point when d.f. $= 17$.
   (c)   The lower .01 point when d.f. $= 10$.
   (d)   The upper .10 point when d.f. $= 15$.

9.2   Name the t percentiles shown and find their values from Appendix B, Table 4.

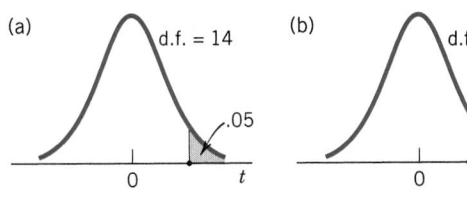

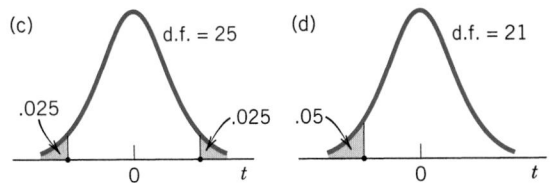

9.3   Using the table for the t distributions find:
   (a)   The 90th percentile of the t distribution when d.f. $= 10$.
   (b)   The 99th percentile of the t distribution when d.f. $= 4$.
   (c)   The 5th percentile of the t distribution when d.f. $= 24$.
   (d)   The lower and upper quartiles of the t distribution when d.f. $= 18$.

9.4   Find the probability of
   (a) $T < -1.761$ when d.f. $= 14$
   (b) $|T| > 2.306$ when d.f. $= 8$
   (c) $-1.734 < T < 1.734$ when d.f. $= 18$
   (d) $-1.812 < T < 2.764$ when d.f. $= 10$

9.5   In each case, find the number b so that
   (a) $P[T < b] = .95$ when d.f. $= 7$
   (b) $P[-b < T < b] = .95$ when d.f. $= 16$
   (c) $P[T > b] = .01$ when d.f. $= 9$
   (d) $P[T > b] = .99$ when d.f. $= 12$

9.6   Record the $t_{.05}$ values for d.f. of 5, 10, 15, 20, and 29. Does this percentile increase or decrease with increasing degrees of freedom?

9.7   Using the table for the t distributions, make an assessment for the probability of the stated event. (The answer to part (a) is provided.)

(a) $T > 2.6$ when   d.f. $= 7$ (Answer: $P[T > 2.6]$ is between .01 and .025 because 2.6 lies between $t_{.05} = 2.365$ and $t_{.01} = 2.998$.)

(b) $T > 1.9$ when d.f. $= 16$

(c) $T < -1.5$ when d.f. $= 11$

(d) $|T| > 1.9$ when d.f. $= 10$

(e) $|T| < 2.8$ when d.f. $= 17$

9.8   What can you say about the number c in each case? Justify your answer. (The answer to part (a) is provided.)

(a) $P[T > c] = .03$ when d.f. $= 5$

(Answer: c is between 2.015 and 2.571 because $t_{.05} = 2.015$ and $t_{.025} = 2.571$.)

(b) $P[T > c] = .016$ when d.f. $= 11$

(c) $P[T < -c] = .004$ when d.f. $= 13$

(d) $P[|T| > c] = .03$ when d.f. $= 6$

(e) $P[|T| < c] = .96$ when d.f. $= 27$

3.   INFERENCES ABOUT μ—SMALL SAMPLE SIZE

3.1.   CONFIDENCE INTERVAL FOR μ

The distribution of

$$T = \frac{\overline{X} - \mu}{S/\sqrt{n}}$$

provides the key for determining a confidence interval for μ, the mean of a normal population. For a $100(1 - \alpha)\%$ confidence interval, we consult the t table (Appendix B, Table 4) and find $t_{\alpha/2}$, the upper $\alpha/2$ point of the t distribution with $n - 1$ degrees of freedom (see Figure 4).

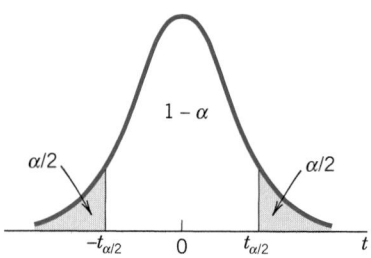

Figure 4 $t_{\alpha/2}$ and the probabilities.

Since $\dfrac{\overline{X} - \mu}{S/\sqrt{n}}$ has the t distribution with d.f. $= n - 1$, we have

$$P\left[-t_{\alpha/2} < \frac{\overline{X} - \mu}{S/\sqrt{n}} < t_{\alpha/2} \right] = 1 - \alpha$$

In order to obtain a confidence interval, let us rearrange the terms inside the brackets so that only the parameter μ remains in the center. The above probability statement then becomes

$$P\left[\overline{X} - t_{\alpha/2}\frac{S}{\sqrt{n}} < \mu < \overline{X} + t_{\alpha/2}\frac{S}{\sqrt{n}}\right] = 1 - \alpha$$

which is precisely in the form required for a confidence statement about μ. The probability is $1 - \alpha$ that the random interval $\overline{X} - t_{\alpha/2}S/\sqrt{n}$ to $\overline{X} + t_{\alpha/2}S/\sqrt{n}$ will cover the true population mean μ. This argument is virtually the same as in Section 3 of Chapter 8. Only now the unknown σ is replaced by the sample standard deviation S, and the t percentage point is used instead of the standard normal percentage point.

**A 100(1 − α)% Confidence Interval
for a Normal Population Mean**

$$\left(\overline{X} - t_{\alpha/2}\frac{S}{\sqrt{n}}, \qquad \overline{X} + t_{\alpha/2}\frac{S}{\sqrt{n}}\right)$$

where $t_{\alpha/2}$ is the upper $\alpha/2$ point of the t distribution with d.f. $= n - 1$.

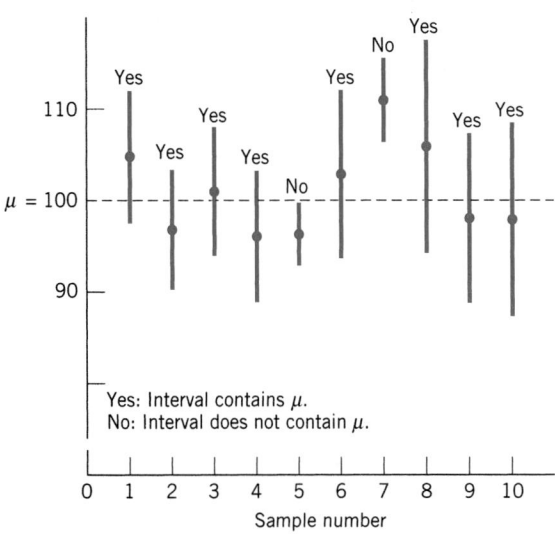

Figure 5   Behavior of confidence intervals based on the t distribution.

Let us review the meaning of a confidence interval in the present context. Imagine that random samples of size n are repeatedly drawn from a normal population and the interval $(\bar{x} - t_{\alpha/2} s/\sqrt{n}, \quad \bar{x} + t_{\alpha/2} s/\sqrt{n})$ calculated in each case. The interval is centered at $\bar{x}$ so the center varies from sample to sample. The length of an interval, $2t_{\alpha/2} s/\sqrt{n}$, also varies from sample to sample because it is a multiple of the sample standard deviation s. (This is unlike the fixed-length situation illustrated in Figure 4 of Chapter 8, which was concerned with a known σ.) Thus, in repeated sampling, the intervals have variable centers and variable lengths. However, our confidence statement means that if the sampling is repeated many times, about $100(1 - \alpha)\%$ of the resulting intervals would cover the true population mean μ.

Figure 5 shows the results of drawing 10 samples of size $n = 7$ from the normal population with $\mu = 100$ and $\sigma = 10$. Selecting $\alpha = .05$, we find that the value of $t_{.025}$ with 6 d.f. is 2.447, so the 95% confidence interval is $\overline{X} \pm 2.447 S/\sqrt{7}$. In the first sample, $\bar{x} = 103.88$ and $s = 7.96$, so the interval is $(96.52, 111.24)$. The 95% confidence intervals are shown by the vertical line segments.

Example 3   Interpreting a Confidence Interval

The weights (pounds) of $n = 8$ female wolves captured in the Yukon-Charley National Reserve (see Table D.9 of the Data Bank) are

$$57 \quad 84 \quad 90 \quad 71 \quad 77 \quad 68 \quad 73 \quad 71$$

Treating these weights as a random sample:
   (a)   Find a 95% confidence interval for the population mean weight of all female wolves living on the Yukon-Charley National Reserve.
   (b)   Is μ included in this interval?

SOLUTION   (a)   If we assume that the weights are normally distributed, a 90% confidence interval for the mean weight μ is given by

$$\left(\overline{X} - t_{.05} \frac{S}{\sqrt{n}}, \quad \overline{X} - t_{.05} \frac{S}{\sqrt{n}} \right)$$

where $n = 8$. The t statistic is based on $n - 1 = 8 - 1 = 7$ degrees of freedom so, consulting the t table, we find $t_{.05} = 1.895$. Beforehand, this is a random interval.

To determine the confidence interval from the given sample, we first obtain the summary statistics

$$\bar{x} = \frac{591.000}{8} = 73.875 \qquad s^2 = \frac{708.875}{7} = 101.268 \qquad \text{so}$$

$$s = \sqrt{101.268} = 10.063$$

The 90% confidence interval for μ is then

$$73.875 \pm 1.895 \times \frac{10.063}{\sqrt{8}} = 73.875 \pm 6.742 \qquad \text{or } (67.13, 80.62)$$

We are 90% confident that the mean weight of all female wolves is between 67.13 and 80.62 pounds. We have this confidence because, over many occasions of sampling, approximately 90% of the intervals calculated using this procedure will contain the true mean.

(b)   We will never know if a single realization of the confidence interval, such as (67.13, 80.62), covers the unknown μ. It is unknown because it is based on every female wolf in the very large reserve. Our confidence in the method is based on the high percentage of times that μ is covered by intervals in repeated samplings.

Remark: When repeated independent measurements are made on the same material and any variation in the measurements is basically due to experimental error (possibly compounded by nonhomogeneous materials), the normal model is often found to be appropriate. It is still necessary to graph the individual data points (too few here) in a dot diagram and normal-scores plot to reveal any wild observations or serious departures from normality. In all small sample situations, it is important to remember that the validity of a confidence interval rests on the reasonableness of the model assumed for the population.

Recall from the previous chapter that the length of a $100(1 - \alpha)\%$ confidence interval for a normal μ is $2 z_{\alpha/2} \sigma/\sqrt{n}$ when σ is known, whereas it is $2 t_{\alpha/2} S/\sqrt{n}$ when σ is unknown. Given a small sample size n and consequently a small number of degrees of freedom $(n - 1)$, the extra variability caused by estimating σ with S makes the t percentage point $t_{\alpha/2}$ much larger than the normal percentage point $z_{\alpha/2}$. For instance, with d.f. $= 4$, we have $t_{.025} = 2.776$, which is considerably larger than $z_{.025} = 1.96$. Thus, when σ is unknown, the confidence estimation of μ based on a very small sample size is expected to produce a much less precise inference (namely, a wide confidence interval) compared to the situation when σ is known. With increasing n, σ can be more closely estimated by S and the difference between $t_{\alpha/2}$ and $z_{\alpha/2}$ tends to diminish.

3.2.   HYPOTHESES TESTS FOR μ

The steps for conducting a test of hypotheses concerning a population mean were presented in the previous chapter. If the sample size is small, basically the same procedure can be followed provided it is reasonable to assume that the population distribution is normal. However, in the small sample situation, our test statistic

$$T = \frac{\overline{X} - \mu_0}{S/\sqrt{n}}$$

has Student's t distribution with $n - 1$ degrees of freedom.

The t table (Appendix B, Table 4) is used to determine the rejection region to test hypotheses about μ.

Hypotheses Tests for μ—Small Samples

To test $H_0: \mu = \mu_0$ concerning the mean of a **normal population**, the test statistic is

$$T = \frac{\overline{X} - \mu_0}{S/\sqrt{n}}$$

which has Student's t distribution with $n - 1$ degrees of freedom:

$$H_1: \mu > \mu_0 \qquad R: T \geq t_\alpha$$
$$H_1: \mu < \mu_0 \qquad R: T \leq -t_\alpha$$
$$H_1: \mu \neq \mu_0 \qquad R: |T| \geq t_{\alpha/2}$$

The test is called a **Student's t test** or simply a **t test**.

Example 4   A Student's t Test to Confirm the Water Is Safe

A city health department wishes to determine if the mean bacteria count per unit volume of water at a lake beach is within the safety level of 200. A researcher collected 10 water samples of unit volume and found the bacteria counts to be

$$
\begin{array}{ccccc}
175 & 190 & 205 & 193 & 184 \\
207 & 204 & 193 & 196 & 180
\end{array}
$$

Do the data strongly indicate that there is no cause for concern? Test with $\alpha = .05$.

SOLUTION   Let μ denote the current (population) mean bacteria count per unit volume of water. Then, the statement "no cause for concern" translates to $\mu < 200$, and the researcher is seeking strong evidence in support of this hypothesis. So the formulation of the null and alternative hypotheses should be

$$H_0: \mu = 200 \qquad \text{versus} \qquad H_1: \mu < 200$$

Since the counts are spread over a wide range, an approximation by a continuous distribution is not unrealistic for inference about the mean. Assuming further that the measurements constitute a sample from a normal population, we employ the t test with

$$T = \frac{\overline{X} - 200}{S/\sqrt{10}} \qquad \text{d.f.} = 9$$

We test at the level of significance $\alpha = .05$. Since H_1 is left-sided, we set the rejection region $T \leq -t_{.05}$. From the t table we find that $t_{.05}$ with d.f. $= 9$ is 1.833, so our rejection region is $R:T \leq -1.833$ as in Figure 6. Computations from the sample data yield

$$\bar{x} = 192.7$$
$$s = 10.81$$
$$t = \frac{192.7 - 200}{10.81/\sqrt{10}} = \frac{-7.3}{3.418} = -2.14$$

Because the observed value $t = -2.14$ is smaller than -1.833, the null hypothesis is rejected at $\alpha = .05$. On the basis of the data obtained from these 10 measurements, there does seem to be strong evidence that the true mean is within the safety level.

Values of T smaller than -2.14 are more extreme evidence in favor of the alternative hypothesis. Since, with 9 degrees of freedom, $t_{.05} = 1.833$ and $t_{.025} = 2.262$, the P-value is between .05 and .025. The accompanying computer output gives

$$P\text{-value} = P(T \leq -2.14) = .031 \quad (\text{see Figure 7})$$

There is strong evidence that the mean bacteria count is within the safety level.

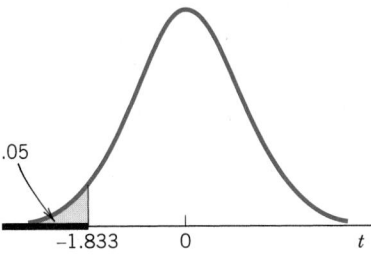

Figure 6   Left-sided rejection region $T \leq -1.833$.

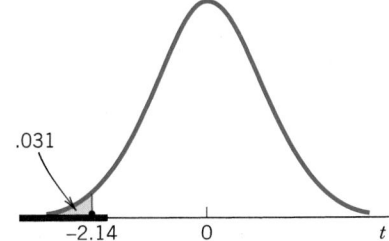

Figure 7 P-value for left-sided rejection region, T statistic.

Computer Solution to Example 4

(see the Techology section for the MINITAB COMMANDS)

T Test of the Mean

Test of mu $= 200.00$ vs mu < 200.00

Variable	N	Mean	StDev	SE Mean	T	P
Bac Coun	10	192.70	10.81	3.42	−2.14	0.031

Exercises

9.9   A random sample of size $n = 20$ from a normal population gives the sample mean 140 and the sample standard deviation 8.

(a)   Construct a 98% confidence interval for the population mean.

(b)   What is the length of this confidence interval? What is its center?

(c)   If a 98% confidence interval were calculated from another random sample of size $n = 20$, would it have the same length as that found in part (b)? Why or why not?

9.10   Recorded here are the germination times (number of days) for seven seeds of a new strain of snap bean.

$$12 \quad 16 \quad 15 \quad 20 \quad 17 \quad 11 \quad 18$$

Stating any assumptions that you make, determine a 95% confidence interval for the true mean germination time for this strain.

9.11   A zoologist collected 20 wild lizards in the southwestern United States. The total length (mm) of each was measured.

179 157 169 146 143 131 159 142 141 130
142 116 130 140 138 137 134 114   90 114

Obtain a 95% confidence interval for the mean length.

9.12   In an investigation on toxins produced by molds that infect corn crops, a biochemist prepares extracts of the mold culture with organic solvents and then measures the amount of the toxic substance per gram of solution. From nine preparations of the mold culture, the following measurements of the toxic substance (in milligrams) are obtained:

$$1.2 \quad .8 \quad .6 \quad 1.1 \quad 1.2 \quad .9 \quad 1.5 \quad .9 \quad 1.0.$$

(a)   Calculate the mean $\bar{x}$ and the standard deviation s from the data.

(b)   Compute a 98% confidence interval for the mean weight of toxic substance per gram of mold culture. State the assumption you make about the population.

9.13   An experimenter studying the feasibility of extracting protein from seaweed to use in animal feed makes 18 determinations of the protein extract, each based on a different 50-kilogram sample of seaweed. The sample mean and the standard deviation are found to be 3.6 and .8 kilograms, respectively. Determine a 95% confidence interval for the mean yield of protein extract per 50 kilograms of seaweed.

9.14   The monthly rent (dollars) for a two-bedroom apartment on the west side of town was recorded for a sample of ten apartments.

550 575 610 620 650 675 720 740 790 860

Obtain a 95% confidence interval for the mean monthly rent for two-bedroom apartments.

9.15   From a random sample of size 12, one has calculated the 95% confidence interval for μ and obtained the result (18.6, 26.2).

(a)   What were the $\bar{x}$ and s for that sample?

(b)   Calculate a 98% confidence interval for μ.

9.16   Suppose that with a random sample of size 18 from a normal population, one has calculated the 90% confidence interval for the population mean μ and obtained the result (122, 146). Using this result, obtain:

(a)   A point estimate of μ and its 90% margin of error.

(b)   A 95% confidence interval for μ.

9.17   Henry Cavendish (1731–1810) provided direct experimental evidence of Newton's law of universal gravitation, which specifies the force of attraction between two masses. In an experiment with known masses determined by weighing, the measured force can also be used to calculate a value for the density of the earth. The values of the earth's density from Cavendish's renowned experiment in time order by column are

5.36	5.62	5.27	5.46
5.29	5.29	5.39	5.30
5.58	5.44	5.42	5.75
5.65	5.34	5.47	5.68
5.57	5.79	5.63	5.85
5.53	5.10	5.34	

[These data were published in *Philosophical Transactions*, **17** (1798), p. 469.] Find a 99% confidence interval for the density of the earth.

9.18   Refer to Exercise 9.14. Do these data support the claim that the mean monthly rent for a two-bedroom apartment differs from 750 dollars? Take $\alpha = .05$.

9.19   Refer to Exercise 9.14. Do these data provide strong evidence for the claim that the monthly rent for a two-bedroom apartment is greater than 620 dollars? Take $\alpha = .05$.

9.20   In a lake pollution study, the concentration of lead in the upper sedimentary layer of a lake bottom is measured from 25 sediment samples of 1000 cubic centimeters each. The sample mean and the standard deviation of the measurements are found to be .38 and .06, respectively. Compute a 99% confidence interval for the mean concentration of lead per 1000 cubic centimeters of sediment in the lake bottom.

9.21   The data on the lengths of anacondas on the front piece of the chapter yield a 95% confidence interval for the population mean length of all anaconda snakes in the area of the study.

```
Variable   N    Mean    StDev    95.0% CI
Length(ft) 21   12.871  2.262  ( 11.842, 13.901)
```

   (a)   Is the population mean length of all anacondas living in the study area contained in this interval?

   (b)   Explain why you are 95% confident that it is contained in the interval.

9.22   Refer to the data on the weight (pounds) of male wolves given in Table D.9 of the Data Bank. A computer calculation gives a 95% confidence interval.

```
Variable   N    Mean    StDev    95.0% CI
Malewt     11   91.91   12.38  ( 83.59, 100.23)
```

   (a)   Is the population mean weight for all wolves in the Yukon-Charley Rivers National Reserve contained in this interval?

   (b)   Explain why you are 95% confident that it is contained in the interval.

9.23   The following measurements of the diameters (in feet) of Indian mounds in southern Wisconsin were gathered by examining reports in *Wisconsin Archeologist* (courtesy of J. Williams).

$$22 \quad 24 \quad 24 \quad 30 \quad 22 \quad 20 \quad 28$$
$$30 \quad 24 \quad 34 \quad 36 \quad 15 \quad 37$$

   (a)   Do these data substantiate the conjecture that the population mean diameter is larger than 21 feet? Test at $\alpha = .01$.

   (b)   Determine a 90% confidence interval for the population mean diameter of Indian mounds.

9.24   Measurements of the acidity (pH) of rain samples were recorded at 13 sites in an industrial region.

$$3.5 \quad 5.1 \quad 5.0 \quad 3.6 \quad 4.8 \quad 3.6 \quad 4.7$$
$$4.3 \quad 4.2 \quad 4.5 \quad 4.9 \quad 4.7 \quad 4.8$$

   Determine a 95% confidence interval for the mean acidity of rain in that region.

9.25   Refer to Exercise 9.11, where a zoologist collected 20 wild lizards in the southwestern United States. Do these data substantiate a claim that the mean length is greater than 128 mm? Test with $\alpha = .05$.

9.26   District court records provided data on sentencing for 19 criminals convicted of negligent homicide. The mean and standard deviation of the sentences were found to be 72.7 and 10.2 months, respectively. Determine a 95% confidence interval for the mean sentence for this crime.

9.27   The data on the weight (lb) of female wolves, from Table D.9 of the Data Bank, are

$$57 \quad 84 \quad 90 \quad 71 \quad 71 \quad 77 \quad 68 \quad 73$$

   Test the null hypothesis that the mean weight of females is 83 pounds versus a two-sided alternative. Take $\alpha = .05$.

9.28   Five years ago, the average size of farms in a state was 160 acres. From a recent survey of 27 farms, the mean and standard deviation were found to be 180 and 36 acres, respectively.

(a) Is there strong evidence that the average farm size is larger than what it was 5 years ago?

(b) Give a 98% confidence interval for the current average size.

9.29 The mean drying time of a brand of spray paint is known to be 90 seconds. The research division of the company that produces this paint contemplates that adding a new chemical ingredient to the paint will accelerate the drying process. To investigate this conjecture, the paint with the chemical additive is sprayed on 15 surfaces and the drying times are recorded. The mean and the standard deviation computed from these measurements are 86 and 4.5 seconds, respectively.

(a) Do these data provide strong evidence that the mean drying time is reduced by the addition of the new chemical?

(b) Construct a 98% confidence interval for the mean drying time of the paint with the chemical additive.

9.30 A few years ago, noon bicycle traffic past a busy section of campus had a mean of $\mu = 300$. To see if any change in traffic has occurred, counts were taken for a sample of 15 weekdays. It was found that $\bar{x} = 340$ and $s = 30$.

(a) Construct an $\alpha = .05$ test of $H_0 : \mu = 300$ against the alternative that some change has occurred.

(b) Obtain a 95% confidence interval for μ.

9.31 Refer to the computer anxiety scores for female accounting students in Table D.4 of the Data Bank. A computer calculation for a test of $H_0 : \mu = 2$ versus $H_1 : \mu \neq 2$ is given below.

```
Test of mu = 2 vs mu not = 2
```

Variable	N	Mean	StDev
FCARS	15	2.514	0.773

Variable	95.0% CI	T	P
FCARS	(2.086, 2.942)	2.58	0.022

(a) What is the conclusion if you test with $\alpha = .05$?

(b) What mistake could you have made in part (a)?

(c) Before you collected the data, what was the probability of making the mistake in part (a)?

(d) Give a long-run relative frequency interpretation of the probability in part (c).

4. RELATIONSHIP BETWEEN TESTS AND CONFIDENCE INTERVALS

By now the careful reader should have observed a similarity between the formulas we use in testing hypotheses and in estimation by a confidence interval. To clarify the link between these two concepts, let us consider again the inferences about the mean μ of a normal population.

A $100(1 - \alpha)\%$ confidence interval for μ is

$$\left(\bar{X} - t_{\alpha/2} \frac{S}{\sqrt{n}}, \quad \bar{X} + t_{\alpha/2} \frac{S}{\sqrt{n}} \right)$$

because before the sample is taken, the probability that

$$\bar{X} - t_{\alpha/2} \frac{S}{\sqrt{n}} < \mu < \bar{X} + t_{\alpha/2} \frac{S}{\sqrt{n}}$$

is $1 - \alpha$. On the other hand, the rejection region of a level α test for $H_0 : \mu = \mu_0$ versus the two-sided alternative $H_1 : \mu \neq \mu_0$ is

$$R : \left| \frac{\bar{X} - \mu_0}{S/\sqrt{n}} \right| \geq t_{\alpha/2}$$

Let us use the name "acceptance region" to mean the opposite (or complement) of the rejection region. Reversing the inequality in R, we obtain

Acceptance region
$$-t_{\alpha/2} < \frac{\overline{X} - \mu_0}{S/\sqrt{n}} < t_{\alpha/2}$$

which can also be written as

Acceptance region
$$\overline{X} - t_{\alpha/2}\frac{S}{\sqrt{n}} < \mu_0 < \overline{X} + t_{\alpha/2}\frac{S}{\sqrt{n}}$$

The latter expression shows that any given null hypothesis μ_0 will be accepted (more precisely, will not be rejected) at level α if μ_0 lies within the $100(1 - \alpha)\%$ confidence interval. Thus, having established a $100(1 - \alpha)\%$ confidence interval for μ, we know at once that all possible null hypotheses values μ_0 lying outside this interval will be rejected at level of significance α and all those lying inside will not be rejected.

Example 5   Relation between a 95% Confidence Interval and Two-Sided $\alpha = .05$ Test

A random sample of size $n = 9$ from a normal population produced the mean $\overline{x} = 8.3$ and the standard deviation $s = 1.2$. Obtain a 95% confidence interval for μ and also test $H_0: \mu = 8.5$ versus $H_1: \mu \neq 8.5$ with $\alpha = .05$.

SOLUTION   A 95% confidence interval has the form

$$\left(\overline{X} - t_{.025}\frac{S}{\sqrt{n}}, \quad \overline{X} + t_{.025}\frac{S}{\sqrt{n}} \right)$$

where $t_{.025} = 2.306$ corresponds to $n - 1 = 8$ degrees of freedom. Here $\overline{x} = 8.3$ and $s = 1.2$, so the interval becomes

$$\left(8.3 - 2.306\frac{1.2}{\sqrt{9}}, \quad 8.3 + 2.306\frac{1.2}{\sqrt{9}} \right) = (7.4, 9.2)$$

Turning now to the problem of testing $H_0: \mu = 8.5$, we observe that the value 8.5 lies in the 95% confidence interval we have just calculated. Using the correspondence between confidence interval and acceptance region, we can at once conclude that $H_0: \mu = 8.5$ should not be rejected at $\alpha = .05$. Alternatively, a formal step-by-step solution can be based on the test statistic

$$T = \frac{\overline{X} - 8.5}{S/\sqrt{n}}$$

The rejection region consists of both large and small values.

Rejection region
$$\left| \frac{\overline{X} - 8.5}{S/\sqrt{n}} \right| \geq t_{.025} = 2.306$$

Now the observed value $|t| = \sqrt{9}|8.3 - 8.5|/1.2 = .5$ does not fall in the rejection region, so the null hypothesis $H_0: \mu = 8.5$ is not rejected at $\alpha = .05$. This conclusion agrees with the one we arrived at from the confidence interval.

This relationship indicates how confidence estimation and tests of hypotheses with two-sided alternatives are really integrated in a common framework. A confidence interval statement is regarded as a more comprehensive inference procedure than testing a single null hypothesis, because a confidence interval statement in effect tests many null hypotheses at the same time.

Exercises

9.32 Based on a random sample of size 18 from a normal distribution, an investigator calculates the 95% confidence interval and gets the result $(27.1, 39.3)$.

  (a) What is the conclusion of the t test for $H_0: \mu = 29$ versus $H_1: \mu \neq 29$ at level $\alpha = .05$?

  (b) What is the conclusion if $H_0: \mu = 26.8$?

9.33 In Example 3, the 95% confidence interval for the mean weight of male wolves was found to be $(67.13, 80.62)$.

  (a) What is the conclusion of testing $H_0: \mu = 81$ versus $H_1: \mu \neq 81$ at level $\alpha = .05$?

  (b) What is the conclusion if $H_0: \mu = 69$?

9.34 Suppose that from a random sample a 90% confidence interval for the population mean has been found to be $(16.8, 19.6)$. Answer each question "yes," "no," or "can't tell," and justify your answer. On the basis of the same sample:

  (a) Would $H_0: \mu = 20$ be rejected in favor of $H_1: \mu \neq 20$ at $\alpha = .10$?

  (b) Would $H_0: \mu = 18$ be rejected in favor of $H_1: \mu \neq 18$ at $\alpha = .10$?

  *(c) Would $H_0: \mu = 17$ be rejected in favor of $H_1: \mu \neq 17$ at $\alpha = .05$?

  *(d) Would $H_0: \mu = 22$ be rejected in favor of $H_1: \mu \neq 22$ at $\alpha = .01$?

9.35 Recorded here are the amounts of decrease in percent body fat for eight participants in an exercise program over three weeks.

  1.8 10.6 −1.2 12.9 15.1 −2.0 6.2 10.8

  (a) Construct a 95% confidence interval for the population mean amount μ of decrease in percent body fat over the three-week program.

  (b) If you were to test $H_0: \mu = 15$ versus $H_1: \mu \neq 15$ at $\alpha = .05$, what would you conclude from your result in part (a)?

  (c) Perform the hypothesis test indicated in part (b) and confirm your conclusion.

9.36 Refer to the data in Exercise 9.35.

  (a) Construct a 90% confidence interval for μ.

  (b) If you were to test $H_0: \mu = 10$ versus $H_1: \mu \neq 10$ at $\alpha = .10$, what would you conclude from your result in part (a)? Why?

  (c) Perform the hypothesis test indicated in part (b) and confirm your conclusion.

9.37 Establish the connection between the large sample Z test, which rejects $H_0: \mu = \mu_0$ in favor of $H_1: \mu \neq \mu_0$, at $\alpha = .05$, if

$$Z = \frac{\overline{X} - \mu_0}{S/\sqrt{n}} \geq 1.96 \qquad \text{or}$$

$$\frac{\overline{X} - \mu_0}{S/\sqrt{n}} \leq -1.96$$

and the 95% confidence interval

$$\overline{X} - 1.96\frac{S}{\sqrt{n}} \qquad \text{to} \qquad \overline{X} + 1.96\frac{S}{\sqrt{n}}$$

5. INFERENCES ABOUT THE STANDARD DEVIATION σ (THE CHI-SQUARE DISTRIBUTION)

Aside from inferences about the population mean, the population variability may also be of interest. Apart from the record of a baseball player's batting average, information on the variability of the player's performance from one game to the next may be an indicator of reliability. Uniformity is often a criterion of production quality for a manufacturing process. The quality control engineer must ensure that the variability of the measurements does not exceed a specified limit. It may also be important to ensure sufficient uniformity of the inputted raw material for trouble-free operation of the machines. In this section, we consider inferences for the standard deviation σ of a population under the assumption that the population distribution is normal. In contrast to the inference procedures concerning the population mean μ, the usefulness of the methods to be presented here is extremely limited when this assumption is violated.

To make inferences about σ^2, the natural choice of a statistic is its sample analog, which is the sample variance,

$$S^2 = \frac{\sum\limits_{i=1}^{n} (X_i - \overline{X})^2}{n-1}$$

We take S^2 as the point estimator of σ^2 and its square root S as the point estimator of σ. To estimate by confidence intervals and test hypotheses, we must consider the sampling distribution of S^2. To do this, we introduce a new distribution, called the χ^2 distribution (read "chi-square distribution"), whose form depends on $n - 1$.

χ^2 Distribution

Let $X_1, \ldots, X_n$ be a random sample from a normal population $N(\mu, \sigma)$. Then the distribution of

$$\chi^2 = \frac{\sum\limits_{i=1}^{n} (X_i - \overline{X})^2}{\sigma^2} = \frac{(n-1)S^2}{\sigma^2}$$

is called the χ^2 distribution with $n - 1$ degrees of freedom.

Unlike the normal or t distribution, the probability density curve of a χ^2 distribution is an asymmetric curve stretching over the positive side of the line and having a long right tail. The form of the curve depends on the value of the degrees of freedom. A typical χ^2 curve is illustrated in Figure 8.

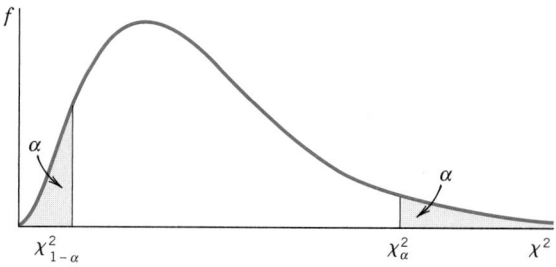

Figure 8   Probability density curve of a χ^2 distribution.

Appendix B, Table 5, provides the upper α points of χ^2 distributions for various values of α and the degrees of freedom. As in both the cases of the t and the normal distributions, the upper α point χ_α^2 denotes the χ^2 value such that the area to the right is α. The lower α point or 100αth percentile, read from the column $\chi_{1-\alpha}^2$ in the table, has an area $1 - \alpha$ to the right. For example, the lower .05 point is obtained from the table by reading the $\chi_{.95}^2$ column, whereas the upper .05 point is obtained by reading the column $\chi_{.05}^2$.

Example 6   Finding Percentage Points of the χ^2 Distribution

Find the upper .05 point of the χ^2 distribution with 17 degrees of freedom. Also find the lower .05 point.

SOLUTION

**Percentage Points of the χ^2 Distributions
(Appendix B, Table 5)**

d.f. $\quad\alpha$	$\cdots$	.95	$\cdots$	.05
.		.		.
.		.		.
.		.		.
17	$\cdots$	8.67	$\cdots$	27.59

The upper .05 point is read from the column labeled $\alpha = .05$. We find $\chi_{.05}^2 = 27.59$ for 17 d.f. The lower .05 point is read from the column $\alpha = .95$. We find $\chi_{.95}^2 = 8.67$, as the lower .05 point.

The χ^2 is the basic distribution for constructing confidence intervals for σ^2 or σ. We outline the steps in terms of a 95% confidence interval for σ^2. Dividing the probability $\alpha = .05$ equally between the two tails of the χ^2 distribution and using the notation just explained, we have

$$P\left[\chi^2_{.975} < \frac{(n-1)S^2}{\sigma^2} < \chi^2_{.025}\right] = .95$$

where the percentage points are read from the χ^2 table at d.f. $= n - 1$. Because

$$\frac{(n-1)S^2}{\sigma^2} < \chi^2_{.025} \quad \text{is equivalent to} \quad \frac{(n-1)S^2}{\chi^2_{.025}} < \sigma^2$$

and

$$\chi^2_{.975} < \frac{(n-1)S^2}{\sigma^2} \quad \text{is equivalent to} \quad \sigma^2 < \frac{(n-1)S^2}{\chi^2_{.975}}$$

we have

$$P\left[\frac{(n-1)S^2}{\chi^2_{.025}} < \sigma^2 < \frac{(n-1)S^2}{\chi^2_{.975}}\right] = .95$$

This last statement, concerning a random interval covering σ^2, provides a 95% confidence interval for σ^2.

A confidence interval for σ can be obtained by taking the square root of the endpoints of the interval. For a confidence level .95, the interval for σ becomes

$$\left(S\sqrt{\frac{n-1}{\chi^2_{.025}}}, \quad S\sqrt{\frac{n-1}{\chi^2_{.975}}}\right)$$

Example 7 A Confidence Interval for the Standard Deviation of Scores

Beginning students in accounting took a test and a computer anxiety score (CARS) was assigned to each student on the basis of their answers to nineteen questions on the test (see Table D.4 of the Data Bank). A small population standard deviation would indicate that computer anxiety is nearly the same for all female beginning accounting students.

The scores for 15 female students are

2.90	1.00	1.90	2.37	3.32	3.79	3.26	1.90
1.84	2.58	1.58	2.90	2.42	3.42	2.53	

Assuming that the distribution of scores can be modeled as a normal distribution, find a 90% confidence interval for the CARS population standard deviation of all female beginning accounting students.

SOLUTION We first obtain the summary statistics

$$\bar{x} = \frac{37.71}{15} = 2.514 \qquad s^2 = \frac{8.3602}{14} = .5972 \qquad \text{so} \quad s = .773$$

Here $n = 15$, so d.f. $= n - 1 = 14$. The χ^2 table gives $\chi^2_{.95} = 6.57$ and $\chi^2_{.05} = 23.68$. Using the preceding probability statement, we determine that a 90% confidence interval for σ^2 is

$$\left(\frac{14 \times .5972}{23.68}, \frac{14 \times .5972}{6.57} \right) = (.3531, 1.2726)$$

and the corresponding interval for σ is $(\sqrt{.3531}, \sqrt{1.2726}) = (.90, 1.13)$. We are 90% confident that the CARS standard deviation is between .90 and 1.13. We have this confidence because 90% of the intervals calculated by this procedure in repeated samples will cover the true σ.

   You may verify that the pattern of a normal-scores plot does not contradict the assumption of normality.

It is instructive to note that the midpoint of the confidence interval for σ^2 in Example 7 is not $s^2 = .5972$, which is the best point estimate. This is in sharp contrast to the confidence intervals for μ, and it serves to accent the difference in logic between interval and point estimation.

   For a test of the null hypothesis $H_0 : \sigma^2 = \sigma_0^2$ it is natural to employ the statistic S^2. If the alternative hypothesis is one-sided, say $H_1 : \sigma^2 > \sigma_0^2$, then the rejection region should consist of large values of S^2 or alternatively large values of the

Test statistic $\qquad \chi^2 = \dfrac{(n - 1)S^2}{\sigma_0^2} \qquad$ d.f. $= n - 1$

To **test hypothesis about** σ, the rejection region of a level α test is,

$$R: \frac{(n - 1)S^2}{\sigma_0^2} \geq \chi^2_\alpha \qquad \text{d.f.} = n - 1$$

For a two-sided alternative $H_1 : \sigma^2 \neq \sigma_0^2$, a level α rejection region is

$$R: \frac{(n - 1)S^2}{\sigma_0^2} \leq \chi^2_{1 - \alpha/2} \qquad \text{or} \qquad \frac{(n - 1)S^2}{\sigma_0^2} \geq \chi^2_{\alpha/2}$$

   Once again, we remind the reader that the inference procedures for σ presented in this section are extremely sensitive to departures from a normal population.

Exercises

9.38   Using the table for the χ^2 distributions, find:

(a)   The upper 5% point when d.f. $= 9$.

(b)   The upper 1% point when d.f. $= 16$.

(c)   The lower 2.5% point when d.f. $= 10$.

(d)   The lower 1% point when d.f. $= 22$.

9.39   Name the χ^2 percentiles shown and find their values from Appendix B, Table 5.

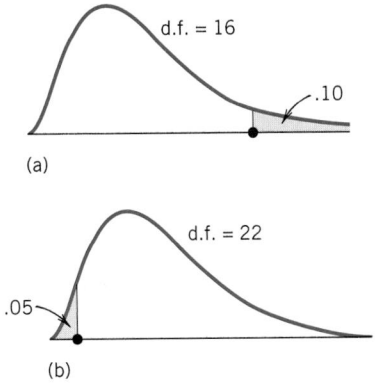

(a)

(b)

(c) Find the percentile in part (a) if d.f. = 30.

(d) Find the percentile in part (b) if d.f. = 9.

9.40 Find the probability of

(a) $\chi^2 > 31.33$ when d.f. = 18

(b) $\chi^2 < 1.15$ when d.f. = 5

(c) $3.24 < \chi^2 < 18.31$ when d.f. = 10

(d) $3.49 < \chi^2 < 20.09$ when d.f. = 8

9.41 Given the sample data

$$10 \quad 16 \quad 7 \quad 13 \quad 12$$

(a) Obtain a point estimate of the population standard deviation σ.

(b) Construct a 95% confidence interval for σ.

(c) Examine whether or not your point estimate is located at the center of the confidence interval.

9.42 Find a 90% confidence interval for σ based on the $n = 40$ measurements of heights of red pine seedlings given in Table 1 of Chapter 8 (Note: $s = .475$ for this data set. State any assumption you make about the population.)

9.43 Refer to Exercise 9.42. A related species has population standard deviation $\sigma = .6$. Do the data provide strong evidence that the red pine population standard deviation is smaller than .6? Test with $\alpha = .05$.

9.44 Plastic sheets produced by a machine are periodically monitored for possible fluctuations in thickness. Uncontrollable heterogeneity in the viscosity of the liquid mold makes some variation in thickness measurements unavoidable. However, if the true standard deviation of thickness exceeds 1.5 millimeters, there is cause to be concerned about the product quality. Thickness measurements (in millimeters) of 10 specimens produced on a particular shift resulted in the following data.

$$226 \quad 228 \quad 226 \quad 225 \quad 232$$
$$228 \quad 227 \quad 229 \quad 225 \quad 230$$

Do the data substantiate the suspicion that the process variability exceeded the stated level on this particular shift? (Test at $\alpha = .05$.) State the assumption you make about the population distribution.

9.45 Refer to Exercise 9.44. Construct a 95% confidence interval for the true standard deviation of the thickness of sheets produced on this shift.

9.46 From a data set of $n = 10$ observations, one has calculated the 95% confidence interval for σ and obtained the result (4.05, 10.75).

(a) What was the standard deviation s for the sample? (*Hint:* Examine how s enters the formula of a confidence interval.)

(b) Calculate a 90% confidence interval for σ.

9.47 Refer to the data of lizard lengths in Exercise 9.11.

(a) Determine a 90% confidence interval for the population standard deviation σ.

(b) Should $H_0 : \sigma = 9$ be rejected in favor of $H_1 : \sigma \neq 9$ at $\alpha = .10$? [Answer by using your result in part (a).]

9.48 Referring to the data in Exercise 9.17, determine a 99% confidence interval for the population standard deviation of the density measurements.

9.49 Referring to Exercise 9.23, construct a 95% confidence interval for the population standard deviation of the diameters of Indian mounds.

9.50 Do the data in Exercise 9.24 substantiate the conjecture that the true standard deviation of the acidity measurements is larger than 0.4? Test at $\alpha = .05$.

6. ROBUSTNESS OF INFERENCE PROCEDURES

The small sample methods for both confidence interval estimation and hypothesis testing presuppose that the sample is obtained from a normal population. Users of these methods would naturally ask:

1. What method can be used to determine if the population distribution is nearly normal?
2. What can go wrong if the population distribution is nonnormal?
3. What procedures should be used if it is nonnormal?
4. If the observations are not independent, is this serious?

1. To answer the first question, we could construct the dot diagram or normal-scores plot. These may indicate a wild observation or a striking departure from normality. If none of these is visible, the investigator would feel more secure using the preceding inference procedures. However, any plot based on a small sample cannot provide convincing justification for normality. Lacking sufficient observations to justify or refute the normal assumption, we are led to a consideration of the second question.

2. Confidence intervals and tests of hypotheses concerning μ are based on Student's t distribution. If the population is nonnormal, the actual percentage points may differ substantially from the tabulated values. When we say that $\overline{X} \pm t_{.025} S/\sqrt{n}$ is a 95% confidence interval for μ, the true probability that this random interval will contain μ may be, say, 85% or 99%. Fortunately, the effects on inferences about μ using the t statistic are not too serious if the sample size is at least moderately large (say, 15). In larger samples, such disturbances tend to disappear due to the central limit theorem. We express this fact by saying that **inferences about μ using the t statistic are reasonably** "robust." However, this qualitative discussion should not be considered a blanket endorsement for t. When the sample size is small, a wild observation or a distribution with long tails can produce misleading results.

Unfortunately, inferences about σ using the χ^2 distribution may be seriously affected by nonnormality even with large samples. We express this by saying that **inferences about σ using the χ^2 distribution are not "robust" against departures of the population distribution from normality.**

3. We cannot give a specific answer to the third question without knowing something about the nature of nonnormality. Dot diagrams or histograms of the original data may suggest some transformations that will bring the shape of the distribution closer to normality. If it is possible to obtain a transformation that leads to reasonably normal data plots, the problem can then be recast in terms of the transformed data. Otherwise, users can benefit from consulting with a statistician.

4. A basic assumption throughout Chapters 8 and 9 is that the sample is drawn at random, so the observations are independent of one another. If the sampling is made in such a manner that the observations are dependent, how-

ever, all the inferential procedures we discussed here for small as well as large samples may be seriously in error. This applies to both the level of significance of a test and a stated confidence level. Concerned with the possible effect of a drug on the blood pressure, suppose that an investigator includes 5 patients in an experiment and makes 4 successive measurements on each. This does *not* yield a random sample of size $5 \times 4 = 20$, because the 4 measurements made on each person are likely to be dependent. This type of sampling requires a more sophisticated method of analysis. An investigator who is sampling opinions about a political issue may choose 100 families at random and record the opinions of both the husband and wife in each family. This also does *not* provide a random sample of size $100 \times 2 = 200$, although it may be a convenient sampling method. When measurements are made closely together in time or distance, there is a danger of losing independence because adjacent observations are more likely to be similar than observations that are made farther apart. Because independence is the most crucial assumption, we must be constantly alert to detect such violations. Prior to a formal analysis of the data, a close scrutiny of the sampling process is imperative.

USING STATISTICS WISELY

1. The inferences in this chapter, based on the t distribution, require that the distribution of the individual observations be normal. With very small sample sizes, say 10 or smaller, it is not possible to check this assumption. The best we can do is make dot diagrams, or normal-scores plots, to make sure there are no obvious outliers. With somewhat larger sample sizes serious asymmetry can be recognized . Recall, however, the central limit result that the distribution of $\overline{X}$ becomes more nearly normal with increasing sample size whatever the population. This tends to make inferences based on the t statistic relatively insensitive to small or moderate departures from a normal population as long as sample size is greater than about 15.

2. Understand the interpretation of a $100(1 - \alpha)\%$ confidence interval. For normal populations, before the data are collected,

$$\left(\overline{X} - t_{\alpha/2} \frac{S}{\sqrt{n}}, \qquad \overline{X} + t_{\alpha/2} \frac{S}{\sqrt{n}} \right)$$

is a random interval that will cover the fixed unknown mean μ with probability $1 - \alpha$. Once a numerical value for the interval is obtained, the interval is fixed and we say we have $100(1 - \alpha)\%$ confidence that the mean is contained in the interval. After many applications of this procedure, to different samples from independent experiments, approximately proportion $1 - \alpha$ of the intervals will cover the respective population mean.

3. Formulate the assertion that the experiment seeks to confirm as the alternative hypothesis. Then, base a test of the null hypothesis $H_0 : \mu = \mu_0$ on the test statistic

$$\frac{\overline{X} - \mu_0}{S/\sqrt{n}}$$

which has a t distribution with $n - 1$ degrees of freedom. The rejection region is one-sided or two-sided corresponding to the alternative hypothesis.

KEY IDEAS AND FORMULAS

When the sample size is small, additional conditions need to be imposed on the population. In this chapter, we **assume that the population distribution is normal**. Inferences about the mean of a normal population are based on

$$T = \frac{\overline{X} - \mu}{S/\sqrt{n}}$$

which has Student's t distribution with $n - 1$ degrees of freedom.

Inferences about the standard deviation of a normal population are based on $(n - 1)S^2/\sigma^2$, which has a χ^2 distribution with $n - 1$ degrees of freedom.

Moderate departures from a normal population distribution do not seriously affect inferences based on Student's t. These procedures are "robust."

Nonnormality can seriously affect inferences about σ.

Inferences about a Normal Population Mean

When n is small, we assume that the population is approximately normal. Inference procedures are derived from Student's t sampling distribution of

$$T = \frac{\overline{X} - \mu}{S/\sqrt{n}}$$

1. A $100(1 - \alpha)$% confidence interval for μ is

$$\left(\overline{X} - t_{\alpha/2} \frac{S}{\sqrt{n}}, \quad \overline{X} + t_{\alpha/2} \frac{S}{\sqrt{n}} \right)$$

2. To test hypotheses about μ, using Student's t test, the test statistic is

$$T = \frac{\overline{X} - \mu_0}{S/\sqrt{n}}$$

Given a level of significance α, the t test will:

Reject $H_0: \mu = \mu_0$ in favor of $H_1: \mu > \mu_0$ if $T \geq t_\alpha$
Reject $H_0: \mu = \mu_0$ in favor of $H_1: \mu < \mu_0$ if $T \leq -t_\alpha$
Reject $H_0: \mu = \mu_0$ in favor of $H_1: \mu \neq \mu_0$ if $|T| \geq t_{\alpha/2}$

Inferences about a Normal Population Standard Deviation

Inferences are derived from the χ^2 distribution for $(n - 1)S^2/\sigma^2$.

1. A point estimator of σ is the sample standard deviation S.

2. A 95% confidence interval for σ is

$$\left(S\sqrt{\frac{n - 1}{\chi^2_{.025}}}, \quad S\sqrt{\frac{n - 1}{\chi^2_{.975}}} \right)$$

3. To **test hypotheses about** σ, the test statistic is

$$\frac{(n - 1)S^2}{\sigma_0^2}$$

Given a level of significance α,

$$\left.\begin{array}{l} \text{Reject } H_0\colon \sigma = \sigma_0 \\ \text{in favor of } H_1\colon \sigma < \sigma_0 \end{array}\right\} \text{ if } \quad \frac{(n - 1)S^2}{\sigma_0^2} \leq \chi^2_{1-\alpha}$$

$$\left.\begin{array}{l} \text{Reject } H_0\colon \sigma = \sigma_0 \\ \text{in favor of } H_1\colon \sigma > \sigma_0 \end{array}\right\} \text{ if } \quad \frac{(n - 1)S^2}{\sigma_0^2} \geq \chi^2_{\alpha}$$

$$\left.\begin{array}{l} \text{Reject } H_0\colon \sigma = \sigma_0 \\ \text{in favor of } H_1\colon \sigma \neq \sigma_0 \end{array}\right\} \begin{array}{l} \text{if} \quad \dfrac{(n - 1)S^2}{\sigma_0^2} \leq \chi^2_{1-\alpha/2} \\[2mm] \text{or} \quad \dfrac{(n - 1)S^2}{\sigma_0^2} \geq \chi^2_{\alpha/2} \end{array}$$

TECHNOLOGY

Confidence intervals and tests concerning a normal mean

MINITAB

Confidence intervals for μ

We illustrate the calculation of a 99% confidence interval for μ based on the t distribution.

Data: C1

Stat > Basic Statistics > 1-Sample t.
Type *C1* in **Samples in Columns.**
Click **Options** and type 99 in **Confidence level.**
Click **OK.** Click **OK.**

Tests of Hypotheses Concerning μ

We illustrate the calculation of an $\alpha = .01$ level test of $H_0 : \mu = 32$ versus a one-sided alternative, $H_1 : \mu > 32$.

Data: *C1*

Stat > Basic Statistics > 1-Sample t.
Type *C1* in **Samples in.**
Following **Test mean,** type *32,* the value of the mean under the null hypothesis. Click **Options** and type *99* in **Confidence level.**
In the **Alternative** cell select **greater than,** the direction of the alternative hypothesis. Click **OK.** Click **OK.**

If the sample size, mean, and standard deviation are available, instead of the second step, type these values in the corresponding cells.

EXCEL

First, obtain the summary data: sample size, $\bar{x}$, and s, as described in Chapter 2 Technology. Then substitute these values into the formulas for confidence intervals or tests.

Note that $t_{a/2}$ can be obtained. For $\alpha = 0.05$ and 6 degrees of freedom:

Select **Insert** and then **Function.** Choose **Statistical** and then **TINV.**
Enter 0.025 in **Probability** and 6 in **Deg_freedom.** Click **OK.**

TI-84/-83 PLUS

Confidence intervals for μ

We illustrate the calculation of a 99% confidence interval for μ. Start with the data entered in **L**₁.

Press **STAT** and select *TESTS* and then **8: Tinterval.**
Select **Data** with **List** set to **L**₁ and **Freq** to **1.**
Enter *.99* following *C-Level:.* Select **Calculate.**
Then press **ENTER.**

If, instead, the sample size, mean, and standard deviation are available, the second step is:

Select **Stats** (instead of **Data**) and enter the sample size, mean, and standard deviation.

Tests of hypotheses concerning μ

We illustrate the calculation of an $\alpha = .01$ level test of $H_0 : \mu = 32$ versus a one-sided alternative $H_1 : \mu > 32$. Start with the data entered in **L**₁.

Press **STAT** and select *TESTS* and then **2: T-Test.**
Select **Data** with **List** set to **L**₁ and **Freq** to **1.**
Enter 32 for μ_0. Select the direction of the alternative hypothesis.
Select **Calculate.** Press **ENTER.**

The calculator will return the *P*–value.

If, instead, the sample size, mean, and standard deviation are available, the second step is:

Select **Stats** (instead of **Data**) and enter the sample size, mean, and standard deviation.

7. REVIEW EXERCISES

9.51 Using the table of percentage points for the t distributions, find

(a) $t_{.05}$ when d.f. $= 3$

(b) $t_{.025}$ when d.f. $= 12$

(c) The lower .05 point when d.f. $= 3$

(d) The lower .05 point when d.f. $= 12$

9.52 Using the table for the t distributions, find the probability of

(a) $T > 2.720$ when d.f. $= 22$

(b) $T < 3.250$ when d.f. $= 9$

(c) $|T| < 2.567$ when d.f. $= 17$

(d) $-1.383 < T < 2.262$ when d.f. $= 9$

9.53 A t distribution assigns more probability to large values than the standard normal.

(a) Find $t_{.05}$ for d.f. $= 12$ and then evaluate $P[Z > t_{.05}]$. Verify that $P[T > t_{.05}]$ is greater than $P[Z > t_{.05}]$.

(b) Examine the relation for d.f. of 5 and 20, and comment.

9.54 Measurements of the amount of suspended solids in river water on 14 Monday mornings yield $\bar{x} = 47$ and $s = 9.4$. Obtain a 95% confidence interval for the mean amount of suspended solids. State any assumption you make about the population.

9.55 Determine a 99% confidence interval for μ using the data in Exercise 9.54.

9.56 The time to blossom of 21 plants has $\bar{x} = 38.4$ days and $s = 5.1$ days. Give a 95% confidence interval for the mean time to blossom.

9.57 Based on a random sample of size 11 from a normal population, one has calculated the 99%

confidence interval for the population mean μ and obtained the result $(62.5, 86.9)$. Using this result, obtain:

(a) A point estimate of μ and its 95% margin of error.

(b) A 90% confidence interval for μ.

9.58 Refer to Exercise 9.54. The water quality is acceptable if the mean amount of suspended solids is less than 49. Construct an $\alpha = .05$ test to establish that the quality is acceptable.

(a) Specify H_0 and H_1.

(b) State the test statistic.

(c) What does the test conclude?

9.59 Refer to Exercise 9.56. Do these data provide strong evidence that the mean time to blossom is less than 42 days? Test with $\alpha = .01$.

9.60 An accounting firm wishes to set a standard time μ required by employees to complete a certain audit operation. Times from 18 employees yield a sample mean of 4.1 hours and a sample standard deviation of 1.6 hours. Test H_0: $\mu = 3.5$ versus H_1: $\mu > 3.5$ using $\alpha = .05$.

9.61 Referring to Exercise 9.60, test H_0: $\mu = 3.5$ versus H_1: $\mu \neq 3.5$ using $\alpha = .02$.

9.62 A random sample of 15 observations provided $\bar{x} = 182$ and $s = 12$.

(a) Test H_0: $\mu = 190$ versus H_1: $\mu < 190$ at $\alpha = .05$. State your assumption about the population distribution.

(b) What can you say about the P–value of the test statistic calculated in part (a)?

9.63 The supplier of a particular brand of vitamin pills claims that the average potency of these pills after a certain exposure to heat and humidity is at least 65. Before buying these pills, a distributor wants to verify the supplier's claim is valid. To this end, the distributor will choose a random sample of 9 pills from a batch and measure their potency after the specified exposure.

(a) Formulate the hypotheses about the mean potency μ.

(b) Determine the rejection region of the test with $\alpha = .05$. State any assumption you make about the population.

(c) The data are 63, 72, 64, 69, 59, 65, 66, 64, 65. Apply the test and state your conclusion.

9.64 A weight loss program advertises "LOSE 40 POUNDS IN 4 MONTHS." A random sample of $n = 25$ customers has $\bar{x} = 32$ pounds lost and $s = 12$. Test $H_0: \mu = 40$ against $H_1: \mu < 40$ with $\alpha = .05$.

9.65 A car advertisement asserts that with the new collapsible bumper system, the average body repair cost for the damages sustained in a collision impact of 10 miles per hour does not exceed $1500. To test the validity of this claim, 5 cars are crashed into a stone barrier at an impact force of 10 miles per hour and their subsequent body repair costs are recorded. The mean and the standard deviation are found to be $1620 and $90, respectively. Do these data strongly contradict the advertiser's claim?

9.66 Combustion efficiency measurements were recorded for 10 home heating furnaces of a new model. The sample mean and standard deviation were found to be 73.2 and 2.74, respectively. Do these results provide strong evidence that the average efficiency of the new model is higher than 70? (Test at $\alpha = .05$. Comment also on the P-value.)

9.67 A person with asthma took measurements by blowing into a peak-flow meter on seven consecutive days.

429   425   471   422   432   444   454

(a) Obtain a 95% confidence interval for the population mean peak-flow.

(b) Conduct an $\alpha = .10$ level test of $H_0 : \mu = 453$ versus $H_1 : \mu \neq 453$.

9.68 The number of days to maturity was recorded for 25 plants grown from seeds of a single stock. The mean and standard deviation were $\bar{x} = 68.4$ days and $s = 6.5$ days.

(a) Do these results contradict the claim that the average maturity time is 65 days for this stock?

(b) Construct a 95% confidence interval for the mean maturity time.

(c) Construct a 90% confidence interval for σ.

9.69 Using the table of percentage points of the χ^2 distributions, find:

(a) $\chi^2_{.05}$ with d.f. $= 6$.

(b) $\chi^2_{.025}$ with d.f. $= 23$.

(c) The lower .05 point with d.f. $= 6$.

(d) The lower .025 point with d.f. $= 23$.

9.70 Using the table for the χ^2 distributions, find:

(a) The 90th percentile of χ^2 when d.f. $= 11$.

(b) The 10th percentile of χ^2 when d.f. $= 8$.

(c) The median of χ^2 when d.f. $= 21$.

(d) The 1st percentile of χ^2 when d.f. $= 50$.

9.71 Test $H_0 : \sigma = 1.0$ versus $H_1 : \sigma > 1.0$ with $\alpha = .05$ in each case.

(a) $n = 25$, $\Sigma(x_i - \bar{x})^2 = 40.16$.

(b) $n = 15$, $s = 1.2$.

(c) $n = 6$ and the sample observations are

11   13   13   15   16   17

9.72 Refer to the data of Exercise 9.66. Is there strong evidence that the standard deviation for the efficiency of the new model is below .30?

9.73 Suppose that based on a random sample of size 10 from a normal population, one has found the 95% confidence interval for the population mean to be (36.2, 45.8). Using this result:

(a) Find the sample standard deviation.

(b) Determine a 98% confidence interval for the population mean.

(c) Determine a 95% confidence interval for the population standard deviation.

9.74 Refer to the data on the head length (cm) of male grizzly bears given in Table D.8 of the Data

Bank. A computer calculation for a test of $H_0 : \mu = 21$ versus $H_1 : \mu \neq 21$ is given below.

```
Test of mu = 21 vs mu not = 21

Variable    N     Mean    StDev
Mhdln      25    18.636   3.697

Variable        95.0% CI          T      P
Mhdln      ( 17.110, 20.162)   -3.20  0.004
```

(a) What is the conclusion if you test with $\alpha = .01$?

(b) What mistake could you have made in part (a)?

(c) Before you collected the data, what was the probability of making the mistake in part (a)?

(d) Give a long-run relative frequency interpretation of the probability in part (c).

9.75 Refer to the computer output concerning the head length (cm) of male grizzly bears in Exercise 9.74.

(a) Is the population mean head length for all male bears in the study area contained in this interval?

(b) Explain why you are 95% confident that it is contained in the interval.

The Following Exercises May Require a Computer

9.76 Refer to the data on the length (cm) of male grizzly bears given in Table D.8 of the Data Bank.

(a) Find a 99% confidence interval for the population mean.

(b) Is the population mean length for all male wolves in the Yukon-Charley Rivers National Reserve mean contained in this interval?

(c) Explain why you are 99% confident that it is contained in the interval.

9.77 Refer to the computer anxiety scores (CARS) for males in Table D.4 of the Data Bank. Conduct an $\alpha = .025$ level of $H_0 : \mu = 2.7$ versus $H_1 : \mu > 2.7$.

9.78 Refer to the data on the body length (cm) of male wolves given in Table D.9 of the Data Bank.

(a) Find a 98% confidence interval for the population mean.

(b) Is the population mean body length for all wolves in the Yukon-Charley Rivers National Reserve mean contained in this interval?

(c) Explain why you are 98% confident that it is contained in the interval.

10

Comparing Two Treatments

Does Playing Action Video Games Modify Ability To See Objects?

Video game playing is now a widespread activity in our society. Two scientists[1] wondered if the game playing itself could alter players' visual abilities and improve performance on related tasks. As one part of their experiment, they briefly flashed different numbers of target squares on the screen. The count of the number of target squares flashed that could be unerringly apprehended was recorded for a sample of 13 regular game players and 13 nonplayers. (Courtesy of G. Green.)

Number of Items Apprehended

	Sample Size	Mean	Standard Deviation
Game players	$n_1 = 13$	$\bar{x}_1 = 4.89$	$s_1 = 1.58$
Non–game players	$n_2 = 13$	$\bar{x}_2 = 3.27$	$s_2 = 1.15$

The difference in sample means, $4.89 - 3.27 = 1.62$ items apprehended, is substantial. A 95% confidence interval for the difference of population means is

$$(0.50, 2.74) \qquad \text{items unerringly apprehended}$$

With 95% confidence we can assert that mean difference is greater than .5 but not greater than 2.74 items unerringly apprehended. These data provide strong evidence that regular game players can unerringly apprehend more items in this new task.

© HIRB/Index Stock Imagery.

[1]*Source:* G. S. Green and D. Bavelier, "Action Video Game Modifies Visual Selective Attention," *Nature* **423** (May 29, 2003), pp. 534–537.

1.   INTRODUCTION

In virtually every area of human activity, new procedures are invented and existing techniques revised. Advances occur whenever a new technique proves to be better than the old. To compare them, we conduct experiments, collect data about their performance, and then draw conclusions from statistical analyses. The manner in which sample data are collected, called an **experimental design** or **sampling design**, is crucial to an investigation. In this chapter, we introduce two experimental designs that are most fundamental to a comparative study: (1) independent samples and (2) a matched pairs sample.

As we shall see, the methods of analyzing the data are quite different for these two processes of sampling. First, we outline a few illustrative situations where a comparison of two methods requires statistical analysis of data.

Example 1   Agricultural Field Trials for Pesticide Residual in Vegetables

The amount of pesticides in the food supply is a major health problem. To ascertain whether a new pesticide will result in less residue in vegetables, field trials must be performed by applying the new and current pesticides to nearly identical farm plots. For instance, the new pesticide could be applied to about one-half of the plot and the current pesticide to the rest but leaving a sufficient buffer zone in between.

Several different fields with different climate and soil conditions and several different crops should be included in the study. A conclusion about pesticide residue in lettuce grown in sandy soil in southern Wisconsin, while of some interest, is not general enough.

Example 2   Drug Evaluation

Pharmaceutical researchers strive to synthesize chemicals to improve their efficiency in curing diseases. New chemicals may result from educated guesses concerning potential biological reactions, but evaluations must be based on their effects on diseased animals or human beings. To compare the effectiveness of two drugs in controlling tumors in mice, several mice of an identical breed may be taken as experimental subjects. After infecting them with cancer cells, some will be subsequently treated with drug 1 and others with drug 2. The data of tumor sizes for the two groups will then provide a basis for comparing the drugs. When testing the drugs on human subjects, the experiment takes a different form. Artificially infecting them with cancer cells is absurd! In fact, it would be criminal. Instead, the drugs will be administered to cancer patients who are available for the study. In contrast, with a pool of mice of an "identical breed," the available subjects may be of varying conditions of general health, prognosis of the disease, and other factors.

When discussing a comparative study, the common statistical term treatment is used to refer to the things that are being compared. The basic unit that is exposed to one treatment or another are called an experimental unit or experimental subject, and the characteristic that is recorded after the application of a treatment to a subject is called the response. For instance, the two treatments in Example 1 are the two varieties of seeds, the experimental subjects are the agricultural plots, and the response is crop yield.

The term experimental design refers to the manner in which subjects are chosen and assigned to treatments. For comparing two treatments, the two basic types of design are:

1. Independent samples (complete randomization).

2. Matched pairs sample (randomization within each matched pair).

The case of independent samples arises when the subjects are randomly divided into two groups, one group is assigned to treatment 1 and the other to treatment 2. The response measurements for the two treatments are then unrelated because they arise from separate and unrelated groups of subjects. Consequently, each set of response measurements can be considered a sample from a population, and we can speak in terms of a comparison between two population distributions.

With the matched pairs design, the experimental subjects are chosen in pairs so that the members in each pair are alike, whereas those in different pairs may be substantially dissimilar. One member of each pair receives treatment 1 and the other treatment 2. Example 3 illustrates these ideas.

Example 3   Independent Samples Versus Matched Pairs Design

To compare the effectiveness of two drugs in curing a disease, suppose that 8 patients are included in a clinical study. Here, the time to cure is the response of interest. Figure 1a portrays a design of independent samples where the 8 patients are randomly split into groups of 4, one group is treated with drug 1, and the other with drug 2. The observations for drug 1 will have no relation to those for drug 2 because the selection of patients in the two groups is left completely to chance.

To conduct a matched pairs design, one would first select the patients in pairs. The two patients in each pair should be as alike as possible in regard to their physiological conditions; for instance, they should be of the same gender and age group and have about the same severity of the disease. These preexisting conditions may be different from one pair to another. Having paired the subjects, we randomly select one member from each pair to be treated with drug 1 and the other with drug 2. Figure 1b shows this matched pairs design.

In contrast with the situation of Figure 1a, here we would expect the responses of each pair to be dependent for the reason that they are governed by the same preexisting conditions of the subjects.

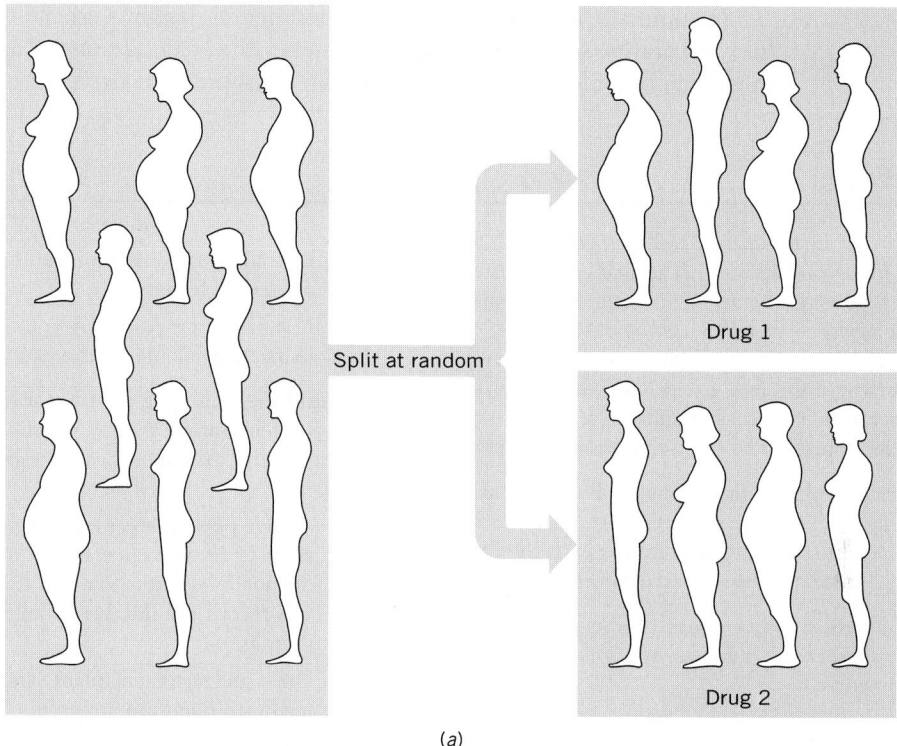

(a)

Figure 1a   Independent samples, each of size 4.

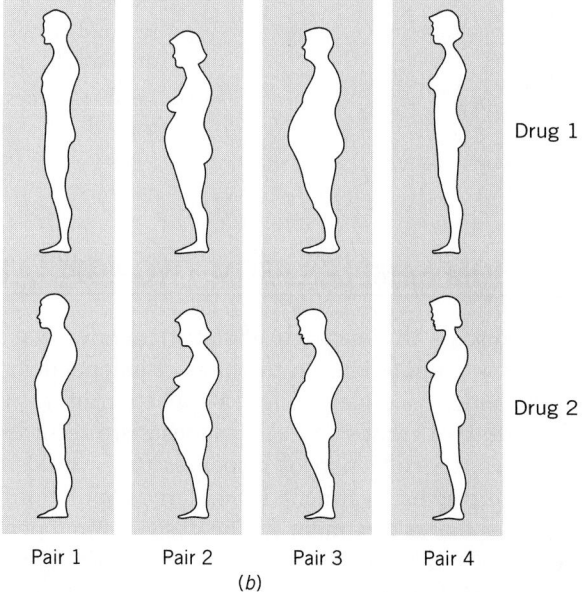

(b)

Figure 1b   Matched pairs design with four pairs of subjects.
Separate random assignment of Drug 1 each pair.

In summary, a carefully planned experimental design is crucial to a successful comparative study. The design determines the structure of the data. In turn, the design provides the key to selecting an appropriate analysis.

Exercises

10.1   Grades for first semester will be compared to those for second semester. The five one-semester courses biology, chemistry, English, history, and psychology must be taken next year. Make a list of all possible ways to split the courses into two groups where the first group has two courses to be taken the first semester and the second group has three courses to be taken the second semester.

10.2   Six mice—Alpha, Tau, Omega, Pi, Beta, and Phi—are to serve as subjects. List all possible ways to split them into two groups with the first having 4 mice and the second 2 mice.

10.3   Six students in a psychology course have volunteered to serve as subjects in a matched pairs experiment.

Name	Age	Gender
Tom	18	M
Sue	20	F
Erik	18	M
Grace	20	F
John	20	M
Roger	18	M

(a)   List all possible sets of pairings if subjects are paired by age.

(b)   If subjects are paired by gender, how many pairs are available for the experiment?

10.4   Identify the following as either matched pair or independent samples. Also identify the experimental units, treatments, and response in each case.

(a)   Twelve persons are given a high-potency vitamin C capsule once a day. Another twelve do not take extra vitamin C. Investigators will record the number of colds in 5 winter months.

(b)   One self-fertilized plant and one cross-fertilized plant are grown in each of 7 pots. Their heights will be measured after 3 months.

(c)   Ten newly married couples will be interviewed. Both the husband and wife will respond to the question, "How many children would you like to have?"

(d)   Learning times will be recorded for 5 dogs trained by a reward method and 3 dogs trained by a reward–punishment method.

2.   INDEPENDENT RANDOM SAMPLES FROM TWO POPULATIONS

Here we discuss the methods of statistical inference for comparing two treatments or two populations on the basis of independent samples. Recall that with the independent samples design, a collection of $n_1 + n_2$ subjects is randomly divided into two groups and the responses are recorded. We conceptualize population 1 as the collection of responses that would result if a vast number of subjects were given treatment 1. Similarly, population 2 refers to the population of responses under treatment 2. The design of independent samples can then be viewed as one that produces unrelated random samples from two populations (see Figure 2). In other situations, the populations to be compared may be quite real entities. For instance, one may wish to compare the residential property values in the east suburb of a city to those in the west suburb. Here the issue of assigning experimental subjects to treatments does not arise. The collection of all residential properties in each suburb constitutes a population from which a sample will be drawn at random.

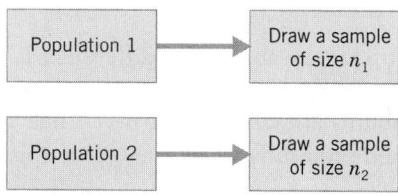

Figure 2   Independent random samples.

With the design of independent samples, we obtain

Sample	Summary Statistics		
$X_1, X_2, \ldots, X_{n_1}$ from population 1	$\overline{X} = \dfrac{1}{n_1} \displaystyle\sum_{i=1}^{n_1} X_i$	$S_1^2 = \dfrac{\displaystyle\sum_{i=1}^{n_1} (X_i - \overline{X})^2}{n_1 - 1}$	
$Y_1, Y_2, \ldots, Y_{n_2}$ from population 2	$\overline{Y} = \dfrac{1}{n_2} \displaystyle\sum_{i=1}^{n_2} Y_i$	$S_2^2 = \dfrac{\displaystyle\sum_{i=1}^{n_2} (Y_i - \overline{Y})^2}{n_2 - 1}$	

To make confidence statements or to test hypotheses, we specify a statistical model for the data.

Statistical Model: Independent Random Samples

1. $X_1, X_2, \ldots, X_{n_1}$ is a random sample of size n_1 from population 1 whose mean is denoted by μ_1 and standard deviation by σ_1.

2. $Y_1, Y_2, \ldots, Y_{n_2}$ is a random sample of size n_2 from population 2 whose mean is denoted by μ_2 and standard deviation by σ_2.

3. The samples are independent. In other words, the response measurements under one treatment are unrelated to the response measurements under the other treatment.

We now set our goal toward drawing a comparison between the mean responses of the two treatments or populations. In statistical language, we are interested in making inferences about the parameter

$$\mu_1 - \mu_2 = \text{(Mean of population 1)} - \text{(Mean of population 2)}$$

INFERENCES FROM LARGE SAMPLES

Inferences about the difference $\mu_1 - \mu_2$ are naturally based on its estimate $\overline{X} - \overline{Y}$, the difference between the sample means. When both sample sizes n_1 and n_2 are large (say, greater than 30), $\overline{X}$ and $\overline{Y}$ are each approximately normal and their difference $\overline{X} - \overline{Y}$ is approximately normal with

Mean

$$E(\overline{X} - \overline{Y}) = \mu_1 - \mu_2$$

Variance

$$\text{Var}(\overline{X} - \overline{Y}) = \frac{\sigma_1^2}{n_1} + \frac{\sigma_2^2}{n_2}$$

Standard error

$$\text{S.E.} (\overline{X} - \overline{Y}) = \sqrt{\frac{\sigma_1^2}{n_1} + \frac{\sigma_2^2}{n_2}}$$

Note: Because the entities $\overline{X}$ and $\overline{Y}$ vary in repeated sampling and independently of each other, the distance between them becomes more variable than the individual members. This explains the mathematical fact that the variance of the difference $\overline{X} - \overline{Y}$ equals the *sum* of the variances of $\overline{X}$ and $\overline{Y}$.

When n_1 and n_2 are both large, the normal approximation remains valid if σ_1^2 and σ_2^2 are replaced by their estimators

$$S_1^2 = \frac{\sum\limits_{i=1}^{n_1} (X_i - \overline{X})^2}{n_1 - 1} \qquad \text{and} \qquad S_2^2 = \frac{\sum\limits_{i=1}^{n_2} (Y_i - \overline{Y})^2}{n_2 - 1}$$

We conclude that, when the sample sizes n_1 and n_2 are large,

$$Z = \frac{(\overline{X} - \overline{Y}) - (\mu_1 - \mu_2)}{\sqrt{\dfrac{S_1^2}{n_1} + \dfrac{S_2^2}{n_2}}} \qquad \text{is approximately } N(0, 1)$$

A confidence interval for $\mu_1 - \mu_2$ is constructed from this sampling distribution. As we did for the single sample problem, we obtain a confidence interval of the form

Estimate of parameter $\pm$ (z value)(estimated standard error)

Large Samples Confidence Interval for $\mu_1 - \mu_2$

When n_1 and n_2 are greater than 30, an approximate $100(1 - \alpha)\%$ confidence interval for $\mu_1 - \mu_2$ is given by

$$\left(\overline{X} - \overline{Y} - z_{\alpha/2} \sqrt{\frac{S_1^2}{n_1} + \frac{S_2^2}{n_2}} \, , \quad \overline{X} - \overline{Y} + z_{\alpha/2} \sqrt{\frac{S_1^2}{n_1} + \frac{S_2^2}{n_2}} \right)$$

where $z_{\alpha/2}$ is the upper $\alpha/2$ point of $N(0, 1)$.

Example 4   Large Sample Confidence Interval for Difference in Mean Age

To compare the age at first marriage of females in two ethnic groups A and B, a random sample of 100 ever-married females is taken from each group and the ages at first marriage are recorded. The means and the standard deviations are found to be

	A	B
Mean	20.7	18.5
sd	6.3	5.8

Construct a 95% confidence interval for $\mu_A - \mu_B$.

SOLUTION   We have

$$n_1 = 100 \qquad \overline{x} = 20.7 \qquad s_1 = 6.3$$
$$n_2 = 100 \qquad \overline{y} = 18.5 \qquad s_2 = 5.8$$

$$\sqrt{\frac{s_1^2}{n_1} + \frac{s_2^2}{n_2}} = \sqrt{\frac{(6.3)^2}{100} + \frac{(5.8)^2}{100}} = .8563$$

For a 95% confidence interval, we use $z_{.025} = 1.96$. We calculate

$$\overline{x} - \overline{y} = 20.7 - 18.5 = 2.2$$

$$z_{.025} \sqrt{\frac{s_1^2}{n_1} + \frac{s_2^2}{n_2}} = 1.96 \times .8563 = 1.68$$

Therefore, a 95% confidence interval for $\mu_A - \mu_B$ is given by

$$2.2 \pm 1.68 \qquad \text{or} \qquad (.52, 3.88)$$

Females in ethnic group B tend, on the average, to marry .52 year to 3.88 years younger than those in ethnic group A.

A test of the null hypothesis that the two population means are the same, $H_0 : \mu_1 - \mu_2 = 0$, employs the test statistic

$$Z = \frac{\overline{X} - \overline{Y}}{\sqrt{\dfrac{S_1^2}{n_1} + \dfrac{S_2^2}{n_2}}}$$

which is approximately $N(0, 1)$ when $\mu_1 - \mu_2 = 0$.

Example 5   Testing Equality of Mean Ages at Marriage

Do the data in Example 4 provide strong evidence that the mean age at first marriage of females is different for the two groups? Test at $\alpha = .02$.

SOLUTION   Because our goal is to establish that the population means μ_A and μ_B are different, we formulate the hypotheses

$$H_0 : \mu_A - \mu_B = 0 \qquad \text{versus} \qquad H_1 : \mu_A - \mu_B \neq 0$$

We use the test statistic

$$Z = \frac{\overline{X} - \overline{Y}}{\sqrt{\dfrac{S_1^2}{n_1} + \dfrac{S_2^2}{n_2}}}$$

and set a two-sided rejection region. Specifically, with $\alpha = .02$, we have $\alpha/2 = .01$ and $z_{\alpha/2} = 2.33$, so the rejection region is $R : |Z| \geq 2.33$.

Using the sample data given in Example 4, we calculate

$$z = \frac{20.7 - 18.5}{\sqrt{\dfrac{(6.3)^2}{100} + \dfrac{(5.8)^2}{100}}} = \frac{2.2}{.8563} = 2.57$$

Because the observed value $z = 2.57$ is in R, we reject H_0 and conclude that, at $\alpha = .02$, the mean ages at first marriage are significantly different. The small P–value

$$P[Z > 2.57] + P[Z < -2.57] = .0102 \quad \text{(see Figure 3)}$$

further strengthens the conclusion.

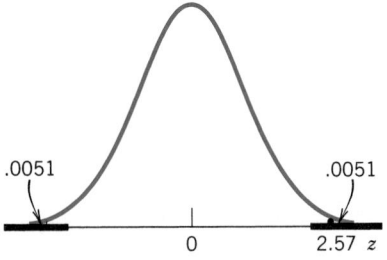

.0051   .0051

0   2.57 z

Figure 3 P–value with two-sided rejection region.

Example 6   Large Samples Test with a One-Sided Alternative

In June two years ago, chemical analyses were made of 85 water samples (each of unit volume) taken from various parts of a city lake, and the measurements of chlorine content were recorded. During the next two winters, the use of road salt was substantially reduced in the catchment areas of the lake. This June, 110 water samples were analyzed and their chlorine contents recorded. Calculations of the mean and the standard deviation for the two sets of data give

	Chlorine Content	
	Two Years Ago	Current Year
Mean	18.3	17.8
Standard deviation	1.2	1.8

Test the claim that lower salt usage has reduced the amount of chlorine in the lake.

SOLUTION   Let μ_1 be the population mean two years ago and μ_2 the population mean in the current year. Because the claim is that μ_2 is less than μ_1, we formulate the hypotheses

$$H_0 : \mu_1 - \mu_2 = 0 \quad \text{versus} \quad H_1 : \mu_1 - \mu_2 > 0$$

With the test statistic

$$Z = \frac{\overline{X} - \overline{Y}}{\sqrt{\dfrac{S_1^2}{n_1} + \dfrac{S_2^2}{n_2}}}$$

the rejection region should be of the form $R : Z \geq c$ because H_1 is right-sided. Using the data

$$n_1 = 85 \quad \overline{x} = 18.3 \quad s_1 = 1.2$$
$$n_2 = 110 \quad \overline{y} = 17.8 \quad s_2 = 1.8$$

we calculate

$$z = \frac{18.3 - 17.8}{\sqrt{\dfrac{(1.2)^2}{85} + \dfrac{(1.8)^2}{110}}} = \frac{.5}{.2154} = 2.32$$

The significant probability of this observed value is (see Figure 4)

$$P\text{-value} = P(Z \geq 2.32) = .0102$$

Because the P-value is very small, we conclude that there is strong evidence in support of H_1.

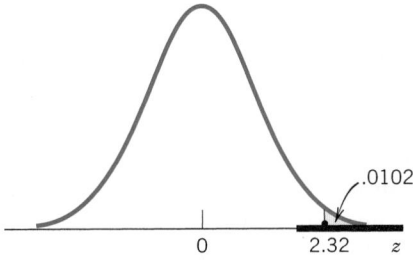

Figure 4   P–value = .0102.

We summarize the procedure for testing $\mu_1 - \mu_2 = \delta_0$ where δ_0 is specified under the null hypothesis. The case $\mu_1 = \mu_2$ corresponds to $\delta_0 = 0$.

Testing $H_0 : \mu_1 - \mu_2 = \delta_0$ with Large Samples

Test statistic:

$$Z = \frac{\overline{X} - \overline{Y} - \delta_0}{\sqrt{\dfrac{S_1^2}{n_1} + \dfrac{S_2^2}{n_2}}}$$

Alternative Hypothesis	Level α Rejection Region
$H_1 : \mu_1 - \mu_2 > \delta_0$	$R : Z \geq z_\alpha$
$H_1 : \mu_1 - \mu_2 < \delta_0$	$R : Z \leq -z_\alpha$
$H_1 : \mu_1 - \mu_2 \neq \delta_0$	$R : \lvert Z \rvert \geq z_{\alpha/2}$

INFERENCES FROM SMALL SAMPLES

Not surprisingly, more distributional structure is required to formulate appropriate inference procedures for small samples. Here we introduce the small samples inference procedures that are valid under the following assumptions about the population distributions. Naturally, the usefulness of such procedures depends on how closely these assumptions are realized.

Additional Assumptions When the Sample Sizes Are Small

1. Both populations are normal.
2. The population standard deviations σ_1 and σ_2 are equal.

A restriction to normal populations is not new. It was previously introduced for inferences about the mean of a single population. The second assumption, requiring equal variability of the populations, is somewhat artificial but we reserve comment until later. Letting σ denote the common standard deviation, we summarize.

Small Samples Assumptions

1. $X_1, X_2, \ldots, X_{n_1}$ is a random sample from $N(\mu_1, \sigma)$.
2. $Y_1, Y_2, \ldots, Y_{n_2}$ is a random sample from $N(\mu_2, \sigma)$.
   (*Note:* σ is the same for both distributions.)
3. $X_1, X_2, \ldots, X_{n_1}$ and $Y_1, Y_2, \ldots, Y_{n_2}$ are independent.

Again, $\overline{X} - \overline{Y}$ is our choice for a statistic.

$$\text{Mean of } (\overline{X} - \overline{Y}) = E(\overline{X} - \overline{Y}) = \mu_1 - \mu_2$$

$$\text{Var}(\overline{X} - \overline{Y}) = \frac{\sigma^2}{n_1} + \frac{\sigma^2}{n_2} = \sigma^2\left(\frac{1}{n_1} + \frac{1}{n_2}\right)$$

The common variance σ^2 can be estimated by combining information provided by both samples. Specifically, the sum $\sum_{i=1}^{n_1}(X_i - \overline{X})^2$ incorporates $n_1 - 1$ pieces of information about σ^2, in view of the constraint that the deviations $X_i - \overline{X}$ sum to zero. Independently of this, $\sum_{i=1}^{n_2}(Y_i - \overline{Y})^2$ contains $n_2 - 1$ pieces of information about σ^2. These two quantities can then be combined,

$$\sum(X_i - \overline{X})^2 + \sum(Y_i - \overline{Y})^2$$

to obtain a pooled estimate of the common σ^2. The proper divisor is the sum of the component degrees of freedom, or $(n_1 - 1) + (n_2 - 1) = n_1 + n_2 - 2$.

Pooled Estimator of the Common σ^2

$$S_{pooled}^2 = \frac{\sum_{i=1}^{n_1}(X_i - \overline{X})^2 + \sum_{i=1}^{n_2}(Y_i - \overline{Y})^2}{n_1 + n_2 - 2}$$

$$= \frac{(n_1 - 1)S_1^2 + (n_2 - 1)S_2^2}{n_1 + n_2 - 2}$$

Example 7   Calculating the Pooled Estimate of Variance

Calculate the observed value s^2_{pooled} from these two samples.

Sample from population 1:	8	5	7	6	9	7
Sample from population 2:	2	6	4	7	6	

SOLUTION   The sample means are

$$\bar{x} = \frac{\sum x_i}{6} = \frac{42}{6} = 7 \qquad \bar{y} = \frac{\sum y_i}{5} = \frac{25}{5} = 5$$

Furthermore,

$$(6 - 1)s^2_1 = \sum (x_i - \bar{x})^2$$
$$= (8 - 7)^2 + (5 - 7)^2 + (7 - 7)^2 + (6 - 7)^2$$
$$+ (9 - 7)^2 + (7 - 7)^2 = 10$$
$$(5 - 1)s^2_2 = \sum (y_i - \bar{y})^2$$
$$= (2 - 5)^2 + (6 - 5)^2 + (4 - 5)^2$$
$$+ (7 - 5)^2 + (6 - 5)^2 = 16$$

Thus, $s^2_1 = 2$, $s^2_2 = 4$, and the pooled variance is

$$s^2_{\text{pooled}} = \frac{\sum (x_i - \bar{x})^2 + \sum (y_i - \bar{y})^2}{n_1 + n_2 - 2} = \frac{10 + 16}{6 + 5 - 2} = 2.89$$

The pooled variance is closer to 2 than 4 because the first sample size is larger.

These arithmetic details serve to demonstrate the concept of pooling. Using a calculator with a "standard deviation" key, one can directly get the sample standard deviations $s_1 = 1.414$ and $s_2 = 2.000$. Noting that $n_1 = 6$ and $n_2 = 5$, we can then calculate the pooled variance as

$$s^2_{\text{pooled}} = \frac{5(1.414)^2 + 4(2.000)^2}{9} = 2.89$$

Employing the pooled estimator $\sqrt{S^2_{\text{pooled}}}$ for the common σ, we obtain a Student's t variable that is basic to inferences about $\mu_1 - \mu_2$.

$$T = \frac{(\bar{X} - \bar{Y}) - (\mu_1 - \mu_2)}{S_{\text{pooled}} \sqrt{\dfrac{1}{n_1} + \dfrac{1}{n_2}}}$$

has Student's t distribution with $n_1 + n_2 - 2$ degrees of freedom.

We can now obtain confidence intervals for $\mu_1 - \mu_2$, which are of the form

Estimate of parameter $\pm$ (t value) $\times$ (Estimated standard error)

Confidence Interval for $\mu_1 - \mu_2$ Small Samples

A $100(1 - \alpha)\%$ confidence interval for $\mu_1 - \mu_2$ is given by

$$\bar{X} - \bar{Y} \pm t_{\alpha/2}\, S_{\text{pooled}} \sqrt{\frac{1}{n_1} + \frac{1}{n_2}}$$

where

$$S_{\text{pooled}}^2 = \frac{(n_1 - 1)S_1^2 + (n_2 - 1)S_2^2}{n_1 + n_2 - 2}$$

and $t_{\alpha/2}$ is the upper $\alpha/2$ point of the t distribution with d.f. $= n_1 + n_2 - 2$.

Example 8   Calculating a Small Samples Confidence Interval

Beginning male and female accounting students were given a test and, on the basis of their answers, were assigned a computer anxiety score (CARS). Using the data given in Table D.4 of the Data Bank, obtain a 95% confidence interval for the difference in mean computer anxiety score between beginning male and female accounting students.

SOLUTION   The dot diagrams of these data, plotted in Figure 5, give the appearance of approximately equal amounts of variation.

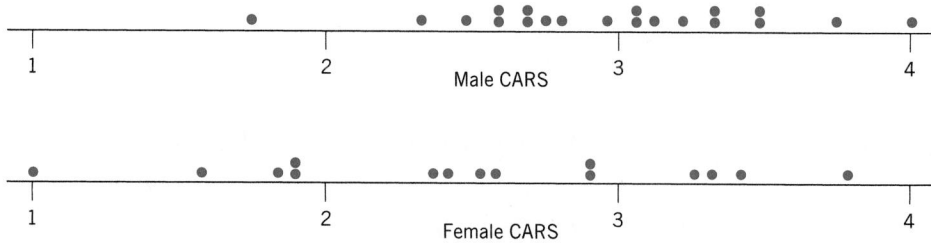

Figure 5   Dot diagrams of the computer anxiety data in Example 8.

We assume that the CARS data for both females and males are random samples from normal populations with means μ_1 and μ_2, respectively, and a common standard deviation σ. Computations from the data provide the summary statistics:

$$\text{Female CARS}$$
$$n_1 = 15 \qquad \bar{x} = 2.514 \qquad s_1 = .773$$

$$\text{Male CARS}$$
$$n_2 = 20 \qquad \bar{y} = 2.963 \qquad s_2 = .525$$

We calculate

$$s_{\text{pooled}} = \sqrt{\frac{14(.773)^2 + 19(.525)^2}{15 + 20 - 2}} = .642$$

With a 95% confidence interval $\alpha/2 = .025$ and consulting the t table, we find (interpolating) that $t_{.025} = 2.035$ for d.f. $= n_1 + n_2 - 2 = 33$. Thus a 95% confidence interval for $\mu_1 - \mu_2$ becomes

$$\bar{x} - \bar{y} \pm t_{0.25} s_{\text{pooled}} \sqrt{\frac{1}{n_1} + \frac{1}{n_2}}$$

$$= 2.514 - 2.963 \pm 2.035 \times .642 \sqrt{\frac{1}{15} + \frac{1}{20}}$$

$$= -.449 \pm .446 \qquad \text{or} \qquad (-.895, -.003)$$

We can be 95% confident that the mean computer anxiety score for female beginning accounting students can be .003 to .895 units lower than the mean score for males.

This interval is quite wide. Certainly the very small values represent technically insignificant differences.

Testing $H_0 : \mu_1 - \mu_2 = \delta_0$ with Small Samples

Test statistic:

$$T = \frac{\bar{X} - \bar{Y} - \delta_0}{S_{\text{pooled}} \sqrt{\frac{1}{n_1} + \frac{1}{n_2}}} \qquad \text{d.f.} = n_1 + n_2 - 2$$

Alternative Hypothesis	Level α Rejection Region
$H_1 : \mu_1 - \mu_2 > \delta_0$	$R : T \geq t_\alpha$
$H_1 : \mu_1 - \mu_2 < \delta_0$	$R : T \leq -t_\alpha$
$H_1 : \mu_1 - \mu_2 \neq \delta_0$	$R : |T| \geq t_{\alpha/2}$

Example 9   Testing the Equality of Mean Computer Anxiety Scores

Refer to the computer anxiety scores (CARS) described in Example 8 and the summary statistics

Female CARS
$$n_1 = 15 \qquad \bar{x} = 2.514 \qquad s_1 = .773$$

Male CARS
$$n_2 = 20 \qquad \bar{y} = 2.963 \qquad s_2 = .525$$

Do these data strongly indicate that the mean score for females is lower than that for males? Test at level $\alpha = .05$.

SOLUTION We are seeking strong evidence in support of the hypothesis that the mean computer anxiety score for females (μ_1) is less than the mean score for males. Therefore the alternative hypothesis should be taken as $H_1 : \mu_1 < \mu_2$ or $H_1 : \mu_1 - \mu_2 < 0$, and our problem can be stated as testing

$$H_0 : \mu_1 - \mu_2 = 0 \qquad \text{versus} \qquad H_1 : \mu_1 - \mu_2 < 0$$

We employ the test statistic

$$T = \frac{\bar{X} - \bar{Y}}{S_{\text{pooled}} \sqrt{\dfrac{1}{n_1} + \dfrac{1}{n_2}}} \qquad \text{d.f.} = n_1 + n_2 - 2$$

and set the left-sided rejection region $R : T \leq -t_{.05}$. For d.f. $= n_1 + n_2 - 2 = 33$, we approximate the tabled value as $t_{.05} = 1.692$, so the rejection region is $R : T \leq -1.692$.

With $S_{\text{pooled}} = .642$ already calculated in Example 8, the observed value of the test statistic T is

$$t = \frac{2.514 - 2.963}{.642 \sqrt{\dfrac{1}{15} + \dfrac{1}{20}}} = \frac{-.449}{.2193} = -2.05$$

This value lies in the rejection region R. Consequently, at the .05 level of significance, we reject the null hypothesis in favor of the alternative hypothesis that males have a higher mean computer anxiety score.

A computer calculation gives a P-value of about .025 so the evidence of H_1 is moderately strong.

DECIDING WHETHER OR NOT TO POOL

Our preceding discussion of large and small sample inferences raises a few questions:

For small sample inference, why do we assume the population standard deviations to be equal when no such assumption was needed in the large samples case?

When should we be wary about this assumption, and what procedure should we use when the assumption is not reasonable?

Learning statistics would be a step simpler if the ratio

$$\frac{(\overline{X} - \overline{Y}) - (\mu_1 - \mu_2)}{\sqrt{\dfrac{S_1^2}{n_1} + \dfrac{S_2^2}{n_2}}}$$

had a t distribution for small samples from normal populations. Unfortunately, statistical theory proves it otherwise. The distribution of this ratio is *not* a t and, worse yet, it depends on the unknown quantity σ_1/σ_2. The assumption $\sigma_1 = \sigma_2$ and the change of the denominator to $S_{\text{pooled}} \sqrt{1/n_1 + 1/n_2}$ allow the t-based inferences to be valid. However, the $\sigma_1 = \sigma_2$ restriction and accompanying pooling are not needed in large samples where a normal approximation holds.

With regard to the second question, the relative magnitude of the two sample standard deviations s_1 and s_2 should be the prime consideration. The assumption $\sigma_1 = \sigma_2$ is reasonable if s_1/s_2 is not very much different from 1. As a working rule, the range of values $\frac{1}{2} \le s_1/s_2 \le 2$ may be taken as reasonable cases for making the assumption $\sigma_1 = \sigma_2$ and hence for pooling. If s_1/s_2 is seen to be smaller than $\frac{1}{2}$ or larger than 2, the assumption $\sigma_1 = \sigma_2$ would be suspect. In that case, some approximate methods of inference about $\mu_1 - \mu_2$ are available, but those will not be discussed here because of their complex forms. Instead, we outline a simple though conservative procedure, which treats the ratio

$$T^* = \frac{(\overline{X} - \overline{Y}) - (\mu_1 - \mu_2)}{\sqrt{\dfrac{S_1^2}{n_1} + \dfrac{S_2^2}{n_2}}}$$

as a t variable with d.f. $=$ smaller of $n_1 - 1$ and $n_2 - 1$.

Small Sample Inferences for $\mu_1 - \mu_2$ When the Populations Are Normal But σ_1 and σ_2 Are Not Assumed to Be Equal

A $100(1 - \alpha)\%$ conservative confidence interval for $\mu_1 - \mu_2$ is given by

$$\overline{X} - \overline{Y} \pm t_{\alpha/2} \sqrt{\dfrac{S_1^2}{n_1} + \dfrac{S_2^2}{n_2}}$$

where $t_{\alpha/2}$ denotes the upper $\alpha/2$ point of the t distribution with d.f. $=$ smaller of $n_1 - 1$ and $n_2 - 1$.

The null hypothesis $H_0 : \mu_1 - \mu_2 = \delta_0$ is tested using the test statistic

$$T^* = \frac{\overline{X} - \overline{Y} - \delta_0}{\sqrt{\dfrac{S_1^2}{n_1} + \dfrac{S_2^2}{n_2}}} \qquad \text{d.f.} = \text{smaller of } n_1 - 1 \text{ and } n_2 - 1$$

Here, the confidence interval is conservative in the sense that the actual level of confidence is at least $1 - \alpha$. Likewise, the level α test is conservative in the sense that the actual Type I error probability is no more than α.

Example 10 Testing Equality of Means When Variances Are Unequal

A new method of storing snap beans is believed to retain more ascorbic acid than an old method. In an experiment, snap beans were harvested under uniform conditions and frozen in 18 equal-size packages. Nine of these packages were randomly selected and stored according to the new method, and the other 9 packages were stored by the old method. Subsequently, ascorbic acid determinations (mg/kg) were made, and the following summary statistics were calculated.

	New Method	Old Method
Mean ascorbic acid	449	410
Standard deviation	19	45

Do these data substantiate the claim that more ascorbic acid is retained under the new method of storing? Test at $\alpha = .05$.

SOLUTION Let μ_1 denote the population mean ascorbic acid under the new method of storing and μ_2 that under the old method. The problem concerns substantiation of the claim that μ_1 is larger than μ_2. Therefore, we formulate the testing problem as

$$H_0 : \mu_1 - \mu_2 = 0 \quad \text{versus} \quad H_1 : \mu_1 - \mu_2 > 0$$

The summary statistics are

$$n_1 = 9 \quad \bar{x} = 449 \quad s_1 = 19$$
$$n_2 = 9 \quad \bar{y} = 410 \quad s_2 = 45$$

We assume that both population distributions are normal. Looking at the observed sample standard deviations, we note that s_2 is more than twice s_1 so the assumption $\sigma_1 = \sigma_2$ is suspect. We therefore use the conservative test based on the test statistic:

$$T^* = \frac{\bar{X} - \bar{Y}}{\sqrt{\dfrac{S_1^2}{n_1} + \dfrac{S_2^2}{n_2}}} \qquad \begin{aligned} \text{d.f.} &= \text{smaller of } n_1 - 1 \text{ and } n_2 - 1 \\ &= 8 \end{aligned}$$

For d.f. $= 8$, the tabled value is $t_{.05} = 1.860$, so we set the rejection region $R : T^* > 1.860$. The observed value of the test statistic is

$$t^* = \frac{449 - 410}{\sqrt{\dfrac{(19)^2}{9} + \dfrac{(45)^2}{9}}} = 2.40$$

and it is in R. Therefore, H_0 is rejected at $\alpha = .05$, and we conclude that the claim is substantiated by the data. In fact, H_0 would be rejected even with $\alpha = .025$ because the observed t^* is larger than $t_{.025} = 2.306$.

Finally, we would like to emphasize that with large samples we can also learn about other differences between the two populations.

Example 11 Large Samples Reveal Additional Differences between Populations

Natural resource managers have attempted to use the Satellite Landsat Multi-spectral Scanner data for improved landcover classification. When the satellite was flying over country known to consist of forest, the following intensities were recorded on the near-infrared band of a thermatic mapper. The sample has already been ordered.

```
77  77  78  78  81  81  82  82  82  82  82  83  83  84  84  84
84  85  86  86  86  86  86  87  87  87  87  87  87  87  89  89
89  89  89  89  89  90  90  90  91  91  91  91  91  91  91  91
91  91  93  93  93  93  93  93  94  94  94  94  94  94  94  94
94  94  94  94  95  95  95  95  95  96  96  96  96  96  96  97
97  97  97  97  97  97  97  97  98  99  100 100 100 100
100 100 100 100 100 101 101 101 101 101 101 102
102 102 102 102 102 103 103 104 104 104 105 107
```

When the satellite was flying over urban areas, the intensities of reflected light on the same near-infrared band were

```
71  72  73  74  75  77  78  79  79  79  79  80  80  80  81  81  81
82  82  82  82  84  84  84  84  84  84  85  85  85  85  85  85  86
86  87  88  90  91  94
```

If the means are different, the readings could be used to tell urban from forest area. Obtain a 95% confidence interval for the difference in mean radiance levels.

SOLUTION Computer calculations give

	Forest	Urban
Number	118	40
Mean	92.932	82.075
Standard deviation	6.9328	4.9789

and, for large sample sizes, the approximate 95% confidence interval for $\mu_1 - \mu_2$ is given by

$$\left(\overline{X} - \overline{Y} - z_{.025} \sqrt{\frac{S_1^2}{n_1} + \frac{S_2^2}{n_2}}, \quad \overline{X} - \overline{Y} + z_{.025} \sqrt{\frac{S_1^2}{n_1} + \frac{S_2^2}{n_2}} \right)$$

Since $z_{.025} = 1.96$, the 95% confidence interval is calculated as

$$92.932 - 82.075 \pm 1.96 \sqrt{\frac{(6.9328)^2}{118} + \frac{(4.9789)^2}{40}} \quad \text{or} \quad (8.87, 12.84)$$

The mean for the forest is 8.87 to 12.84 levels of radiance higher than the mean for the urban areas.

Because the sample sizes are large, we can also learn about other differences between the two populations. The stem-and-leaf displays and boxplots in Figure 6 reveal that there is some difference in the standard deviation as well as the means. The graphs further indicate a range of high readings that are more likely to come from forests than urban areas. This feature has proven helpful in discriminating between forest and urban areas on the basis of near-infrared readings.

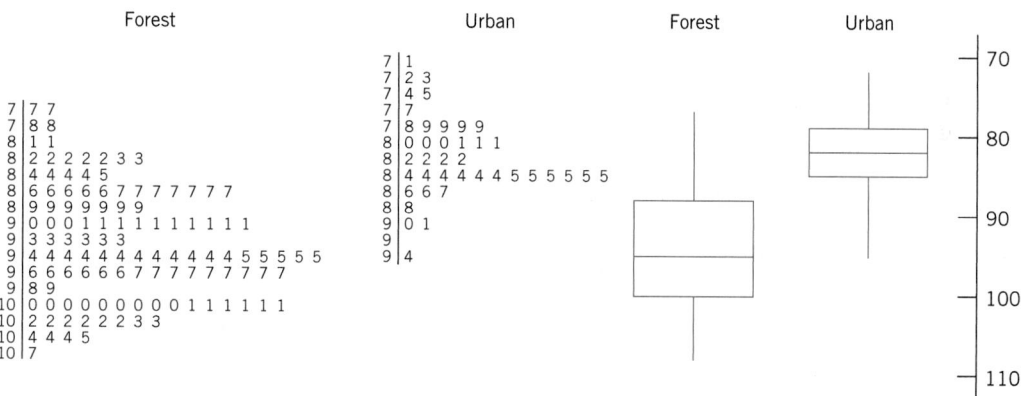

Figure 6   Stem-and-leaf displays and boxplots give additional information about population differences.

Exercises

10.5 Independent random samples from two populations have provided the summary statistics

Sample 1	Sample 2
$n_1 = 52$	$n_2 = 44$
$\bar{x} = 73$	$\bar{y} = 66$
$s_1^2 = 151$	$s_2^2 = 142$

(a) Obtain a point estimate of $\mu_1 - \mu_2$ and calculate the estimated standard error.

(b) Construct a 95% confidence interval for $\mu_1 - \mu_2$.

10.6 Rural and urban students are to be compared on the basis of their scores on a nationwide musical aptitude test. Two random samples of sizes 90 and 100 are selected from rural and urban seventh-grade students. The summary statistics from the test scores are

	Rural	Urban
Sample size	90	100
Mean	76.4	81.2
Standard deviation	8.2	7.6

Establish a 98% confidence interval for the difference in population mean scores between urban and rural students.

10.7   Construct a test to determine if there is a significant difference between the population mean scores in Exercise 10.6. Use $\alpha = .05$.

10.8   A group of 140 subjects is used in an experiment to compare two treatments. Treatment 1 is given to 78 subjects selected at random, and the remaining 62 are given treatment 2. The means and standard deviations of the responses are

	Treatment 1	Treatment 2
Mean	92	118
Standard deviation	46.2	53.4

Determine a 98% confidence interval for the mean difference of the treatment effects.

10.9   Refer to the data in Exercise 10.8. Suppose the investigator wishes to establish that treatment 2 has a higher mean response than treatment 1.
(a)   Formulate H_0 and H_1.
(b)   State the test statistic and the rejection region with $\alpha = .05$.
(c)   Perform the test at $\alpha = .05$. Also, find the P-value and comment.

10.10   A national equal employment opportunities committee is conducting an investigation to determine if women employees are as well paid as their male counterparts in comparable jobs. Random samples of 75 males and 64 females in junior academic positions are selected, and the following calculations are obtained from their salary data.

	Male	Female
Mean	$58,530	$57,620
Standard deviation	780	750

Construct a 95% confidence interval for the difference between the mean salaries of males and females in junior academic positions.

10.11   Refer to the confidence interval obtained in Exercise 10.10. If you were to test the null

hypothesis that the mean salaries are equal versus the two-sided alternatives, what would be the conclusion of your test with $\alpha = .05$? (See Section 4 of Chapter 9.)

10.12   Suppose that, with independent random samples of sizes 52 and 67 from two populations, a 95% confidence interval for $\mu_1 - \mu_2$ has been found to be (6.88, 9.26). Using this result, determine:
(a)   A point estimate of $\mu_1 - \mu_2$ and its estimated standard error.
(b)   A 90% confidence interval for $\mu_1 - \mu_2$.

10.13   In a study of interspousal aggression and its possible effect on child behavior, the behavior problem checklist (BPC) scores were recorded for 47 children whose parents were classified as aggressive. The sample mean and standard deviation were 7.92 and 3.45, respectively. For a sample of 38 children whose parents were classified as nonaggressive, the mean and standard deviation of the BPC scores were 5.80 and 2.87, respectively. Do these observations substantiate the conjecture that the children of aggressive families have a higher mean BPC than those of nonaggressive families? (Answer by calculating the P-value.)

10.14   Suppose that measurements of the size of butterfly wings (cm) for two related species yielded the data

Species 1	6	4	7	3
Species 2	6	9	6	

(a)   Calculate s^2_{pooled}.
(b)   Give an estimate of the common standard deviation for the wing size for the two species.
(c)   Evaluate the t statistic for testing equality of the two population mean wing sizes.

10.15   Suppose that sentence lengths (inches) from two authors yielded the data

Author 1	9	5	7
Author 2	4	7	4

(a) Calculate s^2_{pooled}.

(b) Give an estimate of the common standard deviation for sentence length for the two authors.

(c) Evaluate the t statistic for testing equality of the two population mean sentence lengths.

10.16 Given that

$$n_1 = 13, \bar{x} = 20, \Sigma(x_i - \bar{x})^2 = 28,$$

$$n_2 = 14, \bar{y} = 17, \Sigma(y_i - \bar{y})^2 = 22$$

(a) Obtain s^2_{pooled}.

(b) Test $H_0: \mu_1 = \mu_2$ against $H_1: \mu_1 > \mu_2$ with $\alpha = .05$.

(c) Determine a 95% confidence interval for $\mu_1 - \mu_2$.

10.17 The data on the weight (lb) of male and female wolves, from Table D.9 of the Data Bank, are

Female	57	84	90	71	71	77	68	73		
Male	71	93	101	84	88	117	86	86	93	86	106

(a) Test the null hypothesis that the mean weights of males and females are equal versus a two-sided alternative. Take $\alpha = .05$.

(b) Obtain a 95% confidence interval for the difference of population mean weights.

(c) State any assumptions you make about the populations.

10.18 Given here are the sample sizes and the sample standard deviations for independent random samples from two populations. For each case, state which of the three tests you would use in testing hypotheses about $\mu_1 - \mu_2$: (1) Z test, (2) t test with pooling, (3) conservative t test without pooling. Also, state any assumptions you would make about the population distributions.

(a) $n_1 = 35, \quad s_1 = 12.2$
$\quad n_2 = 50, \quad s_2 = 8.6$

(b) $n_1 = 8, \quad s_1 = 0.86$
$\quad n_2 = 7, \quad s_2 = 1.12$

(c) $n_1 = 8, \quad s_1 = 1.54$
$\quad n_2 = 15, \quad s_2 = 5.36$

(d) $n_1 = 70, \quad s_1 = 6.2$
$\quad n_2 = 60, \quad s_2 = 2.1$

10.19 Psychologists have made extensive studies on the relationship between child abuse and later criminal behavior. Consider a study that consisted of the follow-ups of 52 boys who were abused in their preschool years and 67 boys who were not abused. The data of the number of criminal offenses of those boys in their teens yielded the following summary statistics.

	Abused	Nonabused
Mean	2.48	1.57
Standard deviation	1.94	1.31

Is the mean number of criminal offenses significantly higher for the abused group than that for the nonabused group? Answer by calculating the P-value.

10.20 Referring to the data in Exercise 10.19, determine a 99% confidence interval for the difference between the true means for the two groups.

10.21 To compare two programs for training industrial workers to perform a skilled job, 20 workers are included in an experiment. Of these, 10 are selected at random and trained by method 1; the remaining 10 workers are trained by method 2. After completion of training, all the workers are subjected to a time-and-motion test that records the speed of performance of a skilled job. The following data are obtained.

Time (minutes)										
Method 1	15	20	11	23	16	21	18	16	27	24
Method 2	23	31	13	19	23	17	28	26	25	28

(a) Can you conclude from the data that the mean job time is significantly less after training with method 1 than after training with method 2? (Test with $\alpha = .05$.)

(b) State the assumptions you make for the population distributions.

(c) Construct a 95% confidence interval for the population mean difference in job times between the two methods.

10.22 Based on the data of independent random samples of sizes 6 and 5 from two populations, one reports the following.

"Under the assumption of normal populations with equal but unknown standard deviations, the 90% confidence interval for $\mu_1 - \mu_2$ is (76, 102)."

From this report,

State the conclusion of testing $H_0 : \mu_1 - \mu_2 = 100$ versus $H_1 : \mu_1 - \mu_2 \neq 100$ at $\alpha = .10$. (See Section 4 of Chapter 9.)

10.23 To compare the effectiveness of isometric and isotonic exercise methods, 20 potbellied business executives are included in an experiment: 10 are selected at random and assigned to one exercise method; the remaining 10 are assigned to the other exercise method. After five weeks, the reductions in abdomen measurements are recorded in centimeters, and the following results are obtained.

	Isometric Method A	Isotonic Method B
Mean	2.4	3.2
Standard deviation	0.8	1.0

(a) Do these data support the claim that the isotonic method is more effective?

(b) Construct a 95% confidence interval for $\mu_B - \mu_A$.

10.24 Refer to Exercise 10.23.

(a) Aside from the type of exercise method, identify a few other factors that are likely to have an important effect on the amount of reduction accomplished in a five-week period.

(b) What role does randomization play in achieving a valid comparison between the two exercise methods?

(c) If you were to design this experiment, describe how you would divide the 20 business executives into two groups.

10.25 Given here are four sets of values of the summary statistics for independent random samples from two populations. For each case,

determine a 98% confidence interval for the difference between the population means and state any assumptions required for your procedure.

	Sample 1			Sample 2		
	Sample Size	Mean	sd	Sample Size	Mean	sd
(a)	14	28.1	3.6	12	30.0	5.1
(b)	110	410	26	91	390	38
(c)	13	1.25	0.079	14	1.32	0.326
(d)	65	75.6	18.1	55	62.5	6.8

10.26 The following generic computer output summarizes the data, given in Table D.5 of the Data Bank, on the pretest percent body fat in male and female students.

Gender	N	Mean	StDev
M	40	14.38	7.34
F	43	23.72	5.78

Find a 99% confidence interval for the difference of the two population means.

10.27 Refer to the data on the weight of wolves in Table D.9 of the Data Bank. A computer analysis produces the output

Two-sample T for Male wt vs Female wt

	N	Mean	StDev
Male wt	11	91.9	12.4
Female wt	8	73.9	10.1

T-Test of difference = 0 (vs not =):
T-Value = 3.38 P-Value = 0.004 DF = 17

(a) What is the conclusion to testing the equality of mean weights at level $\alpha = .05$?

(b) Test the null hypothesis that males weigh an average of 5 pounds more than females against a two-sided alternative. Take $\alpha = .05$.

3.   RANDOMIZATION AND ITS ROLE IN INFERENCE

We have presented the methods of drawing inferences about the difference between two population means. Let us now turn to some important questions regarding the design of the experiment or data collection procedure. The manner in which experimental subjects are chosen for the two treatment groups can be crucial. For example, suppose that a remedial-reading instructor has developed a new teaching technique and is permitted to use the new method to instruct half the pupils in the class. The instructor might choose the most alert or the students who are more promising in some other way, leaving the weaker students to be taught in the conventional manner. Clearly, a comparison between the reading achievements of these two groups would not just be a comparison of two teaching methods. A similar fallacy can result in comparing the nutritional quality of a new lunch package if the new diet is given to a group of children suffering from malnutrition and the conventional diet is given to a group of children who are already in good health.

When the assignment of treatments to experimental units is under our control, steps can be taken to ensure a valid comparison between the two treatments. At the core lies the principle of impartial selection, or randomization. The choice of the experimental units for one treatment or the other must be made by a chance mechanism that does not favor one particular selection over any other. It must not be left to the discretion of the experimenters because, even unconsciously, they may be partial to one treatment.

Suppose that a comparative experiment is to be run with N experimental units, of which n_1 units are to be assigned to treatment 1 and the remaining $n_2 = N - n_1$ units are to be assigned to treatment 2. The principle of randomization tells us that the n_1 units for treatment 1 must be chosen at random from the available collection of N units—that is, in a manner such that all $\binom{N}{n_1}$ possible choices are equally likely to be selected.

Randomization Procedure for Comparing Two Treatments

From the available $N = n_1 + n_2$ experimental units, choose n_1 units at random to receive treatment 1 and assign the remaining n_2 units to treatment 2. The random choice entails that all $\binom{N}{n_1}$ possible selections are equally likely to be chosen.

As a practical method of random selection, we can label the available units from 1 to N. Then read random digits from Table 1, Appendix B, until n_1 different numbers between 1 and N are obtained. These n_1 experimental units receive treatment 1 and the remaining units receive treatment 2. For a quicker and more efficient means of random sampling, one can use the computer (see, for instance, the Technology section, page 162, Chapter 4).

Although randomization is not a difficult concept, it is one of the most fundamental principles of a good experimental design. It guarantees that uncontrolled sources of variation have the same chance of favoring the response of treatment 1 as they do of favoring the response of treatment 2. Any systematic effects of uncontrolled variables, such as age, strength, resistance, or intelligence, are chopped up or confused in their attempt to influence the treatment responses.

> **Randomization** prevents uncontrolled sources of variation from influencing the responses in a systematic manner.

Of course, in many cases, the investigator does not have the luxury of randomization. Consider comparing crime rates of cities before and after a new law. Aside from a package of criminal laws, other factors such as poverty, inflation, and unemployment play a significant role in the prevalence of crime. As long as these contingent factors cannot be regulated during the observation period, caution should be exercised in crediting the new law if a decline in the crime rate is observed or discrediting the new law if an increase in the crime rate is observed. When randomization cannot be performed, extreme caution must be exercised in crediting an apparent difference in means to a difference in treatments. The differences may well be due to another factor.

Exercises

10.28 Randomly allocate 2 subjects from among

      Al, Bob, Carol, Ellen, John

to be in the control group. The others will receive a treatment.

10.29 Randomly allocate three subjects from among 6 mice,

      Alpha, Tau, Omega, Pi, Beta, Phi

to group 1.

10.30 Early studies showed a disproportionate number of heavy smokers among lung cancer patients. One scientist theorized that the presence of a particular gene could tend to make a person want to smoke and be susceptible to lung cancer.

(a) How would randomization settle this question?

(b) Would such a randomization be ethical with human subjects?

10.31 Observations on 10 mothers who nursed their babies and 8 who did not revealed that nursing mothers felt warmer toward their babies. Can we conclude that nursing affects a mother's feelings toward her child?

10.32 Suppose that you are asked to design an experiment to study the effect of a hormone injection on the weight gain of pregnant rats during gestation. You have decided to inject 6 of the 12 rats available for the experiment and retain the other 6 as controls.

(a) Briefly explain why it is important to randomly divide the rats into the two groups. What might be wrong with the experimental results if you choose to give the hormone treatment to 6 rats that are easy to grab from their cages?

(b) Suppose that the 12 rats are tagged with serial numbers from 1 through 12 and 12 marbles identical in appearance are also numbered from 1 through 12. How can you use these marbles to randomly select the rats in the treatment and control groups?

4. MATCHED PAIR COMPARISONS

In comparing two treatments, it is desirable that the experimental units or subjects be as alike as possible, so that a difference in responses between the two groups can be attributed to differences in treatments. If some identifiable conditions vary over the units in an uncontrolled manner, they could introduce a large variability in the measurements. In turn, this could obscure a real difference in treatment effects. On the other hand, the requirement that all subjects be alike may impose a severe limitation on the number of subjects available for a comparative experiment. To compare two analgesics, for example, it would be impractical to look for a sizable number of patients who are of the same sex, age, and general health condition and who have the same severity of pain. Aside from the question of practicality, we would rarely want to confine a comparison to such a narrow group. A broader scope of inference can be attained by applying the treatments on a variety of patients of both sexes and different age groups and health conditions.

Matched Pairs Design

Units in each pair are alike, whereas units in different pairs may be dissimilar. In each pair, a unit is chosen at random to receive treatment 1, the other unit receives treatment 2.

The concept of **matching** or **blocking** is fundamental to providing a compromise between the two conflicting requirements that the experimental units be alike and also of different kinds. The procedure consists of choosing units in pairs or blocks so that the units in each block are similar and those in different blocks are dissimilar. One of the units in each block is assigned to treatment 1, the other to treatment 2. This process preserves the effectiveness of a comparison within each block and permits a diversity of conditions to exist in different blocks. Of course, the treatments must be allotted to each pair randomly to

Matched Pairs

Identical twins are the epitome of matched pair experimental subjects. They are matched with respect to not only age but also a multitude of genetic factors. Social scientists, trying to determine the influence of environment and heredity, have been especially interested in studying identical twins who were raised apart. Observed differences in IQ and behavior are then supposedly due to environmental factors.

When the subjects are animals like mice, two from the same litter can be paired. Going one step further, genetic engineers can now provide two identical plants or small animals by cloning these subjects.

avoid selection bias. This design is called a matched pairs design or sampling. For example, in studying how two different environments influence the learning capacities of preschoolers, it is desirable to remove the effect of heredity: Ideally, this is accomplished by working with twins.

In a matched pairs design, the response of an experimental unit is influenced by:

1. The conditions prevailing in the block (pair).
2. A treatment effect.

By taking the difference between the two observations in a block, we can filter out the common block effect. These differences then permit us to focus on the effects of treatments that are freed from undesirable sources of variation.

Pairing (or Blocking)

Pairing like experimental units according to some identifiable characteristic(s) serves to remove this source of variation from the experiment.

The structure of the observations in a paired comparison is given below, where X and Y denote the responses to treatments 1 and 2, respectively. The difference between the responses in each pair is recorded in the last column, and the summary statistics are also presented.

Structure of Data for a Matched Pair Comparison

Pair	Treatment 1	Treatment 2	Difference
1	X_1	Y_1	$D_1 = X_1 - Y_1$
2	X_2	Y_2	$D_2 = X_2 - Y_2$
.	.	.	.
.	.	.	.
.	.	.	.
n	X_n	Y_n	$D_n = X_n - Y_n$

The differences $D_1, D_2, \ldots, D_n$ are a random sample.
Summary statistics:

$$\overline{D} = \frac{1}{n} \sum_{i=1}^{n} D_i \qquad S_D^2 = \frac{\sum_{i=1}^{n} (D_i - \overline{D})^2}{n - 1}$$

Although the pairs (X_i, Y_i) are independent of one another, X_i and Y_i within the ith pair will usually be dependent. In fact, if the pairing of experimental units is effective, we would expect X_i and Y_i to be relatively large or small together. Expressed in another way, we would expect (X_i, Y_i) to have a high positive correlation. Because the differences $D_i = X_i - Y_i, i = 1, 2, \ldots, n$, are freed from the block effects, it is reasonable to assume that they constitute a random sample from a population with mean δ and variance σ_D^2, where δ represents the mean difference of the treatment effects. In other words,

$$E(D_i) = \delta$$
$$\mathrm{Var}(D_i) = \sigma_D^2 \qquad i = 1, \ldots, n$$

If the mean difference δ is zero, then the two treatments can be considered equivalent. A positive δ signifies that treatment 1 has a higher mean response than treatment 2. Considering $D_1, \ldots, D_n$ to be a single random sample from a population, we can immediately apply the techniques discussed in Chapters 8 and 9 to learn about the population mean δ.

As we learned in Chapter 8, the assumption of an underlying normal distribution can be relaxed when the sample size is large. The central limit theorem applied to the differences $D_1, \ldots, D_n$ suggests that when n is large, say, greater than 30,

$$\frac{\overline{D} - \delta}{S_D/\sqrt{n}} \qquad \text{is approximately } N(0, 1)$$

The inferences can then be based on the percentage points of the $N(0, 1)$ distribution or, equivalently, those of the t distribution, with the degrees of freedom marked "infinity."

Example 12   Does Conditioning Reduce Percent Body Fat?

A conditioning class is designed to introduce students to a variety of training techniques to improve fitness and flexibility. The percent body fat was measured at the start of the class and at the end of the semester. The data for 81 students are given in Table D.5 of the Data Bank.

(a) Obtain a 98% confidence interval for the mean reduction in percent body fat.

(b) Test, at $\alpha = .01$, the claim that the conditioning class reduces the mean percent body fat.

SOLUTION   (a) Each subject represents a block which produces one measurement of percent body fat at the start of the semester (x) and one at the end (y). The 81 paired differences $d_i = x_i - y_i$ are summarized by a computer calculation.

	N	Mean	StDev	SE Mean
Difference	81	3.322	2.728	0.303

That is, $\bar{d} = 3.322$ and $s_D = 2.728$. The sample size 81 is large so there is no need to assume that the population is normal. Since, from the normal table, $z_{.01} = 2.33$, the 98% confidence interval becomes

$$\left(\bar{d} - 2.33\frac{s_D}{\sqrt{81}}, \quad \bar{d} + 2.33\frac{s_D}{\sqrt{81}}\right)$$

$$\left(3.322 - 2.33 \times \frac{2.728}{\sqrt{81}}, \quad 3.322 + 2.33 \times \frac{2.728}{\sqrt{81}}\right) =$$

$$(3.322 - .706, 3.322 + .706)$$

or $(2.62, 4.03)$ percent.

(b) Because the claim is that $\delta > 0$, the initial reading tends to be higher than at the end of class, we formulate:

$$H_0 : \delta = 0 \qquad \text{versus} \qquad H_1 : \delta > 0$$

The test statistic

$$Z = \frac{\bar{D}}{S_D/\sqrt{n}}$$

is approximately normally distributed so the rejection region is $R : Z \geq z_{.01} = 2.33$. The observed value of the test statistic

$$z = \frac{\bar{d}}{S_D/\sqrt{81}} = \frac{3.322}{2.728/\sqrt{81}} = \frac{3.322}{.303} = 10.96$$

falls in the rejection region. Consequently H_0 is rejected in favor of H_1 at level $\alpha = .01$. We conclude that the conditioning class does reduce the mean percent body fat. The value of the test statistic is so far in the rejection region that the P-value is .0000 to at least four places. The evidence in support of H_1 is very very strong.

When the sample size is not large, we make the additional assumption that the distribution of the differences is normal.

In summary,

Small Samples Inferences about the Mean Difference δ

Assume that the differences $D_i = X_i - Y_i$ are a random sample from an $N(\delta, \sigma_D)$ distribution. Let

$$\overline{D} = \frac{\sum\limits_{i=1}^{n} D_i}{n} \quad \text{and} \quad S_D = \sqrt{\frac{\sum\limits_{i=1}^{n} (D_i - \overline{D})^2}{n - 1}}$$

Then:

1. A $100(1 - \alpha)\%$ confidence interval for δ is given by

$$\left(\overline{D} - t_{\alpha/2}\frac{S_D}{\sqrt{n}}, \quad \overline{D} + t_{\alpha/2}\frac{S_D}{\sqrt{n}} \right)$$

where $t_{\alpha/2}$ is based on $n - 1$ degrees of freedom.

2. A test of $H_0: \delta = \delta_0$ is based on the test statistic

$$T = \frac{\overline{D} - \delta_0}{S_D/\sqrt{n}} \qquad \text{d.f.} = n - 1$$

Example 13   Does a Pill Incidentally Reduce Blood Pressure?

A medical researcher wishes to determine if a pill has the undesirable side effect of reducing the blood pressure of the user. The study involves recording the initial blood pressures of 15 college-age women. After they use the pill regularly for six months, their blood pressures are again recorded. The researcher wishes to draw inferences about the effect of the pill on blood pressure from the observations given in Table 1.

(a) Calculate a 95% confidence interval for the mean reduction in blood pressure.

(b) Do the data substantiate the claim that use of the pill reduces blood pressure? Test at $\alpha = .01$.

TABLE 1   Blood-Pressure Measurements before and after Use of Pill

	Subject														
	1	2	3	4	5	6	7	8	9	10	11	12	13	14	15
Before (x)	70	80	72	76	76	76	72	78	82	64	74	92	74	68	84
After (y)	68	72	62	70	58	66	68	52	64	72	74	60	74	72	74
$d = x - y$	2	8	10	6	18	10	4	26	18	-8	0	32	0	-4	10

Courtesy of a family planning clinic.

SOLUTION (a) Here each subject represents a block generating a pair of measurements: one before using the pill and the other after using the pill. The paired differences $d_i = x_i - y_i$ are computed in the last row of Table 1, and we calculate the summary statistics

$$\bar{d} = \frac{\sum d_i}{15} = 8.80 \qquad s_D = \sqrt{\frac{\sum (d_i - \bar{d})^2}{14}} = 10.98$$

If we assume that the paired differences constitute a random sample from a normal population $N(\delta, \sigma_D)$, a 95% confidence interval for the mean difference δ is given by

$$\bar{D} \pm t_{.025} \frac{S_D}{\sqrt{15}}$$

where $t_{.025}$ is based on d.f. $= 14$. From the t table, we find $t_{.025} = 2.145$. The 95% confidence interval is then computed as

$$8.80 \pm 2.145 \times \frac{10.98}{\sqrt{15}} = 8.80 \pm 6.08 \qquad \text{or} \qquad (2.72, 14.88)$$

This means that we are 95% confident the mean reduction of blood pressure is between 2.72 and 14.88.

(b) Because the claim is that $\delta > 0$, we formulate

$$H_0 : \delta = 0 \qquad \text{versus} \qquad H_1 : \delta > 0$$

We employ the test statistic $T = \dfrac{\bar{D}}{S_D / \sqrt{n}}$, d.f. $= 14$ and set a right-sided rejection region. With d.f. $= 14$, we find $t_{.01} = 2.624$, so the rejection region is $R : T \geq 2.624$.

The observed value of the test statistic

$$t = \frac{\bar{d}}{S_D / \sqrt{n}} = \frac{8.80}{10.98 / \sqrt{15}} = \frac{8.80}{2.84} = 3.10$$

falls in the rejection region. Consequently, H_0 is rejected in favor of H_1 at $\alpha = .01$. We conclude that a reduction in blood pressure following use of the pill is strongly supported by the data.

Note: To be more convinced that the pill causes the reduction in blood pressure, it is advisable to measure the blood pressures of the same subjects once again after they have stopped using the pill for a period of time. This amounts to performing the experiment in reverse order to check the findings of the first stage.

Example 13 is a typical before–after situation. Data gathered to determine the effectiveness of a safety program or an exercise program would have the same structure. In such cases, there is really no way to choose how to order the experiments within a pair. The before situation must precede the after situation. If something other than the institution of the program causes performance to improve, the improvement will be incorrectly credited to the program. However, when the order of the application of treatments can be determined by the investigator, something can be done about such systematic influences. Suppose that a coin is flipped to select the treatment for the first unit in each pair. Then the other treatment is applied to the second unit. Because the coin is flipped again for each new pair, any uncontrolled variable has an equal chance of helping the performance of either treatment 1 or treatment 2. After eliminating an identified source of variation by pairing, we return to randomization in an attempt to reduce the systematic effects of any uncontrolled sources of variation.

Randomization with Pairing

After pairing, the assignment of treatments should be randomized for each pair.

Randomization within each pair chops up or diffuses any systematic influences that we are unable to control.

Exercises

10.33 Given the following matched pairs sample,

x	y
6	4
4	2
8	4
6	6
9	8
6	9

(a) Evaluate the t statistic
$$t = \frac{\bar{d}}{S_D/\sqrt{n}}.$$

(b) How many degrees of freedom does this t have?

10.34 Given the following matched pairs sample,

x	y
6	8
10	10
8	12
13	12

(a) Evaluate the t statistic
$$t = \frac{\bar{d}}{S_D/\sqrt{n}}.$$

(b) How many degrees of freedom does this t have?

10.35 It is claimed that an industrial safety program is effective in reducing the loss of working hours due to factory accidents. The following data are collected concerning the weekly loss of working hours due to accidents in six plants both before and after the safety program is instituted.

	Plant					
	1	2	3	4	5	6
Before	12	30	15	37	29	15
After	10	29	16	35	26	16

Do the data substantiate the claim? Use $\alpha = .05$.

10.36 A manufacturer claims his boot waterproofing is better than the major brand. Five pairs of shoes are available for a test.

   (a) Explain how you would conduct a paired sample test.

   (b) Write down your assignment of waterproofing to each shoe. How did you randomize?

10.37 A food scientist wants to study whether quality differences exist between yogurt made from skim milk with and without the preculture of a particular type of bacteria, called Psychrotrops (PC). Samples of skim milk are procured from seven dairy farms. One-half of the milk sampled from each farm is inoculated with PC, and the other half is not. After yogurt is made with these milk samples, the firmness of the curd is measured, and those measurements are given below.

Curd Firmness	Dairy Farm						
	A	B	C	D	E	F	G
With PC	68	75	62	86	52	46	72
Without PC	61	69	64	76	52	38	68

   (a) Do these data substantiate the conjecture that the treatment of PC results in a higher degree of curd firmness? Test at $\alpha = .05$.

   (b) Determine a 90% confidence interval for the mean increase of curd firmness due to the PC treatment.

10.38 A study is to be made of the relative effectiveness of two kinds of cough medicines in increasing sleep. Six people with colds are given medicine A the first night and medicine B the second night. Their hours of sleep each night are recorded.

	Subject					
	1	2	3	4	5	6
Medicine A	4.8	4.1	5.8	4.9	5.1	7.4
Medicine B	3.9	4.2	5.0	4.9	5.2	7.1

   (a) Establish a 95% confidence interval for the mean change in hours of sleep when switching from medicine A to medicine B.

   (b) How and what would you randomize in this study? Briefly explain your reason for randomization.

10.39 Two methods of memorizing difficult material are being tested to determine if one produces better retention. Nine pairs of students are included in the study. The students in each pair are matched according to IQ and academic background and then assigned to the two methods at random. A memorization test is given to all the students, and the following scores are obtained:

	Pair								
	1	2	3	4	5	6	7	8	9
Method A	90	86	72	65	44	52	66	38	83
Method B	85	87	70	62	44	53	62	35	86

At $\alpha = .05$, test to determine if there is a significant difference in the effectiveness of the two methods.

10.40 In an experiment conducted to see if electrical pricing policies can affect consumer behavior, 10 homeowners in Wisconsin had to pay a premium for power use during the peak hours. They were offered lower off-peak rates. For each home, the July on-peak usage (kilowatt hours) under the pricing experiment was compared to the previous July usage.

Year	
Previous	Experimental
200	160
180	175
240	210
425	370
120	110
333	298
418	368
380	250
340	305
516	477

(a) Find a 95% confidence interval for the mean decrease.

(b) Test $H_0: \delta = 0$ against $H_1: \delta \neq 0$ at level $\alpha = .05$.

(c) Comment on the feasibility of randomization of treatments.

(d) Without randomization, in what way could the results in parts (a) and (b) be misleading?

(*Hint:* What if air conditioner use is a prime factor, and the July with experimental pricing was cooler than the previous July?)

10.41 To compare the crop yields from two strains of wheat, *A* and *B*, an experiment was conducted at eight farms located in different parts of a state. At each farm, strain *A* was grown on one plot and strain *B* on another; all 16 plots were of equal sizes. Given below are data of yield in pounds per plot.

	Farm							
	1	2	3	4	5	6	7	8
Strain A	23	39	19	43	33	29	28	42
Strain B	18	33	21	34	33	20	21	40

(a) Is there strong evidence that strain *A* has a higher mean yield than strain *B*? Test at $\alpha = .05$.

(b) What should be randomized in this experiment and how?

10.42 Refer to the problem stated in Exercise 10.41, but now suppose that the study was conducted at 16 farms, of which 8 were selected for planting strain *A* and the other 8 for strain *B*. Here also the plots used were all of equal sizes. Recorded below are the data of yields in pounds per plot.

Strain A	23	39	19	28	42	43	33	29
Strain B	20	21	40	34	33	18	33	21

(a) Is there strong evidence that strain *A* has a higher mean yield than strain *B*? State the assumptions you make and use $\alpha = .05$.

(b) What should be randomized in this experiment and how?

(c) Check to see that each data set here is just a scrambled form of the data set in Exercise 10.41. Briefly explain why the conclusion of your test is different in the two situations.

10.43 Measurements of the left- and right-hand gripping strengths of 10 left-handed writers are recorded.

	Person									
	1	2	3	4	5	6	7	8	9	10
Left hand	140	90	125	129	95	121	85	97	131	110
Right hand	138	87	110	131	96	120	86	90	129	100

(a) Do the data provide strong evidence that people who write with the left hand have a greater gripping strength in the left hand than in the right hand?

(b) Construct a 90% confidence interval for the mean difference.

5.   CHOOSING BETWEEN INDEPENDENT SAMPLES AND A MATCHED PAIRS SAMPLE

When planning an experiment to compare two treatments, we often have the option of either designing two independent samples or designing a sample with paired observations. Therefore, some comments about the pros and cons of these two sampling methods are in order here. Because a paired sample with n pairs of observations contains $2n$ measurements, a comparable situation would be two independent samples with n observations in each. First, note that the sample mean difference is the same whether or not the samples are paired. This is because

$$\overline{D} = \frac{1}{n} \sum (X_i - Y_i) = \overline{X} - \overline{Y}$$

Therefore, using either sampling design, the confidence intervals for the difference between treatment effects have the common form

$$(\overline{X} - \overline{Y}) \pm t_{\alpha/2} \text{ (estimated standard error)}$$

However, the estimated standard error as well as the degrees of freedom for t are different between the two situations.

	Independent Samples $(n_1 = n_2 = n)$	Paired Sample $(n$ Pairs$)$
Estimated standard error	$S_{pooled} \sqrt{\dfrac{1}{n} + \dfrac{1}{n}}$	$\dfrac{S_D}{\sqrt{n}}$
d.f. of t	$2n - 2$	$n - 1$

Because the length of a confidence interval is determined by these two components, we now examine their behavior under the two competing sampling schemes.

Paired sampling results in a loss of degrees of freedom and, consequently, a larger value of $t_{\alpha/2}$. For instance, with a paired sample of $n = 10$, we have $t_{.05} = 1.833$ with d.f. $= 9$. But the t value associated with independent samples, each of size 10, is $t_{.05} = 1.734$ with d.f. $= 18$. Thus, if the estimated standard errors are equal, then a loss of degrees of freedom tends to make confidence intervals larger for paired samples. Likewise, in testing hypotheses, a loss of degrees of freedom for the t test results in a loss of power to detect real differences in the population means.

The merit of paired sampling emerges when we turn our attention to the other component. If experimental units are paired so that an interfering factor is held nearly constant between members of each pair, the treatment responses X and Y within each pair will be equally affected by this factor. If the prevailing condition in a pair causes the X measurement to be large, it will also cause the corresponding Y measurement to be large and vice versa. As a result, the variance of the difference $X - Y$ will be smaller in the case of an effective pairing than it will be in the case of independent random variables. The estimated standard deviation will be typically smaller as well. With an effective pairing, the reduction in the standard deviation usually more than compensates for the loss of degrees of freedom.

In Example 13, concerning the effect of a pill in reducing blood pressure, we note that a number of important factors (age, weight, height, general health, etc.)

affect a person's blood pressure. By measuring the blood pressure of the same person before and after use of the pill, these influencing factors can be held nearly constant for each pair of measurements. On the other hand, independent samples of one group of persons using the pill and a separate control group of persons not using the pill are apt to produce a greater variability in blood-pressure measurements if all the persons selected are not similar in age, weight, height, and general health.

In summary, paired sampling is preferable to independent sampling when an appreciable reduction in variability can be anticipated by means of pairing. When the experimental units are already alike or their dissimilarities cannot be linked to identifiable factors, an arbitrary pairing may fail to achieve a reduction in variance. The loss of degrees of freedom will then make a paired comparison less precise.

6. COMPARING TWO POPULATION PROPORTIONS

We are often interested in comparing two populations with regard to the rate of incidence of a particular characteristic. Comparing the jobless rates in two cities, the percentages of female employees in two categories of jobs, and infant mortality in two ethnic groups are just a few examples. Let p_1 denote the proportion of members possessing the characteristic in Population 1 and p_2 that in Population 2. Our goals in this section are to construct confidence intervals for $p_1 - p_2$ and test $H_0 : p_1 = p_2$, the null hypothesis that the rates are the same for two populations. The methods would also apply to the problems of comparison between two treatments, where the response of a subject falls into one of two possible categories that we may technically call "success" and "failure." The success rates for the two treatments can then be identified as the two population proportions p_1 and p_2.

The form of the data is displayed in Table 2, where X and Y denote the numbers of successes in independent random samples of sizes n_1 and n_2 taken from Population 1 and Population 2, respectively.

TABLE 2 Independent Samples from Two Dichotomous Populations

	No. of Successes	No. of Failures	Sample Size
Population 1	X	$n_1 - X$	n_1
Population 2	Y	$n_2 - Y$	n_2

The population proportions of successes p_1 and p_2 are estimated by the corresponding sample proportions

$$\hat{p}_1 = \frac{X}{n_1} \quad \text{and} \quad \hat{p}_2 = \frac{Y}{n_2}$$

Naturally, $\hat{p}_1 - \hat{p}_2$ serves to estimate the difference $p_1 - p_2$. Its standard error is given by

$$S.E.(\hat{p}_1 - \hat{p}_2) = \sqrt{\frac{p_1 q_1}{n_1} + \frac{p_2 q_2}{n_2}}$$

where $q_1 = 1 - p_1$ and $q_2 = 1 - p_2$. This formula of the standard error stems from the fact that because $\hat{p}_1$ and $\hat{p}_2$ are based on independent samples, the variance of their difference equals the sum of their individual variances.

We can calculate the estimated standard error of $\hat{p}_1 - \hat{p}_2$ by using the above expression with the population proportions replaced by the corresponding sample proportions. Moreover, when n_1 and n_2 are large, the estimator $\hat{p}_1 - \hat{p}_2$ is approximately normally distributed. Specifically,

$$Z = \frac{(\hat{p}_1 - \hat{p}_2) - (p_1 - p_2)}{\text{Estimated standard error}} \quad \text{is approximately } N(0, 1)$$

and this can be the basis for constructing confidence intervals for $p_1 - p_2$.

Large Samples Confidence Interval for $p_1 - p_2$

An approximate $100(1 - \alpha)\%$ confidence interval for $p_1 - p_2$ is

$$(\hat{p}_1 - \hat{p}_2) \pm z_{\alpha/2} \sqrt{\frac{\hat{p}_1(1 - \hat{p}_1)}{n_1} + \frac{\hat{p}_2(1 - \hat{p}_2)}{n_2}}$$

provided the sample sizes n_1 and n_2 are large.

Example 14 A Confidence Interval for a Difference in Success Rates

An investigation comparing a medicated patch with the unmedicated control patch for helping smokers quit the habit was discussed on page 88. At the end of the study, the number of persons in each group who were abstinent and who were smoking are repeated in Table 3.

TABLE 3 Quitting Smoking

	Abstinent	Smoking	Total
Medicated patch	21	36	57
Unmedicated patch	11	44	55
	32	80	112

Determine a 95% confidence interval for the difference in success probabilities.

SOLUTION   Let p_1 and p_2 denote the probabilities of quitting smoking with the medicated and unmedicated patches, respectively. We calculate

$$\hat{p}_1 = \frac{21}{57} = .3684 \qquad \hat{p}_2 = \frac{11}{55} = .2000$$

$$\hat{p}_1 - \hat{p}_2 = .1684$$

$$\sqrt{\frac{\hat{p}_1 \hat{q}_1}{n_1} + \frac{\hat{p}_2 \hat{q}_2}{n_2}} = \sqrt{\frac{.3684 \times .6316}{57} + \frac{.2000 \times .8000}{55}} = .0836$$

A 95% confidence interval for $p_1 - p_2$ is

$$(.1684 - 1.96 \times .0836, .1684 + 1.96 \times .0836) \qquad \text{or} \qquad (.005, .332)$$

The confidence interval only covers positive values so we conclude that the success rate with the medicated patch is .005 to .332 higher than for the control group that received the untreated patches. The lower value is so close to 0 that it is still plausible that the medicated patch is not very effective.

Note: A confidence interval for each of the population proportions p_1 and p_2 can be determined by using the method described in Section 5 of Chapter 8. For instance, with the data of Example 14, a 90% confidence interval for p_1 is calculated as

$$\hat{p}_1 \pm 1.645 \sqrt{\frac{\hat{p}_1 \hat{q}_1}{n_1}} = .3684 \pm 1.645 \sqrt{\frac{.3684 \times .6316}{57}}$$

$$= .3684 \pm .1051 \qquad \text{or} \qquad (.263, .474)$$

In order to formulate a test of $H_0: p_1 = p_2$ when the sample sizes n_1 and n_2 are large, we again turn to the fact that $\hat{p}_1 - \hat{p}_2$ is approximately normally distributed. But now we note that under H_0 the mean of this normal distribution is $p_1 - p_2 = 0$ and the standard deviation is

$$\sqrt{pq} \sqrt{\frac{1}{n_1} + \frac{1}{n_2}}$$

where p stands for the common probability of success $p_1 = p_2$ and $q = 1 - p$. The unknown common p is estimated by pooling information from the two samples. The proportion of successes in the combined sample provides

$$\text{Pooled estimate } \hat{p} = \frac{X + Y}{n_1 + n_2}$$

$$\text{or, alternatively,} \qquad \hat{p} = \frac{n_1 \hat{p}_1 + n_2 \hat{p}_2}{n_1 + n_2}$$

$$\text{Also,} \qquad \text{Estimated S.E.}(\hat{p}_1 - \hat{p}_2) = \sqrt{\hat{p}\hat{q}} \sqrt{\frac{1}{n_1} + \frac{1}{n_2}}$$

In summary,

Testing $H_0 : p_1 = p_2$ with Large Samples

Test statistic:

$$Z = \frac{\hat{p}_1 - \hat{p}_2}{\sqrt{\hat{p}(1 - \hat{p})}\sqrt{\dfrac{1}{n_1} + \dfrac{1}{n_2}}} \qquad \text{where } \hat{p} = \frac{X + Y}{n_1 + n_2}$$

The level α rejection region is $|Z| \geq z_{\alpha/2}$, $Z \leq -z_\alpha$, or $Z \geq z_\alpha$ according to whether the alternative hypothesis is $p_1 \neq p_2$, $p_1 < p_2$, or $p_1 > p_2$.

Example 15   Testing Equality of Prevalence of a Virus

A study (courtesy of R. Golubjatnikov) is undertaken to compare the rates of prevalence of CF antibody to parainfluenza I virus among boys and girls in the age group 5 to 9 years. Among 113 boys tested, 34 are found to have the antibody; among 139 girls tested, 54 have the antibody. Do the data provide strong evidence that the rate of prevalence of the antibody is significantly higher in girls than boys? Use $\alpha = .05$. Also, find the P–value.

SOLUTION   Let p_1 denote the population proportion of boys who have the CF antibody and p_2 the population proportion of girls who have the CF antibody. Because we are looking for strong evidence in support of $p_1 < p_2$, we formulate the hypotheses as

$$H_0 : p_1 = p_2 \qquad \text{versus} \qquad H_1 : p_1 < p_2$$

or equivalently as

$$H_0 : p_1 - p_2 = 0 \qquad \text{versus} \qquad H_1 : p_1 - p_2 < 0$$

The sample sizes $n_1 = 113$ and $n_2 = 139$ being large, we will employ the test statistic

$$Z = \frac{\hat{p}_1 - \hat{p}_2}{\sqrt{\hat{p}\hat{q}}\sqrt{\dfrac{1}{n_1} + \dfrac{1}{n_2}}}$$

and set a left-sided rejection region in view of the fact that H_1 is left-sided. With $\alpha = .05$, the rejection region is $R : Z \leq -1.645$. We calculate

$$\hat{p}_1 = \frac{34}{113} = .301 \qquad \hat{p}_2 = \frac{54}{139} = .388$$

$$\text{Pooled estimate } \hat{p} = \frac{34 + 54}{113 + 139} = .349$$

The observed value of the test statistic is then

$$z = \frac{.301 - .388}{\sqrt{.349 \times .651}\sqrt{\dfrac{1}{113} + \dfrac{1}{139}}} = -1.44$$

Because the value $z = -1.44$ is not in R, we do not reject H_0. Consequently, the assertion that the girls have a higher rate of prevalence of the CF antibody than boys is not substantiated at the level of significance $\alpha = .05$.

The significance probability of the observed z is

$$P\text{-value} = P[Z \le -1.44]$$
$$= .0749$$

This means that we must allow an α of at least .0749 in order to consider the result significant.

Exercises

10.44   Given the data

$$n_1 = 100 \qquad \hat{p}_1 = \frac{50}{100} = .50$$

$$n_2 = 200 \qquad \hat{p}_2 = \frac{134}{200} = .72$$

(a) Find a 95% confidence interval for $p_1 - p_2$.

(b) Perform the Z test for the null hypotheses $H_0: p_1 = p_2$ versus $H_1: p_1 < p_2$.

10.45   A sample of 100 females was collected from ethnic group A and a sample of 100 from ethnic group B. Each female was asked, "Did you get married before you were 19?" The following counts were obtained:

	A	B
Yes	61	31
No	39	69

(a) Test for equality of two proportions against a two-sided alternative. Take $\alpha = .05$.

(b) Establish a 95% confidence interval for the difference $p_A - p_B$.

10.46   In a comparative study of two new drugs, A and B, 120 patients were treated with drug A and 150 patients with drug B, and the following results were obtained.

	Drug A	Drug B
Cured	50	88
Not cured	70	62
Total	120	150

(a) Do these results demonstrate a significantly higher cure rate with drug B than drug A? Test at $\alpha = .05$.

(b) Construct a 95% confidence interval for the difference in the cure rates of the two drugs.

10.47   In a study of the relationship between temperament and personality, 49 female high school students who had a high level of reactivity (HRL) and 54 students who had a low level of reactivity (LRL) were classified according to their attitude to group pressure with the following results.

	Attitude		
Reactivity	Submissive	Resistant	Total
HRL	34	15	49
LRL	12	42	54

Is resistance to group pressure significantly lower in the HRL group than the LRL group? Answer by calculating the P-value.

10.48   Refer to the data in Exercise 10.47. Determine a 99% confidence interval for the difference between the proportions of resistant females in the HRL and LRL populations.

10.49   Refer to Exercise 10.19 concerning a study on the relationship between child abuse and later criminal behavior. Suppose that from follow-ups of 85 boys who were abused in their preschool years and 120 boys who were not abused, it was found that 21 boys in the abused group and 11 boys in the nonabused group were chronic offenders in their teens. Do these data substantiate the conjecture that abused boys are more prone to be chronic offenders than nonabused boys? Test at $\alpha = .01$.

10.50   Referring to the data of Exercise 10.49, determine a 95% confidence interval for the difference between the true proportions of chronic offenders in the populations of abused and nonabused boys.

10.51   The popular disinfectant Listerine is named after Joseph Lister, a British physician who pioneered the use of antiseptics. Lister conjectured that human infections might have an organic origin and thus could be prevented by using a disinfectant. Over a period of several years, he performed 75 amputations: 40 using carbolic acid as a disinfectant and 35 without any disinfectant. The following results were obtained.

	Patient Survived	Patient Died	Total
With carbolic acid	34	6	40
Without carbolic acid	19	16	35

Are the survival rates significantly different between the two groups? Test at $\alpha = .05$ and calculate the P-value.

10.52   Referring to the data of Exercise 10.51, calculate a 95% confidence interval for the difference between the survival rates for the two groups.

10.53   Random samples of 250 persons in the 30- to 40-year age group and 250 persons in the 60- to 70-year age group are asked about the average number of hours they sleep per night, and the following summary data are recorded.

Age	Hours of Sleep ≤ 8	> 8	Total
30–40	173	77	250
60–70	120	130	250
Total	293	207	500

Do these data demonstrate that the proportion of persons who have ≤ 8 hours of sleep per night is significantly higher for the age group 30 to 40 than that for the age group 60 to 70? Answer by calculating the P-value.

10.54   Referring to Exercise 10.53, denote by p_1 and p_2 the population proportions in the two groups who have ≤ 8 hours of sleep per night. Construct a 95% confidence interval for $p_1 - p_2$.

10.55   A medical researcher conjectures that smoking can result in wrinkled skin around the eyes. By observing 150 smokers and 250 nonsmokers, the researcher finds that 95 of the smokers and 103 of the nonsmokers have prominent wrinkles around their eyes.

(a)   Do these data substantiate the belief that prominent wrinkles around eyes are more prevalent among smokers than nonsmokers? Answer by calculating the P-value.

(b)   If the results are statistically significant, can the researcher readily conclude that smoking causes wrinkles around the eyes? Why or why not?

10.56   Consider testing the equality of two proportions using independent random samples of sizes n_1 and n_2 from two populations. Suppose that the sample proportions of successes are $\hat{p}_1 = .65$ and $\hat{p}_2 = .52$. For each case, examine whether or not the observed difference $\hat{p}_1 - \hat{p}_2 = .11$ is statistically significant at the specified α.

(a) $n_1 = 40$, $n_2 = 50$, $\alpha = .05$

(b) $n_1 = 200$, $n_2 = 250$, $\alpha = .05$

(c) $n_1 = 200$, $n_2 = 250$, $\alpha = .01$

10.57 A major clinical trial of a new vaccine for type-B hepatitis was conducted with a high-risk group of 1083 male volunteers. From this group, 549 men were given the vaccine and the other 534 a placebo. A follow-up of all these individuals yielded the data:

	Follow-up		Total
	Got Hepatitis	Did Not Get Hepatitis	
Vaccine	11	538	549
Placebo	70	464	534

(a) Do these observations testify that the vaccine is effective? Use $\alpha = .01$.

(b) Construct a 95% confidence interval for the difference between the incidence rates of hepatitis among the vaccinated and nonvaccinated individuals in the high-risk group.

10.58 Records of drivers with a major medical condition (diabetes, heart condition, or epilepsy)

and also a group of drivers with no known health conditions were retrieved from a motor vehicle department. Drivers in each group were classified according to their driving record in the last year.

	Traffic Violations		Total
Medical Condition	None	One or More	
Diabetes	119	41	160
Heart condition	121	39	160
Epilepsy	72	78	150
None (control)	157	43	200

Let p_D, p_H, p_E, and p_C denote the population proportions of drivers having one or more traffic violations in the last year for the four groups "diabetes," "heart condition," "epilepsy," and "control," respectively.

(a) Test $H_0: p_D = p_C$ versus $H_1: p_D > p_C$ at $\alpha = .10$.

(b) Is there strong evidence that p_E is higher than p_C? Answer by calculating the P-value.

10.59 Refer to Exercise 10.58.

(a) Construct a 95% confidence interval for $p_E - p_H$.

(b) Construct a 90% confidence interval for $p_H - p_C$.

(c) Construct 95% confidence intervals for p_D, p_H, p_E, and p_C, individually.

USING STATISTICS WISELY

1. In all cases where a sample size is small and normality is assumed, the data should be graphed in a dot plot to reveal any obvious outliers which could invalidate the inferences based on the normal theory.

2. When comparing two treatments under a matched pairs design, whenever possible, assign the treatments within each pair at random. To analyze the resulting data, use the results for one sample but applied to the differences from each matched pair. For instance, if the difference of paired measure-

ments has a normal distribution, determine a $100(1 - \alpha)\%$ confidence interval for the mean difference μ_D as

$$\left(\bar{d} - t_{\alpha/2} \frac{s_D}{\sqrt{n}}, \ \bar{d} + t_{\alpha/2} \frac{s_D}{\sqrt{n}} \right)$$

where $t_{\alpha/2}$ is based on $n - 1$ degrees of freedom. Otherwise, with large samples use $z_{\alpha/2}$.

3. When comparing two treatments using the independent samples design, randomly assign the treatments to groups whenever possible. With the matched pairs design, randomly assign the treatments within each pair.

4. When sample sizes are large, determine the limits of a $100(1 - \alpha)\%$ confidence interval for the difference of means $\mu_1 - \mu_2$ as

$$\bar{x}_1 - \bar{x}_2 \pm z_{\alpha/2} \sqrt{\frac{s_1^2}{n_1} + \frac{s_2^2}{n_2}}$$

5. When each of the two samples are from normal populations, having the same variance, determine the limits of a $100(1 - \alpha)\%$ confidence interval for the difference of means $\mu_1 - \mu_2$ as

$$\bar{x}_1 - \bar{x}_2 \pm t_{\alpha/2} \, s_{\text{pooled}} \sqrt{\frac{1}{n_1} + \frac{1}{n_2}}$$

where the pooled estimate of variance

$$s_{\text{pooled}}^2 = \frac{(n_1 - 1)s_1^2 + (n_2 - 1)s_2^2}{(n_1 - 1) + (n_2 - 1)}$$

and $t_{\alpha/2}$ is based on $n_1 + n_2 - 2$ degrees of freedom.

6. Do not pool the two sample variances s_1^2 and s_2^2 if they are very different. We suggest a factor of 4 as being too different. There are alternative procedures including the conservative procedures based on page 390.

KEY IDEAS AND FORMULAS

In any comparative study of treatments, products, methods, and so on, the term **treatment** refers to the things being compared. The basic unit or object, to which one of the treatments is applied, is called an **experimental unit** or an **experimental subject**. The **response variable** is the characteristic that is recorded on each unit.

The specification of which treatment to compare and method of assigning experimental units is called the **experimental (or sampling) design**. The choice of appropriate statistical methods for making inferences depends heavily on the experimental design chosen for data collection.

A carefully designed experiment is fundamental to the success of a comparative study.

The most basic experimental designs to compare two treatments are independent samples and matched pairs sample.

The independent samples design require the subjects to be randomly selected for assignment to each treatment. Randomization prevents uncontrolled factors from systematically favoring one treatment over the other.

With a matched pairs design, subjects in each pair are alike, while those in different pairs may be dissimilar. For each pair, the two treatments should be randomly allocated to the members.

The idea of matching or blocking experimental units is to remove a known source of variation from comparisons. Pairing subjects according to some feature prevents that source of variation from interfering with treatment comparisons. By contrast, random allocation of subjects according to the independent random sampling design spreads these variations between the two treatments.

Inferences with Two Independent Random Samples

1. *Large samples.* When n_1 and n_2 are both greater than 30, inferences about $\mu_1 - \mu_2$ are based on the fact that

$$\frac{(\overline{X} - \overline{Y}) - (\mu_1 - \mu_2)}{\sqrt{\dfrac{S_1^2}{n_1} + \dfrac{S_2^2}{n_2}}} \quad \text{is approximately } N(0, 1)$$

A $100(1 - \alpha)\%$ confidence interval for $\mu_1 - \mu_2$ is

$$(\overline{X} - \overline{Y}) \pm z_{\alpha/2} \sqrt{\frac{S_1^2}{n_1} + \frac{S_2^2}{n_2}}$$

To test $H_0 : \mu_1 - \mu_2 = \delta_0$, we use the normal test statistic

$$Z = \frac{(\overline{X} - \overline{Y}) - \delta_0}{\sqrt{\dfrac{S_1^2}{n_1} + \dfrac{S_2^2}{n_2}}}$$

No assumptions are needed in regard to the shape of the population distributions.

2. *Small samples.* When n_1 and n_2 are small, inferences using the t distribution require the assumptions:

   (a) Both populations are normal.

   (b) $\sigma_1 = \sigma_2$.

   The common σ^2 is estimated by

$$S_{\text{pooled}}^2 = \frac{(n_1 - 1)S_1^2 + (n_2 - 1)S_2^2}{n_1 + n_2 - 2}$$

Inferences about $\mu_1 - \mu_2$ are based on

$$T = \frac{(\overline{X} - \overline{Y}) - (\mu_1 - \mu_2)}{S_{\text{pooled}}\sqrt{\dfrac{1}{n_1} + \dfrac{1}{n_2}}} \qquad \text{d.f.} = n_1 + n_2 - 2$$

A $100(1 - \alpha)\%$ confidence interval for $\mu_1 - \mu_2$ is

$$(\overline{X} - \overline{Y}) \pm t_{\alpha/2}\, S_{\text{pooled}}\sqrt{\dfrac{1}{n_1} + \dfrac{1}{n_2}}$$

To test $H_0 : \mu_1 - \mu_2 = \delta_0$, the test statistic is

$$T = \frac{(\overline{X} - \overline{Y}) - \delta_0}{S_{\text{pooled}}\sqrt{\dfrac{1}{n_1} + \dfrac{1}{n_2}}} \qquad \text{d.f.} = n_1 + n_2 - 2$$

Inferences with a Matched Pair Sample

With a paired sample $(X_1, Y_1), \ldots, (X_n, Y_n)$, the first step is to calculate the differences $D_i = X_i - Y_i$, their mean $\overline{D}$, and standard deviation S_D.

If n is small, we assume that the D_i's are normally distributed $N(\delta, \sigma_D)$. Inferences about δ are based on

$$T = \frac{\overline{D} - \delta}{S_D/\sqrt{n}} \qquad \text{d.f.} = n - 1$$

A $100(1 - \alpha)\%$ confidence interval for δ is

$$\overline{D} \pm t_{\alpha/2}\, S_D/\sqrt{n}$$

The test of $H_0 : \delta = \delta_0$ is performed with the test statistic:

$$T = \frac{\overline{D} - \delta_0}{S_D/\sqrt{n}} \qquad \text{d.f.} = n - 1$$

If n is large, the assumption of normal distribution for the D_i's is not needed. Inferences are based on the fact that

$$Z = \frac{\overline{D} - \delta_0}{S_D/\sqrt{n}} \qquad \text{is approximately } N(0, 1)$$

Summary of Inferences about Means

Table 4 summarizes all of the statistical procedures we have considered for making inferences about (1) a single mean, (2) the difference of two means, or (3) a mean difference for a pair of observations.

TABLE 4 General Formulas for Inferences about a Mean (μ), Difference of Two Means ($\mu_1 - \mu_2$)

$$\text{Confidence interval} = \text{Point estimator} \pm (\text{Tabled value})(\text{Estimated or true std. dev.})$$

$$\text{Test statistic} = \frac{\text{Point estimator} - \text{Parameter value at } H_0 \text{ (null hypothesis)}}{(\text{Estimated or true}) \text{ std. dev. of point estimator}}$$

	Ch. 8 General	Ch. 9 Normal with unknown σ	Ch. 10	Independent Samples — Normal $N(\mu_1,\sigma_1), N(\mu_2,\sigma_2)$ $\sigma_1 \neq \sigma_2$	General	Matched Samples — Normal for the difference $D_i = X_i - Y_i$
Population(s)	Mean μ	Mean μ	Normal $N(\mu_1,\sigma_1), N(\mu_2,\sigma_2)$ $\sigma_1 = \sigma_2 = \sigma$		$\mu_1 - \mu_2$	
Inference on	Mean μ	Mean μ	$\mu_1 - \mu_2$	$\mu_1 - \mu_2$	$\mu_1 - \mu_2$	$\delta = \mu_1 - \mu_2$
Sample(s)	$X_1, \ldots, X_n$	$X_1, \ldots, X_n$	$X_1, \ldots, X_{n_1}$ $Y_1, \ldots, Y_{n_2}$	$X_1, \ldots, X_{n_1}$ $Y_1, \ldots, Y_{n_2}$	$X_1, \ldots, X_{n_1}$ $Y_1, \ldots, Y_{n_2}$	$X_1\ Y_1$ $X_2\ Y_2$ $\ldots$ $X_n\ Y_n$ $\quad$ $D_1 = X_1 - Y_1$ $D_2 = X_2 - Y_2$ $\ldots$ $D_n = X_n - Y_n$
Sample size n	Large $n > 30$	$n \geq 2$	$n_1 \geq 2$ $n_2 \geq 2$	$n_1 \geq 2$ $n_2 \geq 2$	$n_1 > 30$ $n_2 > 30$	$n \geq 2$
Point estimator	$\bar{X}$	$\bar{X}$	$\bar{X} - \bar{Y}$	$\bar{X} - \bar{Y}$	$\bar{X} - \bar{Y}$	$\bar{D} = \bar{X} - \bar{Y}$
Variance of point estimator	$\dfrac{\sigma^2}{n}$	$\dfrac{\sigma^2}{n}$	$\sigma^2\left(\dfrac{1}{n_1} + \dfrac{1}{n_2}\right)$	$\dfrac{\sigma_1^2}{n_1} + \dfrac{\sigma_2^2}{n_2}$	$\dfrac{\sigma_1^2}{n_1} + \dfrac{\sigma_2^2}{n_2}$	$\dfrac{\sigma_D^2}{n}$
Std. dev. of point estimator	$\dfrac{\sigma}{\sqrt{n}}$	$\dfrac{\sigma}{\sqrt{n}}$	$\sigma\sqrt{\dfrac{1}{n_1} + \dfrac{1}{n_2}}$	$\sqrt{\dfrac{\sigma_1^2}{n_1} + \dfrac{\sigma_2^2}{n_2}}$	$\sqrt{\dfrac{\sigma_1^2}{n_1} + \dfrac{\sigma_2^2}{n_2}}$	$\dfrac{\sigma_D}{\sqrt{n}}$
Estimated std. dev.	$\dfrac{S}{\sqrt{n}}$	$\dfrac{S}{\sqrt{n}}$	$S_{pooled}\sqrt{\dfrac{1}{n_1} + \dfrac{1}{n_2}}$	$\sqrt{\dfrac{S_1^2}{n_1} + \dfrac{S_2^2}{n_2}}$	$\sqrt{\dfrac{S_1^2}{n_1} + \dfrac{S_2^2}{n_2}}$	$\dfrac{S_D}{\sqrt{n}}$
Distribution	Normal	t with d.f. $= n - 1$	t with d.f. $= n_1 + n_2 - 2$	t with d.f. $=$ smaller of $n_1 - 1$ and $n_2 - 1$	Normal	t with d.f. $= n - 1$
Test statistic	$\dfrac{\bar{X} - \mu_0}{S/\sqrt{n}}$	$\dfrac{\bar{X} - \mu_0}{S/\sqrt{n}}$	$\dfrac{(\bar{X} - \bar{Y}) - \delta_0}{S_{pooled}\sqrt{\dfrac{1}{n_1} + \dfrac{1}{n_2}}}$ $S_{pooled}^2 = \dfrac{(n_1 - 1)S_1^2 + (n_2 - 1)S_2^2}{n_1 + n_2 - 2}$	$\dfrac{(\bar{X} - \bar{Y}) - \delta_0}{\sqrt{\dfrac{S_1^2}{n_1} + \dfrac{S_2^2}{n_2}}}$	$\dfrac{(\bar{X} - \bar{Y}) - \delta_0}{\sqrt{\dfrac{S_1^2}{n_1} + \dfrac{S_2^2}{n_2}}}$	$\dfrac{\bar{D} - \delta_0}{S_D/\sqrt{n}}$ $S_D = $ sample std. dev. of the D_i's

Comparing Two Binomial Proportions—Large Samples

Data:

$$X = \text{No. of successes in } n_1 \text{ trials with success probability } P(S) = p_1$$
$$Y = \text{No. of successes in } n_2 \text{ trials with success probability } P(S) = p_2$$

To test $H_0: p_1 = p_2$ versus $H_1: p_1 \neq p_2$, use the Z test:

$$Z = \frac{\hat{p}_1 - \hat{p}_2}{\sqrt{\hat{p}(1 - \hat{p})} \sqrt{\dfrac{1}{n_1} + \dfrac{1}{n_2}}} \qquad \text{with} \qquad R: |Z| \geq z_{\alpha/2}$$

where

$$\hat{p}_1 = \frac{X}{n_1} \qquad \hat{p}_2 = \frac{Y}{n_2} \qquad \hat{p} = \frac{X + Y}{n_1 + n_2}$$

To test $H_0: p_1 = p_2$ versus $H_1: p_1 > p_2$, use the Z test with $R: Z \geq z_\alpha$.
A $100(1 - \alpha)\%$ confidence interval for $p_1 - p_2$ is

$$(\hat{p}_1 - \hat{p}_2) \pm z_{\alpha/2} \sqrt{\frac{\hat{p}_1(1 - \hat{p}_1)}{n_1} + \frac{\hat{p}_2(1 - \hat{p}_2)}{n_2}}$$

TECHNOLOGY

Confidence intervals and tests for comparing means

MINITAB

Matched pair samples

We illustrate with the calculation of a 98% confidence interval and .02 level test.

Enter the first sample in *C1* and second sample in *C2*.

Dialog box:

Stat > Basic Statistics > Paired t.
Type *C1* in **First Sample** and *C2* in **Second Sample**.
Click **Options.** Type *98* in **Confidence level**.
Enter the value of the mean under the null hypothesis in **Test mean** and choose direction of the **alternative.** Click **OK.** Click **OK.**

Two-sample t tests and confidence intervals

We illustrate with the calculation of a 98% confidence interval for $\mu_1 - \mu_2$ and .02 level test of the null hypothesis of no difference in means.

Data:

C1 First sample
C2 Second sample

Dialog box:
Stat > Basic Statistics > 2-Sample t.
Select **Samples in different columns.** Type *C1* in **First** and *C2* in **Second.**
To pool the estimates of variance, click the box **Assume equal variances.** Click **Options.** Type *98* in **Confidence level,** the null hypothesis value of the mean in **Test mean,** and select the direction of the alternative hypothesis. Click **OK.** Click **OK.**

EXCEL

Matched pairs t tests

We illustrate with the calculation of a test of the null hypothesis $H_0 : \delta = 0$ versus a two-sided alternative. Begin with values for the first variable in column A and the second in column B.

Select **Tools** and then **Data Analysis.**
Select **t-Test: Paired Two-Sample for Means.** Click **OK.**
With the cursor in **Variable 1 Range,** highlight the data in column A.
With the cursor in **Variable 2 Range,** highlight the date in column B.
Type the hypothesized value *0* after **Hypothesized mean difference.** Click **OK.**

The program returns a summary that includes the value of the *t* statistic, the *P*–value for a one-sided test (actually the smallest tail probability), and the *P*–value for a two-sided test.

Two-sample t tests

We illustrate with a test of the null hypothesis of no difference between the two means. Begin with the first sample in column A and the second in column B.

Select **Tools** and then **Data Analysis.**
Select **t-Test: Two-Sample Assuming Unequal Variances.** Click **OK** or, to pool, Select **t-Test: Two-Sample Assuming Equal Variances** and click **OK.**
With the cursor in **Variable 1 Range,** highlight the data in column A.
With the cursor in **Variable 2 Range,** highlight the date in column B.
Type the hypothesized value *0* after **Hypothesized mean difference.**
Enter *.02* for **Alpha.** Click **OK.**

The program returns the one-sided and two-sided *P*–values.

TI-84/-83 PLUS

Matched Pairs Samples

Confidence intervals

We illustrate the calculation of a 98% confidence interval. Start with the values for the first variable entered in L_1 and the second in L_2.

> Press **STAT** and select **TESTS** and then **8: Tinterval.**
> Select **Data** with **List** set to L_3 and **Freq** to **1.**
> Enter *.98* following *C–Level:* Select **Calculate.**
> Then press **ENTER.**

Tests

We illustrate the calculation of a test of the null hypothesis of 0 mean difference. Start with the first sample entered in L_1 and the second in L_2. Then let $L_3 = L_2 - L_1$ or $L_3 = L_1 - L_2$ depending on how the alternative is defined.

> Press **STAT** and select **TESTS** and then **2: T-Test.**
> Select **Data** with **List** set to L_3 and **Freq** to **1.**
> Enter *0* for μ_0. Select the direction of the alternative hypothesis for the mean difference. Select **Calculate.**
> Press **ENTER.**

If, instead, the sample size, mean, and standard deviation are available, the second step is

> Select **Stats** and type in the sample sizes, means, and standard deviations.

Two-sample t tests

We illustrate with the calculation of a test of the null hypothesis of no difference between the two means. Start with the data entered in L_1 and L_2.

> Press **STAT** and select **TESTS** and then **4: 2-SampT-Test.**
> Select **Data** with **List1** set to L_1, **List2** to L_2, **FREQ1** to **1**, and **Freq2** to **1.**
> Select the direction of the alternative hypothesis. Set **Pooled** to *NO* if you do not wish to pool.
> Select **Calculate** or **Draw** and press **ENTER.**

The calculator will return the *P*–value. **Draw** will draw the *t* distribution and shade the area of the *P*–value. If, instead, the sample sizes, means, and standard deviations are available, the second step is

> Select **Stats** and enter the sample sizes, means, and standard deviations.

7. REVIEW EXERCISES

10.60 The following summary is recorded for independent samples from two populations.

Sample 1	Sample 2
$n_1 = 55$	$n_2 = 60$
$\bar{x} = 18.4$	$\bar{y} = 16.5$
$s_1^2 = 8.6$	$s_2^2 = 13.7$

(a) Construct a 98% confidence interval for $\mu_1 - \mu_2$.

(b) Test $H_0:\mu_1 - \mu_2 = 2.5$ versus H_1: $\mu_1 - \mu_2 \neq 2.5$ with $\alpha = .02$.

(c) Test $H_0:\mu_1 - \mu_2 = 2.5$ versus H_1: $\mu_1 - \mu_2 < 2.5$ with $\alpha = .05$.

10.61 Given here are two sets of values for the standard deviations of independent random samples, each of size 52, from two populations. For each case, determine whether or not an observed difference of 8 between the sample means is statistically significant at $\alpha = .05$.

(a) $s_1 = 20, \quad s_2 = 28$

(b) $s_1 = 12, \quad s_2 = 15$

10.62 A group of 88 subjects is used in an experiment to compare two treatments. From this group, 40 subjects are randomly selected to be assigned to treatment 1 and the remaining 48 subjects are assigned to treatment 2. The means and standard deviations of the responses are

	Treatment 1	Treatment 2
Mean	16.21	27.84
Standard deviation	2.88	4.32

Determine a 95% confidence interval for the mean difference of the treatment effects.

10.63 Refer to the data in Exercise 10.62. Suppose the investigator wishes to establish that the mean response of treatment 2 is larger than that of treatment 1 by more than 10 units.

(a) Formulate the null hypothesis and the alternative hypothesis.

(b) State the test statistic and the rejection region with $\alpha = .10$.

(c) Perform the test at $\alpha = .10$. Also, find the P–value and comment.

10.64 A study of postoperative pain relief is conducted to determine if drug A has a significantly longer duration of pain relief than drug B. Observations of the hours of pain relief are recorded for 55 patients given drug A and 58 patients given drug B. The summary statistics are

	A	B
Mean	4.64	4.03
Standard deviation	1.25	1.82

(a) Formulate H_0 and H_1.

(b) State the test statistic and the rejection region with $\alpha = .10$.

(c) State the conclusion of your test with $\alpha = .10$. Also, find the P–value and comment.

10.65 Consider the data of Exercise 10.64.

(a) Construct a 90% confidence interval for $\mu_A - \mu_B$.

(b) Give a 95% confidence interval for μ_A using the data of drug A alone. (*Note:* Refer to Chapter 8.)

10.66 Obtain s_{pooled}^2 for the gaming data in the chapter opening.

10.67 Given the following two samples,

8 11 5 9 7   and   5 3 4 8

obtain (a) s_{pooled}^2 and (b) value of the t statistic for testing $H_0:\mu_1 - \mu_2 = 2$. State the d.f. of the t.

10.68 A fruit grower wishes to evaluate a new spray that is claimed to reduce the loss due to damage by insects. To this end, he performs an experiment with 27 trees in his orchard by treating 12 of those trees with the new spray and the other 15 trees with the standard spray. From the data of fruit yield (in pounds) of those trees, the following summary statistics were found.

	New Spray	Standard Spray
Mean yield	249	233
Standard deviation	19	45

Do these data substantiate the claim that a higher yield should result from the use of the new spray? State the assumptions you make and test at $\alpha = .05$.

10.69 Referring to Exercise 10.68, construct a conservative 95% confidence interval for the difference in mean yields between the new spray and the standard spray.

10.70 Referring to Exercise 10.68, construct 90% confidence intervals for the mean yields under the use of the new spray and the standard spray individually.

10.71 An investigation is conducted to determine if the mean age of welfare recipients differs between two cities A and B. Random samples of 75 and 100 welfare recipients are selected from city A and city B, respectively, and the following computations are made.

	City A	City B
Mean	37.8	43.2
Standard deviation	6.8	7.5

(a) Do the data provide strong evidence that the mean ages are different in city A and city B? (Test at $\alpha = .02$.)

(b) Construct a 98% confidence interval for the difference in mean ages between A and B.

(c) Construct a 98% confidence interval for the mean age for city A and city B individually. (*Note:* Refer to Chapter 8.)

10.72 The following generic computer output summarizes the data, given in Table D.6 of the Data Bank, on the neck size of male and female bears.

Sex	N	Mean	StDev
F	36	52.92	8.83
M	25	59.70	17.5

Test for equality of mean neck size versus a two-sided alternative. Take $\alpha = .03$.

10.73 Refer to the computer attitude scores (CAS) of students given in Table D.4 of the Data Bank. A computer analysis produces the output

Two-sample T for CAS

sex	N	Mean	StDev
F	15	2.643	0.554
M	20	2.945	0.390

```
Difference  = mu M  -  mu F
Estimate for difference: 0.302
90\% CI for the difference
(0.032, 0.571)
T-Test of difference  = 0
(vs not    =):
T-Value   = 1.89 P-Value   = 0.067
DF  = 33
Both use Pooled StDev   = 0.467
```

(a) What is the conclusion to testing the equality of mean computer attitude scores at level $\alpha = .05$?

(b) Find a 95% confidence interval for the mean attitude score for males minus the mean attitude score for females.

(c) Test the null hypothesis that, on average, males score .1 lower than females against a two-sided alternative. Take $\alpha = .05$.

10.74 In each of the following cases, how would you select the experimental units and conduct the experiment—matched pairs or independent samples?

(a) Compare the mileage obtained from two gasolines. Twelve SUVs of various sizes are available.

(b) Compare the drying times of two latex-based interior paints. Ten walls are available.

(c) Compare two methods of teaching swimming. Twenty five-year-old girls are available.

10.75 A sample of river water is divided into two specimens. One is randomly selected to be sent to Lab A and the other is sent to Lab B. This is repeated for a total of nine times. The measurement of suspended solids at Lab B is subtracted from that of Lab A to obtain the differences

     12   10   15   42   11   −4   −2   10   −7

(a) Is there strong evidence that the mean difference is not zero? Test with $\alpha = .02$.

(b) Construct a 90% confidence interval for the mean difference of the suspended solids measurements.

10.76 An experiment is conducted to determine if the use of a special chemical additive with a standard fertilizer accelerates plant growth. Ten locations are included in the study. At each location, two plants growing in close proximity are treated. One is given the standard fertilizer, the other the standard fertilizer with the chemical additive. Plant growth after four weeks is measured in centimeters. Do the following data substantiate the claim that use of the chemical additive accelerates plant growth? State the assumptions that you make and devise an appropriate test of the hypothesis. Take $\alpha = .05$.

	Location									
	1	2	3	4	5	6	7	8	9	10
Without additive	20	31	17	22	19	32	25	18	21	19
With additive	23	34	16	21	22	31	29	20	25	23

10.77 Obtain a 95% confidence interval for δ using the data in Exercise 10.76.

10.78 Referring to Exercise 10.76, suppose that the two plants at each location are situated in the east–west direction. In designing this experiment, you must decide which of the two plants at each location—the one in the east or the one in the west—is to be given the chemical additive.

(a) Explain how, by repeatedly tossing a coin, you can randomly allocate the treatments to the plants at the 10 locations.

(b) Perform the randomization by actually tossing a coin 10 times.

10.79 Students can bike to a park the other side of a lake by going around one side of the lake or the other. After much discussion about which was faster, they decided to perform an experiment. Among the 12 students available, 6 were randomly selected to follow Path A on one side of the lake and the rest followed Path B on the other side. They all went on different days so the conclusion would apply to a variety of conditions.

	Travel Time (minutes)					
Path A	10	12	15	11	16	11
Path B	12	15	17	13	18	16

(a) Is there a significant difference between the mean travel times between the two paths? State the assumptions you have made in performing the test.

(b) Suggest an alternative design for this study that would make the comparison more effective.

10.80 Five pairs of tests are conducted to compare two methods of making rope. Each sample batch contains enough hemp to make two ropes. The tensile strength measurements are

	Test				
	1	2	3	4	5
Method 1	14	12	18	16	15
Method 2	16	15	17	16	14

(a) Treat the data as 5 paired observations and calculate a 95% confidence interval for the mean difference in tensile strengths between ropes made by the two methods.

(b) Repeat the calculation of a 95% confidence interval treating the data as independent random samples.

(c) Briefly discuss the conditions under which each type of analysis would be appropriate.

10.81 An experiment was conducted to study whether cloud seeding reduces the occurrences of hail. At a hail-prone geographical area, seeding was done on 50 stormy days and another 165 stormy days were also observed without seeding. The following counts were obtained.

	Days	
	Seeded	Not Seeded
Hail	7	43
No hail	43	122
Total	50	165

Do these data substantiate the conjecture that seeding reduces the chance of hail? (Answer by determining the P–value.)

10.82 Referring to the data of Exercise 10.81, calculate a 90% confidence interval for the difference between the probabilities of hail with and without seeding.

10.83 An antibiotic for pneumonia was injected into 100 patients with kidney malfunctions (called uremic patients) and 100 patients with no kidney malfunctions (called normal patients). Some allergic reaction developed in 38 of the uremic patients and 21 of the normal patients.

(a) Do the data provide strong evidence that the rate of incidence of allergic reaction to the antibiotic is higher in uremic patients than normal patients?

(b) Construct a 95% confidence interval for the difference between the population proportions.

10.84 An experiment is conducted to compare the viability of seeds with and without a cathodic protection, which consists of subjecting the seeds to a negatively charged conductor. Seeds of a common type are randomly divided into 2 batches of 250 each. One batch is given cathodic protection, and the other is retained as the control group. Both batches are then subjected to a common high temperature to induce artificial aging. Subsequently, all the seeds are soaked in water and left to germinate. It is found that 25% of the control seeds and 10% of the cathodically protected seeds fail to germinate.

(a) Do the data provide strong evidence that the cathodic protection permits a higher germination rate in seeds subjected to artificial aging?

(b) Construct a 98% confidence interval for the difference $p_T - p_C$, where p_T and p_C represent the germination rates of cathodically treated seeds and control seeds, respectively.

The Following Exercises May Require a Computer

10.85 We illustrate the MINITAB commands and output for the two-sample t test.

Data

C1: 5 2 8 3
C2: 8 9 6

Dialog box:

Stat > Basic Statistics > 2-Sample t

Click **Samples in different columns**

Type C1 in **First**. C2 in **Second**.

Click **OK**. Click **Assume equal variances**.
Click **O.K.**

```
Two-Sample T-Test and CI: C1, C2
Two-sample T for C1 vs C2

        N     Mean    StDev    SE Mean
C1      4     4.50    2.65     1.3
C2      3     7.67    1.53     0.88


Difference  =  mu (C1)  -  mu (C2)
Estimate for difference:  -3.16667
95% CI for difference:
(-7.61492, 1.28159)
T-Test of difference  =  0 (vs
not  =):  T-Value  =  -1.83
P-value  =  0.127 DF  =  5
Both use Pooled StDev  =  2.2657
```

(a) From the output, what is the conclusion to testing $H_0 : \mu_1 - \mu_2 = 0$ versus a two-sided alternative at level $\alpha = .05$?

(b) Find a 97% confidence interval for the difference of mean weights of wolves in Exercise 10.17.

10.86 Refer to the alligator data in Table D.11 of the Data Bank. Using the data on testosterone x_4 from the Lake Apopka alligators, find a 95% confidence interval for the difference of means between males and females. There should be a large difference for healthy alligators. Comment on your conclusion.

10.87 Refer to the alligator data in Table D.11 of the Data Bank. Using the data on testosterone x_4 for male alligators, compare the means for the two lake regions.

(a) Should you pool the variances with these data?

(b) Find a 90% confidence interval for the difference of means between the two lakes. Use the conservative procedure on page 390.

(c) Which population has the highest mean and how much higher is it?

10.88 Refer to the bear data in Table D.6 of the Data Bank. Compare the head widths of males and females by obtaining a 95% confidence interval for the difference of means and also test equality versus a two-sided alternative with $\alpha = .05$.

10.89 Refer to the marine growth of salmon data in Table D.7 of the Data Bank. Compare the mean growth of males and females by obtaining a 95% confidence interval for the difference of means and also test equality versus a two-sided alternative with $\alpha = .05$.

10.90 Refer to the physical fitness data in Table D.5 of the Data Bank. Find a 95% confidence interval for the mean difference of the pretest minus posttest number of situps. Also test that the mean difference is zero versus a two-sided alternative with $\alpha = .05$.

10.91 Refer to the physical fitness data in Table D.5 of the Data Bank. Find a 95% confidence interval for the mean difference of the pretest minus posttest time to complete the rowing test. Also test that the mean difference is zero versus a two-sided alternative with $\alpha = .05$.

11

Regression Analysis I
Simple Linear Regression

The Highest Roller Coasters are Fastest

Some roller coasters are designed to twist riders and turn them upside down. Others are designed to provide fast rides over large drops. Among the 12 tallest roller coasters in the world, the maximum height (inches) is related to top speed (miles per hour). Each data point, consisting of the pair of values (height, speed), represents one roller coaster. The fitted line predicts an increase in top speed of .17 miles per hour for each foot of height, or 17 miles per hour for each 100 feet in height.

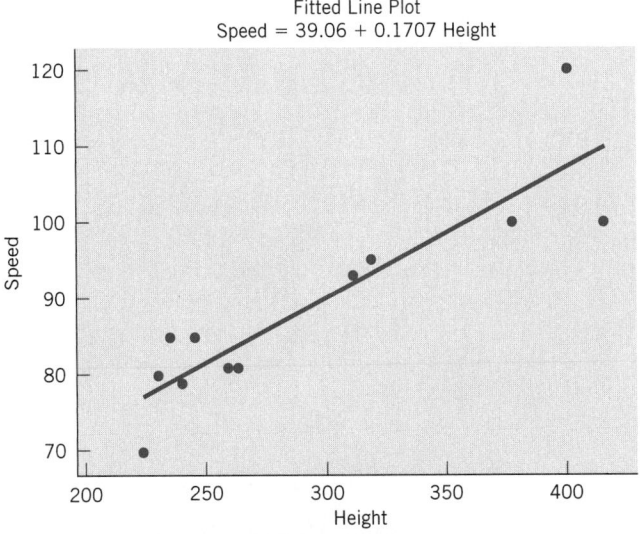

© Rafael Macia/Photo Researchers, Inc.

1.   INTRODUCTION

Except for the brief treatment in Sections 5 to 8 of Chapter 3, we have discussed statistical inferences based on the sample measurements of a single variable. In many investigations, two or more variables are observed for each experimental unit in order to determine:

1.   Whether the variables are related.
2.   How strong the relationships appear to be.
3.   Whether one variable of primary interest can be predicted from observations on the others.

Regression analysis concerns the study of relationships between variables with the object of identifying, estimating, and validating the relationship. The estimated relationship can then be used to predict one variable from the value of the other variable(s). In this chapter, we introduce the subject with specific reference to the straight-line model. Chapter 3 treated the subject of fitting a line from a descriptive statistics viewpoint. Here, we take the additional step of including the omnipresent random variation as an error term in the model. Then, on the basis of the model, we can test whether one variable actually influences the other. Further, we produce confidence interval answers when using the estimated straight line for prediction. The correlation coefficient is shown to measure the strength of the linear relationship.

One may be curious about why the study of relationships of variables has been given the rather unusual name "regression." Historically, the word regression was first used in its present technical context by a British scientist, Sir Francis Galton, who analyzed the heights of sons and the average heights of their parents. From his observations, Galton concluded that sons of very tall (short) parents were generally taller (shorter) than the average but not as tall (short) as their parents. This result was published in 1885 under the title "Regression Toward Mediocrity in Hereditary Stature." In this context, "regression toward mediocrity" meant that the sons' heights tended to revert toward the average rather than progress to more extremes. However, in the course of time, the word regression became synonymous with the statistical study of relation among variables.

Studies of relation among variables abound in virtually all disciplines of science and the humanities. We outline just a few illustrative situations in order to bring the object of regression analysis into sharp focus. The examples progress from a case where beforehand there is an underlying straight-line model that is masked by random disturbances to a case where the data may or may not reveal some relationship along a line or curve.

Example 1   A Straight Line Model Masked by Random Disturbances

A factory manufactures items in batches and the production manager wishes to relate the production cost y of a batch to the batch size x. Certain costs are practically constant, regardless of the batch size x. Building costs and

Regression analysis allows us to predict one variable from the value of another variable.
(By permission of Johnny Hart and Field Enterprises, Inc.)

administrative and supervisory salaries are some examples. Let us denote the fixed costs collectively by F. Certain other costs may be directly proportional to the number of units produced. For example, both the raw materials and labor required to produce the product are included in this category. Let C denote the cost of producing one item. In the absence of any other factors, we can then expect to have the relation

$$y = F + Cx$$

In reality, other factors also affect the production cost, often in unpredictable ways. Machines occasionally break down and result in lost time and added expenses for repair. Variation of the quality of the raw materials may also cause occasional slowdown of the production process. Thus, an ideal relation can be masked by random disturbances. Consequently, the relationship between y and x must be investigated by a statistical analysis of the cost and batch-size data.

Example 2   Expect an Increasing Relation But Not Necessarily a Straight Line

Suppose that the yield y of tomato plants in an agricultural experiment is to be studied in relation to the dosage x of a certain fertilizer, while other contributing factors such as irrigation and soil dressing are to remain as constant as possible. The experiment consists of applying different dosages of the fertilizer, over the range of interest, in different plots and then recording the tomato yield from these plots. Different dosages of the fertilizer will typically produce different yields, but the relationship is not expected to follow a precise mathematical formula. Aside from unpredictable chance variations, the underlying form of the relation is not known.

Example 3   A Scatter Diagram May Reveal an Empirical Relation

The aptitude of a newly trained operator for performing a skilled job depends on both the duration of the training period and the nature of the training program. To evaluate the effectiveness of the training program, we must conduct an experimental study of the relation between growth in skill or learning y and duration x of the training. It is too much to expect a precise mathematical relation simply because no two human beings are exactly alike. However, an analysis of the data of the two variables could help us to assess the nature of the relation and utilize it in evaluating a training program.

These examples illustrate the simplest settings for regression analysis where one wishes to determine how one variable is related to one other variable. In more complex situations several variables may be interrelated, or one variable of major interest may depend on several influencing variables. Regression analysis extends to these multivariate problems. (See Section 3, Chapter 12). Even though randomness is omnipresent, regression analysis allows us to identify it and estimate relationships.

2.   REGRESSION WITH A SINGLE PREDICTOR

A regression problem involving a single predictor (also called simple regression) arises when we wish to study the relation between two variables x and y and use it to predict y from x. The variable x acts as an independent variable whose values are controlled by the experimenter. The variable y depends on x and is also subjected to unaccountable variations or errors.

Notation

x = independent variable, also called predictor variable, **causal variable,** or input variable

y = dependent or response variable

For clarity, we introduce the main ideas of regression in the context of a specific experiment. This experiment, described in Example 4, and the data set of Table 1 will be referred to throughout this chapter. By so doing, we provide a flavor of the subject matter interpretation of the various inferences associated with a regression analysis.

Example 4   Relief from Symptoms of Allergy Related to Dosage

In one stage of the development of a new drug for an allergy, an experiment is conducted to study how different dosages of the drug affect the duration of relief from the allergic symptoms. Ten patients are included in the experiment. Each patient receives a specified dosage of the drug and is asked to report back as soon as the protection of the drug seems to wear off. The observations are recorded in Table 1, which shows the dosage x and duration of relief y for the 10 patients.

TABLE 1   Dosage x (in Milligrams) and the Number of Days of Relief y from Allergy for Ten Patients

Dosage x	Duration of Relief y
3	9
3	5
4	12
5	9
6	14
6	16
7	22
8	18
8	24
9	22

Seven different dosages are used in the experiment, and some of these are repeated for more than one patient. A glance at the table shows that y generally increases with x, but it is difficult to say much more about the form of the relation simply by looking at this tabular data.

For a generic experiment, we use n to denote the sample size or the number of runs of the experiment. Each run gives a pair of observations (x, y) in which x is the fixed setting of the independent variable and y denotes the corresponding response. See Table 2.

We always begin our analysis by plotting the data because the eye can easily detect patterns along a line or curve.

TABLE 2   Data Structure
for a Simple Regression

Setting of the Independent Variable	Response
x_1	y_1
x_2	y_2
x_3	y_3
.	.
.	.
.	.
x_n	y_n

First Step in the Analysis

Plotting a scatter diagram is an important preliminary step prior to undertaking a formal statistical analysis of the relationship between two variables.

The scatter diagram of the observations in Table 1 appears in Figure 1. This scatter diagram reveals that the relationship is approximately linear in nature; that is, the points seem to cluster around a straight line. Because a linear relation is the simplest relationship to handle mathematically, we present the details of the statistical regression analysis for this case. Other situations can often be reduced to this case by applying a suitable transformation to one or both variables.

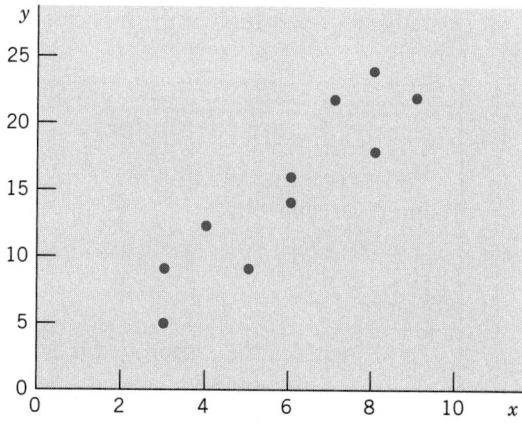

Figure 1   Scatter diagram of the data of Table 1.

3.   A STRAIGHT LINE REGRESSION MODEL

Recall that if the relation between y and x is exactly a straight line, then the variables are connected by the formula

$$y = \beta_0 + \beta_1 x$$

where β_0 indicates the intercept of the line with the y axis and β_1 represents the slope of the line, or the change in y per unit change in x (see Figure 2).

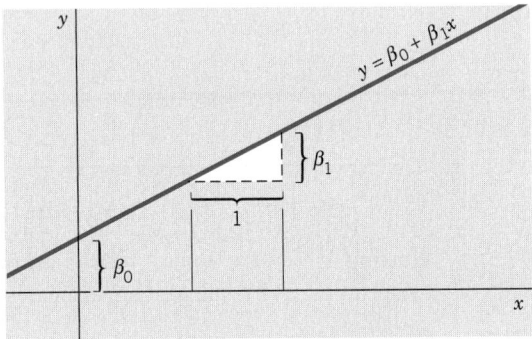

Figure 2   Graph of straight line $y = \beta_0 + \beta_1 x$.

Statistical ideas must be introduced into the study of relation when the points in a scatter diagram do not lie perfectly on a line, as in Figure 1. We think of these data as observations on an underlying linear relation that is being masked by random disturbances or experimental errors due in part to differences in severity of allergy, physical condition of subjects, their environment, and so on. All of variables that influence the response, days of relief, are not even known, yet alone measured. The effects of all these variables are modeled as unobservable random variables. Given this viewpoint, we formulate the following linear regression model as a tentative representation of the mode of relationship between y and x.

Statistical Model for a Straight Line Regression

We assume that the response Y is a random variable that is related to the input variable x by

$$Y_i = \beta_0 + \beta_1 x_i + e_i \qquad i = 1, \ldots, n$$

where:

1. Y_i denotes the response corresponding to the ith experimental run in which the input variable x is set at the value x_i.

2. $e_1, \ldots, e_n$ are the unknown error components that are superimposed on the true linear relation. These are **unobservable random**

variables, which we assume are independently and normally distributed with mean zero and an unknown standard deviation σ.

3. The parameters β_0 and β_1, which together locate the straight line, are unknown.

According to this model, the observation Y_i corresponding to level x_i of the controlled variable is one observation from the normal distribution with mean $\beta_0 + \beta_1 x_i$ and standard deviation σ. One interpretation of this is that as we attempt to observe the true value on the line, nature adds the random error e to this quantity. This statistical model is illustrated in Figure 3, which shows a few normal distributions for the response variable Y for different values of the input variable x. All these distributions have the same standard deviation and their means lie on the unknown true straight line $\beta_0 + \beta_1 x$. Aside from the fact that σ is unknown, the line on which the means of these normal distributions are located is also unknown. In fact, an important objective of the statistical analysis is to estimate this line.

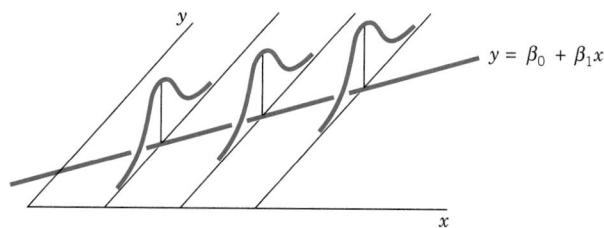

Figure 3   Normal distributions of Y with means on a straight line.

Exercises

11.1   Plot the line $y = 2 + 3x$ on graph paper by locating the points for $x = 1$ and $x = 4$. What is its intercept? What is its slope?

11.2   A store manager has determined that the monthly profit y realized from selling a particular brand of car battery is given by

$$y = 10x - 155$$

where x denotes the number of these batteries sold in a month.

(a)   If 41 batteries were sold in a month, what was the profit?

(b)   At least how many batteries must be sold in a month in order to make a profit?

11.3   Identify the predictor variable x and the response variable y in each of the following situations.

(a)   A training director wishes to study the relationship between the duration of training for new recruits and their performance in a skilled job.

(b)   The aim of a study is to relate the carbon monoxide level in blood samples from smokers with the average number of cigarettes they smoke per day.

(c)   An agronomist wishes to investigate the growth rate of a fungus in relation to the level of humidity in the environment.

(d) A market analyst wishes to relate the expenditures incurred in promoting a product in test markets and the subsequent amount of product sales.

11.4 Identify the values of the parameters β_0, β_1, and σ in the statistical model

$$Y = 2 + 4x + e$$

where e is a normal random variable with mean 0 and standard deviation 5.

11.5 Identify the values of the parameters β_0, β_1, and σ in the statistical model

$$Y = 7 - 5x + e$$

where e is a normal random variable with mean 0 and standard deviation 3.

11.6 Under the linear regression model:

(a) Determine the mean and standard deviation of Y, for $x = 4$, when $\beta_0 = 1$, $\beta_1 = 3$, and $\sigma = 2$.

(b) Repeat part (a) with $x = 2$.

11.7 Under the linear regression model:

(a) Determine the mean and standard deviation of Y, for $x = 1$, when $\beta_0 = 3$, $\beta_1 = -4$, and $\sigma = 4$.

(b) Repeat part (a) with $x = 2$.

11.8 Graph the straight line for the means of the linear regression model

$$Y = \beta_0 + \beta_1 x + e$$

having $\beta_0 = -3$, $\beta_1 = 4$, and the normal random variable e has standard deviation 3.

11.9 Graph the straight line for the means of the linear regression model $Y = \beta_0 + \beta_1 x + e$ having $\beta_0 = 7$ and $\beta_1 = 2$.

11.10 Consider the linear regression model

$$Y = \beta_0 + \beta_1 x + e$$

where $\beta_0 = -2$, $\beta_1 = -1$, and the normal random variable e has standard deviation 3.

(a) What is the mean of the response Y when $x = 3$? When $x = 6$?

(b) Will the response at $x = 3$ always be larger than that at $x = 6$? Explain.

11.11 Consider the following linear regression model

$$Y = \beta_0 + \beta_1 x + e,$$

where $\beta_0 = 4$, $\beta_1 = 3$, and the normal random variable e has the standard deviation 4.

(a) What is the mean of the response Y when $x = 4$? When $x = 5$?

(b) Will the response at $x = 5$ always be larger than that at $x = 4$? Explain.

4. THE METHOD OF LEAST SQUARES

Let us tentatively assume that the preceding formulation of the model is correct. We can then proceed to estimate the regression line and solve a few related inference problems. The problem of estimating the regression parameters β_0 and β_1 can be viewed as fitting the best straight line of the y to x relationship in the scatter diagram. One can draw a line by eyeballing the scatter diagram, but such a judgment may be open to dispute. Moreover, statistical inferences cannot be based on a line that is estimated subjectively. On the other hand, the method of least squares is an objective and efficient method of determining the best fitting straight line. Moreover, this method is quite versatile because its application extends beyond the simple straight line regression model.

Suppose that an arbitrary line $y = b_0 + b_1 x$ is drawn on the scatter diagram as it is in Figure 4. At the value x_i of the independent variable, the y value predicted by this line is $b_0 + b_1 x_i$ whereas the observed value is y_i. The discrepancy

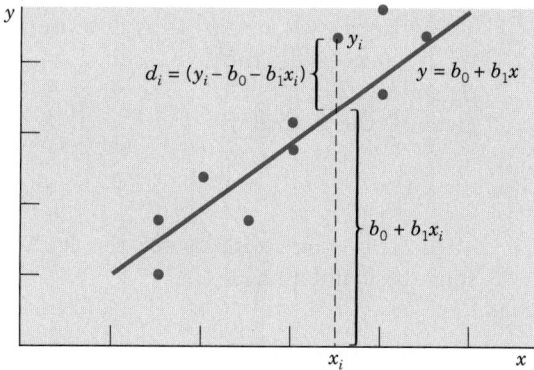

Figure 4   Deviations of the observations from a line
$y = b_0 + b_1 x$.

between the observed and predicted y values is then $y_i - b_0 - b_1 x_i = d_i$, which is the **vertical** distance of the point from the line.

Considering such discrepancies at all the n points, we take

$$D = \sum_{i=1}^{n} d_i^2 = \sum_{i=1}^{n} (y_i - b_0 - b_1 x_i)^2$$

as an overall measure of the discrepancy of the observed points from the trial line $y = b_0 + b_1 x$. The magnitude of D obviously depends on the line that is drawn. In other words, it depends on b_0 and b_1, the two quantities that determine the trial line. A good fit will make D as small as possible. We now state the principle of least squares in general terms to indicate its usefulness to fitting many other models.

The Principle of Least Squares

Determine the values for the parameters so that the overall discrepancy

$$D = \sum (\text{Observed response} - \text{Predicted response})^2$$

is minimized.

The parameter values thus determined are called the least squares estimates.

For the straight line model, the least squares principle involves the determination of b_0 and b_1 to minimize.

$$D = \sum_{i=1}^{n} (y_i - b_0 - b_1 x_i)^2$$

The quantities b_0 and b_1 thus determined are denoted by $\hat{\beta}_0$ and $\hat{\beta}_1$, respectively, and called the least squares estimates of the regression parameters β_0 and β_1. The best fitting straight line or best fitting regression line is then given by the equation

$$\hat{y} = \hat{\beta}_0 + \hat{\beta}_1 x$$

To describe the formulas for the least squares estimators, we first introduce some basic notation.

Basic Notation

$$\bar{x} = \frac{1}{n}\sum x \qquad \bar{y} = \frac{1}{n}\sum y$$

$$S_{xx} = \sum (x - \bar{x})^2 = \sum x^2 - \frac{(\sum x)^2}{n}$$

$$S_{yy} = \sum (y - \bar{y})^2 = \sum y^2 - \frac{(\sum y)^2}{n}$$

$$S_{xy} = \sum (x - \bar{x})(y - \bar{y}) = \sum xy - \frac{(\sum x)(\sum y)}{n}$$

The quantities $\bar{x}$ and $\bar{y}$ are the sample means of the x and y values; S_{xx} and S_{yy} are the sums of squared deviations from the means, and S_{xy} is the sum of the cross products of deviations. These five summary statistics are the key ingredients for calculating the least squares estimates and handling the inference problems associated with the linear regression model. The reader may review Sections 5 and 6 of Chapter 3 where calculations of these statistics were illustrated.

The formulas for the least squares estimators are

Least squares estimator of β_0

$$\hat{\beta}_0 = \bar{y} - \hat{\beta}_1 \bar{x}$$

Least squares estimator of β_1

$$\hat{\beta}_1 = \frac{S_{xy}}{S_{xx}}$$

The estimates $\hat{\beta}_0$ and $\hat{\beta}_1$ can then be used to locate the best fitting line:

Fitted (or estimated) regression line

$$\hat{y} = \hat{\beta}_0 + \hat{\beta}_1 x$$

As we have already explained, this line provides the best fit to the data in the sense that the sum of squares of the deviations, or

$$\sum_{i=1}^{n} (y_i - \hat{\beta}_0 - \hat{\beta}_1 x_i)^2$$

is the smallest.

The individual deviations of the observations y_i from the fitted values $\hat{y}_i = \hat{\beta}_0 + \hat{\beta}_1 x_i$ are called the residuals, and we denote these by $\hat{e}_i$.

Residuals

$$\hat{e}_i = y_i - \hat{\beta}_0 - \hat{\beta}_1 x_i \qquad i = 1, \ldots, n$$

Although some residuals are positive and some negative, a property of the least squares fit is that the **sum of the residuals is always zero.**

In Chapter 12, we will discuss how the residuals can be used to check the assumptions of a regression model. For now, the sum of squares of the residuals is a quantity of interest because it leads to an estimate of the variance σ^2 of the error distributions illustrated in Figure 3. The residual sum of squares is also called the sum of squares due to error and is abbreviated as SSE.

The residual sum of squares or the sum of squares due to error is

$$\text{SSE} = \sum_{i=1}^{n} \hat{e}_i^2 = S_{yy} - \frac{S_{xy}^2}{S_{xx}}$$

The second expression for SSE, which follows after some algebraic manipulations (see Exercise 11.24), is handy for directly calculating SSE. However, we stress the importance of determining the individual residuals for their role in model checking (see Section 4, Chapter 12).

An estimate of variance σ^2 is obtained by dividing SSE by $n - 2$. The reduction by 2 is because two degrees of freedom are lost from estimating the two parameters β_0 and β_1.

Estimate of Variance

The estimator of the error variance σ^2 is

$$S^2 = \frac{\text{SSE}}{n - 2}$$

In applying the least squares method to a given data set, we first compute the basic quantities $\bar{x}, \bar{y}, S_{xx}, S_{yy}$, and S_{xy}. Then the preceding formulas can be used to obtain the least squares regression line, the residuals, and the value of SSE. Computations for the data given in Table 1 are illustrated in Table 3.

TABLE 3   Computations for the Least Squares Line, SSE, and Residuals Using the Data of Table 1

x	y	x^2	y^2	xy	$\hat{\beta}_0 + \hat{\beta}_1 x$	Residual $\hat{e}$
3	9	9	81	27	7.15	1.85
3	5	9	25	15	7.15	−2.15
4	12	16	144	48	9.89	2.11
5	9	25	81	45	12.63	−3.63
6	14	36	196	84	15.37	−1.37
6	16	36	256	96	15.37	.63
7	22	49	484	154	18.11	3.89
8	18	64	324	144	20.85	−2.85
8	24	64	576	192	20.85	3.15
9	22	81	484	198	23.59	−1.59
Total 59	151	389	2651	1003		.04 (rounding error)

$\bar{x} = 5.9, \quad \bar{y} = 15.1$

$\hat{\beta}_1 = \dfrac{112.1}{40.9} = 2.74$

$S_{xx} = 389 - \dfrac{(59)^2}{10} = 40.9$

$\hat{\beta}_0 = 15.1 - 2.74 \times 5.9 = -1.07$

$S_{yy} = 2651 - \dfrac{(151)^2}{10} = 370.9$

$\text{SSE} = 370.9 - \dfrac{(112.1)^2}{40.9} = 63.6528$

$S_{xy} = 1003 - \dfrac{59 \times 151}{10} = 112.1$

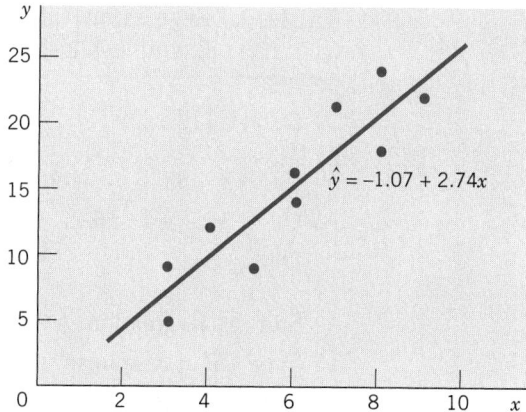

Figure 5   The least squares regression line for the
data given in Table 1.

The equation of the line fitted by the least squares method is then

$$\hat{y} = -1.07 + 2.74x$$

Figure 5 shows a plot of the data along with the fitted regression line.
The residuals $\hat{e}_i = y_i - \hat{y}_i = y_i + 1.07 - 2.74x_i$ are computed in the
last column of Table 3. The sum of squares of the residuals is

$$\sum_{i=1}^{n} \hat{e}_i^2 = (1.85)^2 + (-2.15)^2 + (2.11)^2 + \cdots + (-1.59)^2 = 63.653$$

which agrees with our previous calculations of SSE, except for the error due to
rounding. Theoretically, the sum of the residuals should be zero, and the difference between the sum .04 and zero is also due to rounding.
The estimate of the variance σ^2 is

$$s^2 = \frac{SSE}{n-2} = \frac{63.6528}{8} = 7.96$$

The calculations involved in a regression analysis become increasingly
tedious with larger data sets. Access to a computer proves to be a considerable
advantage. Table 4 illustrates a part of the computer-based analysis of linear
regression using the data of Example 4 and the MINITAB package. For a more
complete regression analysis, see Table 5 in Section 6.4.

TABLE 4   Regression Analysis of the Data in Table 1, Example 4, Using MINITAB

Data: C11T3 DAT

C1: 3 3 4 5 6 6 7 8 8 9

C2: 9 5 12 9 14 16 22 18 24 22

Dialog box:

Stat > Regression > Regression

Type C2 in Response

Type C1 in Predictors. Click OK.

Output:

Regression Analysis

The regression equation is
$y = -1.07 + 2.74x$

Exercises

11.12 Given the five pairs of (x, y) values

x	0	1	6	3	5
y	6	5	2	4	3

(a) Construct a scatter diagram.

(b) Calculate $\bar{x}, \bar{y}, S_{xx}, S_{xy}$, and S_{yy}.

(c) Calculate the least squares estimates $\hat{\beta}_0$ and $\hat{\beta}_1$.

(d) Determine the fitted line and draw the line on the scatter diagram.

11.13 Given these six pairs of (x, y) values,

x	1	2	3	3	4	5
y	8	4	5	2	2	0

(a) Plot the scatter diagram.

(b) Calculate $\bar{x}, \bar{y}, S_{xx}, S_{xy}$, and S_{yy}.

(c) Calculate the least squares estimates $\hat{\beta}_0$ and $\hat{\beta}_1$.

(d) Determine the fitted line and draw the line on the scatter diagram.

11.14 Refer to Exercise 11.12.

(a) Find the residuals and verify that they sum to zero.

(b) Calculate the residual sum of squares SSE by

   (i) Adding the squares of the residuals.

   (ii) Using the formula
   $$SSE = S_{yy} - S_{xy}^2 / S_{xx}$$

(c) Obtain the estimate of σ^2.

11.15 Refer to Exercise 11.13.

(a) Find the residuals and verify that they sum to zero.

(b) Calculate the residual sums of squares SSE by

   (i) Adding the squares of the residuals.

   (ii) Using the formula
   $$SSE = S_{yy} - S_{xy}^2 / S_{xx}$$

(c) Obtain the estimate of σ^2.

11.16 Given the five pairs of (x, y) values

x	0	1	2	3	4
y	3	2	5	6	9

(a) Calculate $\bar{x}, \bar{y}, S_{xx}, S_{xy}$, and S_{yy}.
(b) Calculate the least squares estimates $\hat{\beta}_0$ and $\hat{\beta}_1$.
(c) Determine the fitted line.

11.17 Given the five pairs of (x, y) values

x	0	2	4	6	8
y	4	3	6	8	9

(a) Calculate $\bar{x}, \bar{y}, S_{xx}, S_{xy}$, and S_{yy}.
(b) Calculate the least squares estimates $\hat{\beta}_0$ and $\hat{\beta}_1$.
(c) Determine the fitted line.

11.18 Computing from a data set of (x, y) values, we obtained the following summary statistics.

$$n = 14 \qquad \bar{x} = 3.5 \qquad \bar{y} = 2.84$$
$$S_{xx} = 10.82 \quad S_{xy} = 2.677 \quad S_{yy} = 1.125$$

(a) Obtain the equation of the best fitting straight line.
(b) Calculate the residual sum of squares.
(c) Estimate σ^2.

11.19 Computing from a data set of (x, y) values, we obtained the following summary statistics.

$$n = 20 \qquad \bar{x} = 1.4 \qquad \bar{y} = 5.2$$
$$S_{xx} = 28.2 \quad S_{xy} = 3.25 \quad S_{yy} = 2.01$$

(a) Obtain the equation of the best fitting straight line.
(b) Calculate the residual sum of squares.
(c) Estimate σ^2.

11.20 The data on female wolves in Table D.9 of the Data Bank concerning body weight (lb) and body length (cm) are

Weight	57	84	90	71	77	68	73
Body length	123	129	143	125	122	125	122

(a) Obtain the least squares fit of body weight to the predictor body length.
(b) Calculate the residual sum of squares.
(c) Estimate σ^2.

11.21 Refer to the data on female wolves in Exercise 11.20.

(a) Obtain the least squares fit of body length to the predictor body weight.
(b) Calculate the residual sum of squares.
(c) Estimate σ^2.
(d) Compare your answer in part (a) with your answer to part (a) of Exercise 11.20. Should the two answers be the same? Why or why not?

11.22 Using the formulas of $\hat{\beta}_1$ and SSE, show that SSE can also be expressed as

(a) $\text{SSE} = S_{yy} - \hat{\beta}_1 S_{xy}$
(b) $\text{SSE} = S_{yy} - \hat{\beta}_1^2 S_{xx}$

11.23 Referring to the formulas of $\hat{\beta}_0$ and $\hat{\beta}_1$, show that the point $(\bar{x}, \bar{y})$ lies on the fitted regression line.

11.24 To see why the residuals always sum to zero, refer to the formulas of $\hat{\beta}_0$ and $\hat{\beta}_1$ and verify that

(a) The predicted values are $\hat{y}_i = \bar{y} + \hat{\beta}_1 (x_i - \bar{x})$.
(b) The residuals are

$$\hat{e}_i = y_i - \hat{y}_i = (y_i - \bar{y}) - \hat{\beta}_1(x_i - \bar{x})$$

Then show that $\displaystyle\sum_{i=1}^{n} \hat{e}_i = 0$.

(c) Verify that $\displaystyle\sum_{i=1}^{n} \hat{e}_i^2 = S_{yy} + \hat{\beta}_1^2 S_{xx} -$
$$2\hat{\beta}_1 S_{xy} = S_{yy} - S_{xy}^2 / S_{xx}.$$

5. THE SAMPLING VARIABILITY OF THE LEAST SQUARES ESTIMATORS — TOOLS FOR INFERENCE

It is important to remember that the line $\hat{y} = \hat{\beta}_0 + \hat{\beta}_1 x$ obtained by the principle of least squares is an **estimate** of the unknown true regression line $y = \beta_0 + \beta_1 x$. In our drug evaluation problem (Example 4), the estimated line is

$$\hat{y} = -1.07 + 2.74x$$

Its slope $\hat{\beta}_1 = 2.74$ suggests that the mean duration of relief increases by 2.74 days for each unit dosage of the drug. Also, if we were to estimate the expected duration of relief for a specified dosage $x^* = 4.5$ milligrams, we would naturally use the fitted regression line to calculate the estimate $-1.07 + 2.74 \times 4.5 = 11.26$ days. A few questions concerning these estimates naturally arise at this point.

1. In light of the value 2.74 for $\hat{\beta}_1$, could the slope β_1 of the true regression line be as much as 4? Could it be zero so that the true regression line is $y = \beta_0$, which does not depend on x? What are the plausible values for β_1?

2. How much uncertainty should be attached to the estimated duration of 11.26 days corresponding to the given dosage $x^* = 4.5$?

To answer these and related questions, we must know something about the sampling distributions of the least squares estimators. These sampling distributions will enable us to test hypotheses and set confidence intervals for the parameters β_0 and β_1 that determine the straight line and for the straight line itself. Again, the t distribution is relevant.

---

1. The standard deviations (also called standard errors) of the least squares estimators are

$$\text{S.E.}(\hat{\beta}_1) = \frac{\sigma}{\sqrt{S_{xx}}} \qquad \text{S.E.}(\hat{\beta}_0) = \sigma \sqrt{\frac{1}{n} + \frac{\bar{x}^2}{S_{xx}}}$$

To estimate the standard error, use

$$S = \sqrt{\frac{\text{SSE}}{n-2}} \qquad \text{in place of } \sigma$$

---

2. Inferences about the slope β_1 are based on the t distribution

$$T = \frac{\hat{\beta}_1 - \beta_1}{S/\sqrt{S_{xx}}} \qquad \text{d.f.} = n - 2$$

Inferences about the intercept β_0 are based on the t distribution

$$T = \frac{\hat{\beta}_0 - \beta_0}{S\sqrt{\dfrac{1}{n} + \dfrac{\bar{x}^2}{S_{xx}}}} \qquad \text{d.f.} = n - 2$$

---

3. At a specified value $x = x^*$, the expected response is $\beta_0 + \beta_1 x^*$. This is estimated by $\hat{\beta}_0 + \hat{\beta}_1 x^*$ with

    Estimated standard error

    $$S\sqrt{\frac{1}{n} + \frac{(x^* - \bar{x})^2}{S_{xx}}}$$

    Inferences about $\beta_0 + \beta_1 x^*$ are based on the t distribution

    $$T = \frac{(\hat{\beta}_0 + \hat{\beta}_1 x^*) - (\beta_0 + \beta_1 x^*)}{S\sqrt{\dfrac{1}{n} + \dfrac{(x^* - \bar{x})^2}{S_{xx}}}} \qquad \text{d.f.} = n - 2$$

6. IMPORTANT INFERENCE PROBLEMS

We are now prepared to test hypotheses, construct confidence intervals, and make predictions in the context of straight line regression.

6.1. INFERENCE CONCERNING THE SLOPE β_1

In a regression analysis problem, it is of special interest to determine whether the expected response does or does not vary with the magnitude of the input variable x. According to the linear regression model,

$$\text{Expected response} = \beta_0 + \beta_1 x$$

This does not change with a change in x if and only if $\beta_1 = 0$. We can therefore test the null hypothesis $H_0: \beta_1 = 0$ against a one- or a two-sided alternative, depending on the nature of the relation that is anticipated. If we refer to the boxed statement (2) of Section 5, the null hypothesis $H_0: \beta_1 = 0$ is to be tested using the test statistic

$$T = \frac{\hat{\beta}_1}{S/\sqrt{S_{xx}}} \qquad \text{d.f.} = n - 2$$

Example 5   A Test to Establish That Duration of Relief Increases with Dosage

Do the data given in Table 1 constitute strong evidence that the mean duration of relief increases with higher dosages of the drug?

SOLUTION   For an increasing relation, we must have $\beta_1 > 0$. Therefore, we are to test the null hypothesis $H_0: \beta_1 = 0$ versus the one-sided alternative $H_1: \beta_1 > 0$. We select $\alpha = .05$. Since $t_{.05} = 1.860$, with d.f. $= 8$ we set the rejection region $R: T \geq 1.860$. Using the calculations that follow Table 3, we have

$$\hat{\beta}_1 = 2.74$$

$$s^2 = \frac{SSE}{n-2} = \frac{63.6528}{8} = 7.9566, \qquad s = 2.8207$$

$$\text{Estimated S.E.}(\hat{\beta}_1) = \frac{s}{\sqrt{S_{xx}}} = \frac{2.8207}{\sqrt{40.90}} = .441$$

$$\text{Test statistic} \qquad t = \frac{2.74}{.441} = 6.213$$

The observed t value is in the rejection region, so H_0 is rejected. Moreover, 6.213 is much larger than $t_{.005} = 3.355$, so the P–value is much smaller than .005.

A computer calculation gives $P[T > 6.213] = .0001$. There is strong evidence that larger dosages of the drug tend to increase the duration of relief over the range covered in the study.

A warning is in order here concerning the interpretation of the test of $H_0: \beta_1 = 0$. If H_0 is not rejected, we may be tempted to conclude that y does not depend on x. Such an unqualified statement may be erroneous. First, the absence of a linear relation has only been established over the range of the x values in the experiment. It may be that x was just not varied enough to influence y. Second, the interpretation of lack of dependence on x is valid only if our model formulation is correct. If the scatter diagram depicts a relation on a curve but we inadvertently formulate a linear model and test $H_0: \beta_1 = 0$, the conclusion that H_0 is not rejected should be interpreted to mean "no linear relation," rather than "no relation." We elaborate on this point further in Section 7. Our present viewpoint is to assume that the model is correctly formulated and discuss the various inference problems associated with it.

More generally, we may test whether or not β_1 is equal to some specified value β_{10}, not necessarily zero.

The **test of the null hypothesis**

$$H_0: \beta_1 = \beta_{10}$$

is based on

$$T = \frac{\hat{\beta}_1 - \beta_{10}}{S/\sqrt{S_{xx}}} \qquad \text{d.f.} = n - 2$$

In addition to testing hypotheses, we can provide a confidence interval for the parameter β_1 using the t distribution.

A $100(1 - \alpha)\%$ **confidence interval** for β_1 is

$$\left(\hat{\beta}_1 - t_{\alpha/2} \frac{S}{\sqrt{S_{xx}}}, \qquad \hat{\beta}_1 + t_{\alpha/2} \frac{S}{\sqrt{S_{xx}}} \right)$$

where $t_{\alpha/2}$ is the upper $\alpha/2$ point of the t distribution with d.f. $= n - 2$.

Example 6 A Confidence Interval for β_1

Construct a 95% confidence interval for the slope of the regression line in reference to the data of Table 1.

SOLUTION   In Example 5, we found that $\hat{\beta}_1 = 2.74$ and $s/\sqrt{S_{xx}} = .441$. The required confidence interval is given by

$$2.74 \pm 2.306 \times .441 = 2.74 \pm 1.02 \qquad \text{or} \qquad (1.72, 3.76)$$

We are 95% confident that by adding one extra milligram to the dosage, the mean duration of relief would increase somewhere between 1.72 and 3.76 days.

6.2. INFERENCE ABOUT THE INTERCEPT β_0

Although somewhat less important in practice, inferences similar to those outlined in Section 6.1 can be provided for the parameter β_0. The procedures are again based on the t distribution with d.f. $= n - 2$, stated for $\hat{\beta}_0$ in Section 5. In particular,

---

A $100(1 - \alpha)\%$ **confidence interval** for β_0 is

$$\left(\hat{\beta}_0 - t_{\alpha/2} S \sqrt{\frac{1}{n} + \frac{\bar{x}^2}{S_{xx}}}, \qquad \hat{\beta}_0 + t_{\alpha/2} S \sqrt{\frac{1}{n} + \frac{\bar{x}^2}{S_{xx}}} \right)$$

To illustrate this formula, let us consider the data of Table 1. In Table 3, we have found $\hat{\beta}_0 = -1.07$, $\bar{x} = 5.9$, and $S_{xx} = 40.9$. Also, $s = 2.8207$. Therefore, a 95% confidence interval for β_0 is calculated as

$$-1.07 \pm 2.306 \times 2.8207 \sqrt{\frac{1}{10} + \frac{(5.9)^2}{40.9}}$$

$$= -1.07 \pm 6.34 \quad \text{or} \quad (-7.41, 5.27)$$

Note that β_0 represents the mean response corresponding to the value 0 for the input variable x. In the drug evaluation problem of Example 4, the parameter β_0 is of little practical interest because the range of x values covered in the experiment was 3 to 9 and it would be unrealistic to extend the line to $x = 0$. In fact, the estimate $\hat{\beta}_0 = -1.07$ does not have an interpretation as a (time) duration of relief.

6.3. ESTIMATION OF THE MEAN RESPONSE FOR A SPECIFIED x VALUE

Often, the objective in a regression study is to employ the fitted regression in estimating the expected response corresponding to a specified level of the input variable. For example, we may want to estimate the expected duration of relief for a specified dosage x^* of the drug. According to the linear model described in Section 3, the expected response at a value x^* of the input variable x is given by $\beta_0 + \beta_1 x^*$. The expected response is estimated by $\hat{\beta}_0 + \hat{\beta}_1 x^*$ which is the ordinate of the fitted regression line at $x = x^*$. Referring to statement (3) of Section 5, we determine that the t distribution can be used to construct confidence intervals or test hypotheses.

A $100(1 - \alpha)\%$ **confidence interval for the expected response** $\beta_0 + \beta_1 x^*$ is

$$\hat{\beta}_0 + \hat{\beta}_1 x^* \pm t_{\alpha/2} S \sqrt{\frac{1}{n} + \frac{(x^* - \bar{x})^2}{S_{xx}}}$$

To **test the hypothesis** that $\beta_0 + \beta_1 x^* = \mu_0$, some specified value, we use

$$T = \frac{\hat{\beta}_0 + \hat{\beta}_1 x^* - \mu_0}{S \sqrt{\frac{1}{n} + \frac{(x^* - \bar{x})^2}{S_{xx}}}} \qquad \text{d.f.} = n - 2$$

Example 7   A Confidence Interval for the Expected Duration of Relief

Again consider the data given in Table 1 and the calculations for the regression analysis given in Table 3. Obtain a 95% confidence interval for the expected duration of relief when the dosage is (a) $x^* = 6$ and (b) $x^* = 9.5$.

SOLUTION   The fitted regression line is

$$\hat{y} = -1.07 + 2.74x$$

The expected duration of relief corresponding to the dosage $x^* = 6$ milligrams of the drug is estimated as

$$\hat{\beta}_0 + \hat{\beta}_1 x^* = -1.07 + 2.74 \times 6 = 15.37 \text{ days}$$

$$\text{Estimated standard error} = s\sqrt{\frac{1}{10} + \frac{(6 - 5.9)^2}{40.9}}$$

$$= 2.8207 \times .3166 = .893$$

A 95% confidence interval for the mean duration of relief with the dosage $x^* = 6$ is therefore

$$15.37 \pm t_{.025} \times .893 = 15.37 \pm 2.306 \times .893$$
$$= 15.37 \pm 2.06 \quad \text{or} \quad (13.31, 17.43)$$

We are 95% confident that 6 milligrams of the drug produces an average duration of relief that is between about 13.3 and 17.4 days.

Suppose that we also wish to estimate the mean duration of relief under the dosage $x^* = 9.5$. We follow the same steps to calculate the point estimate.

$$\hat{\beta}_0 + \hat{\beta}_1 x^* = -1.07 + 2.74 \times 9.5 = 24.96 \text{ days}$$

$$\text{Estimated standard error} = 2.8207\sqrt{\frac{1}{10} + \frac{(9.5 - 5.9)^2}{40.9}}$$

$$= 1.821$$

A 95% confidence interval is

$$24.96 \pm 2.306 \times 1.821 = 24.96 \pm 4.20 \quad \text{or} \quad (20.76, 29.16)$$

The formula for the standard error shows that when x^* is close to $\bar{x}$, the standard error is smaller than it is when x^* is far removed from $\bar{x}$. This is confirmed by Example 7, where the standard error at $x^* = 9.5$ can be seen to be more than twice as large as the value at $x^* = 6$. Consequently, the confidence interval for the former is also wider. In general, estimation is more precise near the mean $\bar{x}$ than it is for values of the x variable that lie far from the mean.

Caution: Extreme caution should be exercised in extending a fitted regression line to make long-range predictions far away from the range of *x* values covered in the experiment. Not only does the confidence interval become so wide that predictions based on it can be extremely unreliable, but an even greater danger exists. If the pattern of the relationship between the variables changes drastically at a distant value of *x*, the data provide no information with which to detect such a change. Figure 6 illustrates this situation. We would observe a good linear relationship if we experimented with *x* values in the 5 to 10 range, but if the fitted line were extended to estimate the response at $x^* = 20$, then our estimate would drastically miss the mark.

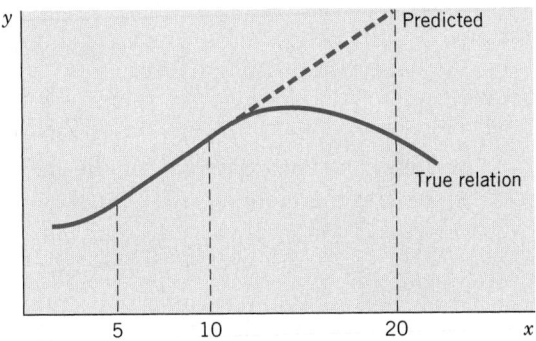

Figure 6   Danger in long-range prediction.

6.4.   PREDICTION OF A SINGLE RESPONSE FOR A SPECIFIED *x* VALUE

Suppose that we give a specified dosage x^* of the drug to a **single** patient and we want to predict the duration of relief from the symptoms of allergy. This problem is different from the one considered in Section 6.3, where we were interested in estimating the mean duration of relief for the population of **all** patients given the dosage x^*. The prediction is still determined from the fitted line; that is, the predicted value of the response is $\hat{\beta}_0 + \hat{\beta}_1 x^*$ as it was in the preceding case. However, the standard error of the prediction here is larger, because a single observation is more uncertain than the mean of the population distribution. We now give the formula of the estimated standard error for this case.

The **estimated standard error when predicting a single observation** *y* at a given x^* is

$$S\sqrt{1 + \frac{1}{n} + \frac{(x^* - \bar{x})^2}{S_{xx}}}$$

The formula for the confidence interval must be modified accordingly. We call the resulting interval a **prediction** interval because it pertains to a future observation.

Example 8   Calculating a Prediction Interval for a Future Trial

Once again, consider the drug trial data given in Table 1. A new trial is to be made on a single patient with the dosage $x^* = 6.5$ milligrams. Predict the duration of relief and give a 95% prediction interval for the duration of relief.

SOLUTION   The predicted duration of relief is

$$\hat{\beta}_0 + \hat{\beta}_1 x^* = -1.07 + 2.74 \times 6.5 = 16.74 \text{ days}$$

Since $t_{.025} = 2.306$ with d.f. $= 8$, a 95% prediction interval for the new patient's duration of relief is

$$16.74 \pm 2.306 \times 2.8207 \sqrt{1 + \frac{1}{10} + \frac{(6.5 - 5.9)^2}{40.9}}$$

$$= 16.74 \pm 6.85 \quad \text{or} \quad (9.89, 23.59)$$

This means we are 95% confident that this particular patient will have relief from symptoms of allergy for about 9.9 to 23.6 days.

In the preceding discussion, we have used the data of Example 4 to illustrate the various inferences associated with a straight-line regression model. Example 9 gives applications to a different data set.

Example 9   Fitting a Straight Line Relation of Skill to the Amount of Training

In a study to determine how the skill in doing a complex assembly job is influenced by the amount of training, 15 new recruits were given varying amounts of training ranging between 3 and 12 hours. After the training, their times to perform the job were recorded. After denoting $x =$ duration of training (in hours) and $y =$ time to do the job (in minutes), the following summary statistics were calculated.

$$\bar{x} = 7.2 \quad S_{xx} = 33.6 \quad S_{xy} = -57.2$$
$$\bar{y} = 45.6 \quad S_{yy} = 160.2$$

(a)   Determine the equation of the best fitting straight line.

(b)   Do the data substantiate the claim that the job time decreases with more hours of training?

(c)   Estimate the mean job time for 9 hours of training and construct a 95% confidence interval.

(d)   Find the predicted y for $x = 35$ hours and comment on the result.

SOLUTION   Using the summary statistics we find:

(a)   The least squares estimates are

$$\hat{\beta}_1 = \frac{S_{xy}}{S_{xx}} = \frac{-57.2}{33.6} = -1.702$$

$$\hat{\beta}_0 = \bar{y} - \hat{\beta}_1\bar{x} = 45.6 - (-1.702) \times 7.2 = 57.85$$

So, the equation of the fitted line is

$$\hat{y} = 57.85 - 1.702x$$

(b)   To answer this question, we are to test $H_0 : \beta_1 = 0$ versus $H_1 : \beta_1 < 0$. The test statistic is

$$T = \frac{\hat{\beta}_1}{S/\sqrt{S_{xx}}}$$

We select $\alpha = .01$. Since $t_{.01} = 2.650$ with d.f. $= 13$, we set the left-sided rejection region $R : T \leq -2.650$. We calculate

$$SSE = S_{yy} - \frac{S_{xy}^2}{S_{xx}} = 160.2 - \frac{(-57.2)^2}{33.6} = 62.824$$

$$s = \sqrt{\frac{SSE}{n - 2}} = \sqrt{\frac{62.824}{13}} = 2.198$$

$$\text{Estimated S.E.} (\hat{\beta}_1) = \frac{s}{\sqrt{S_{xx}}} = \frac{2.198}{\sqrt{33.6}} = .379$$

The t statistic has the value

$$t = \frac{-1.702}{.379} = -4.49$$

Since the observed $t = -4.49$ is less than -2.650, H_0 is rejected with $\alpha = .01$. The P-value is smaller than .010. (See Figure 7.)

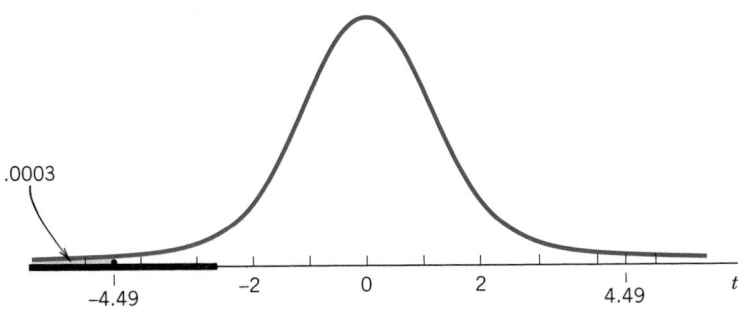

Figure 7 P-value $= .0003$ for one-sided test.

A computer calculation gives

$$P[T \leq -4.49] = .0003$$

We conclude that increasing the duration of training significantly reduces the mean job time within the range covered in the experiment.

(c) The expected job time corresponding to $x^* = 9$ hours is estimated as

$$\hat{\beta}_0 + \hat{\beta}_1 x^* = 57.85 + (-1.702) \times 9$$
$$= 42.53 \text{ minutes}$$

and its

$$\text{Estimated S.E.} = s \sqrt{\frac{1}{15} + \frac{(9 - 7.2)^2}{33.6}} = .888$$

Since $t_{.025} = 2.160$ with d.f. $= 13$, the required confidence interval is

$$42.53 \pm 2.160 \times .888 = 42.53 \pm 1.92 \qquad \text{or} \qquad (40.6, 44.5) \text{ minutes}$$

(d) Since $x = 35$ hours is far beyond the experimental range of 3 to 12 hours, it is not sensible to predict y at $x = 35$ using the fitted regression line. Here a formal calculation gives

$$\text{Predicted job time} = 57.85 - 1.702 \times 35$$
$$= -1.72 \text{ minutes}$$

which is a nonsensical result.

Regression analyses are most conveniently done on a computer. A more complete selection of the output from the computer software package MINITAB, for the data in Example 4, is given in Table 5.

TABLE 5 MINITAB Computer Output for the Data in Example 4

```
THE REGRESSION EQUATION IS
Y  =  -1.07  +  2.74X
```

PREDICTOR	COEF	STDEV	T-RATIO	P
CONSTANT	-1.071	2.751	-0.39	0.707
X	2.7408	0.4411	6.21	0.000

```
S  =  2.821   R-SQ  =  82.8%
```

ANALYSIS OF VARIANCE

SOURCE	DF	SS	MS	F	P
REGRESSION	1	307.25	307.25	38.62	0.000
ERROR	8	63.65	7.96		
TOTAL	9	370.90			

The output of the computer software package SAS for the data in Example 4 is given in Table 6. Notice the similarity of information in Tables 5 and 6. Both include the least squares estimates of the coefficients, their estimated standard deviations, and the t test for testing that the coefficient is zero. The estimate of σ^2 is presented as the mean square error in the analysis of variance table.

TABLE 6   SAS Computer Output for the Data in Example 4

MODEL: MODEL 1
DEPENDENT VARIABLE: Y

ANALYSIS OF VARIANCE

SOURCE	DF	SUM OF SQUARES	MEAN SQUARE	F VALUE	PROB > F
MODEL	1	307.24719	307.24719	38.615	0.0003
ERROR	8	63.65281	7.95660		
C TOTAL	9	370.90000			

ROOT MSE	2.82074	R-SQUARE	0.8284

PARAMETER ESTIMATES

VARIABLE	DF	PARAMETER ESTIMATE	STANDARD ERROR	T FOR HO: PARAMETER = 0	PROB > |T|
INTERCEP	1	−1.070905	2.75091359	−0.389	0.7072
X1	1	2.740831	0.44106455	6.214	0.0003

Example 10   Predicting the Number of Situps after a Semester of Conditioning

Refer to the physical fitness data in Table D.5 of the Data Bank. Using the data on numbers of situps:

(a) Find the least squares fitted line to predict the posttest number of situps from the pretest number at the start of the conditioning class.

(b) Find a 95% confidence interval for the mean number of posttest situps for persons who can perform 35 situps in the pretest. Also find a 95% prediction interval for the number of posttest situps that will be performed by a new person this semester who does 35 situps in the pretest.

(c) Repeat part (b), but replace the number of pretest situps with 20.

SOLUTION   The scatter plot in Figure 8 suggests that a straight line may model the expected value of posttest situps given the number of pretest situps. Here x is the number of pretest situps and y is the number of posttest situps. We use MINITAB statistical software to obtain the output

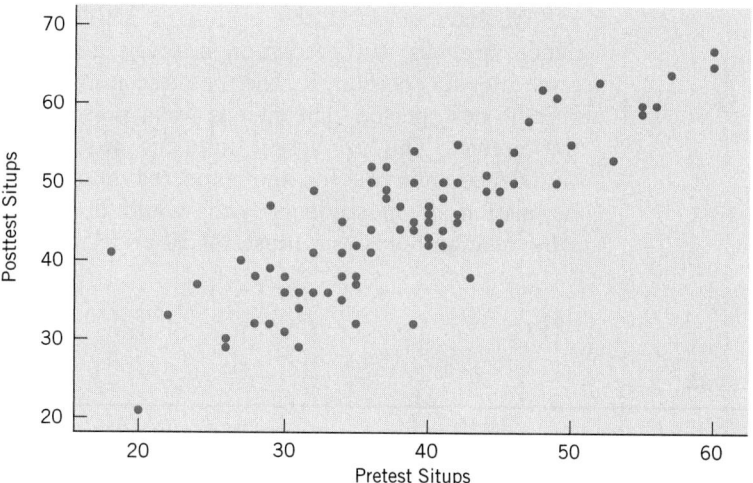

Figure 8   Scatter plot of number of situps.

Regression Analysis: Post Situps versus Pre Situps

```
The regression equation is
Post Situps = 10.3  +  0.899 Pre Situps

Predictor          Coef    SE Coef        T        P
Constant         10.331      2.533     4.08    0.000
Pre Situps      0.89904    0.06388    14.07    0.000


S  =   5.17893     R-Sq  =  71.5%      R-Sq(adj)   =   71.1%
```

Analysis of Variance

```
Source              DF         SS        MS        F        P
Regression           1     5312.9    5312.9   198.09    0.000
Residual Error      79     2118.9      26.8
Total               80     7431.8
```

Predicted Values for New Observations

```
New
Obs   Pre Sit     Fit  SE Fit        95% CI               95% PI
1        35.0  41.797   0.620  (40.563, 43.032) (31.415, 52.179)
2        20.0  28.312   1.321  (25.682, 30.941) (17.673, 38.950)
```

From the output $\hat{y} = \hat{\beta}_0 + \hat{\beta}_1 x = 10.3 + 0.899x$ and $s^2 = (5.1789)^2 = 26.8$ is the estimate of σ^2.

We have selected the option in MINITAB to obtain the two confidence intervals and prediction intervals given in the output. The prediction intervals pertain to the posttest number of situps performed by a specific new person. The first is for a person who performed 35 situps in the pretest. The prediction intervals are wider than the corresponding confidence intervals for the expected number of posttest situps for the population of all students who would do 35 situps in the pretest. The same relation holds, as it must, for 20 pretest situps.

Exercises

11.25 Given the five pairs of (x, y) values

x	0	1	6	3	5
y	5	4	1	3	2

(a) Calculate the least squares estimates $\hat{\beta}_0$ and $\hat{\beta}_1$. Also estimate the error variance σ^2.

(b) Test $H_0: \beta_1 = 0$ versus $H_1: \beta_1 \neq 0$ with $\alpha = .05$.

(c) Estimate the expected y value corresponding to $x = 2.5$ and give a 90% confidence interval.

11.26 Refer to Exercise 11.25. Construct a 90% confidence interval for the intercept β_0.

11.27 Refer to Exercise 11.25. Obtain a 95% confidence interval for β_1.

11.28 Given these five pairs of (x, y) values,

x	2	3	4	5	6
y	.9	2.1	2.4	3.3	3.8

(a) Calculate the least squares estimates $\hat{\beta}_0$ and $\hat{\beta}_1$. Also estimate the error variance σ^2.

(b) Test $H_0: \beta_1 = 1$ versus $H_1: \beta_1 \neq 1$ with $\alpha = .05$.

(c) Estimate the expected y value corresponding to $x = 3.5$ and give a 95% confidence interval.

(d) Construct a 90% confidence interval for the intercept β_0.

11.29 For a random sample of seven homes that are recently sold in a city suburb, the assessed values x and the selling prices y are

($1000)		($1000)	
x	y	x	y
183.5	188.0	210.2	211.0
190.0	191.2	194.6	199.0
170.5	176.2	220.0	218.0
200.8	207.0		

(a) Plot the scatter diagram.

(b) Determine the equation of the least squares regression line and draw this line on the scatter diagram.

(c) Construct a 95% confidence interval for the slope of the regression line.

11.30 Refer to the data in Exercise 11.29.

(a) Estimate the expected selling price of homes that were assessed at $190,000 and construct a 95% confidence interval.

(b) For a single home that was assessed at $190,000, give a 95% prediction interval for the selling price.

11.31 In an experiment designed to determine the relationship between the doses of a compost fertilizer x and the yield of a crop y, the following summary statistics are recorded:

$$n = 15 \qquad \bar{x} = 1.1 \qquad \bar{y} = 4.6$$
$$S_{xx} = 4.2 \qquad S_{yy} = 12.2 \qquad S_{xy} = 6.7$$

Assume a linear relationship.

(a) Find the equation of the least squares regression line.

(b) Compute the error sum of squares and estimate σ^2.

(c) Do the data establish the experimenter's conjecture that, over the range of x values covered in the study, the average increase in yield per unit increase in the compost dose is more than 1.3?

11.32 Refer to Exercise 11.31.

(a) Construct a 95% confidence interval for the expected yield corresponding to $x = 1.2$.

(b) Construct a 95% confidence interval for the expected yield corresponding to $x = 1.5$.

11.33 According to the computer output in Table 7:

(a) What model is fitted?

(b) Test, with $\alpha = .05$, if the x term is needed in the model.

11.34 According to the computer output in Table 7:

(a) Predict the mean response when $x = 5000$.

(b) Find a 90% confidence interval for the mean response when $x = 5000$. You will need the additional information $n = 30$, $\bar{x} = 8354$, and $\Sigma(x_i - \bar{x})^2 = 97,599,296$.

11.35 According to the computer output in Table 8 on page 460:

(a) What model is fitted?

(b) Test, with $\alpha = .05$, if the x term is needed in the model.

11.36 According to the computer output in Table 8:

(a) Predict the mean response when $x = 3$.

(b) Find a 90% confidence interval for the mean response when $x = 3$. You will need the additional information $n = 25$, $\bar{x} = 1.793$, and $\Sigma(x_i - \bar{x})^2 = 1.848$.

(c) Find a 90% confidence interval for the mean response when $x = 2$.

11.37 Consider the data on male wolves in Table D.9 of the Data Bank concerning age (years) and canine length (mm).

(a) Obtain the least squares fit of canine length to the predictor age.

(b) Test $H_0: \beta_1 = 0$ versus $H_1: \beta_1 \neq 0$ with $\alpha = .05$.

(c) Obtain a 90% confidence interval for the canine length when age is $x = 4$.

(d) Obtain a 90% prediction interval for the canine length of an individual wolf when the age is $x = 4$.

TABLE 7 Computer Output for Exercises 11.33 and 11.34

```
THE REGRESSION EQUATION IS
Y = 994 + 0.104X
```

PREDICTOR	COEF	STDEV	T-RATIO	P
CONSTANT	994.0	254.7	3.90	0.001
X	0.10373	0.02978	3.48	0.002

```
S = 299.4    R-SQ = 30.2%
```

ANALYSIS OF VARIANCE

SOURCE	DF	SS	MS	F	P
REGRESSION	1	1087765	1087765	12.14	0.002
ERROR	28	2509820	89636		
TOTAL	29	3597585			

TABLE 8   Computer Output for Exercises 11.35 and 11.36

```
THE REGRESSION EQUATION IS
Y  =  0.338  +  0.831X
```

PREDICTOR	COEF	STDEV	T-RATIO	P
CONSTANT	0.3381	0.1579	2.14	0.043
X	0.83099	0.08702	9.55	0.000

```
S  =  0.1208    R-SQ  =  79.9%
```

ANALYSIS OF VARIANCE

SOURCE	DF	SS	MS	F	P
REGRESSION	1	1.3318	1.3318	91.20	0.000
ERROR	23	0.3359	0.0146		
TOTAL	24	1.6676			

7.   THE STRENGTH OF A LINEAR RELATION

To arrive at a measure of adequacy of the straight line model, we examine how much of the variation in the response variable is explained by the fitted regression line. To this end, we view an observed y_i as consisting of two components.

$$y_i \quad = \quad (\hat{\beta}_0 + \hat{\beta}_1 x_i) \quad + \quad (y_i - \hat{\beta}_0 - \hat{\beta}_1 x_i)$$

Observed	Explained by	Residual or
y value	linear relation	deviation from
		linear relation

In an ideal situation where all the points lie exactly on the line, the residuals are all zero, and the y values are completely accounted for or **explained** by the linear dependence on x.

We can consider the sum of squares of the residuals

$$\text{SSE} = \sum_{i=1}^{n} (y_i - \hat{\beta}_0 - \hat{\beta}_1 x_i)^2 = S_{yy} - \frac{S_{xy}^2}{S_{xx}}$$

to be an overall measure of the discrepancy or departure from linearity. The total variability of the y values is reflected in the **total sum of squares**

$$S_{yy} = \sum_{i=1}^{n} (y_i - \bar{y})^2$$

of which SSE forms a part. The difference

$$S_{yy} - SSE = S_{yy} - \left(S_{yy} - \frac{S_{xy}^2}{S_{xx}} \right)$$

$$= \frac{S_{xy}^2}{S_{xx}}$$

forms the other part. Motivated by the decomposition of the observation y_i, just given, we can now consider a decomposition of the variability of the y values.

Decomposition of Variability

$$S_{yy} \qquad = \qquad \frac{S_{xy}^2}{S_{xx}} \qquad + \qquad SSE$$

| Total variability of y | Variability explained by the linear relation | Residual or unexplained variability |

The first term on the right-hand side of this equality is called the **sum of squares (SS) due to regression**. Likewise, the total variability S_{yy} is also called the **total SS** of y. In order for the straight line model to be considered as providing a good fit to the data, the SS due to the linear regression should comprise a major portion of S_{yy}. In an ideal situation in which all points lie on the line, SSE is zero, so S_{yy} is completely explained by the fact that the x values vary in the experiment. That is, the linear relationship between y and x is solely responsible for the variability in the y values.

As an index of how well the straight line model fits, it is then reasonable to consider the **proportion of the y variability explained by the linear relation**

$$\frac{\text{SS due to linear regression}}{\text{Total SS of } y} = \frac{S_{xy}^2 / S_{xx}}{S_{yy}} = \frac{S_{xy}^2}{S_{xx} S_{yy}}$$

From Section 6 of Chapter 3, recall that the quantity

$$r = \frac{S_{xy}}{\sqrt{S_{xx} S_{yy}}}$$

is named the **sample correlation coefficient**. Thus, the square of the sample correlation coefficient represents the proportion of the y variability explained by the linear relation.

The **strength of a linear relation** is measured by

$$r^2 = \frac{S_{xy}^2}{S_{xx} S_{yy}}$$

which is the square of the sample correlation coefficient r.

Example 11   The Proportion of Variability in Duration Explained by Dosage

Let us consider the drug trial data in Table 1. From the calculations provided in Table 3,

$$S_{xx} = 40.9 \qquad S_{yy} = 370.9 \qquad S_{xy} = 112.1$$

Fitted regression line

$$\hat{y} = -1.07 + 2.74x$$

How much of the variability in y is explained by the linear regression model?

SOLUTION   To answer this question, we calculate

$$r^2 = \frac{S_{xy}^2}{S_{xx}S_{yy}} = \frac{(112.1)^2}{40.9 \times 370.9} = .83$$

This means that 83% of the variability in y is explained by linear regression, and the linear model seems satisfactory in this respect.

Example 12   Proportion of Variation Explained in Number of Situps

Refer to physical fitness data in Table D.5 of the Data Bank. Using the data on numbers of situps, find the proportion of variation in the posttest number of situps explained by the pretest number that was obtained at the beginning of the conditioning class.

SOLUTION   Repeating the relevant part of the computer output from Example 10,

```
The regression equation is
Post Situps  =  10.3  +  0.899 Pre Situps

Predictor        Coef     SE Coef         T        P
Constant       10.331       2.533      4.08    0.000
Pre Situps    0.89904     0.06388     14.07    0.000

S  =  5.17893     R-Sq  =  71.5%     R-Sq(adj)  =  71.1%

Analysis of Variance

Source            DF          SS         MS        F        P
Regression         1      5312.9     5312.9   198.09    0.000
Residual Error    79      2118.9       26.8
Total             80      7431.8
```

we find R-Sq $=$ 71.5%, or proportion .715. From the analysis-of-variance table we could also have calculated

$$\frac{\text{Sum of squares regression}}{\text{Total sum of squares}} = \frac{5312.9}{7431.8} = .715$$

Using a person's pretest number of situps to predict their posttest number of situps explains that 71.5% of the variation is the posttest number.

When the value of r^2 is small, we can only conclude that a straight line relation does not give a good fit to the data. Such a case may arise due to the following reasons.

1. There is little relation between the variables in the sense that the scatter diagram fails to exhibit any pattern, as illustrated in Figure 9a. In this case, the use of a different regression model is not likely to reduce the SSE or explain a substantial part of S_{yy}.

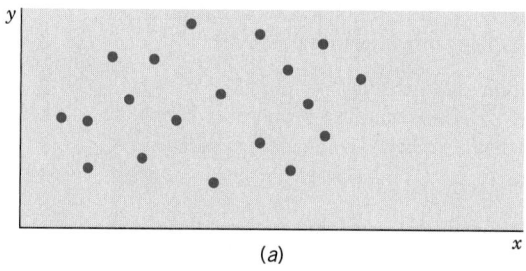

(a)

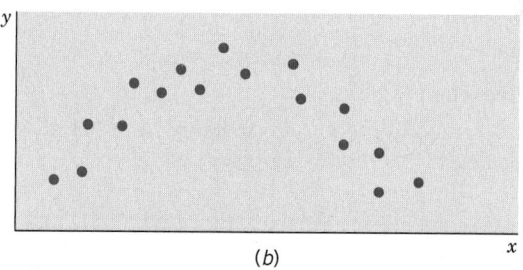

(b)

Figure 9   Scatter diagram patterns:
(a) No relation. (b) A nonlinear relation.

2. There is a prominent relation but it is nonlinear in nature; that is, the scatter is banded around a curve rather than a line. The part of S_{yy} that is explained by straight line regression is small because the model is inappropriate. Some other relationship may improve the fit substantially. Figure 9b illustrates such a case, where the SSE can be reduced by fitting a suitable curve to the data.

Exercises

11.38 Computing from a data set of (x, y) values, the following summary statistics are obtained.

$$n = 12 \qquad \bar{x} = 1.2 \qquad \bar{y} = 5.1$$
$$S_{xx} = 15.10 \qquad S_{xy} = 2.31 \qquad S_{yy} = 2.35$$

Determine the proportion of variation in y that is explained by linear regression.

11.39 Computing from a data set of (x, y) values, the following summary statistics are obtained:

$$n = 16 \qquad \bar{x} = 7.3 \qquad \bar{y} = 2.1$$
$$S_{xx} = 43.2 \qquad S_{xy} = 9.4 \qquad S_{yy} = 6.7$$

Determine the proportion of variation in y that is explained by linear regression.

11.40 Given $S_{xx} = 14.2$, $S_{yy} = 18.3$, and $S_{xy} = 10.3$, determine the proportion of variation in y that is explained by linear regression.

11.41 A calculation shows that $S_{xx} = 9.2$, $S_{yy} = 49$, and $S_{xy} = 16$. Determine the proportion of variation in y that is explained by linear regression.

11.42 Refer to Exercise 11.25.

(a) What proportion of the y variability is explained by the linear regression on x?

(b) Find the sample correlation coefficient.

(c) Calculate the residual sum of squares.

(d) Estimate σ^2.

11.43 Refer to Exercise 11.28.

(a) What proportion of y variability is explained by the linear regression on x?

(b) Find the sample correlation coefficient.

11.44 Refer to Exercise 11.33. According to the computer output in Table 7, find the proportion of y variability explained by x.

11.45 Refer to Exercise 11.35. According to the computer output in Table 8, find the proportion of y variability explained by x.

11.46 Consider the data on wolves in Table D.9 of the Data Bank concerning body length (cm) and weight (lb). Calculate the correlation coefficient r and r^2 for

(a) all wolves.

(b) male wolves.

(c) female wolves.

(d) Comment on the differences in your answers. Make a multiple scatter diagram (see Chapter 3) to clarify the situation.

*11.47 (a) Show that the sample correlation coefficient r and the slope $\hat{\beta}_1$ of the fitted regression line are related as

$$r = \frac{\hat{\beta}_1 \sqrt{S_{xx}}}{\sqrt{S_{yy}}}$$

(b) Show that SSE $= (1 - r^2) S_{yy}$.

*11.48 Show that the SS due to regression, S_{xy}^2 / S_{xx}, can also be expressed as $\hat{\beta}_1^2 S_{xx}$.

8. REMARKS ABOUT THE STRAIGHT LINE MODEL ASSUMPTIONS

A regression study is not completed by performing a few routine hypothesis tests and constructing confidence intervals for parameters on the basis of the formulas given in Section 5. Such conclusions can be seriously misleading if the assumptions made in the model formulations are grossly incompatible with the data. It is therefore essential to check the data carefully for indications of any violation of the assumptions. To review, the assumptions involved in the formulation of our straight line model are briefly stated again.

1. The underlying relation is linear.

2. Independence of errors.

3. Constant variance.

4. Normal distribution.

Of course, when the general nature of the relationship between y and x forms a curve rather than a straight line, the prediction obtained from fitting a straight line model to the data may produce nonsensical results. Often, a suitable transformation of the data reduces a nonlinear relation to one that is approximately linear in form. A few simple transformations are discussed in Chapter 12. Violating the assumption of independence is perhaps the most serious matter, because this can drastically distort the conclusions drawn from the t tests and the confidence statements associated with interval estimation. The implications of assumptions 3 and 4 were illustrated earlier in Figure 3. If the scatter diagram shows different amounts of variability in the y values for different levels of x, then the assumption of constant variance may have been violated. Here, again, an appropriate transformation of the data often helps to stabilize the variance. Finally, using the t distribution in hypothesis testing and confidence interval estimation is valid as long as the errors are approximately normally distributed. A moderate departure from normality does not impair the conclusions, especially when the data set is large. In other words, a violation of assumption 4 alone is not as serious as a violation of any of the other assumptions. Methods of checking the residuals to detect any serious violation of the model assumptions are discussed in Chapter 12.

USING STATISTICS WISELY

1. As a first step, plot the response variable versus the predictor variable. Examine the plot to see if a linear or other relationship exists.

2. Apply the principal of least squares to obtain estimates of the coefficients when fitting a straight line model.

3. Determine the $100(1 - \alpha)\%$ confidence intervals for the slope and intercept parameters. You can also look at P–values to decide whether or not they are zero. If not, you can use the fitted line for prediction.

4. Don't use the fitted line to make predictions beyond the range of the data. The model may be different over that range.

KEY IDEAS AND FORMULAS

In its simplest form, **regression analysis** deals with studying the manner in which the **response variable** y depends on a **predictor variable** x. Sometimes, the response variable is called the **dependent variable** and predictor variable is called the **independent** or **input variable**.

The first important step in studying the relation between the variables y and x is to plot the scatter diagram of the data (x_i, y_i), $i = 1, \ldots, n$. If this plot indicates an approximate linear relation, a straight line regression model is formulated:

$$\text{Response} = \text{A straight line in } x + \text{Random error}$$
$$Y_i = \beta_0 + \beta_1 x_i + e_i$$

The random errors are assumed to be independent, normally distributed, and have mean 0 and equal standard deviations σ.

The least squares estimate of $\hat{\beta}_0$ and least squares estimate of $\hat{\beta}_1$ are obtained by the method of least squares, which minimizes the sum of squared deviations $\Sigma\, (y_i - \beta_0 - \beta_1 x_i)^2$. The least squares estimates $\hat{\beta}_0$ and $\hat{\beta}_1$ determine the best fitting regression line $\hat{y} = \hat{\beta}_0 + \hat{\beta}_1 x$, which serves to predict y from x.

The differences $y_i - \hat{y}_i = $ Observed response $-$ Predicted response are called the residuals.

The adequacy of a straight line fit is measured by r^2, which represents the proportion of y variability that is explained by the linear relation between y and x. A low value of r^2 only indicates that a linear relation is not appropriate—there may still be a relation on a curve.

Least squares estimators

$$\hat{\beta}_1 = \frac{S_{xy}}{S_{xx}} \qquad \hat{\beta}_0 = \bar{y} - \hat{\beta}_1 \bar{x}$$

Best fitting straight line

$$\hat{y} = \hat{\beta}_0 + \hat{\beta}_1 x$$

Residuals

$$\hat{e}_i = y_i - \hat{y}_i = y_i - \hat{\beta}_0 - \hat{\beta}_1 x_i$$

Residual sum of squares

$$\text{SSE} = \sum_{i=1}^{n} \hat{e}_i^2 = S_{yy} - \frac{S_{xy}^2}{S_{xx}}$$

Estimate of variance σ^2

$$S^2 = \frac{\text{SSE}}{n - 2}$$

Inferences

1. Inferences concerning the slope β_1 are based on the

Estimator $\hat{\beta}_1$

$$\text{Estimated S.E.} = \frac{S}{\sqrt{S_{xx}}}$$

and the sampling distribution

$$T = \frac{\hat{\beta}_1 - \beta_1}{S/\sqrt{S_{xx}}} \qquad \text{d.f.} = n - 2$$

A $100(1 - \alpha)\%$ confidence interval for β_1 is

$$\hat{\beta}_1 \pm t_{\alpha/2} \frac{S}{\sqrt{S_{xx}}}$$

To test $H_0: \beta_1 = \beta_{10}$, the test statistic is

$$T = \frac{\hat{\beta}_1 - \beta_{10}}{S/\sqrt{S_{xx}}} \qquad \text{d.f.} = n - 2$$

2. Inferences concerning the intercept β_0 are based on the

Estimator $\hat{\beta}_0$

$$\text{Estimated S.E.} = S \sqrt{\frac{1}{n} + \frac{\bar{x}^2}{S_{xx}}}$$

and the sampling distribution

$$T = \frac{\hat{\beta}_0 - \beta_0}{S \sqrt{\dfrac{1}{n} + \dfrac{\bar{x}^2}{S_{xx}}}} \qquad \text{d.f.} = n - 2$$

A $100(1 - \alpha)\%$ confidence interval for β_0 is

$$\hat{\beta}_0 \pm t_{\alpha/2} S \sqrt{\frac{1}{n} + \frac{\bar{x}^2}{S_{xx}}}$$

3. At a specified $x = x^*$, the expected response is $\beta_0 + \beta_1 x^*$. Inferences about the expected response are based on the

Estimator $\hat{\beta}_0 + \hat{\beta}_1 x^*$

$$\text{Estimated S.E.} = S \sqrt{\frac{1}{n} + \frac{(x^* - \bar{x})^2}{S_{xx}}}$$

A $100(1 - \alpha)\%$ confidence interval for the expected response at x^* is given by

$$\hat{\beta}_0 + \hat{\beta}_1 x^* \pm t_{\alpha/2} S \sqrt{\frac{1}{n} + \frac{(x^* - \bar{x})^2}{S_{xx}}}$$

4. A single response at a specified $x = x^*$ is predicted by $\hat{\beta}_0 + \hat{\beta}_1 x^*$ with

$$\text{Estimated S.E.} = S \sqrt{1 + \frac{1}{n} + \frac{(x^* - \bar{x})^2}{S_{xx}}}$$

A $100(1 - \alpha)\%$ prediction interval for a single response is

$$\hat{\beta}_0 + \hat{\beta}_1 x^* \pm t_{\alpha/2} S \sqrt{1 + \frac{1}{n} + \frac{(x^* - \bar{x})^2}{S_{xx}}}$$

Decomposition of Variability

The total sum of squares S_{yy} is the sum of two components, the sum of squares due to regression S_{xy}^2/S_{xx} and the sum of squares due to error

$$S_{yy} = \frac{S_{xy}^2}{S_{xx}} + \text{SSE}$$

Variability explained by the linear relation $= \dfrac{S_{xy}^2}{S_{xx}} = \hat{\beta}_1^2 S_{xx}$

Residual or unexplained variability $=$ SSE

Total y variability $= S_{yy}$

The strength of a linear relation, or proportion of y variability explained by linear regression

$$r^2 = \frac{S_{xy}^2}{S_{xx} S_{yy}}$$

Sample correlation coefficient

$$r = \frac{S_{xy}}{\sqrt{S_{xx} S_{yy}}}$$

TECHNOLOGY

Fitting a straight line and calculating the correlation coefficient

MINITAB

Fitting a straight line—regression analysis

Begin with the values for the predictor variable x in $C1$ and the response variable y in $C2$.

> Stat > Regression < Regression.
> Type C2 in **Response**. Type $C1$ in **Predictors**.
> Click **OK**.

To calculate the correlation coefficient, start as above with data in $C1$ and $C2$.

> Stat > Basic Statistics > Correlation.
> Type $C1$ $C2$ in **Variables**. Click **OK**.

EXCEL

Fitting a straight line — regression analysis

Begin with the values of the predictor variable in column A and the values of the response variable in column B. To plot,

> Highlight the data and go to **Insert** and then **Chart.**
> Select **XY(Scatter)** and click **Finish.**
> Go to **Chart** and then **Add Trendline.**
> Click on the **Options** tab and check **Display equation on chart.**
> Click **OK.**

To obtain a more complete statistical analysis and diagnostic plots, instead use the following steps:

> Select **Tools** and then **Data Analysis.**
> Select **Regression.** Click **OK.**
> With the cursor in the **Y Range,** highlight the data in column B.
> With the cursor in the **X Range,** highlight the data in column A.
> Check boxes for **Residuals, Residual Plots,** and **Line Fit Plot.** Click **OK.**

To calculate the correlation coefficient, begin with the first variable in column A and the second in column B.

> Click on a blank cell. Select **Insert** and then **Function**
> (or click on the f_x icon).
> Select **Statistical** and then **CORREL.**
> **Highlight the data** in column A for **Array1** and highlight the data in column B for **Array2.** Click **OK.**

TI-84/-83 PLUS

Fitting a straight line — regression analysis

Enter the values of the predictor variable in **L₁** and those of the response variable in **L₂.**

> Select **STAT,** then **CALC,** and then **4 : LinReg (ax + b).**
> With **LinReg** on the Home screen Press **Enter.**

The calculator will return the intercept a, slope b, and correlation coefficient r. If r is not shown, go to the **2nd 0 : CATALOG** and select **Diagnostic.** Press **ENTER** twice. Then go back to **LinReg.**

9. REVIEW EXERCISES

11.49 Concerns that were raised for the environment near a government facility led to a study of plants. Since leaf area is difficult to measure, the leaf area (cm^2) was fit to

$$x = \text{Leaf length} \times \text{Leaf width}$$

using a least squares approach. For data collected one year, the fitted regression line is

$$\hat{y} = .2 + 0.5x$$

and $s^2 = (0.3)^2$. Comment on the size of the slope. Should it be positive or negative, less than one, equal to one, or greater than one?

11.50 Given these nine pairs of (x, y) values:

x	1	1	1	2	3	3	4	5	5
y	9	7	8	10	15	12	19	24	21

(a) Plot the scatter diagram.

(b) Calculate $\bar{x}$, $\bar{y}$, S_{xx}, S_{yy}, and S_{xy}.

(c) Determine the equation of the least squares fitted line and draw the line on the scatter diagram.

(d) Find the predicted y corresponding to $x = 3$.

11.51 Refer to Exercise 11.50.

(a) Find the residuals.

(b) Calculate the SSE by (i) summing the squares of the residuals and also (ii) using the formula SSE $= S_{yy} - S_{xy}^2/S_{xx}$.

(c) Estimate the error variance.

11.52 Refer to Exercise 11.50.

(a) Construct a 95% confidence interval for the slope of the regression line.

(b) Obtain a 90% confidence interval for the expected y value corresponding to $x = 4$.

11.53 An experiment is conducted to determine how the strength y of plastic fiber depends on the size x of the droplets of a mixing polymer in suspension. Data of (x, y) values, obtained

from 15 runs of the experiment, have yielded the following summary statistics.

$$\bar{x} = 8.3 \qquad \bar{y} = 54.8$$
$$S_{xx} = 5.6 \qquad S_{xy} = -12.4 \qquad S_{yy} = 38.7$$

(a) Obtain the equation of the least squares regression line.

(b) Test the null hypothesis $H_0: \beta_1 = -2$ against the alternative $H_1: \beta_1 < -2$, with $\alpha = .05$.

(c) Estimate the expected fiber strength for droplet size $x = 10$ and set a 95% confidence interval.

11.54 Refer to Exercise 11.53.

(a) Obtain the decomposition of the total y variability into two parts: one explained by linear relation and one not explained.

(b) What proportion of the y variability is explained by the straight line regression?

(c) Calculate the sample correlation coefficient between x and y.

11.55 A recent graduate moving to a new job collected a sample of monthly rent (dollars) and size (square feet) of 2-bedroom apartments in one area of a midwest city.

Size	Rent	Size	Rent
900	550	1000	650
925	575	1033	675
932	620	1050	715
940	620	1100	840

(a) Plot the scatter diagram and find the least squares fit of a straight line.

(b) Do these data substantiate the claim that the monthly rent increases with the size of the apartment? (Test with $\alpha = .05$).

(c) Give a 95% confidence interval for the expected increase in rent for one additional square foot.

(d) Give a 95% prediction interval for the monthly rent of a specific apartment having 1025 square feet.

11.56 Refer to Exercise 11.55.

    (a) Calculate the sample correlation coefficient.

    (b) What proportion of the y variability is explained by the fitted regression line?

11.57 A Sunday newspaper lists the following used-car prices for a foreign compact, with age x measured in years and selling price y measured in thousands of dollars.

x	y	x	y
1	16.9	5	8.9
2	12.9	7	5.6
2	13.9	7	5.7
4	13.0	8	6.0
4	8.8		

    (a) Plot the scatter diagram.

    (b) Determine the equation of the least squares regression line and draw this line on the scatter diagram.

    (c) Construct a 95% confidence interval for the slope of the regression line.

11.58 Refer to Exercise 11.57.

    (a) From the fitted regression line, determine the predicted value for the average selling price of a 5-year-old car and construct a 95% confidence interval.

    (b) Determine the predicted value for a 5-year-old car to be listed in next week's paper. Construct a 90% prediction interval.

    (c) Is it justifiable to predict the selling price of a 15-year-old car from the fitted regression line? Give reasons for your answer.

11.59 Again referring to Exercise 11.57, find the sample correlation coefficient between age and selling price. What proportion of the y variability is explained by the fitted straight line? Comment on the adequacy of the straight line fit.

11.60 Given

$$n = 20 \qquad \Sigma x = 17 \qquad \Sigma y = 31$$
$$\Sigma x^2 = 19 \qquad \Sigma xy = 21 \qquad \Sigma y^2 = 73$$

    (a) Find the equation of the least squares regression line.

    (b) Calculate the sample correlation coefficient between x and y.

    (c) Comment on the adequacy of the straight line fit.

The Following Exercises Require a Computer

11.61 *Using the computer.* The calculations involved in a regression analysis become increasingly tedious with larger data sets. Access to a computer proves to be of considerable advantage. We repeat here a computer-based analysis of linear regression using the data of Example 4 and the MINITAB package.

The sequence of steps in MINITAB:

> **Data: C11T3.DAT**
>
> C1: 3 3 4 5 6 6 7 8 8 9
> C2: 9 5 12 9 14 16 22 18 24 22
> **Dialog box:**
>
> **Stat > Regression > Regression**
> Type C2 in **Response**
> Type C1 in **Predictors**. Click **OK**.

produces all the results that are basic to a linear regression analysis. The important pieces in the output are shown in Table 9 on page 472.

Compare Table 9 with the calculations illustrated in Sections 4 to 7. In particular, identify:

    (a) The least squares estimates.

    (b) The SSE.

    (c) The estimated standard errors of $\hat{\beta}_0$ and $\hat{\beta}_1$.

    (d) The t statistics for testing $H_0 : \beta_0 = 0$ and $H_0 : \beta_1 = 0$.

    (e) r^2.

    (f) The decomposition of the total sum of squares into the sum of squares explained by the linear regression and the residual sum of squares.

11.62 Consider the data on all of the wolves in Table D.9 of the Data Bank concerning body length

TABLE 9   MINITAB Regression Analysis of the Data in Example 4

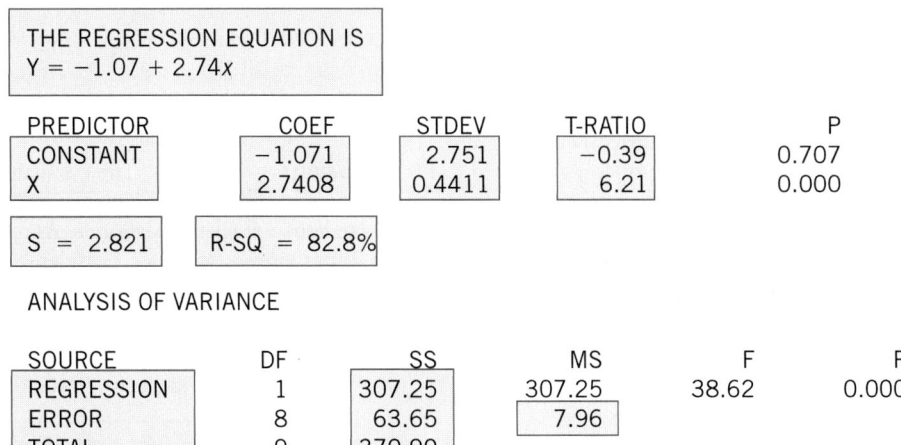

THE REGRESSION EQUATION IS
$Y = -1.07 + 2.74x$

PREDICTOR	COEF	STDEV	T-RATIO	P
CONSTANT	-1.071	2.751	-0.39	0.707
X	2.7408	0.4411	6.21	0.000

$S = 2.821$ R-SQ $= 82.8\%$

ANALYSIS OF VARIANCE

SOURCE	DF	SS	MS	F	P
REGRESSION	1	307.25	307.25	38.62	0.000
ERROR	8	63.65	7.96		
TOTAL	9	370.90			

(cm) and weight (lb). Using MINITAB or some other software program:

(a) Plot weight versus body length.

(b) Obtain the least squares fit of weight to the predictor variable body length.

(c) Test $H_0 : \beta_1 = 0$ versus $H_1 : \beta_1 > 0$ with $\alpha = .05$.

11.63   Refer to Exercise 11.62 and a least squares fit using the data on all of the wolves in Table D.9 of the Data Bank concerning body length (cm) and weight (lb). There is one obvious outlier, row 18 with body length 123 and weight 106, indicated in the MINITAB output. Drop this observation.

(a) Obtain the least squares fit of weight to the predictor variable body length.

(b) Test $H_0 : \beta_1 = 0$ versus $H_1 : \beta_1 > 0$ with $\alpha = .05$.

(c) Comment on any important differences between your answers to parts (a) and (b) and the answer to Exercise 11.62.

11.64   Many college students obtain college degree credits by demonstrating their proficiency on exams developed as part of the College Level Examination Program (CLEP). Based on their scores on the College Qualification Test (CQT), it would be helpful if students could predict their scores on a corresponding portion

of the CLEP exam. The following data (courtesy of R. W. Johnson) are for x = Total CQT score and y = Mathematical CLEP score.

x	y	x	y
170	698	174	645
147	518	128	578
166	725	152	625
125	485	157	558
182	745	174	698
133	538	185	745
146	485	171	611
125	625	102	458
136	471	150	538
179	798	192	778

(a) Find the least squares fit of a straight line.

(b) Construct a 95% confidence interval for the slope.

(c) Construct a 95% prediction interval for the CLEP score of a student who obtains a CQT score of 150.

(d) Repeat part (c) with $x = 175$ and $x = 195$.

11.65   Crickets make a chirping sound with their wing covers. Scientists have recognized that there is a relationship between the frequency of chirps and the temperature. (There is some truth to the cartoon on p. 432.) Use the 15 measurements for the striped ground cricket to:

(a) Fit a least squares line.

(b) Obtain a 95% confidence interval for the slope.

(c) Predict the temperature when $x = 15$ chirps per second.

Chirps (per second) (x)	Temperature (°F) (y)
20.0	88.6
16.0	71.6
19.8	93.3
18.4	84.3
17.1	80.6
15.5	75.2
14.7	69.7
17.1	82.0
15.4	69.4
16.3	83.3
15.0	79.6
17.2	82.6
16.0	80.6
17.0	83.5
14.4	76.3

Source: G. Pierce, *The Songs of Insects*, Cambridge, MA: Harvard University Press, 1949, pp. 12–21.

11.66 Use MINITAB or some other software to obtain the scatter diagram, correlation coefficient, and the regression line of the final time to run 1.5 miles on the initial times given in Table D.5 of the Data Bank.

11.67 Use MINITAB or some other software program to regress the marine growth on freshwater growth for the fish growth data in Table D.7 of the Data Bank. Do separate regression analyses for:

(a) All fish.

(b) Males.

(c) Females.

Your analysis should include (i) a scatter diagram, (ii) a fitted line, (iii) a determination if β_1 differs from zero. Also find a 95% confidence interval for the population mean when the freshwater growth is 100.

11.68 The data on the maximum height and top speed of the 12 highest roller coasters, displayed in the chapter opener, are

Height	Speed
400	120
415	100
377	100
318	95
310	93
263	81
259	81
245	85
240	79
235	85
230	80
224	70

(a) Use MINITAB or some other software program to determine the proportion of variation in speed due to regression on height.

(b) What top speed is predicted for a new roller coaster of height 325 feet?

(c) What top speed is predicted for a new roller coaster of height 480 feet? What additional danger is there in this prediction?

12

Regression Analysis II
Multiple Linear Regression and Other Topics

TABLE 4   Data Structure for Multiple Regression
with Two Input Variables

Experimental Run	Input Variables		Response y
	x_1	x_2	
1	x_{11}	x_{12}	y_1
2	x_{21}	x_{22}	y_2
.	.	.	.
.	.	.	.
.	.	.	.
i	x_{i1}	x_{i2}	y_i
.	.	.	.
.	.	.	.
.	.	.	.
n	x_{n1}	x_{n2}	y_n

By analogy with the simple linear regression model, we can then tentatively formulate:

A Multiple Regression Model

$$Y_i = \beta_0 + \beta_1 x_{i1} + \beta_2 x_{i2} + e_i \qquad i = 1, \ldots, n$$

where x_{i1} and x_{i2} are the values of the input variables for the ith experimental run and Y_i is the corresponding response.

The error components e_i are assumed to be independent normal variables with mean 0 and variance σ^2.

The regression parameters β_0, β_1, and β_2 are unknown and so is σ^2.

This model suggests that aside from the random error, the response varies linearly with each of the independent variables when the other remains fixed.

The principle of least squares is again useful in estimating the regression parameters. For this model, we are required to vary b_0, b_1, and b_2 simultaneously to minimize the sum of squared deviations

$$\sum_{i=1}^{n} (y_i - b_0 - b_1 x_{i1} - b_2 x_{i2})^2$$

The least squares estimates $\hat{\beta}_0, \hat{\beta}_1$, and $\hat{\beta}_2$ are the solutions to the following equations, which are extensions of the corresponding equations for fitting the straight line model (see Section 4 of Chapter 11.)

12

Regression Analysis II
Multiple Linear Regression and Other Topics

Micronutrients and Kelp Cultures: Evidence for Cobalt and Manganese Deficiency in Southern California Deep Seawater

Abstract. *It has been suggested that naturally occurring copper and zinc concentrations in deep seawater are toxic to marine organisms when the free ion forms are overabundant. The effects of micronutrients on the growth of gametophytes of the ecologically and commercially significant giant kelp* Macrocystis pyrifera *were studied in defined media. The results indicate that toxic copper and zinc ion concentrations as well as cobalt and manganese deficiencies may be among the factors controlling the growth of marine organisms in nature.*

A least squares fit of gametophytic growth data in the defined medium generated the expression

$$Y = 136 + 8x_{Mn} - 5x_{Cu} + 7x_{Co}$$
$$- 7x_{Zn}x_{Cu} - 15x_{Zn}^2 - 27x_{Mn}^2 - 12x_{Cu}^2$$
$$- 18x_{Co}^2 - 6x_{Cu}x_{Zn}^2 - 6x_{Cu}x_{Mn}^2 \tag{1}$$

where Y is mean gametophytic length in micrometers. The fit of the experimental data to Eq. (1) was considered excellent.

Here, several variables are important for predicting growth.

Source: J. S. Kuwabara, "Micronutrients and Kelp Cultures: Evidence for Cobalt and Manganese Deficiency in Southern California Deep Sea Waters," *Science,* **216** (June 11, 1982), pp. 1219–1221. Copyright © 1982 by AAAS.

© David Hall/Photo Researchers, Inc.

1. INTRODUCTION

The basic ideas of regression analysis have a much broader scope of application than the straight line model of Chapter 11. In this chapter, our goal is to extend the ideas of regression analysis in two important directions.

1. To handle nonlinear relations by means of appropriate transformations applied to one or both variables.
2. To accommodate several predictor variables into a regression model.

These extensions enable the reader to appreciate the breadth of regression techniques that are applicable to real-life problems. We then discuss some graphical procedures that are helpful in detecting any serious violation of the assumptions that underlie a regression analysis.

2. NONLINEAR RELATIONS AND LINEARIZING TRANSFORMATIONS

When studying the relation between two variables y and x, a scatter plot of the data often indicates that a relationship, although present, is far from linear. This can be established on a statistical basis by checking that the value of r^2 is small so a straight line fit is not adequate.

Statistical procedures for handling nonlinear relationships are more complicated than those for handling linear relationships, with the exception of a specific type of model called the polynomial regression model, which is discussed in Section 3. In some situations, however, it may be possible to transform the variables x and/or y in such a way that the new relationship is close to being linear. A linear regression model can then be formulated in terms of the transformed variables, and the appropriate analysis can be based on the transformed data.

Transformations are often motivated by the pattern of data. Sometimes, when the scatter diagram exhibits a relationship on a curve in which the y values increase too fast in comparison with the x values, a plot of $\sqrt{y}$ or some other fractional power of y can help to linearize the relation. This situation is illustrated in Example 1.

Example 1   Transforming the Response to Approximate a Linear Relation

To determine the maximum stopping ability of cars when their brakes are fully applied, 10 cars are driven each at a specified speed and the distance each requires to come to a complete stop is measured. The various initial speeds selected for each of the 10 cars and the stopping distances recorded are given in Table 1. Can the data be transformed to a nearly straight line relationship?

TABLE 1   Data on Speed and Stopping Distance

Initial speed x (mph)	20	20	30	30	30	40	40	50	50	60
Stopping distance y (ft)	16.3	26.7	39.2	63.5	51.3	98.4	65.7	104.1	155.6	217.2

SOLUTION   The scatter diagram for the data appears in Figure 1. The relation deviates from a straight line most markedly in that y increases at a much faster rate at large x than at small x. This suggests that we can try to linearize the relation by plotting $\sqrt{y}$ or some other fractional power of y with x.

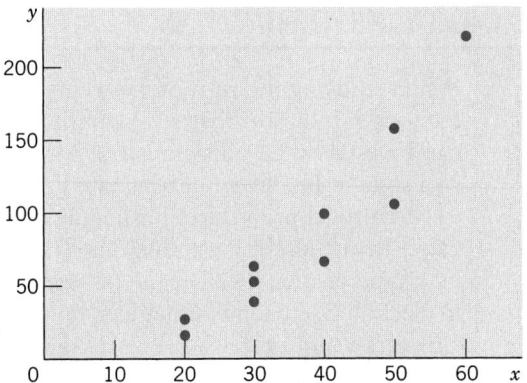

Figure 1   Scatter diagram of the data given in Table 1.

We try the transformed data $\sqrt{y}$ given in Table 2. The scatter diagram for these data, which exhibits an approximate linear relation, appears in Figure 2.

TABLE 2   Data on Speed and Square Root of Stopping Distance

x	20	20	30	30	30	40	40	50	50	60
$y' = \sqrt{y}$	4.037	5.167	6.261	7.969	7.162	9.920	8.106	10.203	12.474	14.738

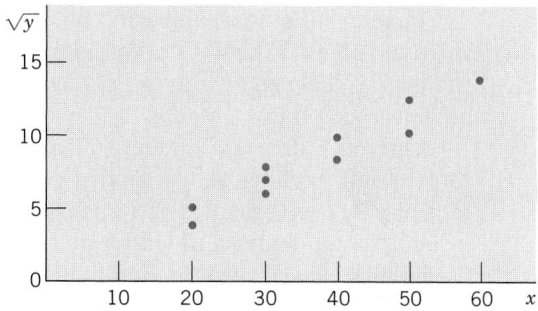

Figure 2   Scatter diagram of the transformed data given in Table 2.

With the aid of a standard computer program for regression analysis (see Exercise 12.27), the following results are obtained by transforming the original data.

$$\bar{x} = 37 \qquad \bar{y}' = 8.604$$
$$S_{xx} = 1610 \qquad S_{y'y'} = 97.773 \qquad S_{xy'} = 381.621$$
$$\hat{\beta}_0 = -.167 \qquad \hat{\beta}_1 = .237$$

Thus, the equation of the fitted line is

$$\hat{y}' = -.167 + .237x$$

The proportion of the y' variation that is explained by the straight line model is

$$r^2 = \frac{(381.621)^2}{(1610)(97.773)} = .925$$

A few common nonlinear models and their corresponding linearizing transformations are given in Table 3.

TABLE 3   Some Nonlinear Models and Their Linearizing Transformations

Nonlinear Model	Transformation		Transformed Model $y' = \beta_0 + \beta_1 x'$	
(a) $y = ae^{bx}$	$y' = \log_e y$	$x' = x$	$\beta_0 = \log_e a$	$\beta_1 = b$
(b) $y = ax^b$	$y' = \log y$	$x' = \log x$	$\beta_0 = \log a$	$\beta_1 = b$
(c) $y = \dfrac{1}{a + bx}$	$y' = \dfrac{1}{y}$	$x' = x$	$\beta_0 = a$	$\beta_1 = b$
(d) $y = a + b\sqrt{x}$	$y' = y$	$x' = \sqrt{x}$	$\beta_0 = a$	$\beta_1 = b$

In some situations, a specific nonlinear relation is strongly suggested by either the data or a theoretical consideration. Even when initial information about the form is lacking, a study of the scatter diagram often indicates the appropriate linearizing transformation.

Once the data are entered on a computer, it is easy to obtain the transformed data $1/y$, $\log_e y$, $y^{1/2}$, and $y^{1/4}$. Note $y^{1/4}$ is obtained by taking the square root of $y^{1/2}$. A scatter plot of $\sqrt{y}$ versus $\log_e x$ or any number of others can then be constructed and examined for a linear relation. Under relation (a) in Table 3, the graph of $\log_e y$ versus x would be linear.

We must remember that all inferences about the transformed model are based on the assumptions of a linear relation and independent normal errors with constant variance. Before we can trust these inferences, this transformed model must be scrutinized to determine whether any serious violation of these assumptions may have occurred (see Section 4).

Exercises

12.1 Given the pairs of (x, y) values

x	.5	1	2	4	5	6	7
y	4.6	3.2	2.1	1.7	.9	.7	.8

(a) Plot the scatter diagram.

(b) Obtain the best fitting straight line and draw it on the scatter diagram.

(c) What proportion of the y variability is explained by the fitted line?

12.2 Refer to the data of Exercise 12.1.

(a) Consider the reciprocal transformation $y' = 1/y$ and plot the scatter diagram of y' versus x.

(b) Fit a straight line regression to the transformed data.

(c) Calculate r^2 and comment on the adequacy of the fit.

12.3 Find a linearizing transformation in each case.

(a) $y = \dfrac{1}{(a + bx)^3}$

(b) $\dfrac{1}{y} = a + \dfrac{b}{1 + x}$

12.4 An experiment was conducted for the purpose of studying the effect of temperature on the life-length of an electrical insulation. Specimens of the insulation were tested under fixed temperatures, and their times to failure recorded.

Temperature x (°C)	Failure Time y (thousand hours)
180	7.3, 7.9, 8.5, 9.6, 10.3
210	1.7, 2.5, 2.6, 3.1
230	1.2, 1.4, 1.6, 1.9
250	.6, .7, 1.0, 1.1, 1.2

(a) Fit a straight line regression to the transformed data

$$x' = \frac{1}{x} \quad \text{and} \quad y' = \log y$$

(b) Is there strong evidence that an increase in temperature reduces the life of the insulation?

(c) Comment on the adequacy of the fitted line.

12.5 In an experiment (courtesy of W. Burkholder) involving stored-product beetles (*Trogoderma glabrum*) and their sex-attractant pheromone, the pheromone is placed in a pit-trap in the centers of identical square arenas. Marked

Release Distance (centimeters)	No. of Beetles Captured out of 8
6.25	5, 3, 4, 6
12.5	5, 2, 5, 4
24	4, 5, 3, 0
50	3, 4, 2, 2
100	1, 2, 2, 3

beetles are then released along the diagonals of each square at various distances from the pheromone source. After 48 hours, the pit-traps are inspected. Control pit-traps containing no pheromone capture no beetles.

(a) Plot the original data with y = number of beetles captured. Also plot y with $x = \log_e$ (distance).

(b) Fit a straight line by least squares to the appropriate graph in part (a).

(c) Construct a 95% confidence interval for β_1.

(d) Establish a 95% confidence interval for the mean at a release distance of 18 cm.

3. MULTIPLE LINEAR REGRESSION

A response variable y may depend on a predictor variable x but, after a straight line fit, it may turn out that the unexplained variation is large, so r^2 is small and a poor fit is indicated. At the same time, an attempt to transform one or both of the variables may fail to dramatically improve the value of r^2. This difficulty may well be due to the fact that the response depends on not just x but other factors as well. When used alone, x fails to be a good predictor of y because of the effects of those other influencing variables. For instance, the yield of a crop depends on not only the amount of fertilizer but also on the rainfall and average temperature during the growing season. Cool weather and no rain could completely cancel the choice of a correct fertilizer.

To obtain a useful prediction model, one should record the observations of all variables that may significantly affect the response. These other variables may then be incorporated explicitly into the regression analysis. The name **multiple regression** refers to a model of relationship where the response depends on two or more predictor variables. Here, we discuss the main ideas of a multiple regression analysis in the setting of two predictor variables.

Suppose that the response variable y in an experiment is expected to be influenced by two input variables x_1 and x_2, and the data relevant to these input variables are recorded along with the measurements of y. With n runs of an experiment, we would have a data set of the form shown in Table 4.

TABLE 4   Data Structure for Multiple Regression
with Two Input Variables

Experimental	Input Variables		Response
Run	x_1	x_2	y
1	x_{11}	x_{12}	y_1
2	x_{21}	x_{22}	y_2
.	.	.	.
.	.	.	.
.	.	.	.
i	x_{i1}	x_{i2}	y_i
.	.	.	.
.	.	.	.
.	.	.	.
n	x_{n1}	x_{n2}	y_n

By analogy with the simple linear regression model, we can then tentatively formulate:

A Multiple Regression Model

$$Y_i = \beta_0 + \beta_1 x_{i1} + \beta_2 x_{i2} + e_i \qquad i = 1, \ldots, n$$

where x_{i1} and x_{i2} are the values of the input variables for the ith experimental run and Y_i is the corresponding response.

The error components e_i are assumed to be independent normal variables with mean 0 and variance σ^2.

The regression parameters $\beta_0, \beta_1,$ and β_2 are unknown and so is σ^2.

This model suggests that aside from the random error, the response varies linearly with each of the independent variables when the other remains fixed.

The principle of least squares is again useful in estimating the regression parameters. For this model, we are required to vary $b_0, b_1,$ and b_2 simultaneously to minimize the sum of squared deviations

$$\sum_{i=1}^{n} (y_i - b_0 - b_1 x_{i1} - b_2 x_{i2})^2$$

The least squares estimates $\hat{\beta}_0, \hat{\beta}_1,$ and $\hat{\beta}_2$ are the solutions to the following equations, which are extensions of the corresponding equations for fitting the straight line model (see Section 4 of Chapter 11.)

$$\hat{\beta}_1 S_{11} + \hat{\beta}_2 S_{12} = S_{1y}$$
$$\hat{\beta}_1 S_{12} + \hat{\beta}_2 S_{22} = S_{2y}$$
$$\hat{\beta}_0 = \bar{y} - \hat{\beta}_1 \bar{x}_1 - \hat{\beta}_2 \bar{x}_2$$

where S_{11}, S_{12}, and so on, are the sums of squares and cross products of deviations of the variables indicated in the suffix. They are computed just as in a straight line regression model. Methods are available for interval estimation, hypothesis testing, and examining the adequacy of fit. In principle, these methods are similar to those used in the simple regression model, but the algebraic formulas are more complex and hand computations become more tedious. However, a multiple regression analysis is easily performed on a computer with the aid of the standard packages such as MINITAB, SAS, or SPSS. We illustrate the various aspects of a multiple regression analysis with the data of Example 2 and computer-based calculations.

Example 2   Interpreting the Regression of Blood Pressure on Weight and Age

We are interested in studying the systolic blood pressure y in relation to weight x_1 and age x_2 in a class of males of approximately the same height. From 13 subjects preselected according to weight and age, the data set listed in Table 5 was obtained.

TABLE 5   The Data of x_1 = Weight in Pounds, x_2 = Age, and y = Blood Pressure of 13 Males

x_1	x_2	y
152	50	120
183	20	141
171	20	124
165	30	126
158	30	117
161	50	129
149	60	123
158	50	125
170	40	132
153	55	123
164	40	132
190	40	155
185	20	147

Use a computer package to perform a regression analysis using the model

$$Y_i = \beta_0 + \beta_1 x_{i1} + \beta_2 x_{i2} + e_i$$

SOLUTION   To use MINITAB, we first enter the data of x_1, x_2, and y in three different columns and then use the regression command,

> **Data:** C12T5.DAT
> C1: 152   183   171 $\cdots$ 185
> C2: 50   20   20 $\cdots$ 20
> C3: 120   141   124 $\cdots$ 147
> **Dialog box:**
> **Stat > Regression > Regression**
> Type C3 in **Response.**
> Type C1 and C2 in **Predictors.**
> Click **OK.**

With the last command, the computer executes a multiple regression analysis. We focus our attention on the principal aspects of the output, as shown in Table 6.

TABLE 6   Regression Analysis of the Data in Table 5: Selected MINITAB Output

① THE REGRESSION EQUATION IS
Y = − 65.1 + 1.08 X1 + 0.425 X2

PREDICTOR	COEF	STDEV	T-RATIO	P
CONSTANT	− 65.10	14.94	− 4.36	0.001
X1	② 1.07710	0.07707	13.98	0.000
X2	0.42541	0.07315	④ 5.82	0.000

③ S = 2.509   ⑤ R-SQ = 95.8%

ANALYSIS OF VARIANCE

SOURCE	DF	SS	MS	F	P
REGRESSION	2	1423.84	711.92	113.13	0.000
ERROR	10	⑥ 62.93	6.29		
TOTAL	12	1486.77			

SOURCE	DF	SEQ SS
X1	1	1211.01
X2	1	212.83

We now proceed to interpret the results in Table 6 and use them to make further statistical inferences.

(i)   The equation of the fitted linear regression is

   ① $$\hat{y} = -65.1 + 1.08x_1 + .425x_2$$

This means that the mean blood pressure increases by 1.08 if weight x_1 increases by one pound and age x_2 remains fixed. Similarly, a 1-year increase in age with the weight held fixed will only increase the mean blood pressure by .425.

(ii)   The estimated regression coefficient and the corresponding estimated standard errors are

②
$$\hat{\beta}_0 = -65.10 \qquad \text{Estimated S.E. } (\hat{\beta}_0) = 14.94$$
$$\hat{\beta}_1 = 1.07710 \qquad \text{Estimated S.E. } (\hat{\beta}_1) = .07707$$
$$\hat{\beta}_2 = .42541 \qquad \text{Estimated S.E. } (\hat{\beta}_2) = .07315$$

③   Further, the error standard deviation σ is estimated by $s = 2.509$ with

$$\begin{aligned} \text{Degrees of freedom} &= n - (\text{No. of input variables}) - 1 \\ &= 13 - 2 - 1 \\ &= 10 \end{aligned}$$

These results are useful in interval estimation and hypothesis tests about the regression coefficients. In particular, a $100(1 - \alpha)\%$ confidence interval for a coefficient β is given by

$$\text{Estimated coefficient} \pm t_{\alpha/2} \text{ (Estimated S.E.)}$$

where $t_{\alpha/2}$ is the upper $\alpha/2$ point of the distribution with d.f. $= 10$. For instance, a 95% confidence interval for β_1 is

$$\begin{aligned} 1.07710 &\pm 2.228 \times .07707 \\ = 1.07710 &\pm .17171 \qquad \text{or} \qquad (.905, 1.249) \end{aligned}$$

To test the null hypothesis that a particular coefficient β is zero, we employ the test statistic

$$t = \frac{\text{Estimated coefficient} - 0}{\text{Estimated S.E.}} \qquad \text{d.f.} = 10$$

These t-ratios appear in Table 6. Suppose that we wish to examine whether the mean blood pressure significantly increases with age. In the language of hypothesis testing, this problem translates to one of testing $H_0 : \beta_2 = 0$ versus $H_1 : \beta_2 > 0$. The observed value of the
④   test statistic is $t = 5.82$ with d.f. $= 10$. Since this is larger than the tabulated value $t_{.01} = 2.764$, the null hypothesis is rejected in favor of H_1, with $\alpha = .01$. In fact, it is rejected even with $\alpha = .005$.

(iii)   In Table 6, the result "R-SQ $=$ 95.8%" or

$$⑤ \quad R^2 \ = \ .958$$

tells us that 95.8% of the variability of y is explained by the fitted multiple regression of y on x_1 and x_2. The "analysis of variance" shows the decomposition of the total variability $\Sigma(y_i - \bar{y})^2 = 1486.77$ into the two components.

$$⑥ \quad 1486.77 \quad = \quad 1423.84 \quad + \quad 62.93$$

Total variability of y	Variability explained by the regression of y on x_1 and x_2	Residual or unexplained variability

Thus,

$$R^2 \ = \ \frac{1423.84}{1486.77} \ = \ .958$$

and σ^2 is estimated by $s^2 \ = \ 62.93/10 \ = \ 6.293$, so $s \ = \ 2.509$ [checks with s from (ii)].

TABLE 7   A Regression Analysis of the Data in Example 2 Using SAS

MODEL: MODEL 1
DEPENDENT VARIABLE: Y

ANALYSIS OF VARIANCE

SOURCE	DF	SUM OF SQUARES	MEAN SQUARE	F VALUE	PROB>F
MODEL	2	1423.83797	711.91898	113.126	0.0001
ERROR	10	⑥ 62.93126	6.29313		
C TOTAL	12	1486.76923			

③ ROOT MSE   2.50861   ⑤ R-SQUARE   0.9577

PARAMETER ESTIMATES

VARIABLE	DF	PARAMETER ESTIMATE	STANDARD ERROR	T FOR H0: PARAMETER = 0	PROB > |T|
INTERCEP	1	− 65.099678	14.94457547	− 4.356	0.0014
X1	1	② 1.077101	0.07707220	13.975	0.0001
X2	1	0.425413	0.07315231	④ 5.815	0.0002

The square of the multiple correlation coefficient R^2 gives the proportion of variability in y explained by the fitted multiple regression.

The output from the SAS package is given in Table 7. The quantities needed in our analysis have been labeled with the same circled numbers as in the MINITAB output.

Example 3   Computer-Aided Regression Analysis — Two Predictors

The times for 81 students to complete a rowing test both before and after completing a one-semester conditioning course are given in Table D.5 in the Data Bank. It may be that not only the pretest rowing time but also gender would be useful for predicting the posttest rowing time. Perform a regression analysis.

SOLUTION   We use MINITAB to obtain the output

```
Regression Analysis: Post row versus Pre row, Gender


The regression equation is
Post row = 97.3 + 0.726 Pre row + 32.1 Gender

Predictor       Coef    SE Coef       T       P
Constant       97.33      31.68    3.07   0.003
Pre row      0.72573    0.05487   13.23   0.000
Gender        32.083      9.756    3.29   0.002


S  =   31.8137 R-Sq  =   85.7%       R-Sq(adj)  =   85.3%

Analysis of Variance

Source             DF       SS       MS       F       P
Regression          2   471547   235774  232.95   0.000
Residual Error     78    78945     1012
Total              80   550492

Source      DF   Seq SS
Pre row      1   460602
Gender       1    10945
```

Which variables should be used to predict posttest rowing time? Reading from the column of P–values for the individual coefficients, the largest is only .003. The constant term and the coefficients of pretest rowing time and gender are significantly different from 0. All three terms are needed in the model.

The plot of residuals versus fit in Figure 3 on page 488 reveals a constant width band so there is no evidence against the assumption of constant variance. The one large negative residual is case 17 and the two large positive residuals are cases 29 and 70.

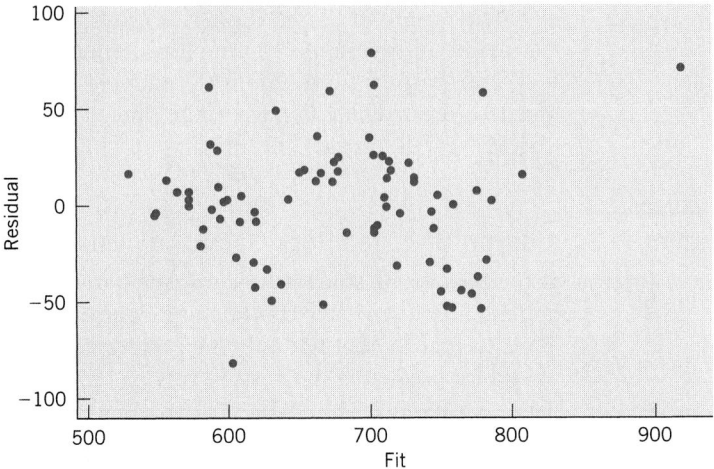

Figure 3   Residuals of posttest row times versus fits.

POLYNOMIAL REGRESSION

A scatter diagram may exhibit a relationship on a curve for which a suitable linearizing transformation cannot be constructed. Another method of handling such a nonlinear relation is to include terms with higher powers of x in the model $Y = \beta_0 + \beta_1 x + e$. In this instance, by including the second power of x, we obtain the model

$$Y_i = \beta_0 + \beta_1 x_i + \beta_2 x_i^2 + e_i \qquad i = 1, \ldots, n$$

which states that aside from the error components e_i, the response y is a quadratic function (or a second-degree polynomial) of the independent variable x. Such a model is called a polynomial regression model of y with x, and the highest power of x that occurs in the model is called the degree or the order of the polynomial regression. It is interesting to note that the analysis of a polynomial regression model does not require any special techniques other than those used in multiple regression analysis. By identifying x and x^2 as the two variables x_1 and x_2, respectively, this second-degree polynomial model reduces to the form of a multiple regression model:

$$Y_i = \beta_0 + \beta_1 x_{i1} + \beta_2 x_{i2} + e_i \qquad i = 1, \ldots, n$$

where $x_{i1} = x_i$ and $x_{i2} = x_i^2$. In fact, both these types of models and many more types are special cases of a class called general linear models [1, 2].

GENERAL LINEAR MODEL

By virtue of its wide applicability, the multiple linear regression model plays a prominent role in the portfolio of a statistician. Although a complete analysis cannot be given here, the general structure of a multiple regression model merits further attention. We have already mentioned that most least squares analyses of multiple linear regression models are carried out with the aid of a computer. All the programs for implementing the analysis require the investigator to provide the values of the response y_i and the p input variables $x_{i1}, \ldots, x_{ip}$ for each run $i = 1, 2, \ldots, n$. In writing $1 \cdot \beta_0$, where 1 is the known value of an extra "dummy" input variable corresponding to β_0, the model is

Observation   Input variables   Error

$$Y_i \quad = 1 \cdot \beta_0 + x_{i1}\beta_1 + x_{i2}\beta_2 + \cdots + x_{ip}\beta_p + \quad e_i$$

The basic quantities can be arranged in the form of these arrays, which are denoted by boldface letters.

Observation   Input variables

$$\mathbf{y} = \begin{bmatrix} y_1 \\ y_2 \\ \cdot \\ \cdot \\ \cdot \\ y_i \\ \cdot \\ \cdot \\ \cdot \\ y_n \end{bmatrix} \qquad \mathbf{X} = \begin{bmatrix} 1 & x_{11} & \cdots & x_{1p} \\ 1 & x_{21} & \cdots & x_{2p} \\ \cdot & \cdot & \cdots & \cdot \\ \cdot & \cdot & \cdots & \cdot \\ \cdot & \cdot & \cdots & \cdot \\ 1 & x_{i1} & \cdots & x_{ip} \\ \cdot & \cdot & \cdots & \cdot \\ \cdot & \cdot & \cdots & \cdot \\ \cdot & \cdot & \cdots & \cdot \\ 1 & x_{n1} & \cdots & x_{np} \end{bmatrix}$$

Only the arrays $\mathbf{y}$ and $\mathbf{X}$ are required to obtain the least squares estimates of β_0, $\beta_1, \ldots, \beta_p$ that minimize

$$\sum_{i=1}^{n} (y_i - b_0 - x_{i1}b_1 - \cdots - x_{ip}b_p)^2$$

The input array $\mathbf{X}$ is called the **design matrix.**

In the same vein, setting

$$\mathbf{e} = \begin{bmatrix} e_1 \\ e_2 \\ \cdot \\ \cdot \\ \cdot \\ e_n \end{bmatrix} \qquad \text{and} \qquad \boldsymbol{\beta} = \begin{bmatrix} \beta_0 \\ \beta_1 \\ \cdot \\ \cdot \\ \cdot \\ \beta_p \end{bmatrix}$$

we can write the model in the suggestive form

	Observation		Design matrix	Parameter		Error
	y	=	X	$\boldsymbol{\beta}$	+	e

which forms the basis for a thorough but more advanced treatment of regression.

Exercises

12.6 The regression model $Y = \beta_0 + \beta_1 x_1 + \beta_2 x_2 + e$ was fitted to a data set obtained from 20 runs of an experiment in which two predictors x_1 and x_2 were observed along with the response y. The least squares estimates were

$$\hat{\beta}_0 = 3.80 \qquad \hat{\beta}_1 = 7.90 \qquad \hat{\beta}_2 = -4.20$$

Predict the response for

(a) $x_1 = 2, \qquad x_2 = 3$

(b) $x_1 = 2, \qquad x_2 = 5$

(c) $x_1 = 3, \qquad x_2 = 5$

12.7 Consider the multiple linear regression model

$$Y = \beta_0 + \beta_1 x_1 + \beta_2 x_2 + e$$

where $\beta_0 = -2, \beta_1 = -1, \beta_2 = 3$, and the normal random variable e has standard deviation 3. What is the mean of the response Y when $x_1 = 3$ and $x_2 = -2$?

12.8 In Exercise 12.6, suppose that the residual sum of squares (SSE) was 55.94 and the SS due to regression was 421.38.

(a) Estimate the error standard deviation σ. State the degrees of freedom.

(b) Find R^2 and interpret the result.

12.9 Refer again to Exercise 12.6. Suppose that the estimated standard errors of $\hat{\beta}_0, \hat{\beta}_1$, and $\hat{\beta}_2$ were 1.345, .907, and .4056, respectively.

(a) Determine 95% confidence intervals for β_0 and β_2.

(b) Test $H_0: \beta_1 = 6$ versus $H_1: \beta_1 > 6$, with $\alpha = .05$.

12.10 Consider the data on all of the wolves in Table D.9 of the Data Bank concerning age (years) and canine length (mm).

(a) Obtain the least squares fit of the straight line regression model $Y = \beta_0 + \beta_1 x + e$ to predict canine length from age.

(b) Obtain the least squares fit of the multiple regression model $Y = \beta_0 + \beta_1 x_1 + \beta_2 x_2 + e$ to predict canine length using age x_1 and body length x_2.

(c) What is the predicted canine length for a wolf of age 2.5 and body length 127?

(d) What proportion of the y variability is explained by the fitted model?

(e) Obtain 95% confidence intervals for β_0, β_1, and β_2.

12.11 Refer to the computer output in Table 8.

(a) Identify the least squares estimates $\hat{\beta}_0$, $\hat{\beta}_1$, and $\hat{\beta}_2$.

(b) What model is suggested from this analysis?

(c) What is the proportion of y variability explained by the regression on x_1 and x_2?

(d) Estimate σ^2.

12.12 Refer to the computer output in Table 9.

(a) Identify the least squares estimates $\hat{\beta}_0$, $\hat{\beta}_1$, and $\hat{\beta}_2$.

(b) What model is suggested from this analysis?

(c) What is the proportion of y variability explained by the regression on x_1 and x_2?

12.13 With reference to Exercise 12.11:

(a) Test $H_0: \beta_1 = 0$ versus $H_1: \beta_1 \neq 0$ with $\alpha = .05$.

(b) Test $H_0: \beta_2 = 0$ versus $H_1: \beta_2 \neq 0$ with $\alpha = .05$.

TABLE 8   Computer Output of a Regression Analysis to Be Used for Exercise 12.11

```
THE REGRESSION EQUATION IS
Y = -0.081 + 0.646 X1 + 0.805 X2

PREDICTOR        COEF     STDEV    T-RATIO       P
CONSTANT      -0.0810    0.1652     -0.49    0.629
X1             0.64588   0.09206     7.02    0.000
X2             0.8046    0.2447      3.29    0.003

S = 0.1012 R-SQ = 86.5%

ANALYSIS OF VARIANCE

SOURCE        DF      SS        MS       F      P
REGRESSION     2   1.44246   0.72123  70.47  0.000
ERROR         22   0.22517   0.01023
TOTAL         24   1.66763

SOURCE        DF  SEQ  SS
X1             1   1.33177
X2             1   0.11069
```

(c) Estimate the expected y value correspond-
     ing to $x_1 = 1.9$ and $x_2 = 1.0$.

(d) Construct a 90% confidence interval for
     the intercept β_0.

12.14 With reference to Exercise 12.12:

(a) Test $H_0 : \beta_1 = 0$ versus $H_1 : \beta_1 \neq 0$
     with $\alpha = .05$.

(b) Test $H_0 : \beta_2 = 0$ versus $H_1 : \beta_2 \neq 0$
     with $\alpha = .05$.

(c) Estimate the expected y value correspond-
     ing to $x_1 = 2.0$ and $x_2 = 3.5$.

(d) Construct a 90% confidence interval for
     the intercept β_0.

TABLE 9   Computer Output of a Regression Analysis to Be Used for Exercise 12.12

```
The regression equation is
Y = 19.6 - 0.0564 X1 - 1.06 X2

PREDICTOR       Coef    SE Coef        T       P
CONSTANT      19.5778   0.8918     21.95   0.000
mil(acc)      -0.05644  0.02314     -2.44   0.033
yr(acc)       -1.0569   2.556       -4.14   0.002

S = 1.574 R-Sq = 89.2%

ANALYSIS OF VARIANCE

Source           DF      SS       MS       F      P
Regression        2    224.05   112.03   45.21  0.000
Residual Error   11     27.26     2.48
Total            13    251.31

Source      DF   Seq SS
X1           1   181.67
X2           1    42.38
```

4. RESIDUAL PLOTS TO CHECK THE ADEQUACY OF A STATISTICAL MODEL

> ### General Attitude Toward a Statistical Model
>
> A regression analysis is not completed by fitting a model by least squares, providing confidence intervals, and testing various hypotheses. These steps tell only half the story: the statistical inferences that can be made when the postulated model is adequate. In most studies, we cannot be sure that a particular model is correct. Therefore, we should adopt the following strategy.
>
> 1. **Tentatively entertain a model.**
> 2. **Obtain least squares estimates and compute the residuals.**
> 3. **Review the model by examining the residuals.**

Step 3 often suggests methods of appropriately modifying the model. We then return to step 1, where the modified model is entertained, and this **iteration** is continued until a model is obtained for which the data do not seem to contradict the assumptions made about the model.

Once a model is fitted by least squares, all the information on variation that cannot be explained by the model is contained in the residuals

$$\hat{e}_i = y_i - \hat{y}_i \qquad i = 1, 2, \ldots, n$$

where y_i is the observed value and $\hat{y}_i$ denotes the corresponding value predicted by the fitted model. For example, in the case of a simple linear regression model, $\hat{y}_i = \hat{\beta}_0 + \hat{\beta}_1 x_i$.

Recall from our discussion of the straight line model in Chapter 11 that we have made the assumptions of independence, constant variance, and a normal distribution for the error components e_i. The inference procedures are based on these assumptions. When the model is correct, the residuals can be considered as estimates of the errors e_i that are distributed as $N(0, \sigma)$.

To determine the merits of the tentatively entertained model, we can examine the residuals by plotting them in various ways. Then if we recognize any systematic pattern formed by the plotted residuals, we would suspect that some assumptions regarding the model are invalid. There are many ways to plot the residuals, depending on what aspect is to be examined. We mention a few of these here to illustrate the techniques. A more comprehensive discussion can be found in Chapter 3 of Draper and Smith [2].

HISTOGRAM OR DOT DIAGRAM OF RESIDUALS

To picture the overall behavior of the residuals, we can plot a histogram for a large number of observations or a dot diagram for fewer observations. For example, in a dot diagram like the one in Figure 4a, the residuals seem to behave like a sample

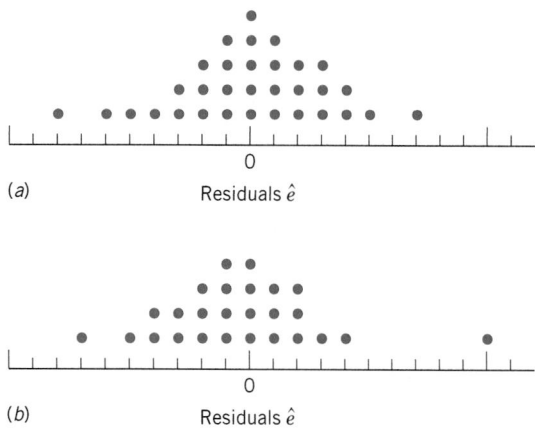

Figure 4   Dot diagram of residuals.
(*a*) Normal pattern. (*b*) One large residual.

from a normal population and there do not appear to be any "wild" observations. In contrast, Figure 4*b* illustrates a situation in which the distribution appears to be quite normal except for a single residual that lies far to the right of the others. The circumstances that produced the associated observation demand a close scrutiny.

PLOT OF RESIDUAL VERSUS PREDICTED VALUE

A plot of the residuals $\hat{e}_i$ versus the predicted value $\hat{y}_i$ often helps to detect the inadequacies of an assumed relation or a violation of the assumption of constant error variance. Figure 5 illustrates some typical phenomena. If the points form a horizontal band around zero, as in Figure 5*a*, then no abnormality is indicated. In Figure 5*b*, the width of the band increases noticeably with increasing values of $\hat{y}$. This indicates that the error variance σ^2 tends to increase with an increasing level of response. We would then suspect the validity of the assumption of constant variance in the model. Figure 5*c* shows residuals that form a systematic pattern. Instead of being randomly distributed around the $\hat{y}$ axis, they tend first to increase steadily and then decrease. This would lead us to suspect that the model is inadequate and a squared term or some other nonlinear x term should be considered.

PLOT OF RESIDUAL VERSUS TIME ORDER

The most crucial assumption in a regression analysis is that the errors e_i are independent. Lack of independence frequently occurs in business and economic applications, where the observations are collected in a time sequence with the intention of using regression techniques to predict future trends. In many other experiments, trials are conducted successively in time. In any event, a plot of the residuals versus time order often detects a violation of the assumption of independence. For example, the plot in Figure 6 exhibits a systematic pattern in that a string of high values is followed by a string of low values. This indicates that consecutive residuals are (positively) correlated, and we

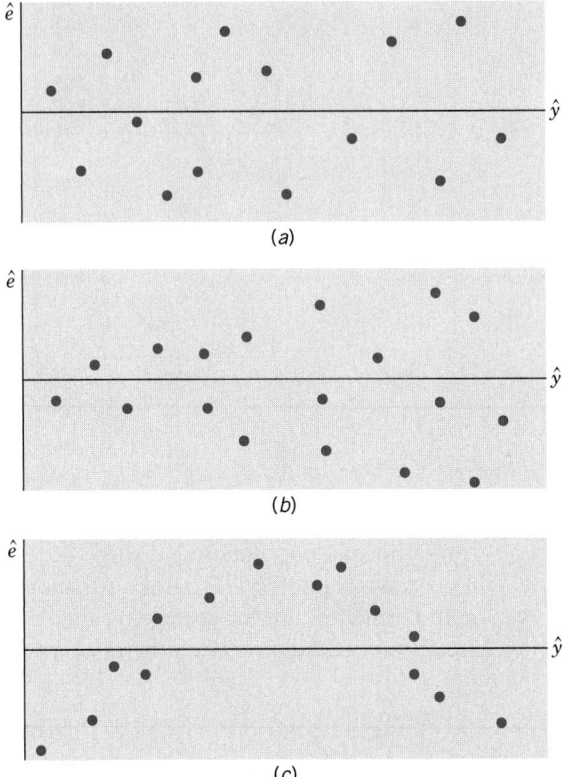

Figure 5   Plot of residual versus predicted value.
(a) Constant spread. (b) Increasing spread.
(c) Curved pattern.

would suspect a violation of the independence assumption. Independence can also be checked by plotting the successive pairs $(\hat{e}_i, \hat{e}_{i-1})$, where $\hat{e}_1$ indicates the residual from the first y value observed, $\hat{e}_2$ indicates the second, and so on. Independence is suggested if the scatter diagram is a patternless cluster, whereas points clustered along a line suggest a lack of independence between adjacent observations.

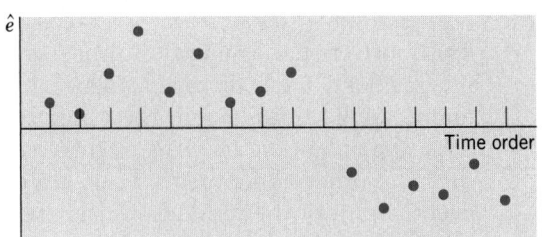

Figure 6   Plot of residual versus time order.

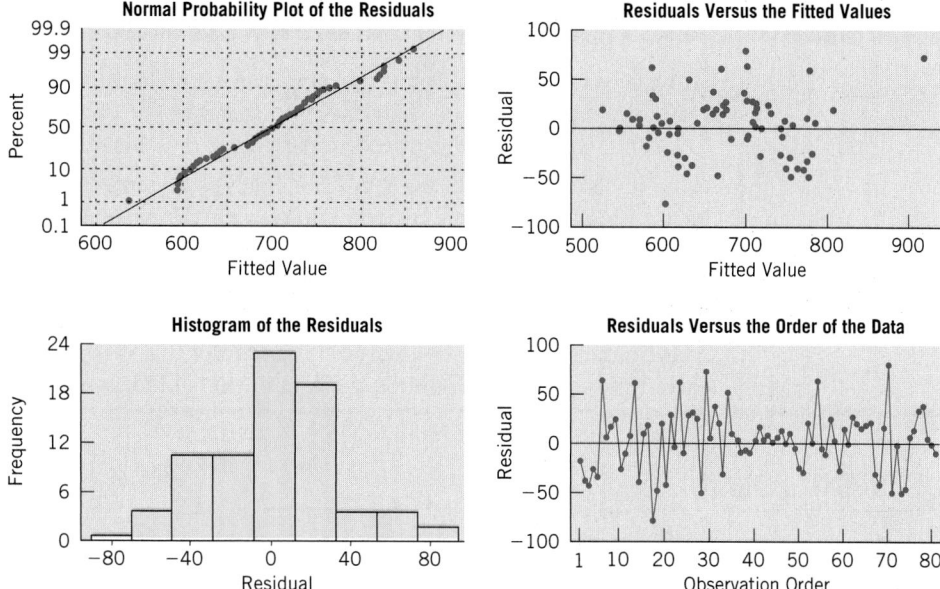

Figure 7   Four residual plots for posttest rowing time using MINITAB.

The MINITAB regression *four-in-one* graphics option created the four residual plots in Figure 7 using the data and fit from Example 3.

It is important to remember that our confidence in statistical inference procedures is related to the validity of the assumptions about them. A mechanically made inference may be misleading if some model assumption is grossly violated. An examination of the residuals is an important part of regression analysis, because it helps to detect any inconsistency between the data and the postulated model.

If no serious violation of the assumption is exposed in the process of examining residuals, we consider the model adequate and proceed with the relevant inferences. Otherwise, we must search for a more appropriate model.

References

1. S. Chatterjee and A. Hadi, *Regression Analysis by Example*, 3rd ed., John Wiley & Sons: New York, 2000.

2. N. R. Draper and H. Smith, *Applied Regression Analysis*, 3rd ed., John Wiley & Sons: New York, 1998.

USING STATISTICS WISELY

1. Always, as a first step, plot the response variable versus the predictor variable. If there is more than one predictor variable, make separate plots for each. Examine the plot to see if a linear or other relationship exists.

2. Do not routinely accept the regression analysis presented in computer output. Instead, criticize the model by inspecting the residuals for outliers and moderate to severe lack of normality. A normal-scores plot is useful if there are more than 20 or so residuals. That plot may suggest a transformation.

3. Plot the residuals versus predicted value to check the assumption of constant variance. Plot the residuals in time order if that is appropriate. A trend over time would cast doubt on the assumption of independent errors.

KEY IDEAS AND FORMULAS

When a scatter diagram shows relationship on a curve, it may be possible to choose a transformation of one or both variables such that the transformed data exhibit a linear relation. A simple linear regression analysis can then be performed on the transformed data.

Multiple regression analysis is a versatile technique of building a prediction model with several input variables. In addition to obtaining the least squares fit, we can construct confidence intervals and test hypotheses about the influence of each input variable.

A polynomial regression model is a special case of multiple regression where the powers x, x^2, x^3, and so on, of a single predictor x play the role of the individual predictors.

The highest power of x that occurs in the model is called the degree or order of the regression model. A quadratic function, or second-degree polynomial, is commonly fit as an alternative to a straight line.

Both the polynomial regression model and the multiple regression with several predictors are special cases of general linear models.

The measure R^2, called the square of the multiple correlation coefficient, represents the proportion of y variability that is explained by the fitted multiple regression model.

$$R^2 = \frac{\text{Regression SS}}{\text{Total SS}}$$

To safeguard against a misuse of regression analysis, we must scrutinize the data for agreement with the model assumptions. An examination of the residuals, especially by graphical plots, including a dot diagram or histogram, a plot versus predicted value, and a plot versus time order, is essential for detecting possible violations of the assumptions and also identifying the appropriate modifications of an initial model.

TECHNOLOGY

Regression with two or more predictors and quadratic regression

MINITAB

Regression with two or more predictors

Begin with the values for the two predictor variables in C2 and C3 and the response variable y in *C1*.

> **Stat > Regression > Regression.**
> Type *C1* in **Response.** Type C2 and C3 in **Predictors.**
> Select **Graphs.** Click **Four in one.** Click **OK.**
> Click **OK.**

The graphics step produces four residual plots: histogram, normal probability plot, residual versus fit, and residual versus order.

Transforming data

We illustrate with the predictor variable x in C2 and transforming to $\log(x)$, where the logarithm is base 10 in C3.

> **Calc > Calculator.**
> Type C3 in **Store results in variable** and *LOGT(C2)* in **Expression.**
> Click **OK.**

Fitting a quadratic regression model

> With the values of x in *C1* and the y values in C2, you must select:

> **Stat > Regression > Fitted Line Plot.**
> Enter C2 in **Response (Y)** and enter *C1* in **Predictor (X).**
> Under **Type of Regression Model** choose **Quadratic.** Click **OK.**

TI-84/-83 PLUS

Fitting a quadratic regression model

Enter the values of the predictor variable in **L1** and those of the response variable in **L2**.

> Select **STAT,** then **CALC,** and then **5: QuadReg (ax + b).**
> With **LinReg** on the Home screen press **Enter.**

The calculator will give a, b, and c in the equation

$$y = ax^2 + bx + c$$

5. REVIEW EXERCISES

12.15 Given the pairs of (x, y) values

x	100	150	250	250	400	650	1000	1600
y	21	20	24	17	18	10	11	9

(a) Transform the x values to $x' = \log_{10} x$ and plot the scatter diagram of y versus x'.

(b) Fit a straight line regression to the transformed data.

(c) Obtain a 90% confidence interval for the slope of the regression line.

(d) Estimate the expected y value corresponding to $x = 300$ and give a 95% confidence interval.

*12.16 Obtain a linearizing transformation in each case.

(a) $y = \dfrac{1}{(1 + ae^{bx})^2}$

(b) $y = e^{ax^b}$

12.17 A genetic experiment is undertaken to study the competition between two types of female *Drosophila melanogaster* in cages with one male genotype acting as a substrate. The independent variable x is the time spent in cages, and the dependent variable y is the ratio of the numbers of type 1 to type 2 females. The following data (courtesy of C. Denniston) are recorded.

Time x (days)	No. Type 1	No. Type 2	$y = \dfrac{\text{No. Type 1}}{\text{No. Type 2}}$
17	137	586	.23
31	278	479	.58
45	331	167	1.98
59	769	227	3.39
73	976	75	13.01

(a) Plot the scatter diagram of y versus x and determine if a linear model of relation is appropriate.

(b) Determine if a linear relation is plausible for the transformed data $y' = \log_{10} y$.

(c) Fit a straight line regression to the transformed data.

12.18 A multiple linear regression was fitted to a data set obtained from 27 runs of an experiment, in which four predictors x_1, x_2, x_3, and x_4 were observed along with the response y. The following results were obtained.

$$\hat{\beta}_0 = -8.51 \qquad \hat{\beta}_1 = 1.93$$
$$\hat{\beta}_2 = 20.2 \qquad \hat{\beta}_3 = -.828 \qquad \hat{\beta}_4 = 5.91$$

$$\text{SS due to regression} = 920.15$$
$$\text{SSE} = 88.21$$

(a) Predict the response for

   (i) $x_1 = 16, x_2 = 5, x_3 = 5, x_4 = 4.6$

   (ii) $x_1 = 25, x_2 = .8, x_3 = 1, x_4 = 2.3$

(b) Estimate the error standard deviation σ and state the degrees of freedom.

(c) What proportion of the y variability is explained by the fitted regression?

12.19 Refer to Exercise 12.18. The estimated standard errors of $\hat{\beta}_1$ and $\hat{\beta}_2$ were .062 and 2.43, respectively.

(a) Obtain a 90% confidence interval for β_1.

(b) Test $H_0 : \beta_2 = 25$ versus $H_1 : \beta_2 < 25$ with $\alpha = .05$.

12.20 A second-degree polynomial $\hat{y} = \hat{\beta}_0 + \hat{\beta}_1 x + \hat{\beta}_2 x^2$ is fitted to a response y, and the following predicted values and residuals are obtained.

$\hat{y}$	Residuals
4.01	.28
5.53	$-.33$
6.21	$-.21$
6.85	.24
8.17	$-.97$
8.34	.46
8.81	.79
9.62	-1.02
10.05	1.35
10.55	-1.55
10.77	.63
10.77	1.73
10.94	-2.14
10.98	1.92
10.98	-1.18

Do the assumptions appear to be violated?

12.21 The following predicted values and residuals are obtained in an experiment conducted to determine the degree to which the yield of an important chemical in the manufacture of penicillin is dependent on sugar concentration (the time order of the experiments is given in parentheses).

Predicted	Residual
2.2(9)	− 1
3.1(6)	− 2
2.5(13)	3
3.3(1)	− 3
2.3(7)	− 1
3.6(14)	5
2.6(8)	0
2.5(3)	0
3.0(12)	3
3.2(4)	− 2
2.9(11)	2
3.3(2)	− 5
2.7(10)	0
3.2(5)	1

(a) Plot the residuals against the predicted values and also against the time order.

(b) Do the basic assumptions appear to be violated?

12.22 An experimenter obtains the following residuals after fitting a quadratic expression in x.

$x = 1$	$x = 2$	$x = 3$	$x = 4$	$x = 5$
− .1	1.3	− .1	0	− .2
0	− .2	− .3	.2	0
− .2	− .1	.1	− .1	− .2
.6	− .3	.4	0	− .2
− .1	.1	− .1	− .2	− .3
		.1		− .1

Do the basic assumptions appear to be violated?

12.23 An interested student used the method of least squares to fit the straight line $\hat{y} = 264.3 + 18.77x$ to gross national product, y, in real dollars. The results for 26 recent years, $x = 1, 2, \ldots , 26$, appear below. Which assumption(s) for a linear regression model appear to be seriously violated by the data? (*Note:* Regression methods are usually not appropriate for this type of data.)

Year	y	$\hat{y}$	Residual
1	309.9	283.1	26.8
2	323.7	301.9	21.8
3	324.1	320.6	3.5
4	355.3	339.4	15.9
5	383.4	358.2	25.2
6	395.1	376.9	18.2
7	412.8	395.7	17.1
8	407	414.5	− 7.5
9	438	433.2	4.8
10	446.1	452.0	− 5.9
11	452.5	470.8	− 18.3
12	447.3	489.5	− 42.2
13	475.9	508.3	− 32.4
14	487.7	527.1	− 39.4
15	497.2	545.8	− 48.6
16	529.8	564.6	− 34.8
17	551	583.4	− 32.4
18	581.1	602.1	− 21.0
19	617.8	620.9	− 3.1
20	658.1	639.7	18.4
21	675.2	658.4	16.8
22	706.6	677.2	29.4
23	725.6	696.0	29.6
24	722.5	714.7	7.8
25	745.4	733.5	11.9
26	790.7	752.3	38.4

The Following Exercises Require a Computer

12.24 Consider the data on male wolves in Table D.9 of the Data Bank concerning age (years) and canine length (mm).

(a) Obtain the least squares fit of canine length to the predictor age.

(b) Obtain the least squares fit of canine length to a quadratic function of the predictor age. The MINITAB commands are

---

Data: DBT9.DAT
C2:   4   2   4 · · · 0
C6: 28.7   27.0   27.2 · · · 24.5
Dialog box:

Stat > Regression > Fitted line plot
Type C6 in **Response.**
Type C2 in **Predictor.**
Click **Quadratic.** Click **OK.**

(c) What proportion of the y variability is explained by the quadratic regression model?

(d) Compare the estimated standard deviations, s, of the random error term in parts (a) and (b).

12.25 Given the pairs of (x, y) values

x	3.2	3.8	5.0	6.2	6.9	7.0	8.0	8.2	8.9	9.8
y	98	118	130	158	185	195	220	233	263	322

(a) Fit a quadratic regression model $Y = \beta_0 + \beta_1 x + \beta_2 x^2 + e$ to these data.

(b) What proportion of the y variability is explained by the quadratic regression model?

(c) Test $H_0: \beta_1 = 0$ versus $H_1: \beta_1 < 0$ with $\alpha = .05$.

12.26 Listed below are the price quotations for a midsize foreign used car along with their age and odometer mileage.

Age (years) x_1	Mileage x_2 (thousand miles)	Price y (thousand miles)
1	14	16.9
2	44	12.9
2	20	13.9
4	36	13.0
4	66	8.8
5	59	8.9
7	100	5.6
7	95	5.7
8	38	6.0

Perform a multiple regression analysis of these data. In particular

(a) Determine the equation for predicting the price from age and mileage. Interpret the meaning of the coefficients $\hat{\beta}_1$ and $\hat{\beta}_2$.

(b) Give 95% confidence intervals for $\hat{\beta}_1$ and $\hat{\beta}_2$.

(c) Obtain R^2 and interpret the result.

12.27 Refer to the data of speed x and stopping distance y given in Table 1. The MINITAB commands for fitting a straight line regression to $y' = \sqrt{y}$ and x are

---

Data: C12T1.DAT

C1: 20   20   30   30   30   40   40   50   50   60

C2: 16.3 26.7 39.2 63.5 51.3 98.4 65.7 104.1 155.6 217.2

Dialog box:

Calc > Calculator

Type *SQRT(C2)* in the **Expression** box.

Type C3 in **Store** box. Click **OK**.

Stat > Regression > Regression

Type C3 in **Response**.

Type C1 in **Predictors**. Click **OK**.

(a) Obtain the computer output and identify the equation of the fitted line and the value of r^2 (see Example 1).

(b) Give a 95% confidence interval for the slope.

(c) Obtain a 95% confidence interval for the expected y' value at $x = 45$.

12.28 A forester seeking information on basic tree dimensions obtains the following measurements of the diameters 4.5 feet above the ground and the heights of 12 sugar maple trees (courtesy of A. Ek). The forester wishes to determine if the diameter measurements can be used to predict the tree height.

Diameter x (inches)	Height y (feet)
.9	18
1.2	26
2.9	32
3.1	36
3.3	44.5
3.9	35.6
4.3	40.5
6.2	57.5
9.6	67.3
12.6	84
16.1	67
25.8	87.5

(a) Plot the scatter diagram and determine if a straight line relation is appropriate.

(b) Determine an appropriate linearizing transformation. In particular, try $x' = \log x$, $y' = \log y$.

(c) Fit a straight-line regression to the transformed data.

(d) What proportion of variability is explained by the fitted model?

12.29 Recorded here are the scores x_1 and x_2 in two midterm examinations, the GPA x_3, and the final examination score y for 20 students in a statistics class.

x_1	x_2	x_3	y	x_1	x_2	x_3	y
87	25	2.9	60	93	60	3.2	44
100	84	3.3	80	92	69	3.1	53
91	52	3.5	73	100	86	3.6	86
85	60	3.7	83	80	67	3.5	59
56	76	2.8	33	100	96	3.8	81
81	28	3.1	65	69	51	2.8	20
85	67	3.1	53	80	75	3.6	64
96	83	3.0	68	74	70	3.1	38
79	60	3.7	88	79	66	2.9	77
96	69	3.7	89	95	83	3.3	47

(a) Ignoring the data of GPA and the first midterm score, fit a simple linear regression of y on x_2. Compute r^2.

(b) Fit a multiple linear regression to predict the final examination score from the GPA and the scores in the midterms. Compute R^2.

(c) Interpret the values of r^2 and R^2 obtained in parts (a) and (b).

12.30 Refer to Exercise 11.64.

(a) Fit a quadratic model $Y = \beta_0 + \beta_1 x + \beta_2 x^2 + e$ to the data for CLEP scores y and CQT scores x.

(b) Use the fitted regression to predict the expected CLEP score when $x = 160$.

(c) Compute r^2 for fitting a line and R^2 for fitting a quadratic expression. Interpret these values and comment on the improvement of fit.

*12.31 Write the design matrix X for fitting a multiple regression model to the data of Exercise 12.26.

*12.32 Write the design matrix X for fitting a quadratic regression model using the data of Exercise 12.25.

12.33 Refer to the physical conditioning data given in Table D.5 of the Data Bank. Use MINITAB or some other package to fit a regression of the final number of situps on the initial number of situps and the gender of the student.

12.34 Refer to the physical fitness data in Table D.5 of the Data Bank. Use both the data on the pretest run time and gender for predicting the posttest run time. Obtain the least squares fit and plot the residuals versus fitted value.

13

Analysis of Categorical Data

Are Firm Mattresses Really Best?

Although it is commonly believed that firm mattresses can help persons with low-back pain, experimental evidence was lacking. Medical researchers[1] designed a carefully designed experiment. Of the patients admitted to the study, 158 were randomly selected to sleep on a new firm mattress and the others were given new medium-firm mattresses. The patients were not told about the firmness of their new mattresses although most perceived the correct firmness. The persons that installed the new beds and the research assistants who collected the data from patients were also kept unaware of the firmness of the mattresses.

This design includes meritable features: (1) the random assignment of treatments (mattress firmness) and (2) the *double blind* feature where patient and researcher are not told which treatment has been assigned.

The patients were asked to report whether or not they had low-back pain upon arising during the 90-day follow-up period.

Firmness	No	Yes	Total
Firm	36	122	158
Medium firm	55	95	150

The proportion with no low-back pain upon arising is .228 = 36/158 for those who slept on firm mattresses and the proportion is .367 = 55/150 for those who slept on medium-firm mattresses. These data provide convincing evidence that medium-firm mattresses reduce low-back pain upon arising. The *P*–value for the test of equality of the proportions, persons in each population who would answer *No*, versus a two-sided alternative is .008. In less than 1 in 100 times would chance variation give an observed difference this large or larger.

[1]*Source:* F. Kovacs et al., "Effect of firmness of mattress on chronic non-specific low-back pain: randomized, double-blind, controlled, multicentre trial," The *Lancet* **362** (November 15, 2003), pp. 1599–1604

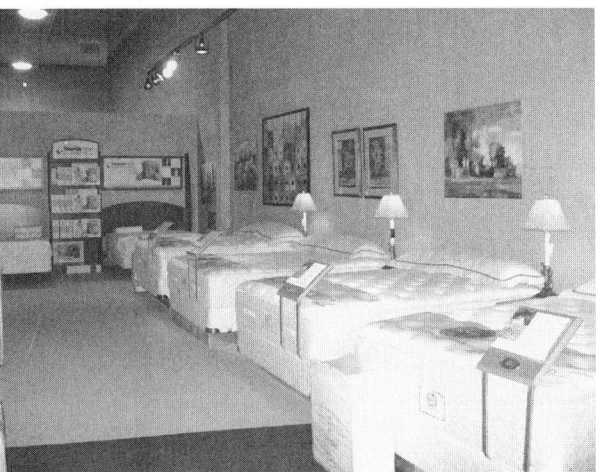

Courtesy of Roberta Johnson.

1.   INTRODUCTION

The expression categorical data refers to observations that are only classified into categories so that the data set consists of frequency counts for the categories. Such data occur abundantly in almost all fields of quantitative study, particularly in the social sciences. In a study of religious affiliations, people may be classified as Catholic, Protestant, Jewish, or other; in a survey of job compatibility, employed persons may be classified as being satisfied, neutral, or dissatisfied with their jobs; in plant breeding, the offspring of a cross-fertilization may be grouped into several genotypes; manufactured items may be sorted into such categories as "free of defects," "slightly blemished," and "rejects." In all these examples, each category is defined by a qualitative trait. Categories can also be defined by specifying ranges of values on an original numerical measurement scale, such as income that is categorized high, medium, or low and rainfall that is classified heavy, moderate, or light.

The next three examples present the testing problems addressed in this chapter under the umbrella term of chi-square, or χ^2, tests.

Example 1   One Sample Classified in Several Categories

The offspring produced by a cross between two given types of plants can be any of the three genotypes denoted by A, B, and C. A theoretical model of gene inheritance suggests that the offspring of types A, B, and C should be in the ratio $1:2:1$. For experimental verification, 100 plants are bred by crossing the two given types. Their genetic classifications are recorded in Table 1. Do these data contradict the genetic model?

TABLE 1   Classification of Crossbred Plants

Genotype	A	B	C	Total
Observed frequency	18	55	27	100

Let us denote the population proportions or the probabilities of the genotypes A, B, and C by p_A, p_B, and p_C, respectively. Since the genetic model states that these probabilities are in the ratio $1:2:1$, our object is to test the null hypothesis

$$H_0 : p_A = \frac{1}{4} \qquad p_B = \frac{2}{4} \qquad p_C = \frac{1}{4}$$

Here the data consist of frequency counts of a random sample classified in three categories or cells, the null hypothesis specifies the numerical values of the cell probabilities, and we wish to examine if the observed frequencies contradict the null hypothesis.

Example 2   Independent Samples Classified in Several Categories

To compare the effectiveness of two diets A and B, 150 infants were included in a study. Diet A was given to 80 randomly selected infants and diet B to the other 70 infants. At a later time, the health of each infant was observed and classified into one of the three categories "excellent," "average," and "poor." From the frequency counts recorded in Table 2, we wish to test the null hypothesis that there is no difference between the quality of the two diets.

TABLE 2   Health under Two Different Diets

	Excellent	Average	Poor	Sample Size
Diet A	37	24	19	80
Diet B	17	33	20	70
Total	54	57	39	150

The two rows of Table 2 have resulted from independent samples. For a descriptive summary of these data, it is proper to compute the relative frequencies for each row. These are given in Table 2(a).

TABLE 2(a)   Relative Frequencies (from Table 2)

	Excellent	Average	Poor	Total
Diet A	.46	.30	.24	1
Diet B	.24	.47	.29	1

The (unknown) population proportions or probabilities are entered in Table 2(b). They allow us to describe the null hypothesis more clearly. The null hypothesis of "no difference" is equivalent to the statement that, for each response category, the probability is the same for diet A and diet B. Consequently, we formulate

$$H_0: \quad p_{A1} = p_{B1} \qquad p_{A2} = p_{B2} \qquad p_{A3} = p_{B3}$$

Note that although H_0 specifies a structure for the cell probabilities, it does not give the numerical value of the common probability in each column.

TABLE 2(b)   Population Proportions or Probabilities

	Excellent	Average	Poor	Total
Diet A	p_{A1}	p_{A2}	p_{A3}	1
Diet B	p_{B1}	p_{B2}	p_{B3}	1

Example 3   One Sample Simultaneously Classified According to Two Characteristics

A random sample of 500 persons is questioned regarding political affiliation and attitude toward a tax reform program. From the observed frequency table given in Table 3, we wish to answer the following question: Do the data indicate that the pattern of opinion is different between the two political groups?

TABLE 3   Political Affiliation and Opinion

	Favor	Indifferent	Opposed	Total
Democrat	138	83	64	285
Republican	64	67	84	215
Total	202	150	148	500

Unlike Example 2, here we have a single random sample, but each sampled individual elicits two types of responses: political affiliation and attitude. In the present context, the null hypothesis of "no difference" amounts to saying that the two types of responses are independent. In other words, attitude to the program is unrelated to or independent of a person's political affiliation. A formal specification of this null hypothesis, in terms of the cell probabilities, is deferred until Section 4.

Frequency count data that arise from a classification of the sample observations according to two or more characteristics are called **cross-tabulated data** or a **contingency table**. If only two characteristics are observed and the contingency table has r rows and c columns, it is designated as an $r \times c$ table. Thus, Tables 2 and 3 are both 2×3 contingency tables.

Although Tables 2 and 3 have the same appearance, there is a fundamental difference in regard to the method of sampling. The row totals 80 and 70 in Table 2 are the predetermined sample sizes; these are not outcomes of random sampling, as are the column totals. By contrast, both sets of marginal totals in Table 3 are outcomes of random sampling—none were fixed beforehand. To draw the distinction, one often refers to Table 2 as a 2×3 contingency table with fixed row totals. In Sections 3 and 4, we will see that the formulation of the null hypothesis is different for the two situations.

2. PEARSON'S χ^2 TEST FOR GOODNESS OF FIT

We first consider the type of problem illustrated in Example 1, where the data consist of frequency counts observed from a random sample and the null hypothesis specifies the unknown cell probabilities. Our primary goal is to test if the model given by the null hypothesis fits the data, and this is appropriately called testing for goodness of fit.

For general discussion, suppose a random sample of size n is classified into k categories or cells labeled $1, 2, \ldots, k$ and let $n_1, n_2, \ldots, n_k$ denote the respective cell frequencies. If we denote the cell probabilities by $p_1, p_2, \ldots, p_k$, a null hypothesis that completely specifies the cell probabilities is of the form

$$H_0: \quad p_1 = p_{10}, \ldots, p_k = p_{k0}$$

where $p_{10}, \ldots, p_{k0}$ are given numerical values that satisfy $p_{10} + \cdots + p_{k0} = 1$.

From Chapter 5 recall that if the probability of an event is p, then the expected number of occurrences of the event in n trials is np. Therefore, once the cell probabilities are specified, the expected cell frequencies can be readily computed by **multiplying** these probabilities by the sample size n. A goodness of fit test attempts to determine if a conspicuous discrepancy exists between the observed cell frequencies and those expected under H_0. (See Table 4.)

TABLE 4   The Basis of a Goodness of Fit Test

Cells	1	2	$\cdots$	k	Total
Observed frequency O	n_1	n_2	$\cdots$	n_k	n
Probability under H_0	p_{10}	p_{20}	$\cdots$	p_{k0}	1
Expected frequency E under H_0	np_{10}	np_{20}	$\cdots$	np_{k0}	n

A useful measure for the overall discrepancy between the observed and expected frequencies is given by the chi-square or χ^2 statistic

$$\chi^2 = \sum_{i=1}^{k} \frac{(n_i - np_{i0})^2}{np_{i0}} = \sum_{\text{cells}} \frac{(O - E)^2}{E}$$

where O and E symbolize an observed frequency and the corresponding expected frequency. The discrepancy in each cell is measured by the squared difference between the observed and the expected frequencies divided by the expected frequency. The χ^2 measure is the sum of these quantities for all cells.

The χ^2 statistic was originally proposed by Karl Pearson (1857–1936), who found the distribution for large n to be approximately a χ^2 distribution with d.f. $= k - 1$. Due to this distribution, the statistic is denoted by χ^2 and called *Pearson's χ^2 statistic for goodness of fit*. Because a large value of the overall discrepancy indicates a disagreement between the data and the null hypothesis, the upper tail of the χ^2 distribution constitutes the rejection region.

Pearson's χ^2 Test for Goodness of Fit (Based on Large n)

Null hypothesis

$$H_0: \quad p_1 = p_{10}, \ldots, p_k = p_{k0}$$

Test statistic

$$\chi^2 = \sum_{i=1}^{k} \frac{(n_i - np_{i0})^2}{np_{i0}} = \sum_{\text{cells}} \frac{(O - E)^2}{E}$$

Rejection region

$$\chi^2 \geq \chi_\alpha^2$$

where χ_α^2 is the upper α point of the χ^2 distribution with

$$\text{d.f.} = k - 1 = (\text{Number of cells}) - 1$$

It should be remembered that Pearson's χ^2 test is an approximate test that is valid only for large samples. As a rule of thumb, n should be large enough so that the expected frequency of each cell is at least 5.

Example 4 χ^2 Goodness of Fit for a Genetic Model

Referring to Example 1, test the goodness of fit of the genetic model to the data in Table 1. Take $\alpha = .05$.

SOLUTION   Following the structure of Table 4, the computations for the χ^2 statistic are exhibited in Table 5.

TABLE 5   The χ^2 Goodness of Fit Test for the Data in Table 1

Cell	A	B	C	Total
Observed frequency O	18	55	27	100
Probability under H_0	.25	.50	.25	1.0
Expected frequency E	25	50	25	100
$\dfrac{(O - E)^2}{E}$	1.96	.50	.16	2.62 = χ^2 d.f. = 2

We use the χ^2 statistic with rejection region $R : \chi^2 \geq 5.99$ since $\chi^2_{.05} = 5.99$ with d.f. = 2 (Appendix B, Table 5). Because the observed $\chi^2 = 2.62$ is smaller than this value, the null hypothesis is not rejected at $\alpha = .05$. We conclude that the data in Table 1 do not contradict the genetic model.

The χ^2 statistic measures the overall discrepancy between the observed frequencies and those expected under a given null hypothesis. Example 4 demonstrates its application when the frequency counts arise from a single random sample and the categories refer to only one characteristic—namely, the genotype of the offspring. Basically, the same principle extends to testing hypotheses with more complex types of categorical data such as the contingency tables illustrated in Examples 2 and 3. In preparation for these developments, we state two fundamental properties of the χ^2 statistic:

Properties of Pearson's χ^2 Statistic

1. **Additivity:** If χ^2 statistics are computed from independent samples, then their sum is also a χ^2 statistic whose d.f. equals the sum of the d.f.'s of the components.

2. **Loss of d.f. due to estimation of parameters:** If H_0 does not completely specify the cell probabilities, then some parameters have to be estimated in order to obtain the expected cell frequencies. In that case, the d.f. of χ^2 is reduced by the number of parameters estimated.

    d.f. of χ^2 = (No. of cells) − 1 − (No. of parameters estimated)

Exercises

13.1 Given below are the frequencies observed from 320 tosses of a die. Do these data cast doubt on the fairness of the die?

Face No.	1	2	3	4	5	6	Total
Frequency	39	63	56	67	57	38	320

13.2 Recorded here is the frequency distribution of the blood types of 100 persons who have volunteered to donate blood at a plasma center.

Blood type	O	A	B	AB	Total
Frequency	40	44	10	6	100

Test the goodness of fit of the model, which assumes that the four blood types are equally likely in the population of plasma donors. Use $\alpha = .05$.

13.3 Referring to the data in Exercise 13.2, test the null hypothesis that the probability of the blood types O, A, B, and AB is in the ratios 4:4:1:1. Use $\alpha = .05$.

13.4 A market researcher wishes to assess consumers' preference among three different colors available on name-brand dishwashers. The following frequencies were observed from a random sample of 150 recent sales.

Color	White	Stainless	Black	Total
Frequency	63	56	31	150

Test the null hypothesis, at $\alpha = .05$, that all three colors are equally popular.

13.5 A shipment of assorted nuts is labeled as having 45% walnuts, 20% hazelnuts, 20% almonds, and 15% pistachios. By randomly picking several scoops of nuts from this shipment, an inspector finds the counts shown at the bottom.
Could these findings be a strong basis for an accusation of mislabeling? Test at $\alpha = .025$.

	Counts
Walnuts	95
Hazelnuts	70
Almonds	33
Pistachios	42
Total	240

13.6 Cross-fertilizing a pure strain of red flowers with a pure strain of white flowers produces pink hybrids that have one gene of each type. Crossing these hybrids can lead to any one of four possible gene pairs. Under Mendel's theory, these four are equally likely, so

$$P(\text{white}) = \tfrac{1}{4} \quad P(\text{pink}) = \tfrac{1}{2} \quad P(\text{red}) = \tfrac{1}{4}$$

An experiment carried out by Correns, one of Mendel's followers, resulted in the frequencies 141, 291, and 132 for the white, pink, and red flowers, respectively. (*Source*: W. Johannsen, 1909, *Elements of the Precise Theory of Heredity*, G. Fischer, Jena.)

Do these observations appear to contradict the probabilities suggested by Mendel's theory?

13.7 According to the records of the National Safety Council, accidental deaths in the United States during a recent year had the following distribution according to the principal types of accidents.

Motor Vehicle	Falls	Drowning	Burns	Poison	Other
46%	15%	4%	4%	8%	23%

In a particular geographical region, the accidental deaths, classified according to the principal types of accidents, yielded the following frequency distribution.

Motor Vehicle	Falls	Drowning	Burns	Poison	Other	Total
462	171	76	57	92	74	932

Do these data show a significantly different distribution of accidental deaths as compared to that for the entire United States? Test at $\alpha = .01$.

13.8 The following table, based on government data, shows the frequency distribution of births by day of the week for all registered births in the United States in a recent year. Test the null hypothesis, at $\alpha = .01$, that all seven days of the week are equally likely for childbirth.

Number of Births (in 10,000) by Day of the Week, United States

	Number of births
Mon.	57.18
Tues.	64.44
Wed.	62.67
Thurs.	61.74
Fri.	60.84
Sat.	45.38
Sun.	40.71
All Days	392.96

Source: S. Ventura et al, "Births: Final Data for 1998," *National Vital Statistics Report* **48,** No. 3 (2000), p. 43, Table 16.

13.9 Observations of 80 litters, each containing 3 rabbits, reveal the following frequency distribution of the number of male rabbits per litter.

Number of males in litter	0	1	2	3	Total
Number of litters	19	32	22	7	80

Under the model of Bernoulli trials for the sex of rabbits, the probability distribution of the number of males per litter should be binomial with 3 trials and p = probability of a male birth. From these data, the parameter p is estimated as

$$\hat{p} = \frac{\text{Total number of males in 80 litters}}{\text{Total number of rabbits in 80 litters}}$$

$$= \frac{97}{240} \approx .4$$

(a) Using the binomial table for three trials and $p = .4$, determine the cell probabilities.

(b) Perform the χ^2 test for goodness of fit. (In determining the d.f., note that one parameter has been estimated from the data.)

*13.10 *An alternative expression for Pearson's χ^2.* By expanding the square on the right-hand side of

$$\chi^2 = \sum_{\text{cells}} \frac{(n_i - np_{i0})^2}{np_{i0}}$$

show that the χ^2 statistic can also be expressed as

$$\chi^2 = \sum_{\text{cells}} \frac{n_i^2}{np_{i0}} - n \qquad \text{that is,}$$

$$\chi^2 = \sum_{\text{cells}} \frac{O^2}{E} - n$$

3. CONTINGENCY TABLE WITH ONE MARGIN FIXED (TEST OF HOMOGENEITY)

From each population, we draw a random sample of a predetermined sample size and classify each response in categories. These data form a two-way contingency table where one classification refers to the populations and the other to the response under study. Our objective is to test whether the populations are alike, or **homogeneous,** with respect to cell probabilities. To do so, we will determine if the observed proportions in each response category are nearly the same for all populations.

Let us pursue our development of the χ^2 test of homogenity with the data of Table 2.

Example 5 Developing a χ^2 Test to Compare Two Diets

Referring to Example 2, test the null hypothesis that there is no difference between the quality of the two diets.

SOLUTION   For ease of reference, the data in Table 2 are reproduced in Table 6. Here the populations correspond to the two diets and the response is recorded in three categories. The row totals 80 and 70 are the fixed sample sizes.

TABLE 6 A 2 × 3 Contingency Table with Fixed Row Totals

	Excellent	Average	Poor	Total
Diet A	37	24	19	80
Diet B	17	33	20	70
Total	54	57	39	150

We have already formulated the null hypothesis of "homogeneity" or "no difference between the diets" as [see Table 2(b)]

$$H_0: \quad p_{A1} = p_{B1} \qquad p_{A2} = p_{B2} \qquad p_{A3} = p_{B3}$$

If we denote these common probabilities under H_0 by p_1, p_2, and p_3, respectively, the expected cell frequencies in each row would be obtained by multiplying these probabilities by the sample size. In particular, the expected frequencies in the first row are $80p_1$, $80p_2$, and $80p_3$, and those in the second row are $70p_1$, $70p_2$, and $70p_3$. However, the p_i's are not specified by H_0. Therefore, we have to estimate these parameters in order to obtain the numerical values of the expected frequencies.

The column totals 54, 57, and 39 in Table 6 are the frequency counts of the three response categories in the combined sample of size 150. Under H_0, the estimated probabilities are

$$\hat{p}_1 = \frac{54}{150} \qquad \hat{p}_2 = \frac{57}{150} \qquad \hat{p}_3 = \frac{39}{150}$$

We use these estimates to calculate the expected frequencies in the first row as

$$80 \times \frac{54}{150} = \frac{80 \times 54}{150} \qquad \frac{80 \times 57}{150} \qquad \frac{80 \times 39}{150}$$

and similarly for the second row. Referring to Table 6, notice the interesting pattern in these calculations:

$$\text{Expected cell frequency} = \frac{\text{Row total} \times \text{Column total}}{\text{Grand total}}$$

Table 7(a) presents the observed frequencies O along with the expected frequencies E. The latter are given in parentheses. Table 7(b) computes the discrepancy measure $(O - E)^2/E$ for the individual cells. Adding these over all the cells, we obtain the value of the χ^2 statistic.

TABLE 7(a) The Observed and Expected Frequencies of the Data in Table 6

	Excellent	Average	Poor
Diet A	37	24	19
	(28.8)	(30.4)	(20.8)
Diet B	17	33	20
	(25.2)	(26.6)	(18.2)

TABLE 7(b) The Values of $(O - E)^2/E$

	Excellent	Average	Poor	
Diet A	2.335	1.347	.156	
Diet B	2.668	1.540	.178	
				$8.224 = \chi^2$

In order to determine the degrees of freedom, we employ the properties of the χ^2 statistic stated in Section 2. Our χ^2 statistic has been computed from two independent samples; each contributes $3 - 1 = 2$ d.f. because there are three categories. The added d.f. $= 2 + 2 = 4$ must now be reduced by the number of parameters we have estimated. Since p_1, p_2, and p_3 satisfy the relation $p_1 + p_2 + p_3 = 1$, there are really two undetermined parameters among them. Therefore, our χ^2 statistic has d.f. $= 4 - 2 = 2$.

With d.f. $= 2$, the tabulated upper 5% point of χ^2 is 5.99 (Appendix B, Table 5). Since the observed $\chi^2 = 8.224$ is larger, the null hypothesis is rejected at $\alpha = .05$. It would also be rejected at $\alpha = .025$. Therefore, a significant difference between the quality of the two diets is indicated by the data.

Having obtained a significant χ^2, we should now examine Tables 7(a) and 7(b) and try to locate the source of the significance. We find that large contributions to χ^2 come from the "excellent" category, where the relative frequency is 37/80, or 46%, for diet A and 17/70, or 24%, for diet B. These data indicate that A is superior.

Motivated by Example 5, we are now ready to describe the χ^2 test procedure for an $r \times c$ contingency table that has independent samples from r populations that are classified in c response categories. As we have seen before, the expected frequency of a cell is given by (row total $\times$ column total)/grand total. With regard to the d.f. of the χ^2 for an $r \times c$ table, we note that each of the r rows contributes $c - 1$ d.f.'s so the total contribution is $r(c - 1)$. Since $c - 1$ number of parameters have to be estimated,

$$
\begin{aligned}
\text{d.f. of } \chi^2 &= r(c - 1) - (c - 1) \\
&= (r - 1)(c - 1) \\
&= (\text{No. of rows} - 1) \times (\text{No. of columns} - 1)
\end{aligned}
$$

The χ^2 Test of Homogeneity in a Contingency Table

Null hypothesis

In each response category, the probabilities are equal for all the populations.

Test statistic

$$
\chi^2 = \sum_{\text{cells}} \frac{(O - E)^2}{E} \qquad
\begin{cases}
O = \text{Observed cell frequency} \\
E = \dfrac{\text{Row total} \times \text{Column total}}{\text{Grand total}}
\end{cases}
$$

$$
\text{d.f.} = (\text{No. of rows} - 1) \times (\text{No. of columns} - 1)
$$

Rejection region

$$
\chi^2 \geq \chi_\alpha^2
$$

Example 6 Conducting a χ^2 Test of Homogeneity

A survey is undertaken to determine the incidence of alcoholism in different professional groups. Random samples of the clergy, educators, executives, and merchants are interviewed, and the observed frequency counts are given in Table 8.

TABLE 8 Contingency Table of Alcoholism versus Profession

	Alcoholic	Nonalcoholic	Sample Size
Clergy	32(58.25)	268(241.75)	300
Educators	51(48.54)	199(201.46)	250
Executives	67(58.25)	233(241.75)	300
Merchants	83(67.96)	267(282.04)	350
Total	233	967	1200

Construct a test to determine if the incidence rate of alcoholism appears to be the same in all four groups.

SOLUTION Let us denote the proportions of alcoholics in the populations of the clergy, educators, executives, and merchants by p_1, p_2, p_3, and p_4, respectively. Based on independent random samples from four binomial populations, we want to test the null hypothesis

$$H_0: \quad p_1 = p_2 = p_3 = p_4$$

The expected cell frequencies, shown in parentheses in Table 8, are computed by multiplying the row and column totals and dividing the results by 1200. The χ^2 statistic is computed in Table 9. With d.f. $= 3$, the tabulated upper 5% point of χ^2 is 7.81 so that the null hypothesis is rejected at $\alpha = .05$. It would be rejected also at $\alpha = .01$, so the P-value is less than .01.

TABLE 9 The Values of $(O - E)^2/E$
for the Data in Table 8

	Alcoholic	Nonalcoholic
Clergy	11.83	2.85
Educators	.12	.03
Executives	1.31	.32
Merchants	3.33	.80

$$20.59 = \chi^2$$
$$\text{d.f. of } \chi^2 = (4 - 1)(2 - 1) = 3$$

Examining Table 9, we notice that a large contribution to the χ^2 statistic has come from the first row. This is because the relative frequency of alcoholics among the clergy is quite low in comparison to the others, as one can see from Table 8.

Example 7 The χ^2 Test for 2 × 2 Contingency Table

To determine the possible effect of a chemical treatment on the rate of seed germination, 100 chemically treated seeds and 150 untreated seeds are sown. The numbers of seeds that germinate are recorded in Table 10. Do the data provide strong evidence that the rate of germination is different for the treated and untreated seeds?

TABLE 10 Germination of Seeds

	Germinated	Not Germinated	Total
Treated	84(86.40)	16(13.60)	100
Untreated	132(129.60)	18(20.40)	150
Total	216	34	250

SOLUTION   Letting p_1 and p_2 denote the probabilities of germination for the chemically treated seeds and the untreated seeds, respectively, we wish to test the null hypothesis $H_0: p_1 = p_2$ versus $H_1: p_1 \neq p_2$. For the χ^2 test, we calculate the expected frequencies in the usual way. These are given in parentheses in Table 10. The computed value of χ^2 is

$$\chi^2 = .067 + .424 + .044 + .282$$
$$= .817$$
$$\text{d.f.} = (2 - 1)(2 - 1) = 1$$

The tabulated 5% value of χ^2 with d.f. $= 1$ is 3.84. Because the observed $\chi^2 = .817$ is smaller, the null hypothesis is not rejected at $\alpha = .05$. The rate of germination is not significantly different between the treated and untreated seeds.

ANOTHER METHOD OF ANALYZING A 2 × 2 CONTINGENCY TABLE

In light of Example 7, we note that a 2 × 2 contingency table, with one margin fixed, is essentially a display of independent random samples from two dichotomous (i.e., two-category) populations. This structure is shown in Table 11, where we have labeled the two categories "success" and "failure." Here X and Y denote the numbers of successes in independent random samples of sizes n_1 and n_2 taken from population 1 and population 2, respectively.

TABLE 11   Independent Samples from Two Dichotomous Populations

	No. of Successes	No. of Failures	Sample Size
Population 1	X	$n_1 - X$	n_1
Population 2	Y	$n_2 - Y$	n_2

If we let p_1 and p_2 denote the probabilities of success for populations 1 and 2, respectively, our objective is to test the null hypothesis $H_0: p_1 = p_2$. The sample proportions

$$\hat{p}_1 = \frac{X}{n_1} \quad \text{and} \quad \hat{p}_2 = \frac{Y}{n_2}$$

provide estimates of p_1 and p_2. When the sample sizes are large, a test of $H_0: p_1 = p_2$ can be based on (see Section 6 of Chapter 10)

Test statistic

$$Z = \frac{\hat{p}_1 - \hat{p}_2}{\sqrt{\hat{p}(1 - \hat{p})}\sqrt{\dfrac{1}{n_1} + \dfrac{1}{n_2}}} \qquad \text{where } \hat{p} = \frac{X + Y}{n_1 + n_2}$$

The level α rejection region is $|Z| \geq z_{\alpha/2}$, $Z \leq -z_\alpha$, or $Z \geq z_\alpha$ according to whether the alternative hypothesis is $p_1 \neq p_2$, $p_1 < p_2$, or $p_1 > p_2$. Here z_α denotes the upper α point of the $N(0, 1)$ distribution.

Although the test statistics Z and

$$\chi^2 = \sum_{\text{cells}} \frac{(O - E)^2}{E}$$

appear to have quite different forms, there is an exact relation between them—namely,

$$Z^2 = \chi^2 \quad \text{(for a 2 × 2 contingency table)}$$

Also, $z_{\alpha/2}^2$ is the same as the upper α point of χ^2 with d.f. $= 1$. For instance, with $\alpha = .05$, $z_{.025}^2 = (1.96)^2 = 3.8416$, which is also the upper 5% point of χ^2 with d.f. $= 1$ (see Appendix B, Table 5). Thus, the two test procedures are equivalent when the alternative hypothesis is two-sided. However, if the alternative hypothesis is one-sided, such as $H_1 : p_1 > p_2$, only the Z test is appropriate.

Example 8   Conducting a Z test to Compare Two Proportions

Use the Z test with the data of Example 7. Also, determine a 95% confidence interval for $p_1 - p_2$.

SOLUTION   We calculate

$$\hat{p}_1 = \frac{84}{100} = .84 \qquad \hat{p}_2 = \frac{132}{150} = .88$$

$$\text{Pooled estimate } \hat{p} = \frac{84 + 132}{100 + 150} = .864$$

$$Z = \frac{\hat{p}_1 - \hat{p}_2}{\sqrt{\hat{p}(1 - \hat{p})}\sqrt{\dfrac{1}{n_1} + \dfrac{1}{n_2}}}$$

$$= \frac{-.04}{\sqrt{.864 \times .136}\sqrt{\dfrac{1}{100} + \dfrac{1}{150}}} = -.904$$

Because the observed $|Z|$ is smaller than $z_{.025} = 1.96$, the null hypothesis is not rejected at $\alpha = .05$. Note that $Z^2 = (-.904)^2 = .817$ agrees with the result $\chi^2 = .817$ found in Example 7.

For the confidence interval, we calculate

$$\hat{p}_1 - \hat{p}_2 = -.04$$

$$\sqrt{\frac{\hat{p}_1(1 - \hat{p}_1)}{n_1} + \frac{\hat{p}_2(1 - \hat{p}_2)}{n_2}} = \sqrt{\frac{.84 \times .16}{100} + \frac{.88 \times .12}{150}}$$

$$= .045$$

A 95% confidence interval for $p_1 - p_2$ is

$$-.04 \pm 1.96 \times .045 = -.04 \pm .09 \quad \text{or} \quad (-.13, .05)$$

Exercises

13.11 Among a sample of 800 adult males, 414 said they usually open all of their mail. Among 900 adult females, 532 said they usually open all of their mail.[2]

(a) Construct a two-way table based on these frequencies.

(b) Formulate the null hypothesis.

(c) Conduct a χ^2 test of your null hypothesis. Use $\alpha = .05$.

(d) Comment on the form of any departure from the null hypothesis.

13.12 A sample of 250 students majoring in business and a sample of 500 executives were asked to respond to the question. Should corporations become more directly involved with social issues such as homelessness, education, and drugs?[3]

	More Involved	Not More Involved	Not Sure
Executives	345	135	20
Business students	222	20	8

(a) Formulate the null hypothesis of no difference of opinion about involvement in social issues.

[2]These proportions are close to those obtained in a recent Gallup survey.
[3]The resulting proportions are close to those obtained in a recent Harris poll.

(b) Conduct a χ^2 test of your null hypothesis. Use $\alpha = .05$.

(c) Comment on the form of any departure from the null hypothesis.

13.13 Nausea from air sickness affects some travelers. In a comparative study of the effectiveness of two brands of motion sickness pills, brand A pills were given to 45 persons randomly selected from a group of 90 air travelers, while the other 45 persons were given brand B pills. The following results were obtained.

	Degree of Nausea				
	None	Slight	Moderate	Severe	Total
Brand A	18	17	6	4	45
Brand B	11	14	14	6	45

Do these observations demonstrate that the two brands of pills are significantly different in quality? Test at $\alpha = .05$.

13.14 Refer to the hardness of mattresses data in the chapter front piece. Confirm that these data establish a difference in the proportions who did not have lower back pain using:

(a) The χ^2 test with level $\alpha = .01$.

(b) The Z test with level $\alpha = .01$. Calculate the P-value.

13.15 A community paper in the Spanish language was delivered to many sites in the San Francisco Bay area. As a check on the circulation numbers that are important to advertisers, a survey was conducted at four drop sites. The number of papers delivered to each site and the number remaining after 3 days was recorded

	Papers Delivered	Papers Remaining
Site 1	50	17
Site 2	47	12
Site 3	48	7
Site 4	50	21

(a) Formulate a null hypothesis of no difference in the proportions of papers taken from the sites.

(b) Conduct a χ^2 test of your null hypothesis. Use $\alpha = .05$.

(c) Let $p_1, p_2, p_3,$ and p_4 denote the probabilities that a paper will be taken from drop sites 1, 2, 3, and 4, respectively. Construct the four individual 95% confidence intervals and plot these intervals.

13.16 Using the data for site 1 and site 3 in Exercise 13.15, make a 2 × 2 table and test $H_0 : p_1 = p_3$ versus $H_1 : p_1 \neq p_3$ at $\alpha = .05$ using:

(a) The χ^2 test.

(b) The Z test.

13.17 Referring to the data for site 3 and site 4 in Exercise 13.15, make a 2 × 2 table and test $H_0 : p_3 = p_4$ versus $H_1 : p_3 > p_4$ at $\alpha = .05$:

(a) Is there strong evidence that the probability a paper will be taken from the drop site is higher for site 3 than for site 4? Answer by calculating the P-value.

(b) Construct a 95% confidence interval for $p_3 - p_4$.

13.18 Osteoporosis, or a loss of bone minerals, is a common cause of broken bones in the elderly. A researcher on aging conjectures that bone mineral loss can be reduced by regular physical therapy or certain kinds of physical activity. A study is conducted on 200 elderly subjects of approximately the same age divided into control, physical therapy, and physical activity groups. After a suitable period of time, the nature of change in bone mineral content is observed.

	Change in Bone Mineral			
	Appreciable Loss	Little Change	Appreciable Increase	Total
Control	38	15	7	60
Therapy	22	32	16	70
Activity	15	30	25	70
Total	75	77	48	200

Do these data indicate that the change in bone mineral varies for the different groups?

The Following Exercises May Require a Computer

13.19 *Using the computer.* The analysis of a contingency table can be conveniently done on a computer. For an illustration, we present here a MINITAB analysis of the data in Table 6, Example 5.

Data: C13T6.DAT

C1: 37 17
C2: 24 33
C3: 19 20

Dialog box:

Stat > Tables > Chisquare Test
Type C1-C3 in Columns containing the table. Click OK.

The **output** is as follows:

```
Chi-Square Test: C1, C2, C3

Expected    counts    are    printed    below
observed counts
Chi-Square    contributions    are    printed
below expected counts

            C1        C2        C3     Total
    1       37        24        19        80
         28.80     30.40     20.80
         2.335     1.347     0.156

    2       17        33        20        70
         25.20     26.60     18.20
         2.668     1.540     0.178

Total      54        57        39       150

Chi-Sq = 8.224, DF = 2, P-Value = 0.016
```

(a) Compare this output with the calculations presented in Example 5.

(b) Do Exercise 13.11 on a computer.

(c) Do Exercise 13.18 on a computer.

13.20 With reference to the sleep data in Table D.10 of the Data Bank, make two categories of snorers on the basis of the last variable: those that responded three or more times a week so their score is coded 4 or 5 and those that responded less than three times a week. Test the equality of proportions for males and females using the χ^2 test with $\alpha = .05$.

4. CONTINGENCY TABLE WITH NEITHER MARGIN FIXED (TEST OF INDEPENDENCE)

When two traits are observed for each element of a random sample, the data can be simultaneously classified with respect to these traits. We then obtain a two-way contingency table in which both sets of marginal totals are random. An illustration was already provided in Example 3. To cite a few other examples: A random sample of employed persons may be classified according to educational attainment and type of occupation; college students may be classified according to the year in college and attitude toward a dormitory regulation; flowering plants may be classified with respect to type of foliage and size of flower.

A typical inferential aspect of cross-tabulation is the study of whether the two characteristics appear to be manifested independently or certain levels of one characteristic tend to be associated or contingent with some levels of another.

Example 9 **Developing a χ^2 Test of Independence**

Analyze the data in Example 3.

SOLUTION   The 2×3 contingency table of Example 3 is given in Table 12. Here a single random sample of 500 persons is classified into six cells.

Dividing the cell frequencies by the sample size 500, we obtain the relative frequencies shown in Table 12(a). Its row marginal totals, .570 and .430, represent the sample proportions of Democrats and Republicans, respectively. Likewise, the column marginal totals show the sample proportions in the three categories of attitude.

TABLE 12   Contingency Table for Political Affiliation and Opinion

	Favor	Indifferent	Opposed	Total
Democrat	138	83	64	285
Republican	64	67	84	215
Total	202	150	148	500

TABLE 12(a)   Proportion of Observations in Each Cell

	Favor	Indifferent	Opposed	Total
Democrat	.276	.166	.128	.570
Republican	.128	.134	.168	.430
Total	.404	.300	.296	1

Imagine a classification of the entire population. The unknown population proportions (i.e., the probabilities of the cells) are represented by the entries in Table 12(b), where the suffixes D and R stand for Democrat and Republican, and 1, 2, and 3 refer to the "favor," "indifferent," and "opposed" categories. Table 12(b) is the population analogue of Table 12(a), which shows the sample proportions. For instance,

$$\text{Cell probability } p_{D1} = P(\text{Democrat } \textbf{and} \text{ in favor})$$
$$\text{Row marginal probability } p_D = P(\text{Democrat})$$
$$\text{Column marginal probability } p_1 = P(\text{in favor})$$

TABLE 12(b)   Cell Probabilities

	Favor	Indifferent	Opposed	Row Marginal Probability
Democrat	p_{D1}	p_{D2}	p_{D3}	p_D
Republican	p_{R1}	p_{R2}	p_{R3}	p_R
Column marginal probability	p_1	p_2	p_3	1

We are concerned with testing the null hypothesis that the two classifications are independent. Recall from Chapter 4 that the probability of the intersection of independent events is the product of their probabilities. Thus, the independence of the two classifications means that $p_{D1} = p_D p_1$, $p_{D2} = p_D p_2$, and so on.

Therefore, the null hypothesis of independence can be formalized as

H_0:   Each cell probability is the product of the
corresponding pair of marginal probabilities.

To construct a χ^2 test, we need to determine the expected frequencies. Under H_0, the expected cell frequencies are

$$500 p_D p_1 \quad 500 p_D p_2 \quad 500 p_D p_3$$
$$500 p_R p_1 \quad 500 p_R p_2 \quad 500 p_R p_3$$

These involve the unknown marginal probabilities that must be estimated from the data. Referring to Table 12, we calculate the estimates as

$$\hat{p}_D = \frac{285}{500} \qquad \hat{p}_R = \frac{215}{500}$$

$$\hat{p}_1 = \frac{202}{500} \qquad \hat{p}_2 = \frac{150}{500} \qquad \hat{p}_3 = \frac{148}{500}$$

Then, the expected frequency in the first cell is estimated as

$$500\hat{p}_D\hat{p}_1 = 500 \times \frac{285}{500} \times \frac{202}{500}$$

$$= \frac{285 \times 202}{500} = 115.14$$

Notice that the expected frequency for each cell of Table 12 is of the form

$$\frac{\text{Row total} \times \text{Column total}}{\text{Grand total}}$$

Table 13(a) presents the observed cell frequencies along with the expected frequencies shown in parentheses. The quantities $(O - E)^2/E$ and the χ^2 statistic are then calculated in Table 13(b).

TABLE 13(a) The Observed and Expected Cell
Frequencies for the Data in Table 12

	Favor	Indifferent	Opposed
Democrat	138 (115.14)	83 (85.50)	64 (84.36)
Republican	64 (86.86)	67 (64.50)	84 (63.64)

TABLE 13(b) The Values of $(O - E)^2/E$

	Favor	Indifferent	Opposed	Total
Democrat	4.539	.073	4.914	
Republican	6.016	.097	6.514	
				$22.153 = \chi^2$

Having calculated the χ^2 statistic, it now remains to determine its d.f. by invoking the properties we stated in Section 2. Because we have a single random sample, only the property (b) is relevant to this problem. Since $p_D + p_R = 1$ and $p_1 + p_2 + p_3 = 1$, we have really estimated $1 + 2 = 3$ parameters. Hence,

$$\text{d.f. of } \chi^2 = (\text{No. of cells}) - 1 - (\text{No. of parameters estimated})$$
$$= 6 - 1 - 3$$
$$= 2$$

We choose level of significance $\alpha = .05$ and the tabulated upper 5% point of χ^2 with d.f. $= 2$ is 5.99. Because the observed χ^2 is larger than the tabulated value, the null hypothesis of independence is rejected at $\alpha = .05$. In fact, it would be rejected even for $\alpha = .01$.

An inspection of Table 13(b) reveals that large contributions to the value of χ^2 have come from the corner cells. Moreover, comparing the observed and expected frequencies in Table 13(a), we see that the support for the program draws more from the Democrats than Republicans.

From our analysis of the contingency table in Example 9, the χ^2 test of independence in a general $r \times c$ contingency table is readily apparent. In fact, it is much the same as the test of homogeneity described in Section 3. The expected cell frequencies are determined in the same way—namely,

$$\text{Expected cell frequency} = \frac{\text{Row total} \times \text{Column total}}{\text{Grand total}}$$

and the test statistic is again

$$\chi^2 = \sum_{\text{cells}} \frac{(O - E)^2}{E}$$

With regard to the d.f. of χ^2 in the present case, we initially have $rc - 1$ d.f. because there are rc cells into which a single random sample is classified. From this, we must subtract the number of estimated parameters. This number is $(r - 1) + (c - 1)$ because there are $r - 1$ parameters among the row marginal probabilities and $c - 1$ parameters among the column marginal probabilities. Therefore,

$$\text{d.f. of } \chi^2 = rc - 1 - (r - 1) - (c - 1)$$
$$= rc - r - c + 1$$
$$= (r - 1)(c - 1)$$
$$= (\text{No. of rows} - 1) \times (\text{No. of columns} - 1)$$

which is identical to the d.f. of χ^2 for testing homogeneity. In summary, the χ^2 test statistic, its d.f., and the rejection region for testing independence are the same as when testing homogeneity. It is only the statement of the null hypothesis that is different between the two situations.

The Null Hypothesis of Independence

H_0: Each cell probability equals the product of the corresponding row and column marginal probabilities.

SPURIOUS DEPENDENCE

When the χ^2 test leads to a rejection of the null hypothesis of independence, we conclude that the data provide evidence of a **statistical association** between the two characteristics. However, we must refrain from making the hasty interpretation that these characteristics are directly related. A claim of casual relationship must draw from common sense, which statistical evidence must not be allowed to supersede.

Two characteristics may appear to be strongly related due to the common influence of a third factor that is not included in the study. In such cases, the dependence is called a **spurious dependence**. For instance, if a sample of individuals is classified in a 2×2 contingency table according to whether or not they are heavy drinkers and whether or not they suffer from respiratory trouble, we would probably find a high value for χ^2 and conclude that a strong statistical association exists between drinking habit and lung condition. But the reason for the association may be that most heavy drinkers are also heavy smokers and the smoking habit is a direct cause of respiratory trouble. This discussion should remind the reader of a similar warning given in Chapter 3 regarding the interpretation of a correlation coefficient between two sets of measurements. In the context of contingency tables, examples of spurious dependence are sometimes called **Simpson's paradox**, which is discussed in Chapter 3, Section 2.

Exercises

13.21 Applicants for public assistance are allowed an appeals process when they feel unfairly treated. At such a hearing, the applicant may choose self-representation or representation by an attorney. The appeal may result in an increase, decrease, or no change of the aid recommendation. Court records of 320 appeals cases provided the following data.

Type of Representation	Amount of Aid		
	Increased	Unchanged	Decreased
Self	59	108	17
Attorney	70	63	3

Test the null hypothesis that the appeals decision and the type of representation are independent. Test at $\alpha = .05$.

13.22 A consultant to all kinds of retailers suggests that they have plenty of baskets available for customers. He bases his suggestion on data collected by watching video tapes from hidden cameras. Suppose that out of 200 customers, 80 picked up a basket when they entered the store. Among those who picked up baskets 60 persons made purchases while only 41 of those without baskets made a purchase.[4]

---

[4]The proportions of purchasers are similar to those given in *Fortune*, July 19, 1999, 131–134.

(a) Conduct a χ^2 test of independence between purchasing and the decision to pick up a basket. Use $\alpha = .05$.

(b) The consultant may be confusing association with cause and effect. For instance, do you think the decision to pick up a basket has anything to do with the intent to purchase when a person enters the store? Comment.

13.23 A survey was conducted by sampling 400 persons who were questioned regarding union membership and attitude toward decreased national spending on social welfare programs. The cross-tabulated frequency counts are presented.

	Support	Indifferent	Opposed	Total
Union	112	36	28	176
Nonunion	84	68	72	224
Total	196	104	100	400

Can these observed differences be explained by chance or are attitude and membership status associated?

13.24 A survey was conducted to study people's attitude toward television programs that show violence. A random sample of 1200 adults was selected and classified according to gender and response to the question: Do you think there is a link between violence on TV and crime?

	Response		
	Yes	No	Not Sure
Male	361	228	17
Female	433	141	20

Do the survey data show a significant association between attitude and gender?

13.25 In a study of factors that regulate behavior, three kinds of subjects are identified: overcontrollers, average controllers, and undercontrollers, with the first group being most inhibited. Each subject is given the routine task of filling a box with buttons and all subjects are told they can stop whenever they wish. Whenever a subject indicates he or she wishes to stop, the experimenter asks, "Don't you really want to continue? The number of subjects in each group who stop and the number who continue are given in the following table.

Controller	Continue	Stop	Total
Over	9	9	18
Average	8	12	20
Under	3	14	17
Total	20	35	55

Are controller group and continue/stop decision associated?

13.26 Out of 120 members of a club who responded to a survey, 80 said that the golfing facilities were influential, 53 said the dining facilities were influential, and 25 said both were influential in their decision to join the club.

(a) Construct a two-way table of frequencies with two categories influential/not influential for each facility, golf and dining.

(b) Formulate the null hypothesis.

(c) Conduct a χ^2 test of your null hypothesis. Use $\alpha = .05$.

(d) Comment on the form of any departure from the null hypothesis.

The Following Exercises May Require a Computer

13.27 *Using the computer.* The analysis of a contingency table becomes tedious especially when the size of the table is large. Using a computer makes the task quite easy. We illustrate this by using MINITAB to analyze the data in Table 12, Example 9.

Data: C13T12.DAT

C1: 138 64

C2: 83 67

C3: 64 84

Dialog box:

Stat > Tables > Chisquare Test
Type C1-C3 in **Columns** containing
the table. Click **OK**.

The **output** is as follows:

```
Chi-Square Test: C1, C2, C3
```

```
Expected counts are printed below ob-
served counts
```

```
Chi-Square contributions are printed be-
low expected counts
```

	C1	C2	C3	Total
1	138	83	64	285
	115.14	85.50	84.36	
	4.539	0.073	4.914	
2	64	67	84	215
	86.86	64.50	63.64	
	6.016	0.097	6.514	
Total	202	150	148	500

```
Chi-Sq = 22.152, DF = 2,
P-Value = 0.000
```

(a) Compare this output with the calculations presented in Example 9.

(b) Do Exercise 13.24 on a computer.

13.28 Do Exercise 13.26 on a computer.

USING STATISTICS WISELY

1. Moderately large sample sizes are required to detect differences in proportions. Usually, 50 to 100 observations from each population are needed.

2. Although the χ^2 test statistic is the same for testing independence and for testing equality of proportions, you should be clear which null hypothesis you are testing. When you sample from separate populations, the χ^2 test concerns the equality of proportions. If a single sample is cross-classified according to two characteristics, the χ^2 tests concerns independence.

3. Don't routinely apply the inference procedures for comparing proportions when it is obvious that the outcomes have a time order dependence.

KEY IDEAS AND FORMULAS

The term categorical data refers to observations that are only classified into categories so the data consist of frequency counts for the categories. When the frequency counts arise because observations are classified according to two or more characteristics, they are called cross-tabulated data or a contingency table. A chi-square (χ^2) statistic compares the observed frequencies with those expected under a null hypothesis.

When a chi-square test of independence leads to the rejection of the null hypothesis, we say we have established a statistical association. There is no assertion of a causal relation as it may be a spurious dependence caused by a third variable. Simpson's paradox is an example of spurious dependence caused when two categorical data sets collected from very different populations are combined.

Pearson's χ^2 Test for Goodness of Fit

Data: Observed cell frequencies $n_1, \ldots, n_k$ from a random sample of size n classified into k cells.

The null hypothesis specifies the cell probabilities

$$H_0: \quad p_1 = p_{10}, \ldots, p_k = p_{k0}$$

Test statistic

$$\chi^2 = \sum_{\text{cells}} \frac{(n_i - np_{i0})^2}{np_{i0}} \qquad \text{d.f.} = k - 1$$

Rejection region $\quad \chi^2 \geq \chi^2_\alpha$

χ^2 Test of Homogeneity in an $r \times c$ Contingency Table

Data: Independent random samples from r populations, each sample classified in c response categories.

Null hypothesis: In each response category, the probabilities are equal for all the populations.

Test statistic

$$\chi^2 = \sum_{\text{cells}} \frac{(O - E)^2}{E} \qquad \text{d.f.} = (r - 1)(c - 1)$$

where for each cell

$$O = \text{Observed cell frequency}$$

$$E = \frac{\text{Row total} \times \text{Column total}}{\text{Grand total}}$$

Rejection region $\quad \chi^2 \geq \chi^2_\alpha$

χ^2 Test of Independence in an $r \times c$ Contingency Table

Data: A random sample of size n is simultaneously classified with respect to two characteristics, one has r categories and the other c categories.

Null hypothesis: The two classifications are independent; that is, each cell probability is the product of the row and column marginal probabilities.

Test statistic and rejection region: Same as when testing homogeneity.

Limitation

All inference procedures of this chapter require large samples. The χ^2 tests are appropriate if no expected cell frequency is too small (≥ 5 is normally required).

TECHNOLOGY

Conducting a χ^2 test

MINITAB

Conducting a χ^2 test

We illustrate the calculation of the χ^2 statistic with an example. Enter the counts

30	42	28
9	10	31

in columns 1 to 3. Select the MINITAB commands

> **Stat > Tables > Chisquare Test.**
> Type C1 – C4 in **columns containing the table.**
> Click **OK.**

EXCEL

Calculating a χ^2 statistic

Enter observed values in a rectangular range of cells.
Enter the expected values in another rectangular range of cells.

> Select **Insert,** then **Function.** Select **Statistical,** and then **CHITEST.**
> Click **OK.**
> With the cursor in the textbox for **Actual_range,** highlight the observed values.
> With the cursor in the textbox for **Expected_range,** highlight the expected values.
> Click **OK.**

The software will return the P-value.

TI-84/-83 PLUS

Calculating a χ^2 statistic

Enter the observed counts and expected values as matrices.

> Select **Matrix** and then **EDIT.**
> Press **Enter** and enter the number of rows, the number of columns, and the entries for [A], the observed values.
> Press 2*nd* **Quit** and select **Matrix,** then **EDIT** again.

Select **Matrix [B]**, press **Enter** and enter the number of rows, the number of columns, and the expected values. Press **2nd Quit** again.
Select **STAT** then **TESTS** and then $> \chi^2$ **test**.
Arrow down to select **Calculate**. Enter the appropriate matrix names (matrices [A] and [B] are the defaults) and then press **Enter**.

The software returns the value of the χ^2 statistic, the P–value, and the degrees of freedom.

5. REVIEW EXERCISES

13.29 To examine the quality of a random number generator, frequency counts of the individual integers are recorded from an output of 500 integers. The concept of randomness implies that the integers 0, 1, . . . , 9 are equally likely. Based on the observed frequency counts, would you suspect any bias of the random number generator? Answer by performing the χ^2 test.

Integer	0	1	2	3	4	5	6	7	8	9	Total
Frequency	41	58	51	61	39	56	45	35	62	52	500

13.30 The following record shows a classification of 41,208 births in Wisconsin (courtesy of Professor Jerome Klotz). Test the goodness of fit of the model that births are uniformly distributed over all 12 months of the year. Use $\alpha = .01$.

Jan.	3,478	July	3,476
Feb.	3,333	Aug.	3,495
Mar.	3,771	Sept.	3,490
Apr.	3,542	Oct.	3,331
May	3,479	Nov.	3,188
June	3,304	Dec.	3,321
		Total	41,208

13.31 The following table records the observed number of births at a hospital in four consecutive quarterly periods.

Quarters	Jan.–March	April–June	July–Sept.	Oct.–Dec.
Number of births	55	29	26	41

It is conjectured that twice as many babies are born during the Jan.–March quarter than any of the other three quarters. At $\alpha = .10$, test if these data strongly contradict the stated conjecture.

13.32 Refer to Exercise 13.12, where 250 students majoring in business and 500 executives were asked to respond to this question: Should corporations become more directly involved with social issues such as homelessness, education, and drugs?

(a) Let p_1 be the proportion of executives who would answer "More involved," and let p_2 be the same probability for students. Test $H_0: p_1 = p_2$ versus $H_1: p_1 < p_2$. Use $\alpha = .05$.

(b) Obtain a 95% confidence interval for the difference in proportions.

13.33 Refer to Exercise 3.48 and the data concerning a vaccine for type B hepatitis.

	Hepatitis	No Hepatitis	Total
Vaccinated	11	538	549
Not vaccinated	70	464	534
Total	81	1002	1083

Do these data indicate that there is a different rate of incidence of hepatitis between the vaccinated and nonvaccinated participants? Use the χ^2 test for homogeneity in a contingency table.

13.34 Refer to the data in Exercise 13.33.

(a) Use the Z test for testing the equality of two population proportions with a two-sided alternative. Verify the relation $\chi^2 = Z^2$ by comparing their numerical values.

(b) If the alternative is that the incidence rate is lower for the vaccinated group, which of the two tests should be used?

13.35 To compare the effectiveness of four drugs in relieving postoperative pain, an experiment was done by randomly assigning 195 surgical patients to the drugs under study. Recorded here are the number of patients assigned to each drug and the number of patients who were free of pain for a period of five hours.

	Free of Pain	No. of Patients Assigned
Drug 1	23	53
Drug 2	30	47
Drug 3	19	51
Drug 4	29	44

(a) Make a 4 × 2 contingency table showing the counts of patients who were free of pain and those who had pain, and test the null hypothesis that all four drugs are equally effective. (Use $\alpha = .05$.)

(b) Let p_1, p_2, p_3, and p_4 denote the population proportions of patients who would be free of pain under the use of drugs 1, 2, 3, and 4, respectively. Calculate a 90% confidence interval for each of these probabilities individually.

13.36 Using the data for drugs 1 and 3 in Exercise 13.35, make a 2 × 2 contingency table and test $H_0: p_1 = p_3$ versus $H_1: p_1 \neq p_3$ at $\alpha = .05$ employing:

(a) The χ^2 test.

(b) The Z test.

13.37 Refer to the data for drugs 3 and 4 in Exercise 13.35.

(a) Is there strong evidence that drug 4 is more effective in controlling postoperative pain than drug 3? Answer by calculating the P-value.

(b) Construct a 95% confidence interval for the difference $p_4 - p_3$.

13.38 In a study on the effect of diet and life-style on heart disease, 96 patients with severe coronary blockage were randomly assigned, 49 to an experimental group and 47 to a control group. The patients in the experimental group had a low-fat vegetarian diet, regular exercise, and stress-management training, whereas those in the control group had a low-fat diet and moderate exercise. The condition of their coronary blockage was monitored throughout the study period, and the following results were noted.

Group	Coronary Blockage			Total
	Worsened	No Change	Improved	
Experimental	4	8	37	49
Control	8	25	14	47
				96

Analyze the data to determine if the changes in coronary blockage were significantly different between the two groups of patients.

13.39 Based on interviews of couples seeking divorces, a social worker compiles the following data related to the period of acquaintanceship before marriage and the duration of marriage.

Acquaintanceship before Marriage	Duration of Marriage		Total
	≤ 4 years	> 4 years	
Under $\frac{1}{2}$ year	11	8	19
$\frac{1}{2} - 1\frac{1}{2}$ years	28	24	52
Over $1\frac{1}{2}$ years	21	19	40
Total	60	51	111

Perform a test to determine if the data substantiate an association between the stability of a marriage and the period of acquaintanceship prior to marriage.

13.40   By polling a random sample of 350 undergraduate students, a campus press obtains the following frequency counts regarding student attitude toward a proposed change in dormitory regulations.

	Favor	Indifferent	Oppose	Total
Male	95	72	19	186
Female	53	79	32	164
Total	148	151	51	350

Are attitude toward the proposal and gender associated?

13.41   In a genetic study of chromosome structures, 132 individuals are classified according to the type of structural chromosome aberration and carriers in their parents. The following counts are obtained.

	Carrier		
Type of Aberration	One Parent	Neither Parent	Total
Presumably innocuous	27	20	47
Substantially unbalanced	36	49	85
Total	63	69	132

Test the null hypothesis that type of aberration is independent of parental carrier.

13.42   A random sample of 130 business executives was classified according to age and the degree of risk aversion as measured by a psychological test.

	Degree of Risk Aversion			
Age	Low	Medium	High	Total
Below 45	14	22	7	43
45–55	16	33	12	61
Over 55	4	15	7	26
				130

Do these data demonstrate an association between risk aversion and age?

13.43   *Pooling contingency tables can produce spurious association.* A large organization is being investigated to determine if its recruitment is sex-biased. Tables 14 and 15, respectively, show the classification of applicants for secretarial and for sales positions according to gender and result of interview. Table 16 is an aggregation of the corresponding entries of Table 14 and Table 15.

TABLE 14   Secretarial Positions

	Offered	Denied	Total
Male	25	50	75
Female	75	150	225
Total	100	200	300

TABLE 15   Sales Positions

	Offered	Denied	Total
Male	150	50	200
Female	75	25	100
Total	225	75	300

TABLE 16   Secretarial and Sales Positions

	Offered	Denied	Total
Male	175	100	275
Female	150	175	325
Total	325	275	600

(a) Verify that the χ^2 statistic for testing independence is zero for each of the data sets given in Tables 14 and 15.

(b) For the pooled data given in Table 16, compute the value of the χ^2 statistic and test the null hypothesis of independence.

(c) Explain the paradoxical result that there is no sex bias in any job category, but the combined data indicate sex discrimination.

14

Analysis of Variance (ANOVA)

Which Brand TV Has the Clearest Picture?

A good experimental design is to collect samples of several sets of each brand and measure their picture clarity. The statistical technique called "analysis of variance" enables us to verify differences among the brands.
© Ed Lallo/Index Stock Imagery.

1. INTRODUCTION

In Chapter 10, we introduced methods for comparing two population means. When several means must be compared, more general methods are required. We now become acquainted with the powerful technique called analysis of variance (ANOVA) that allows us to analyze and interpret observations from several populations. This versatile statistical tool partitions the total variation in a data set according to the sources of variation that are present. In the context of comparing k population means, the two sources of variation are (1) differences between means or treatments and (2) within population variation (error). We restrict our discussion to this case, although ANOVA techniques apply to much more complex situations.

In this chapter, you will learn how to test for differences among several means and to make confidence statements about pairs of means.

2. COMPARISON OF SEVERAL TREATMENTS— THE COMPLETELY RANDOMIZED DESIGN

It is usually more expedient in terms of both time and expense to simultaneously compare several treatments than it is to conduct several comparative trials two at a time. The term completely randomized design is synonymous with independent random sampling from several populations when each population is identified as the population of responses under a particular treatment. Let treatment 1 be applied to n_1 experimental units, treatment 2 to n_2 units, . . . , treatment k to n_k units. In a completely randomized design, n_1 experimental units selected at random from the available collection of $n = n_1 + n_2 + \cdots + n_k$ units are to receive treatment 1, n_2 units randomly selected from the remaining units are to receive treatment 2, and proceeding in this manner, treatment k is to be applied to the remaining n_k units. The special case of this design for a comparison of $k = 2$ treatments has already been discussed in Section 2 of Chapter 10. The data structure for the response measurements can be represented by the format shown in Table 1, where y_{ij} is the jth observation on treatment i. The summary statistics appear in the last two columns.

Before proceeding with the general case of k treatments, it would be instructive to explain the reasoning behind the analysis of variance and the associated calculations in terms of a numerical example.

Example 1   The Structure of Data from an Experiment for Comparing Four Means

In an effort to improve the quality of recording tapes, the effects of four kinds of coatings A, B, C, D on the reproducing quality of sound are compared. Suppose that the measurements of sound distortion given in Table 2 are obtained from tapes treated with the four coatings. Two questions immediately come to mind.

TABLE 1   Data Structure for the Completely Randomized Design with k Treatments

	Observations	Mean	Sum of Squares
Treatment 1	$y_{11}, y_{12}, \ldots, y_{1n_1}$	$\bar{y}_1$	$\sum\limits_{j=1}^{n_1} (y_{1j} - \bar{y}_1)^2$
Treatment 2	$y_{21}, y_{22}, \ldots, y_{2n_2}$	$\bar{y}_2$	$\sum\limits_{j=1}^{n_2} (y_{2j} - \bar{y}_2)^2$
.	.	.	.
.	.	.	.
.	.	.	.
Treatment k	$y_{k1}, y_{k2}, \ldots, y_{kn_k}$	$\bar{y}_k$	$\sum\limits_{j=1}^{n_k} (y_{kj} - \bar{y}_k)^2$

$$\text{Grand mean } \bar{y} = \frac{\text{Sum of all observations}}{n_1 + n_2 + \cdots + n_k} = \frac{n_1\bar{y}_1 + \cdots + n_k\bar{y}_k}{n_1 + \cdots + n_k}$$

TABLE 2   Sound Distortion Obtained with Four Types of Coating

Coating	Observations	Mean	Sum of Squares
A	10, 15, 8, 12, 15	$\bar{y}_1 = 12$	$\sum\limits_{j=1}^{5} (y_{1j} - \bar{y}_1)^2 = 38$
B	14, 18, 21, 15	$\bar{y}_2 = 17$	$\sum\limits_{j=1}^{4} (y_{2j} - \bar{y}_2)^2 = 30$
C	17, 16, 14, 15, 17, 15, 18	$\bar{y}_3 = 16$	$\sum\limits_{j=1}^{7} (y_{3j} - \bar{y}_3)^2 = 12$
D	12, 15, 17, 15, 16, 15	$\bar{y}_4 = 15$	$\sum\limits_{j=1}^{6} (y_{4j} - \bar{y}_4)^2 = 14$
	Grand mean $\bar{y} = 15$		

Does any significant difference exist among the mean distortions obtained using the four coatings? Can we establish confidence intervals for the mean differences between coatings?

An analysis of the results essentially consists of decomposing the observations into contributions from different sources. We reason that the deviation of an individual observation from the grand mean, $y_{ij} - \bar{y}$, is partly due to differences among the mean qualities of the coatings and partly due to random variation in measurements within the same group. This suggests the following decomposition.

$$\text{Observation} = \begin{pmatrix} \text{Grand} \\ \text{mean} \end{pmatrix} + \begin{pmatrix} \text{Deviation due} \\ \text{to treatment} \end{pmatrix} + (\text{Residual})$$

$$y_{ij} \quad = \quad \bar{y} \quad + \quad (\bar{y}_i - \bar{y}) \quad + \quad (y_{ij} - \bar{y}_i)$$

For the data given in Table 2, the decomposition of all the observations can be presented in the form of the following arrays:

Observations

$$y_{ij}$$

$$\begin{bmatrix} 10 & 15 & 8 & 12 & 15 \\ 14 & 18 & 21 & 15 \\ 17 & 16 & 14 & 15 & 17 & 15 & 18 \\ 12 & 15 & 17 & 15 & 16 & 15 \end{bmatrix}$$

	Grand mean	**Treatment effects**

$$\bar{y} \qquad\qquad\qquad (\bar{y}_i - \bar{y})$$

$$= \begin{bmatrix} 15 & 15 & 15 & 15 & 15 \\ 15 & 15 & 15 & 15 \\ 15 & 15 & 15 & 15 & 15 & 15 & 15 \\ 15 & 15 & 15 & 15 & 15 & 15 \end{bmatrix} + \begin{bmatrix} -3 & -3 & -3 & -3 & -3 \\ 2 & 2 & 2 & 2 \\ 1 & 1 & 1 & 1 & 1 & 1 & 1 \\ 0 & 0 & 0 & 0 & 0 & 0 \end{bmatrix}$$

Residuals

$$(y_{ij} - \bar{y}_i)$$

$$+ \begin{bmatrix} -2 & 3 & -4 & 0 & 3 \\ -3 & 1 & 4 & -2 \\ 1 & 0 & -2 & -1 & 1 & -1 & 2 \\ -3 & 0 & 2 & 0 & 1 & 0 \end{bmatrix}$$

For instance, the upper left-hand entries of the arrays show that

$$10 = 15 + (-3) + (-2)$$

$$y_{11} = \bar{y} + (\bar{y}_1 - \bar{y}) + (y_{11} - \bar{y}_1)$$

If there is really no difference in the mean distortions obtained using the four tape coatings, we can expect the entries of the second array on the right-hand side of the equation, whose terms are $\bar{y}_i - \bar{y}$, to be close to zero. As an overall measure of the amount of variation due to differences in the treatment means, we calculate the sum of squares of all the entries in this array, or

$$\underbrace{(-3)^2 + \cdots + (-3)^2}_{n_1 = 5} + \underbrace{2^2 + \cdots + 2^2}_{n_2 = 4} + \underbrace{1^2 + \cdots + 1^2}_{n_3 = 7} + \underbrace{0^2 + \cdots + 0^2}_{n_4 = 6}$$

$$= 5(-3)^2 + 4(2)^2 + 7(1)^2 + 6(0)^2$$

$$= 68$$

Thus, the sum of squares due to differences in the treatment means, also called the **treatment sum of squares**, is given by

$$\text{Treatment sum of squares} = \sum_{i=1}^{4} n_i (\bar{y}_i - \bar{y})^2 = 68$$

The last array consists of the entries $y_{ij} - \bar{y}_i$ that are the deviations of individual observations from the corresponding treatment mean. These deviations reflect inherent variabilities in the experimental material and the measuring device and are called the **residuals**. The overall variation due to random errors is measured by the sum of squares of all these residuals

$$(-2)^2 + 3^2 + (-4)^2 + \cdots + 1^2 + 0^2 = 94$$

Thus, we obtain

$$\text{Error sum of squares} = \sum_{i=1}^{4} \sum_{j=1}^{n_i} (y_{ij} - \bar{y}_i)^2 = 94$$

The double summation indicates that the elements are summed within each row and then over different rows. Alternatively, referring to the last column in Table 2, we obtain

$$\text{Error sum of squares} = \sum_{j=1}^{5} (y_{1j} - \bar{y}_1)^2 + \sum_{j=1}^{4} (y_{2j} - \bar{y}_2)^2$$
$$+ \sum_{j=1}^{7} (y_{3j} - \bar{y}_3)^2 + \sum_{j=1}^{6} (y_{4j} - \bar{y}_4)^2$$
$$= 38 + 30 + 12 + 14 = 94$$

Finally, the deviations of individual observations from the grand mean $y_{ij} - \bar{y}$ are given by the array

$$\begin{bmatrix} -5 & 0 & -7 & -3 & 0 & & \\ -1 & 3 & 6 & 0 & & & \\ 2 & 1 & -1 & 0 & 2 & 0 & 3 \\ -3 & 0 & 2 & 0 & 1 & 0 & \end{bmatrix}$$

The total variation present in the data is measured by the sum of squares of all these deviations.

$$\text{Total sum of squares} = \sum_{i=1}^{4} \sum_{j=1}^{n_i} (y_{ij} - \bar{y})^2$$
$$= (-5)^2 + 0^2 + (-7)^2 + \cdots + 0^2$$
$$= 162$$

Note that the total sum of squares is the sum of the treatment sum of squares and the error sum of squares.

It is time to turn our attention to another property of this decomposition, the degrees of freedom associated with the sums of squares. In general terms:

$$\begin{pmatrix} \text{Degrees of} \\ \text{freedom} \\ \text{associated with a} \\ \text{sum of squares} \end{pmatrix} = \begin{pmatrix} \text{Number of} \\ \text{elements} \\ \text{whose squares} \\ \text{are summed} \end{pmatrix} - \begin{pmatrix} \text{Number of linear} \\ \text{constraints} \\ \text{satisfied by the} \\ \text{elements} \end{pmatrix}$$

In our present example, the treatment sum of squares is the sum of four terms $n_1(\bar{y}_1 - \bar{y})^2 + n_2(\bar{y}_2 - \bar{y})^2 + n_3(\bar{y}_3 - \bar{y})^2 + n_4(\bar{y}_4 - \bar{y})^2$, where the elements satisfy the single constraint

$$n_1(\bar{y}_1 - \bar{y}) + n_2(\bar{y}_2 - \bar{y}) + n_3(\bar{y}_3 - \bar{y}) + n_4(\bar{y}_4 - \bar{y}) = 0$$

This equality holds because the grand mean $\bar{y}$ is a weighted average of the treatment means, or

$$\bar{y} = \frac{n_1\bar{y}_1 + n_2\bar{y}_2 + n_3\bar{y}_3 + n_4\bar{y}_4}{n_1 + n_2 + n_3 + n_4}$$

Consequently, the number of degrees of freedom associated with the treatment sum of squares is $4 - 1 = 3$. To determine the degrees of freedom for the error sum of squares, we note that the entries $y_{ij} - \bar{y}_i$ in each row of the residual array sum to zero and there are 4 rows. The number of degrees of freedom for the error sum of squares is then $(n_1 + n_2 + n_3 + n_4) - 4 = 22 - 4 = 18$. Finally, the number of degrees of freedom for the total sum of squares is $(n_1 + n_2 + n_3 + n_4) - 1 = 22 - 1 = 21$, because the 22 entries $(y_{ij} - \bar{y})$ whose squares are summed satisfy the single constraint that their total is zero. Note that the degrees of freedom for the total sum of squares is the sum of the degrees of freedom for treatment and error.

We summarize the calculations thus far in Table 3.

TABLE 3   ANOVA Table for Distortion Data

Source	Sum of Squares	d.f.
Treatment	68	3
Error	94	18
Total	162	21

Guided by this numerical example, we now present the general formulas for the analysis of variance for a comparison of k treatments using the data structure given in Table 1. Beginning with the basic decomposition

$$(y_{ij} - \bar{y}) = (\bar{y}_i - \bar{y}) + (y_{ij} - \bar{y}_i)$$

and squaring each side of the equation, we obtain

$$(y_{ij} - \bar{y})^2 = (\bar{y}_i - \bar{y})^2 + (y_{ij} - \bar{y}_i)^2 + 2(\bar{y}_i - \bar{y})(y_{ij} - \bar{y}_i)$$

When summed over $j = 1, \ldots, n_i$ the last term on the right-hand side of this equation reduces to zero due to the relation $\sum_{j=1}^{n_i} (y_{ij} - \bar{y}_i) = 0$. Therefore, summing each side of the preceding relation over $j = 1, \ldots, n_i$ and $i = 1, \ldots, k$ provides the decomposition

$$\sum_{i=1}^{k} \sum_{j=1}^{n_i} (y_{ij} - \bar{y})^2 = \sum_{i=1}^{k} n_i (\bar{y}_i - \bar{y})^2 + \sum_{i=1}^{k} \sum_{j=1}^{n_i} (y_{ij} - \bar{y}_i)^2$$

$$\uparrow \qquad\qquad\qquad \uparrow \qquad\qquad\qquad \uparrow$$

| Total SS | Treatment SS | Residual SS or error SS |

$$\text{d.f.} = \sum_{i=1}^{k} n_i - 1 \qquad \text{d.f.} = k - 1 \qquad \text{d.f.} = \sum_{i=1}^{k} n_i - k$$

It is customary to present the decomposition of the sum of squares and the degrees of freedom in a tabular form called the analysis of variance table, abbreviated as ANOVA table. This table contains the additional column for the mean square associated with a component, which is defined as

$$\text{Mean square} = \frac{\text{Sum of squares}}{\text{d.f.}}$$

The ANOVA table for comparing k treatments appears in Table 4.

TABLE 4   ANOVA Table for Comparing k Treatments

Source	Sum of Squares	d.f.	Mean Square
Treatment	$SS_T = \sum_{i=1}^{k} n_i (\bar{y}_i - \bar{y})^2$	$k - 1$	$MS_T = \dfrac{SS_T}{k - 1}$
Error	$SSE = \sum_{i=1}^{k} \sum_{j=1}^{n_i} (y_{ij} - \bar{y}_i)^2$ $\quad \sum_{i=1}^{k} n_i - k$		$MSE = \dfrac{SSE}{\sum_{i=1}^{k} n_i - k}$
Total	$\sum_{i=1}^{k} \sum_{j=1}^{n_i} (y_{ij} - \bar{y})^2$	$\sum_{i=1}^{k} n_i - 1$	

GUIDE TO HAND CALCULATION

When performing an ANOVA on a calculator, it is convenient to express the sums of squares in an alternative form. These employ the treatment totals

$$T_i = \sum_{j=1}^{n_i} y_{ij} = \text{Sum of all responses under treatment } i$$

$$T = \sum_{i=1}^{k} T_i = \sum_{i=1}^{k} \sum_{j=1}^{n_i} y_{ij} = \text{Sum of all observations}$$

to calculate the sums of squares:

$$\text{Total SS} = \sum_{i=1}^{k} \sum_{j=1}^{n_i} y_{ij}^2 - \frac{T^2}{n} \qquad \text{where} \qquad n = \sum_{i=1}^{k} n_i$$

$$SS_T = \sum_{i=1}^{k} \frac{T_i^2}{n_i} - \frac{T^2}{n}$$

$$SSE = \text{Total SS} - SS_T$$

Notice that the SSE can be obtained by subtraction.

Example 2   Calculating Sums of Squares Using the Alternative Formulas

Obtain the Total SS, SS_T, and SSE for the data in Example 1 using the alternative form of calculation.

SOLUTION
$$T_1 = 10 + 15 + 8 + 12 + 15 = 60 \qquad\qquad n_1 = 5$$
$$T_2 = 14 + 18 + 21 + 15 = 68 \qquad\qquad n_2 = 4$$
$$T_3 = 17 + 16 + 14 + 15 + 17 + 15 + 18 = 112 \qquad n_3 = 7$$
$$T_4 = 12 + 15 + 17 + 15 + 16 + 15 = 90 \qquad\qquad n_4 = 6$$

and

$$T = T_1 + T_2 + T_3 + T_4 \qquad\qquad n = n_1 + n_2 + n_3 + n_4$$
$$= 60 + 68 + 112 + 90 = 330 \qquad = 5 + 4 + 7 + 6 = 22$$

Since

$$\sum_{i=1}^{4} \sum_{j=1}^{n_i} y_{ij}^2 = (10)^2 + (15)^2 + \cdots + (16)^2 + (15)^2 = 5112$$

$$\text{Total SS} = 5112 - \frac{(330)^2}{22} = 162$$

$$SS_T = \frac{(60)^2}{5} + \frac{(68)^2}{4} + \frac{(112)^2}{7} + \frac{(90)^2}{6} - \frac{(330)^2}{22} = 68$$

$$SSE = \text{Total SS} - SS_T = 162 - 68 = 94$$

Exercises

14.1 (a) Obtain the arrays that show a decomposition for the following observations.

(b) Find the sum of squares for each array.

(c) Determine the degrees of freedom for each sum of squares.

(d) Summarize by an ANOVA table.

Treatment	Observations
1	5 9
2	8 4
3	4 2
4	7 9

14.2 (a) Obtain the arrays that show a decomposition for the following observations.

(b) Find the sum of squares for each array.

(c) Determine the degrees of freedom for each sum of squares.

(d) Summarize by an ANOVA table.

Treatment	Observations
1	35, 24, 28, 21
2	19, 14, 14, 13
3	21, 16, 21, 14

14.3 Repeat Exercise 14.1 for the data

Treatment	Observations
1	7 5 4 4
2	6 1 2
3	2 1 0 1

14.4 Use the relations for sums of squares and d.f. to complete the following ANOVA table:

Source	Sum of Squares	d.f.
Treatment	36	5
Error		
Total	76	25

14.5 Provide a decomposition of the following observations and obtain the ANOVA table.

Treatment	Observations
1	2 1 3
2	1 5
3	9 5 6 4
4	3 4 5

14.6 Given the summary statistics from three samples,

$$n_1 = 10 \quad \bar{y}_1 = 5$$

$$(n_1 - 1)s_1^2 = \sum_{j=1}^{10} (y_{1j} - \bar{y}_1)^2 = 30$$

$$n_2 = 6 \quad \bar{y}_2 = 2$$

$$(n_2 - 1)s_2^2 = \sum_{j=1}^{6} (y_{2j} - \bar{y}_2)^2 = 16$$

$$n_3 = 9 \quad \bar{y}_3 = 7$$

$$(n_3 - 1)s_3^2 = \sum_{j=1}^{9} (y_{3j} - \bar{y}_3)^2 = 25$$

create the ANOVA table.

14.7 Reading levels vary between different magazines. To avoid difficulties caused by different typefaces and sizes, an investigator just counted the number of letters and punctuation signs. Random samples of 20 sentences were selected from *The New Yorker*, *Sports Illustrated*, and *National Geographic*. The resulting summary statistics are:

The New Yorker	Sports Illustrated	National Geographic
$\bar{y}_1 = 94.4$	$\bar{y}_2 = 92.9$	$\bar{y}_3 = 75.5$
$s_1 = 58.4$	$s_2 = 54.2$	$s_3 = 38.1$
$n_1 = 20$	$n_2 = 20$	$n_3 = 20$

where $s_i^2 = \sum_{j=1}^{n_i} (y_{ij} - \bar{y}_i)^2 / (n_i - 1)$

Present the ANOVA table for these data.

3. POPULATION MODEL AND INFERENCES FOR A COMPLETELY RANDOMIZED DESIGN

To implement a formal statistical test for no difference among treatment effects, we need to have a population model for the experiment. To this end, we assume that the response measurements with the ith treatment constitute a random sample from a normal population with a mean of μ_i and a common variance of σ^2. The samples are assumed to be mutually independent.

Population Model for Comparing k Treatments

$$Y_{ij} = \mu_i + e_{ij} \qquad j = 1, \ldots, n_i \qquad \text{and} \qquad i = 1, \ldots, k$$

where $\mu_i = i$th treatment mean. The errors e_{ij} are all independently distributed as $N(0, \sigma)$.

F DISTRIBUTION

The F test will determine if significant differences exist between the k sample means. The null hypothesis that no difference exists among the k population means can now be phrased as follows:

$$H_0 : \mu_1 = \mu_2 = \cdots = \mu_k$$

The alternative hypothesis is that not all the μ_i's are equal. Seeking a criterion to test the null hypothesis, we observe that when the population means are all equal, $\bar{y}_i - \bar{y}$ is expected to be small, and consequently, the treatment mean square $\sum n_i (\bar{y}_i - \bar{y})^2 / (k - 1)$ is expected to be small. On the other hand, it is likely to be large when the means differ markedly. The error mean square, which provides an estimate of σ^2, can be used as a yardstick for determining how large a treatment mean square should be before it indicates significant differences. Statistical distribution theory tells us that under H_0 the ratio

$$F = \frac{\text{Treatment mean square}}{\text{Error mean square}} = \frac{\text{Treatment SS} / (k - 1)}{\text{Error SS} \Big/ \left(\sum\limits_{i=1}^{k} n_i - k \right)}$$

has an F distribution with d.f. $= (k - 1, n - k)$, where $n = \sum n_i$.

Notice that an F distribution is specified in terms of its numerator degrees of freedom $v_1 = k - 1$ and denominator degrees of freedom $v_2 = n - k$. We denote

$$F_\alpha(v_1, v_2) = \text{Upper } \alpha \text{ point of the } F \text{ distribution with } (v_1, v_2) \text{ d.f.}$$

which is also called the upper 100α-th percentage point.

The upper $\alpha = .05$ and $\alpha = .10$ points are given in Appendix B, Table 6, for several pairs of d.f. With $v_1 = 7$ and $v_2 = 15$, for $\alpha = .05$, we read from column $v_1 = 7$ and row $v_2 = 15$ to obtain $F_{.05}(7,15) = 2.71$ (see Table 5).

TABLE 5   Percentage Points
of $F(v_1, v_2)$ Distributions
$\alpha = .05$

v_1			
	$\cdots$	7	$\cdots$
v_2			
.		.	
.		.	
.		.	
15	$\cdots$	2.71	
.			
.			
.			

We summarize the F test introduced above.

F Test for Equality of Means

Reject $H_0: \mu_1 = \mu_2 = \cdots = \mu_k$ if

$$F = \frac{\text{Treatment SS}/(k - 1)}{\text{Error SS}/(n - k)} \geq F_\alpha(k - 1, n - k)$$

where $n = \sum\limits_{i=1}^{k} n_i$ and $F_\alpha(k - 1, n - k)$ is the upper α point of the F distribution with d.f. $= (k - 1, n - k)$.

The computed value of the F-ratio is usually presented in the last column of the ANOVA table.

Example 3   The F Test for Testing the Null Hypothesis of No Difference in Sound Distortion Means

Construct the ANOVA table for the data given in Example 1 concerning a comparison of four tape coatings. Test the null hypothesis that the means are equal. Use $\alpha = .05$.

SOLUTION   Using our earlier calculations for the component sums of squares, we construct the ANOVA table that appears in Table 6.

TABLE 6   ANOVA Table for the Data Given in Example 1

Source	Sum of Squares	d.f.	Mean Square	F-ratio
Treatment	68	3	22.67	$\dfrac{22.67}{5.22} = 4.34$
Error	94	18	5.22	
Total	162	21		

A test of the null hypothesis $H_0: \mu_1 = \mu_2 = \mu_3 = \mu_4$ is performed by comparing the observed F value 4.34 with the tabulated value of F with d.f. $= (3, 18)$. At a .05 level of significance, the tabulated value is found to be 3.16. Because this is exceeded by the observed value, we conclude that there is a significant difference among the four mean tape sound qualities.

Table 7 gives some typical output from a computer program where the term factor is used instead of treatment. The MINITAB commands for obtaining these results are given in Exercise 14.36.

TABLE 7   Computer Output: One-Way Analysis of Variance
for Distortion Data

```
One-way Analysis of Variance

Analysis of Variance
Source     DF         SS        MS        F        P
Factor      3      68.00     22.67     4.34    0.018
Error      18      94.00      5.22
Total      21     162.00
```

Exercises

14.8 Using the table of percentage points for the F distribution, find

(a) The upper 5% point when $v_1 = 7$ and $v_2 = 10$.

(b) The upper 5% point when $v_1 = 10$ and $v_2 = 7$.

14.9 Using Appendix B, Table 6, find the upper 10% point of F for

(a) d.f. $= (3, 5)$ (b) d.f. $= (3, 10)$

(c) d.f. $= (3, 15)$ (d) d.f. $= (3, 30)$

(e) What effect does increasing the denominator d.f. have?

14.10   Given the following ANOVA table,

Source	Sum of Squares	d.f.
Treatment	97	5
Error	101	20
Total	198	25

carry out the F test for equality of means taking $\alpha = .10$.

14.11   Given the following ANOVA table,

Source	Sum of Squares	d.f.
Treatment	23	5
Error	56	30
Total	79	35

carry out the F test for equality of means taking $\alpha = .05$.

14.12   Using the data from Exercise 14.1, test for equality of means using $\alpha = .05$.

14.13   Test for equality of means based on the data in Exercise 14.2. Take $\alpha = .05$.

14.14   Three bread recipes are to be compared with respect to density of the loaf. Five loaves will be baked using each recipe.

(a)   If one loaf is made and baked at a time, how would you select the order?

(b)   Given the following data, conduct an F test for equality of means. Take $\alpha = .05$.

Recipe	Observation				
1	.95	.86	.71	.72	.74
2	.71	.85	.62	.72	.64
3	.69	.68	.51	.73	.44

14.15   Test for equality of means based on the data in Exercise 14.3. Take $\alpha = .05$.

14.16   Refer to the data on reading levels in Exercise 14.7. Test for equality of means. Take $\alpha = .05$.

4.   SIMULTANEOUS CONFIDENCE INTERVALS

The ANOVA F test is only the initial step in our analysis. It determines if significant differences exist among the treatment means. Our goal should be more than to merely conclude that treatment differences are indicated by the data. Rather, we must detect likenesses and differences among the treatments. Thus, the problem of estimating differences in treatment means is of even greater importance than the overall F test.

Referring to the comparison of k treatments using the data structure given in Table 1, let us examine how a confidence interval can be established for $\mu_1 - \mu_2$, the mean difference between treatment 1 and treatment 2. The statistic

$$T = \frac{(\bar{Y}_1 - \bar{Y}_2) - (\mu_1 - \mu_2)}{\sqrt{\dfrac{SSE}{n-k}}\sqrt{\dfrac{1}{n_1} + \dfrac{1}{n_2}}}$$

has a t distribution with d.f. $= n - k$, and this can be employed to construct a confidence interval for $\mu_1 - \mu_2$. More generally:

Confidence Interval for a Single Difference

A $100(1 - \alpha)\%$ confidence interval for $\mu_i - \mu_{i'}$, the difference of means for treatment i and treatment i' is given by

$$(\overline{Y}_i - \overline{Y}_{i'}) \pm t_{\alpha/2} S \sqrt{\frac{1}{n_i} + \frac{1}{n_{i'}}}$$

where

$$S = \sqrt{MSE} = \sqrt{\frac{SSE}{n - k}}$$

and $t_{\alpha/2}$ is the upper $\alpha/2$ point of t with d.f. $= n - k$.

If the F test first shows a significant difference in means, then some statisticians feel that it is reasonable to compare means pairwise according to the preceding intervals. However, many statisticians prefer a more conservative procedure based on the following reasoning.

Without the provision that the F test is significant, the preceding method provides **individual** confidence intervals for pairwise differences. However, with $k = 4$ treatments, there are $\binom{4}{2} = 6$ pairwise differences $\mu_i - \mu_{i'}$, and this procedure applied to all pairs yields six confidence statements, each having a $100(1 - \alpha)\%$ level of confidence. It is difficult to determine what level of confidence will be achieved for claiming that all six of these statements are correct. To overcome this dilemma, procedures have been developed for several confidence intervals to be constructed in such a manner that the joint probability that all the statements are true is guaranteed not to fall below a predetermined level. Such intervals are called multiple confidence intervals or simultaneous confidence intervals. Numerous methods proposed in the statistical literature have achieved varying degrees of success. We present one that can be used simply and conveniently in general applications.

The procedure, called the multiple-t confidence intervals, consists of setting confidence intervals for the differences $\mu_i - \mu_{i'}$ in much the same way we just did for the individual differences, except that a different percentage point is read from the t table.

Operationally, the construction of these confidence intervals does not require any new concepts or calculations, but it usually involves some nonstandard percentage point of t. For example, with $k = 3$ and $1 - \alpha = .95$, if we want to set simultaneous intervals for all $m = \binom{k}{2} = 3$ pairwise differences, we require that the upper $\alpha/(2m) = .05/6 = .00833$ point of a t distribution.

Multiple-t Confidence Intervals

A set of $100(1 - \alpha)\%$ simultaneous confidence intervals for m number of pairwise differences $\mu_i - \mu_{i'}$ is given by

$$(\overline{Y}_i - \overline{Y}_{i'}) \pm t_{\alpha/2m} S \sqrt{\frac{1}{n_i} + \frac{1}{n_{i'}}}$$

where $S = \sqrt{\text{MSE}}$, $m =$ the number of confidence statements, and $t_{\alpha/2m} =$ the upper $\alpha/(2m)$ point of t with d.f. $= n - k$.

Prior to sampling, the probability of all the m statements being correct is at least $1 - \alpha$.

Example 4   Calculating Multiple-t Confidence Intervals to Reveal Which Means Differ

An experiment is conducted to determine the soil moisture deficit resulting from varying amounts of residual timber left after cutting trees in a forest. The three treatments are treatment 1: no timber left; treatment 2: 2000 bd ft left; treatment 3: 8000 bd ft left. (Board feet is a particular unit of measurement of timber volume.) The measurements of moisture deficit are given in Table 8. Perform the ANOVA test and construct 95% multiple-t confidence intervals for the treatment differences.

SOLUTION   Our analysis employs convenient alternative forms of the expressions for sums of squares involving totals.

TABLE 8   Moisture Deficit in Soil

Treatment	Observations	Total	Mean
1	1.52 1.38 1.29 1.48 1.63	$T_1 = 7.30$	$\overline{y}_1 = 1.460$
2	1.63 1.82 1.35 1.03 2.30 1.45	$T_2 = 9.58$	$\overline{y}_2 = 1.597$
3	2.56 3.32 2.76 2.63 2.12 2.78	$T_3 = 16.17$	$\overline{y}_3 = 2.695$
		Grand total	Grand mean
		$T = 33.05$	$\overline{y} = 1.944$

The total number of observations $n = 5 + 6 + 6 = 17$.

$$\text{Total SS} = \sum_{i=1}^{3} \sum_{j=1}^{n_i} (y_{ij} - \overline{y})^2 = \sum_{i=1}^{3} \sum_{j=1}^{n_i} y_{ij}^2 - \frac{T^2}{n}$$

$$= 71.3047 - 64.2531 = 7.0516$$

$$\text{Treatment SS} = \sum_{i=1}^{3} n_i(\bar{y}_i - \bar{y})^2 = \sum_{i=1}^{3} \frac{T_i^2}{n_i} - \frac{T^2}{n}$$

$$= 69.5322 - 64.2531 = 5.2791$$

$$\text{Error SS} = \text{Total SS} - \text{Treatment SS} = 1.7725$$

The ANOVA table appears in Table 9.

TABLE 9   ANOVA Table for Comparison of Moisture Deficit

Source	Sum of Squares	d.f.	Mean Square	F-ratio
Treatment	5.2791	2	2.640	20.8
Error	1.7725	14	.127	
Total	7.0516	16		

Because the observed value of F is larger than the tabulated value $F_{.05}(2, 14) = 3.74$, the null hypothesis of no difference in the treatment effects is rejected at $\alpha = .05$. In fact, this would be true at almost any significance level. In constructing a set of 95% multiple-t confidence intervals for pairwise differences, note that there are $\binom{3}{2} = 3$ pairs, so

$$\frac{\alpha}{2m} = \frac{.05}{(2 \times 3)} = .00833$$

From Appendix B, Table 4, the upper .00833 point of t with d.f. $= 14$ is 2.718. The simultaneous confidence intervals are calculated as follows:

$$\mu_2 - \mu_1: \quad (1.597 - 1.460) \pm 2.718 \times .356 \times \sqrt{\frac{1}{6} + \frac{1}{5}}$$

$$= (-.45, .72)$$

$$\mu_3 - \mu_2: \quad (2.695 - 1.597) \pm 2.718 \times .356 \times \sqrt{\frac{1}{6} + \frac{1}{6}}$$

$$= (.54, 1.66)$$

$$\mu_3 - \mu_1: \quad (2.695 - 1.460) \pm 2.718 \times .356 \times \sqrt{\frac{1}{6} + \frac{1}{5}}$$

$$= (.65, 1.82)$$

These confidence intervals indicate that treatments 1 and 2 do not differ appreciably, but the mean for treatment 3 is considerably higher than the means for treatments 1 and 2.

Exercises

14.17 Taking $\alpha = .05$ and $n - k = 26$, determine the appropriate percentile of the t distribution when calculating the multiple-t confidence intervals with (a) $m = 3$ and (b) $m = 5$.

14.18 Construct the 90% multiple-t confidence intervals using the sound distortion data in Example 1.

14.19 Given the following summary statistics,

$$
\begin{aligned}
n_1 &= 20 & \bar{y}_1 &= 10.2 \\
n_2 &= 18 & \bar{y}_2 &= 8.1 & s &= 3.2 \\
n_3 &= 24 & \bar{y}_3 &= 9.7 \\
n_4 &= 8 & \bar{y}_4 &= 6.2
\end{aligned}
$$

use $\alpha = .10$ and determine:

(a) t intervals for each of the six differences of means.

(b) The six multiple-t intervals.

14.20 Refer to the data on reading levels in Exercise 14.7.

(a) Calculate simultaneous confidence intervals for the differences in means.

(b) Are you surprised by the conclusion regarding reading levels? Give another variable that might better quantify reading levels.

14.21 Determine the expression for the length of the t interval for $\mu_1 - \mu_2$ and the multiple-t interval for $\mu_1 - \mu_2$ when $m = 10$. The ratio of lengths does not depend on the data. Evaluate this ratio for $\alpha = .10$ and $n - k = 15$.

5. GRAPHICAL DIAGNOSTICS AND DISPLAYS TO SUPPLEMENT ANOVA

In addition to testing hypotheses and setting confidence intervals, an analysis of data must include a critical examination of the assumptions involved in a model. As in regression analysis, of which analysis of variance is a special case, the residuals must be examined for evidence of serious violations of the assumptions. This aspect of the analysis is ignored in the ANOVA table summary.

Example 5 Plotting Residuals

Determine the residuals for the moisture data given in Example 4 (see Table 10) and graphically examine them for possible violations of the assumptions.

SOLUTION **TABLE 10** Residuals $y_{ij} - \bar{y}_i$ for the Data Given In Table 8

Treatment	Residuals					
1	.06	− .08	− .17	.02	.17	
2	.03	.22	− .25	− .57	.70	− .15
3	− .14	.63	.07	− .07	− .58	.09

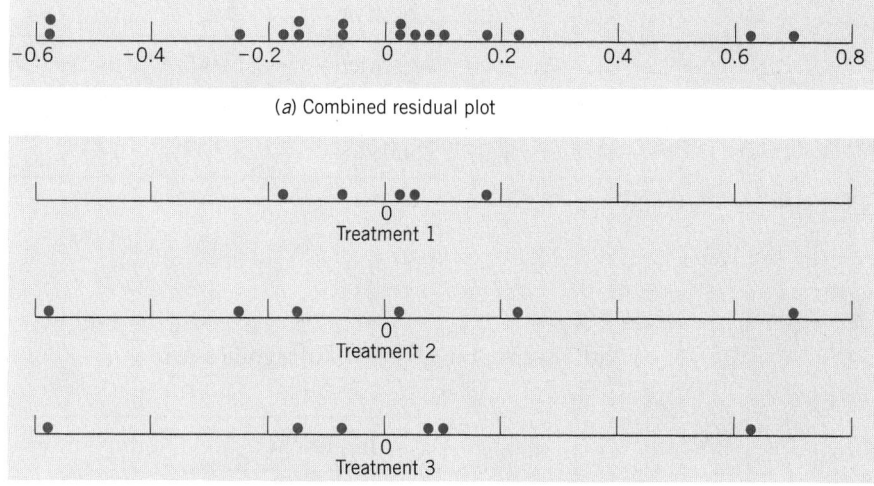

(a) Combined residual plot

Treatment 1

Treatment 2

Treatment 3

(b) Residuals with individual treatment

Figure 1   Residual plots for the data given in Example 4.

The residual plots of these data are shown in Figure 1, where the combined dot diagram is presented in (a) and the dot diagrams of residuals corresponding to individual treatments appear in (b).

From an examination of the dot diagrams, the variability in the points for treatment 1 appears to be somewhat smaller than the variabilities in the points for treatments 2 and 3. However, given so few observations, it is difficult to determine if this has occurred by chance or if treatment 1 actually has a smaller variance. A few more observations are usually necessary to obtain a meaningful pattern for the individual treatment plots.

Fortunately, the ANOVA testing procedure is **robust** in the sense that small or moderate departures from normality and constant variance do not seriously affect its performance.

In addition to the ANOVA a graphical portrayal of the data, as a box plot for each treatment, conveys important information available for making comparisons of populations.

Example 6   Box Plots Reveal Differences between Populations

The sepal width was measured on 50 iris flowers for each of three varieties, *Iris setosa, Iris versicolor,* and *Iris virginica.* A computer calculation produced the summary shown in the following ANOVA table.

Source	SS	d.f.	F
Treatment	11.345	2	49.16
Error	16.962	147	
Total	28.307	149	

Since $F_{.05}(2, 147) \approx 3.05$, we reject the null hypothesis of equal sepal width means at the 5% level of significance.

Treatment	Sample Mean
Iris setosa	3.428
Iris versicolor	2.770
Iris virginica	2.974

A calculation of multiple-t confidence intervals shows that all population means differ from one another (see Exercise 14.36). Display the data on the three varieties in boxplots.

SOLUTION   The data are given in Exercise 14.36. Boxplots graphically display the variation in the sepal width measurements. From Figure 2, we see that the *Iris setosa* typically has larger sepal width. R. A. Fisher, who developed analysis of variance, used these data along with other lengths and widths to introduce a statistical technique for identifying varieties of plants.

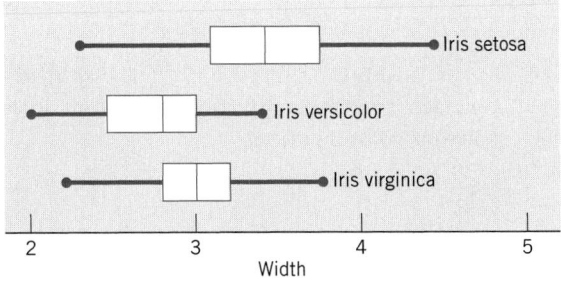

Figure 2   Boxplots for the three iris samples.

6. RANDOMIZED BLOCK EXPERIMENTS FOR COMPARING k TREATMENTS

Just as we can pair like subjects or experimental units to improve upon the procedure of taking two independent samples, we can also arrange, or **block** subjects into homogeneous groups of size k when comparing k treatments. Then if each treatment is applied to exactly one unit in the block and comparisons are only drawn between treatment responses from the same block, extraneous variability should be greatly reduced. It is this concept of blocking that underlies the randomized block design.

The term "block design" originated from the design of agricultural field trials, where "block" refers to a group of adjacent plots. A few typical examples for which the block design may be appropriate are clinical trials to compare several competing drugs, where the experimental subjects are grouped in blocks according to age group and severity of symptoms; psychological experiments comparing several stimuli, where subjects may be blocked according to socioeconomic background; and comparison of several techniques for storing fruit or vegetables, where each incoming shipment is regarded as a block.

As its name implies, *randomization* is a basic part of the block design. This time, once the grouping of like experimental subjects in blocks is accomplished, we randomly select one subject from the first block to receive treatment 1, one of the remaining subjects to receive treatment 2, and so on. The same procedure is repeated with a new randomization for each of the remaining blocks.

Once the data are obtained, they can be arranged in rows according to the treatments and in columns according to the blocks. If we designate the measurement corresponding to treatment i and block j by y_{ij}, the data structure of a randomized block design with b blocks and k treatments is shown in Table 11. The

TABLE 11 Data Structure of a Randomized Block Design with b Blocks and k Treatments

	Block 1	Block 2	$\cdots$	Block b	Treatment Means
Treatment 1	y_{11}	y_{12}	$\cdots$	y_{1b}	$\bar{y}_{1\cdot}$
Treatment 2	y_{21}	y_{22}	$\cdots$	y_{2b}	$\bar{y}_{2\cdot}$
$\cdot$	$\cdot$	$\cdot$	$\cdots$	$\cdot$	$\cdot$
$\cdot$	$\cdot$	$\cdot$	$\cdots$	$\cdot$	$\cdot$
$\cdot$	$\cdot$	$\cdot$	$\cdots$	$\cdot$	$\cdot$
Treatment k	y_{k1}	y_{k2}	$\cdots$	y_{kb}	$\bar{y}_{k\cdot}$
Block means	$\bar{y}_{\cdot 1}$	$\bar{y}_{\cdot 2}$	$\cdots$	$\bar{y}_{\cdot b}$	$\bar{y}_{\cdot\cdot}$

row and column means are denoted by

$$\text{ith treatment (row) mean } \bar{y}_{i.} = \frac{1}{b} \sum_{j=1}^{b} y_{ij}$$

$$\text{jth block (column) mean } \bar{y}_{.j} = \frac{1}{k} \sum_{i=1}^{k} y_{ij}$$

These means are shown in the margins of the table. Here an overbar on y indicates an average and a dot in the subscript denotes that the average is taken over the subscript appearing in that place.

We now discuss the analysis of variance for a randomized block design with illustrative calculations based on the data given in Example 7.

Example 7   The Structure of Data from a Randomized Block Experiment

The cutting speeds of four types of tools are being compared in an experiment. Five materials of varying degrees of hardness are to be used as experimental blocks. The data pertaining to measurements of cutting time in seconds appear in Table 12.

TABLE 12   Measurements of Cutting Time According to Types of Tool (Treatments) and Hardness of Material (Blocks)

Treatment	Block 1	Block 2	Block 3	Block 4	Block 5	Treatment Means
1	12	2	8	1	7	$\bar{y}_{1.} = 6$
2	20	14	17	12	17	$\bar{y}_{2.} = 16$
3	13	7	13	8	14	$\bar{y}_{3.} = 11$
4	11	5	10	3	6	$\bar{y}_{4.} = 7$
Block means	$\bar{y}_{.1} = 14$	$\bar{y}_{.2} = 7$	$\bar{y}_{.3} = 12$	$\bar{y}_{.4} = 6$	$\bar{y}_{.5} = 11$	$\bar{y}_{..} = \dfrac{200}{20} = 10$

The observations form a two-way table, and their decomposition indicates both a term for row (treatment) deviations and column (block) deviations.

Decomposition of Observations

Observation = Grand mean + Deviation due to treatment + Deviation due to block + Residual

$$y_{ij} = \bar{y}_{..} + (\bar{y}_{i.} - \bar{y}_{..}) + (\bar{y}_{.j} - \bar{y}_{..}) + (y_{ij} - \bar{y}_{i.} - \bar{y}_{.j} + \bar{y}_{..})$$

Taking the observation $y_{11} = 12$ in Example 7, we obtain

$$12 = 10 + (6 - 10) + (14 - 10) + (12 - 14 - 6 + 10)$$
$$= 10 + (-4) + (4) + (2)$$

Tables 13 and 14 contain the results of the decomposition of all the observations in two different formats.

TABLE 13 Decomposition of Observations
for the Randomized Block Experiment in Table 12

Observation y_{ij}	=	Grand Mean $\bar{y}..$	+	Treatment Effect $\bar{y}_{i\cdot} - \bar{y}..$

$$\begin{bmatrix} 12 & 2 & 8 & 1 & 7 \\ 20 & 14 & 17 & 12 & 17 \\ 13 & 7 & 13 & 8 & 14 \\ 11 & 5 & 10 & 3 & 6 \end{bmatrix} = \begin{bmatrix} 10 & 10 & 10 & 10 & 10 \\ 10 & 10 & 10 & 10 & 10 \\ 10 & 10 & 10 & 10 & 10 \\ 10 & 10 & 10 & 10 & 10 \end{bmatrix} + \begin{bmatrix} -4 & -4 & -4 & -4 & -4 \\ 6 & 6 & 6 & 6 & 6 \\ 1 & 1 & 1 & 1 & 1 \\ -3 & -3 & -3 & -3 & -3 \end{bmatrix}$$

+	Block Effect $\bar{y}_{\cdot j} - \bar{y}..$	+	Residual $y_{ij} - \bar{y}_{i\cdot} - \bar{y}_{\cdot j} + \bar{y}..$

$$+ \begin{bmatrix} 4 & -3 & 2 & -4 & 1 \\ 4 & -3 & 2 & -4 & 1 \\ 4 & -3 & 2 & -4 & 1 \\ 4 & -3 & 2 & -4 & 1 \end{bmatrix} + \begin{bmatrix} 2 & -1 & 0 & -1 & 0 \\ 0 & 1 & -1 & 0 & 0 \\ -2 & -1 & 0 & 1 & 2 \\ 0 & 1 & 1 & 0 & -2 \end{bmatrix}$$

TABLE 14 An Alternative Format of the Decomposition Table
for a Randomized Block Experiment

Treatment	Block 1	$\cdots$	j	$\cdots$	b	Deviation of Treatment Mean from Grand Mean
1						$(\bar{y}_{1\cdot} - \bar{y}..)$
.						.
.						.
i		$\cdots$	$(y_{ij} - \bar{y}_{i\cdot} - \bar{y}_{\cdot j} + \bar{y}..)$	$\cdots$		$(\bar{y}_{i\cdot} - \bar{y}..)$
.						.
.						.
k						$(\bar{y}_{k\cdot} - \bar{y}..)$
Deviation of block mean from grand mean	$(\bar{y}_{\cdot 1} - \bar{y}..)$ $\cdots$		$(\bar{y}_{\cdot j} - \bar{y}..)$	$\cdots$ $(\bar{y}_{\cdot b} - \bar{y}..)$		Grand mean $\bar{y}..$

The number of distinct entries $\bar{y}_{i\cdot} - \bar{y}_{\cdot\cdot}$ in the treatment effects array is k, and the single constraint is that they must sum to zero. Thus, $k - 1$ degrees of freedom are associated with the treatment sum of squares.

Sum of Squares Due to Treatment

$$SS_T = \sum_{i=1}^{k} \sum_{j=1}^{b} (\bar{y}_{i\cdot} - \bar{y}_{\cdot\cdot})^2$$

$$= b \sum_{i=1}^{k} (\bar{y}_{i\cdot} - \bar{y}_{\cdot\cdot})^2$$

with d.f. $= k - 1$.

In the array of treatment effects in Table 13, each entry appears $b = 5$ times, once in each block. The treatment sum of squares for this example is then

$$SS_T = 5(-4)^2 + 5(6)^2 + 5(1)^2 + 5(-3)^2 = 310$$

with d.f. $= (4 - 1) = 3$.

In a similar manner, we obtain the block sum of squares.

Sum of Squares Due to Block

$$SS_B = \sum_{i=1}^{k} \sum_{j=1}^{b} (\bar{y}_{\cdot j} - \bar{y}_{\cdot\cdot})^2$$

$$= k \sum_{j=1}^{b} (\bar{y}_{\cdot j} - \bar{y}_{\cdot\cdot})^2$$

with d.f. $= b - 1$.

Referring to the array of block effects in Table 13, we find the block sum of squares for our example to be

$$SS_B = 4(4)^2 + 4(-3)^2 + 4(2)^2 + 4(-4)^2 + 4(1)^2 = 184$$

with d.f. $= 5 - 1 = 4$.

The number of degrees of freedom associated with the residual array is $(b - 1)(k - 1)$. To understand why this is so, note that among the $b \times k$ residuals, the following constraints are satisfied. One constraint is that the sum of all entries must be zero. The fact that all row sums are zero introduces $k - 1$ additional constraints. This is so because having fixed any $k - 1$ row totals and the grand total, the remaining row total is automatically fixed. By the same reasoning, $b - 1$ additional constraints arise from the fact that all column totals are zero. Consequently, the number of degrees of freedom for the residual is

$$bk - 1 - (k - 1) - (b - 1) = (b - 1)(k - 1)$$

Residual Sum of Squares or Error Sum of Squares

$$\text{SSE} = \sum_{i=1}^{k} \sum_{j=1}^{b} (y_{ij} - \bar{y}_{i\cdot} - \bar{y}_{\cdot j} + \bar{y}_{\cdot\cdot})^2$$

with d.f. $= (b - 1)(k - 1)$.

In our example,

$$\text{SSE} = 2^2 + 0^2 + (-2)^2 + 0^2 + \cdots + (-2)^2 = 24$$

with d.f. $= (5 - 1)(4 - 1) = 12$.

Finally, the total sum of squares equals the sum of squares of each observation about the grand mean, or

$$\text{Total sum of squares} = \sum_{i=1}^{k} \sum_{j=1}^{b} (y_{ij} - \bar{y}_{\cdot\cdot})^2 = \sum_{i=1}^{k} \sum_{j=1}^{b} y_{ij}^2 - bk\bar{y}_{\cdot\cdot}^2$$

with d.f. $= bk - 1$. In our example, we sum the square of each entry in the array for y_{ij} and subtract the sum of squares of entries in the $\bar{y}_{\cdot\cdot}$ array.

$$\begin{aligned}
\text{Total sum of squares} &= (12)^2 + (2)^2 + \cdots + (6)^2 \\
&\quad - [(10)^2 + (10)^2 + \cdots + (10)^2] \\
&= 518
\end{aligned}$$

With d.f. $= (5)(4) - 1 = 19$. This provides a check on our previous calculations, because those sums of squares and degrees of freedom must sum to these totals.

These calculations are conveniently summarized in the ANOVA tables shown in Table 15 for the general case and Table 16 for the data in our numerical example. The last column of F-ratios will be explained after we discuss the population model.

TABLE 15   ANOVA Table for a Randomized Block Design

Source	Sum of Squares	d.f.	Mean Square	F-Ratio
Treatments	$SS_T = b \sum_{i=1}^{k} (\bar{y}_{i\cdot} - \bar{y}_{\cdot\cdot})^2$	$k - 1$	$MS_T = \dfrac{SS_T}{k - 1}$	$\dfrac{MS_T}{MSE}$
Blocks	$SS_B = k \sum_{j=1}^{b} (\bar{y}_{\cdot j} - \bar{y}_{\cdot\cdot})^2$	$b - 1$	$MS_B = \dfrac{SS_B}{b - 1}$	$\dfrac{MS_B}{MSE}$
Residual	$SSE = \sum_{i=1}^{k} \sum_{j=1}^{b} (y_{ij} - \bar{y}_{i\cdot} - \bar{y}_{\cdot j} + \bar{y}_{\cdot\cdot})^2$	$(b - 1)(k - 1)$	$MSE = \dfrac{SSE}{(b - 1)(k - 1)}$	
Total	$\sum_{i=1}^{k} \sum_{j=1}^{b} (y_{ij} - \bar{y}_{\cdot\cdot})^2$	$bk - 1$		

TABLE 16   ANOVA Table for the Data Given in Example 7

Source	Sum of Squares	d.f.	Mean Square	F-Ratio
Treatments	310	3	103.3	51.7
Blocks	184	4	46	23
Residual	24	12	2	
Total	518	19		

Again, a statistical test of treatment differences is based on an underlying population model.

Population Model for a Randomized Block Experiment

$$\text{Observation} = \underset{\text{mean}}{\text{Overall}} + \underset{\text{effect}}{\text{Treatment}} + \underset{\text{effect}}{\text{Block}} + \text{Error}$$

$$Y_{ij} = \mu + \alpha_i + \beta_j + e_{ij}$$

for $i = 1, \ldots, k$ and $j = 1, \ldots, b$, where the parameters satisfy

$$\sum_{i=1}^{k} \alpha_i = 0 \qquad \sum_{j=1}^{b} \beta_j = 0$$

and the e_{ij} are random errors independently distributed as $N(0, \sigma)$.

Tests for the absence of treatment differences or differences in block effects can now be performed by comparing the corresponding mean square with the yardstick of the error mean square by using an F test.

Reject $H_0: \alpha_1 = \cdots = \alpha_k = 0$ (no treatment differences) if

$$\frac{MS_T}{MSE} > F_\alpha(k - 1, (b - 1)(k - 1))$$

Reject $H_0: \beta_1 = \cdots = \beta_b = 0$ (no block differences) if

$$\frac{MS_B}{MSE} > F_\alpha(b - 1, (b - 1)(k - 1))$$

To test the hypothesis of no treatment differences for the analysis of variance in Table 16, we find that the tabulated .05 point of $F(3, 12)$ is 3.49, a value far exceeded by the observed F-ratio for treatment effect. We therefore conclude that a highly significant treatment difference is indicated by the data. The block effects are also highly significant, because the observed F value of 23 is much larger than the tabulated value $F_{05}(4, 12) = 3.26$.

Again, we stress that a serious violation of the model assumptions is likely to jeopardize the conclusions drawn from the preceding analyses and a careful examination of the residuals should be an integral part of the analysis. In addition to plotting the whole set of residuals in a graph, separate plots for individual treatments and individual blocks should also be studied. When observations are collected over time, a plot of the residuals versus the time order is also important.

CONFIDENCE INTERVALS FOR TREATMENT DIFFERENCES

In addition to performing the overall F test for detecting treatment differences, the experimenter typically establishes confidence intervals to compare specific pairs of treatments. This is particularly important when the F test leads to a rejection of the null hypothesis, thus signifying the presence of treatment differences.

We next estimate the difference $\alpha_i - \alpha_{i'}$ of the mean responses of treatments i and i'. Because $\overline{Y}_{i\cdot} - \overline{Y}_{i'\cdot}$ is normally distributed with a mean of $\alpha_i - \alpha_{i'}$ and a variance of

$$\sigma^2 \left(\frac{1}{b} + \frac{1}{b} \right) = \sigma^2 \left(\frac{2}{b} \right)$$

the ratio

$$T = \frac{(\overline{Y}_{i\cdot} - \overline{Y}_{i'\cdot}) - (\alpha_i - \alpha_{i'})}{\sqrt{MSE} \ \sqrt{(2/b)}}$$

has a t distribution with d.f. $= (b - 1)(k - 1)$. This result can be used to construct a confidence interval for an individual difference $\alpha_i - \alpha_{i'}$.

When several such pairwise comparisons are to be integrated into a combined confidence statement, the concept of simultaneous confidence intervals, discussed in Section 4, is again applied.

Exercises

14.22 Suppose you wish to compare three different brands of tick collars for dogs. You have available three of each of the breeds Poodle, Lab, Collie, and Dachshund. Explain how you would assign a brand of tick collar to each of the 12 dogs in order to conduct a randomized block experiment.

14.23 (a) Provide a decomposition for the following observations from a randomized block experiment.

(b) Find the sum of squares for each array.

(c) Determine the degrees of freedom by checking the constraints for each array.

Treatment	Block 1	2	3	4
1	11	10	7	0
2	7	8	7	2
3	15	6	13	10

14.24 (a) Provide a decomposition for the following observations from a randomized block experiment.

(b) Find the sum of squares for each array.

(c) Determine the degrees of freedom by checking the constraints for each array.

Treatment	Block 1	2	3	4
1	35	24	28	21
2	19	14	14	13
3	21	16	21	14

14.25 Refer to Exercise 14.23. Present the ANOVA table. What conclusions can you draw from the two F tests? Take $\alpha = .05$.

14.26 Refer to Exercise 14.24. Present the ANOVA table. What conclusions can you draw from the two F tests? Take $\alpha = .05$.

14.27 Three loaves of bread, each made according to a different recipe, are baked in one oven at the same time. Because of possible uncontrolled variations in oven performance, each baking is treated as a block. This procedure is repeated five times, and the following measurements of density are obtained.

Recipe	Block 1	2	3	4	5
1	.95	.86	.71	.72	.74
2	.71	.85	.62	.72	.64
3	.69	.88	.51	.73	.44

(a) How should the three oven positions of the three loaves be selected for each trial?

(b) Perform an analysis of variance for these data.

14.28 Referring to Exercise 14.27:

(a) Obtain simultaneous confidence intervals for the pairwise differences in mean density for the three recipes. Take $\alpha = .05$.

(b) Calculate the residuals and make a normal-scores plot.

14.29 As part of a cooperative study on the nutritional quality of oats, 6 varieties of oat kernels with their hulls removed are subjected to a mineral analysis. The plants are grown according to a randomized block design, and the following measurements of protein by percent of dry weight are recorded.

(a) Perform an analysis of variance for these data.

(b) Calculate and plot the residuals. Does the model appear to be adequate?

14.30   Referring to Exercise 14.29, suppose that variety 6 is of special interest. Construct simultaneous 90% confidence intervals for the differences between the mean of variety 6 and each of the other means.

Treatment	Block					
	1	2	3	4	5	6
1	19.09	20.29	20.31	19.60	18.62	20.10
2	16.28	17.88	16.88	17.57	16.72	17.32
3	16.31	18.17	17.38	17.53	16.34	17.88
4	17.50	18.05	17.59	17.64	17.38	18.04
5	16.25	16.92	15.88	14.78	15.97	16.66
6	21.09	21.37	21.38	20.52	21.09	21.58

Data courtesy of D. Peterson, L. Schrader, and V. Youngs.

USING STATISTICS WISELY

1. When collecting data according to a one-way ANOVA design, conduct the trials in random order if at all possible.

2. Do not routinely accept the analysis of variance generated by statistical software. Instead, inspect the residuals for outliers or patterns indicating that the variance is not constant across treatments.

KEY IDEAS AND FORMULAS

Several populations can be compared using the Analysis of variance (ANOVA).

A completely randomised design specifies taking independent random samples from each population.

All of the observations are assumed to have a common variance. Then the ANOVA consists of a separation of the total sum of squares into components due to different sources of variation.

When samples are taken from k populations the two sources of variation are the within population variation or error and the differences between population means or treatments. The jth observation on the ith treatment is y_{ij}. The error variation is estimated using the residuals or $y_{ij} - \bar{y}_i$ which are the deviations of the observations from their respective sample means.

The total variation in the observations y_{ij} is expressed as

$$\text{Total sum of squares} = \sum_{i=1}^{k} \sum_{j=1}^{n_i} (y_{ij} - \bar{y})^2$$

$\sum_{i=1}^{k} \sum_{j=1}^{n_i} (y_{ij} - \bar{y})^2$ is partitioned into the two components:

1. Treatment sum of squares:

$$SS_T = \sum_{i=1}^{k} n_i(\bar{y}_i - \bar{y})^2$$

2. Error sum of squares:

$$SSE = \sum_{i=1}^{k} \sum_{j=1}^{n_i} (y_{ij} - \bar{y}_i)^2$$

This decomposition of sums of squares is summarized in an analysis of variance (ANOVA) table. This table includes the mean square for each sum of squares.

$$\text{Mean square} = \frac{\text{Sum of squares}}{\text{Degrees of freedom}}$$

The F statistic $= (\text{Treatment mean square})/(\text{Error mean square})$

which has an F distribution with $k - 1$ and $n - k$ degrees of freedom.

To ensure an overall confidence level for all confidence intervals, it is desirable to use a simultaneous (multiple) confidence intervals procedure when calculating confidence intervals for the many mean differences. The multiple-t confidence intervals for all $m = k(k - 1)/2$ statements about the mean differences $\mu_i - \mu_{i'}$ are

$$\left(\bar{Y}_i - \bar{Y}_{i'} - t_{\alpha/2m} \sqrt{\frac{MSE}{n - k}}, \quad \bar{Y}_i - \bar{Y}_{i'} + t_{\alpha/2m} \sqrt{\frac{MSE}{n - k}} \right)$$

Fortunately, the one-way analysis of variance is robust with respect to small or moderate departures from the assumptions of normal errors with common variance.

The technique of blocking, or grouping experimental units into homogeneous sets, can reduce extraneous variation and sharpen comparisons among treatments. The analysis of variance of a randomized block experiment is based on the partition of the total sum of squares $\sum_{i=1}^{k} \sum_{j=1}^{b} (y_{ij} - \bar{y}..)^2$ into three components:

1. Treatment sum of squares:

$$SS_T = b \sum_{i=1}^{k} (\bar{y}_{i\cdot} - \bar{y}..)^2$$

2. Block sum of squares:

$$SS_B = k \sum_{j=1}^{b} (\bar{y}_{\cdot j} - \bar{y}..)^2$$

3. Error sum of squares:

$$SSE = \sum_{i=1}^{k} \sum_{j=1}^{b} (y_{ij} - \bar{y}_{i\cdot} - \bar{y}_{\cdot j} + \bar{y}..)^2$$

TECHNOLOGY

One-way analysis of variance (ANOVA)

MINITAB

We illustrate with the data from different populations in separate columns. For example, with data from 3 populations in columns $C1$, $C2$, and $C3$:

Data

    $C1$:   6   10   8
    $C2$: 13   14   15   14
    $C3$:   3   7   5

Stat > ANOVA > One-way (Unstacked)
Type $C1 - C3$ in **Responses (in separate columns:)**
Click on **Graphs**. Select **Three in one**. Click **OK**.
Click **OK**.

In addition to the analysis, this produces a normal probability plot and histogram of the residuals as well as a plot of residuals versus fitted values.

EXCEL

Enter the data from the different populations in separate columns, for instance A, B, C, etc., with a label in the first cell of each column.

Select **Tools**, then **Data Analysis**. Select **Statistical** and then **Anova:Single Factor.**
Click **OK**. Highlight the columns and check **Labels in First Row.**
Enter **Alpha** and click **OK**.

The output includes the ANOVA table and P-value for test of equality of the means.

TI-84/-83 PLUS

Store the data in lists, one for each population. With 3 populations enter data in L_1, L_2, and L_3.

Select STAT > TESTS > ANOVA.
Enter L_1, L_2, L_3 and close the parenthesis to obtain ANOVA(L_1, L_2, L_3).
Press **Enter**.

The software returns the value of the F statistic and the P-value, along with other values from the ANOVA table.

7. REVIEW EXERCISES

14.31 Provide a decomposition for the following observations from a completely randomized design with three treatments.

Treatment 1	Treatment 2	Treatment 3
19	16	13
18	11	16
21	13	18
18	14	11
	11	15
		11

14.32 Compute the sums of squares and construct the ANOVA table for the data given in Exercise 14.31.

14.33 Using the table of percentage points for the F distribution, find

(a) The upper 5% point when d.f. $=$ (7, 13).

(b) The upper 5% point when d.f. $=$ (7, 20).

(c) The upper 10% point when d.f. $=$ (7, 12).

14.34 As part of the multilab study, four fabrics are tested for flammability at the National Bureau of Standards. The following burn times in seconds are recorded after a paper tab is ignited on the hem of a dress made of each fabric.

Fabric 1	Fabric 2	Fabric 3	Fabric 4
17.8	11.2	11.8	14.9
16.2	11.4	11.0	10.8
17.5	15.8	10.0	12.8
17.4	10.0	9.2	10.7
15.0	10.4	9.2	10.7

(a) State the statistical model and present the ANOVA table. With $\alpha = .05$, test the null hypothesis of no difference in the degree of flammability for the four fabrics.

(b) If the null hypothesis is rejected, construct simultaneous confidence intervals to determine the fabric(s) with the lowest mean burn time.

(c) Plot the residuals and comment on the plausibility of the assumptions.

(d) If the tests had been conducted one at a time on a single mannequin, how would you have randomized the fabrics tested in this experiment?

The Following Exercises Require a Computer

14.35 *Using the computer.* MINITAB can be used for ANOVA. Start with the data on each treatment, from Example 1, set in separate columns. The sequence of commands and output is:

Data

C1: 10 15 8 12 15
C2: 14 18 21 15
C3: 17 16 14 15 17 15 18
C4: 12 15 17 15 16 15

Dialog box:

Stat > ANOVA > One-way (Unstacked)
Type *C1-C4* in **Responses**. Click **OK**.

The **Output** is as follows:

```
One-way ANOVA: C1, C2, C3, C4

SOURCE   DF      SS      MS      F       P
FACTOR    3   68.00   22.67   4.34   0.018
ERROR    18   94.00    5.22
TOTAL    21  162.00

S = 2.285     R-Sq= 41.98%

LEVEL    N     MEAN    STDEV
C1       5   12.000    3.082
C2       4   17.000    3.162
C3       7   16.000    1.414
C4       6   15.000    1.673

POOLED STDEV = 2.285
```

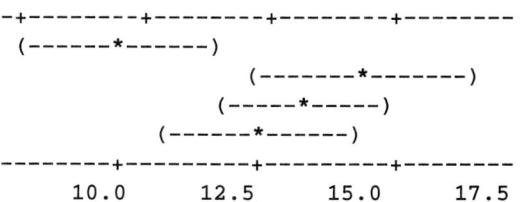

```
INDIVIDUAL 95% CIS FOR MEAN
BASED ON POOLED STDEV


-+---------+---------+---------+---------
 (------*------)
                    (-------*-------)
               (-----*-----)
          (------*------)
--------+---------+---------+---------
      10.0      12.5      15.0      17.5
```

Use computer software to analyze the moisture data in Table 8.

14.36 The iris data described in Example 6 are given in the stem-and-leaf diagrams below.

```
LEAF UNIT = 0.10

SETOSA              VERSICOLOR      VIRGINICA

                    2 0
2 3                 2 22333         2 2
2                   2 4445555       2 5555
2                   2 66677777      2 667777
2 9                 2 8888889999999 2 8888888899
3 0000001111        3 00000000111   3 0000000000001111
3 2222233           3 2223          3 22222333
3 444444444555555 3 4             3 44
3 666777                            3 6
3 888899                            3 88
4 01
4 2
4 4
```

The MINITAB output for the analysis of the iris data is given below.

```
One-way ANOVA:

SOURCE  DF      SS      MS      F       P
IRIS     2   11.345   5.672  49.16   0.000
ERROR  147   16.962   0.115
TOTAL  149   28.307


S = 0.3397   R-Sq = 40.08%
```

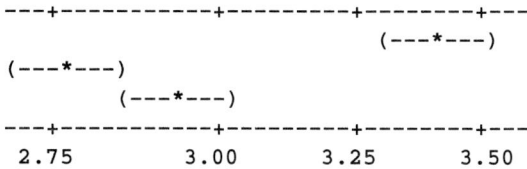

```
LEVEL   N     MEAN    STDEV
1      50    3.4280  0.3791
2      50    2.7700  0.3138
3      50    2.9740  0.3225
```

```
POOLED STDEV = 0.3397
```

```
INDIVIDUAL 95% CIs FOR MEAN
BASED ON POOLED STDEV


---+-----------+---------+---------+---
                                 (---*---)
(---*---)
          (---*---)
---+-----------+---------+---------+---
  2.75        3.00      3.25      3.50
```

(a) Identify the SSE and its degrees of freedom. Also locate s.

(b) Check the calculation of F from the given sums of squares and d.f.

(c) Is there one population with highest mean or are two or more alike? Use multiple-t confidence intervals with $\alpha = .05$.

14.37 (a) Provide a decomposition for the following observations from a randomized block experiment.

(b) Find the sum of squares for each array.

(c) Determine the degrees of freedom by checking the constraints for each array.

		Block		
Treatment	1	2	3	4
1	8	9	1	6
2	5	12	0	11
3	8	15	8	13

14.38 Refer to the output below concerning the time (min) it took four different persons (blocks) to complete three different tasks.

(a) Identify the SSE and its degrees of freedom.

(b) Are the block means different? Check the calculation of F for blocks from the given sums of squares and degrees of freedom.

```
Two-way Analysis of Variance

Analysis of Variance for Time
Source        DF        SS          MS        F        P
Block          3       90.92       30.31     7.96     0.016
Treatment      2       71.17       35.58     9.35     0.014
Error          6       22.83        3.81
Total         11      184.92


Block             Mean
1                  9.0
2                 13.7
3                  6.7
4                 12.3


Treatment         Mean
1                 8.00
2                 9.50
3                13.75
```

(c) Are the mean task times different? Check the calculation of F for treatments (tasks) from the given sums of squares and degrees of freedom.

(d) Use multiple-t 95% confidence intervals to investigate differences between mean task times.

15

Nonparametric Inference

Selecting the Best Vintage

Wines can be ranked without reference to a quantitative scale of measurement. Individuals use non-quantitative characteristics to help select their favorite wines.
© Peter Beck/Corbis Stock Market.

1. INTRODUCTION

Nonparametric refers to inference procedures that do not require the population distribution to be normal or some other form specified in terms of parameters. Nonparametric procedures continue to gain popularity because they apply to a very wide variety of population distributions. Typically, they utilize simple aspects of the sample data, such as the signs of the measurements, order relationships, or category frequencies. Stretching or compressing the scale of measurement does not alter them. As a consequence, the null distribution of a nonparametric test statistic can be determined without regard to the shape of the underlying population distribution. For this reason, these tests are also called distribution-free tests. This distribution-free property is their strongest advantage.

What type of observations are especially suited to a nonparametric analysis? Characteristics like degree of apathy, taste preference, and surface gloss cannot be evaluated on an objective numerical scale, and an assignment of numbers is, therefore, bound to be arbitrary. Also, when people are asked to express their views on a five-point rating scale,

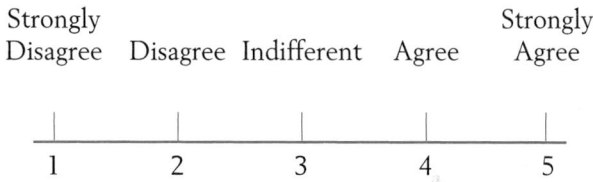

the numbers have little physical meaning beyond the fact that higher scores indicate greater agreement. Data of this type are called ordinal data, because only the order of the numbers is meaningful and the distance between the two numbers does not lend itself to practical interpretation. Nonparametric procedures that utilize information only on order or rank are particularly suited to measurements on an ordinal scale.

2. THE WILCOXON RANK-SUM TEST FOR COMPARING TWO TREATMENTS

The problem of comparing two populations based on independent random samples has already been discussed in Section 2 of Chapter 10. Under the assumption of normality and equal standard deviations, the parametric inference procedures were based on Student's t statistic. Here we describe a useful nonparametric procedure named after its proposer F. Wilcoxon (1945). An equivalent alternative version was independently proposed by H. Mann and D. Whitney (1947).

For a comparative study of two treatments A and B, a set of $n = n_A + n_B$ experimental units is randomly divided into two groups of sizes n_A and n_B, respectively. Treatment A is applied to the n_A units, and treatment B to the other n_B units. The response measurements, recorded in a slightly different notation than before, are

$$\text{Treatment } A \quad X_{11}, \quad X_{12}, \quad \ldots, \quad X_{1n_A}$$
$$\text{Treatment } B \quad X_{21}, \quad X_{22}, \quad \ldots, \quad X_{2n_B}$$

These data constitute independent random samples from two populations. Assuming that larger responses indicate a better treatment, we wish to test the null hypothesis that there is no difference between the two treatment effects versus the one-sided alternative that treatment A is more effective than treatment B. In the present nonparametric setting, we only assume that the distributions are continuous.

Model: Both Population Distributions Are Continuous

Hypotheses

H_0: The two population distributions are identical.

H_1: The distribution of population A is **shifted** to the right of the distribution of population B.

Note that no assumption is made regarding the *shape* of the population distribution. This is in sharp contrast to our t test in Chapter 10, where we assumed that the population distributions were normal with equal standard deviations. Figure 1 illustrates the above hypotheses H_0 and H_1.

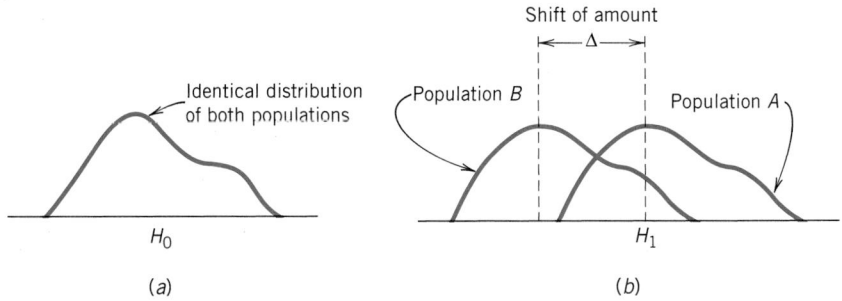

Figure 1   (*a*) Null distribution. (*b*) A shift alternative.

The basic concept underlying the rank-sum test can now be explained by the following intuitive line of reasoning. Suppose that the two sets of observations are plotted on the same diagram using different markings A and B to identify their sources. Under H_0, the samples come from the same population, so that the two sets of points should be well mixed. However, if the larger observations are more often associated with the first sample, for example, we can infer that population A is possibly shifted to the right of population B. These two situations are diagrammed in Figure 2, where the combined set of points in each case is serially numbered from left to right. These numbers are called the combined sample ranks. In Figure 2a, large as well as small ranks are associated with each sample, whereas in Figure 2b, most of the larger ranks are associated with the first sample. Therefore, if we consider the sum of the ranks associated with the first sample as a test statistic, a large value of this statistic should reflect that the first population is located to the right of the second.

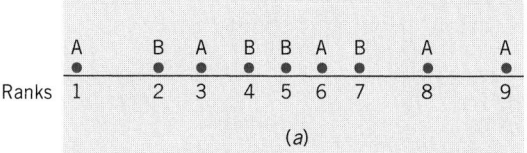
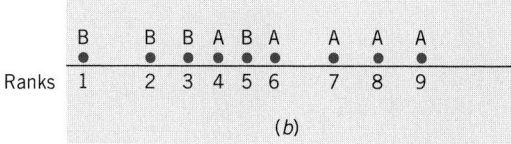

Figure 2   Combined plot of the two samples and the combined sample ranks. (a) Mixed ranks. (b) Higher ranks are mostly A.

To establish a rejection region with a specified level of significance, we must consider the distribution of the rank-sum statistic under the null hypothesis. This concept is explored in Example 1, where small sample sizes are investigated for easy enumeration.

Example 1   Determining the Null Distribution of the Rank-Sum Statistic

To determine if a new hybrid seedling produces a bushier flowering plant than a currently popular variety, a horticulturist plants 2 new hybrid seedlings and 3 currently popular seedlings in a garden plot. After the plants mature, the following measurements of shrub girth in inches are recorded.

	Shrub Girth (in inches)		
Treatment A (new hybrid)	31.8	39.1	
Treatment B (current variety)	35.5	27.6	21.3

Do these data strongly indicate that the new hybrid produces larger shrubs than the current variety?

SOLUTION   We wish to test the null hypothesis

$$H_0: A \text{ and } B \text{ populations are identical}$$

versus the alternative hypothesis

$$H_1: \text{Population } A \text{ is shifted from } B \text{ toward larger values}$$

For the rank-sum test, the two samples are placed together and ranked from smallest to largest:

Combined sample ordered observations	21.3	27.6	31.8	35.5	39.1
Ranks	1	2	3	4	5
Treatment	B	B	A	B	A

$$\text{Rank sum for } A \qquad W_A = 3 + 5 = 8$$
$$\text{Rank sum for } B \qquad W_B = 1 + 2 + 4 = 7$$

Because larger measurements and therefore higher ranks for treatment A tend to support H_1, the rejection region of our test should consist of large values for W_A:

$$\text{Reject } H_0 \quad \text{if} \quad W_A \geq c$$

To determine the critical value c so that the Type I error probability is controlled at a specified level α, we evaluate the probability distribution of W_A under H_0. When the two samples come from the same population, every pair of integers out of $\{1, 2, 3, 4, 5\}$ is equally likely to be the ranks for the two A measurements. There are $\binom{5}{2} = 10$ potential pairs, so that each collection of possible ranks has a probability of $\frac{1}{10} = .1$ under H_0. These rank collections are listed in Table 1 with their corresponding W_A values. The null distribution of W_A can be obtained immediately from Table 1 by collecting the probabilities of identical values (see Table 2). The observed value $W_A = 8$

TABLE 1   Rank Collections for
Treatment A with Sample
Sizes $n_A = 2, n_B = 3$

Ranks of A	Rank Sum W_A	Probability
1,2	3	.1
1,3	4	.1
1,4	5	.1
1,5	6	.1
2,3	5	.1
2,4	6	.1
2,5	7	.1
3,4	7	.1
3,5	8	.1
4,5	9	.1
		Total 1.0

TABLE 2 Distribution of the Rank Sum W_A for Sample Sizes $n_A = 2, n_B = 3$

Value of W_A	3	4	5	6	7	8	9
Probability	.1	.1	.2	.2	.2	.1	.1

has the significance probability $P_{H_0}(W_A \geq 8) = .1 + .1 = .2$. In other words, we must tolerate a Type I error probability of .2 in order to reject H_0. The rank-sum test leads us to conclude that the evidence is not sufficiently strong to reject H_0. Note that even if the A measurements did receive the highest ranks of 4 and 5, a significance level of $\alpha = .1$ would be required to reject H_0.

Guided by Example 1, we now state the rank-sum test procedure in a general setting.

Wilcoxon Rank-Sum Test

Let $X_{11}, \ldots, X_{1n_A}$ and $X_{21}, \ldots, X_{2n_B}$ be independent random samples from continuous populations A and B, respectively. To test H_0: The populations are identical:

1. Rank the combined sample of $n = n_A + n_B$ observations in increasing order of magnitude.
2. Find the rank sum W_A of the first sample.
3. (a) For H_1: Population A is shifted to the right of population B; set the rejection region at the upper tail of W_A.
   (b) For H_1: Population A is shifted to the left of population B; set the rejection region at the lower tail of W_A.
   (c) For H_1: Populations are different; set the rejection region at both tails of W_A having equal probabilities.

A determination of the null distribution of the rank-sum statistic by direct enumeration becomes more tedious as the sample sizes increase. However, tables for the null distribution of this statistic have been prepared for small samples, and an approximation is available for large samples. To explain the use of Appendix B, Table 7, first we note some features of the rank sums W_A and W_B.

The total of the two ranks sums $W_A + W_B$ is a constant, which is the sum of the integers $1, 2, \ldots, n$, where n is the combined sample size. For instance, in Example 1,

$$W_A + W_B = (3 + 5) + (1 + 2 + 4)$$
$$= 1 + 2 + 3 + 4 + 5 = 15$$

Therefore, a test that rejects H_0 for large values of W_A is equivalent to a test that rejects H_0 for small values of W_B. We can just as easily designate W_B the test statistic and set the rejection region at the lower tail. Consequently, we can always concentrate on the rank sum of the smaller sample and set the rejection region at the lower (or upper) tail, depending on whether the alternative hypothesis states that the corresponding population distribution is shifted to the left (or right).

Second, the distribution of each of the rank-sum statistics W_A and W_B is symmetric. In fact, W_A is symmetric about $n_A(n_A + n_B + 1)/2$ and W_B is symmetric about $n_B(n_A + n_B + 1)/2$. Table 2 illustrates the symmetry of the W_A distribution for the case $n_A = 2$, $n_B = 3$. This symmetry also holds for the test statistic calculated from the larger sample size.

THE USE OF APPENDIX B, TABLE 7

The Wilcoxon rank-sum test statistic is taken as

$$W_S = \text{sum of ranks of the smaller sample in the combined sample ranking}$$

When the sample sizes are equal, take the sum of ranks for either of the samples. Appendix B, Table 7 gives the upper- as well as the lower-tail probabilities:

$$\begin{aligned}
&\text{Upper-tail probability} && P[W_S \geq x] \\
&\text{Lower-tail probability} && P[W_S \leq x^*]
\end{aligned}$$

By the symmetry of the distribution, these probabilities are equal when x and x^* are at equal distances from the center. The table includes the x^* values corresponding to the x's at the upper tail.

Example 2   Using Table 7 in Appendix B to Set the Rejection Region
Find $P[W_S \geq 25]$ and $P[W_S \leq 8]$ when

$$\begin{aligned}
\text{Smaller sample size} &= 3 \\
\text{Larger sample size} &= 7
\end{aligned}$$

SOLUTION   From Table 7, we read $P = P[W_S \geq x]$ opposite the entry $x = 25$, so $.033 = P[W_S \geq 25]$.

The lower tail entry $P[W_S \leq 8]$ is obtained by reading $P[W_S \leq x^*]$ opposite $x^* = 8$. We find $P[W_S \leq 8] = .033$ illustrating the symmetry of W_S.

$$P = P[W_S \geq x] = P[W_S \leq x^*]$$

Smaller Sample Size = 3, Larger
Sample Size = 7

x	P	x^*
22	—	11
23	—	10
24	—	9
25	.033	8
26	—	7
27	—	6

The steps to follow when using Appendix B, Table 7 in performing a rank-sum test are:

Use the rank-sum W_S of the smaller sample as the test statistic. (If the sample sizes are equal, take either rank sum as W_S.)

1. If H_1 states that the population corresponding to W_S is shifted to the right of the other population, set a rejection region of the form $W_S \geq c$ and take c as the smallest x value for which $P \leq \alpha$.

2. If H_1 states that the population corresponding to W_S is shifted to the left, set a rejection region of the form $W_S \leq c$ and take c as the largest x^* value for which $P \leq \alpha$.

3. If H_1 states that the population corresponding to W_S is shifted in either direction, set a rejection region of the form $W_S \leq c_1$ or $W_S \geq c_2$ and read c_1 from the x^* column and c_2 from the x column, so that $P \leq \alpha/2$.

Example 3 Apply the Rank-Sum Test to Compare Two Geological Formations

Two geological formations are compared with respect to richness of mineral content. The mineral contents of 7 specimens of ore collected from formation 1 and 5 specimens collected from formation 2 are measured by chemical analysis. The following data are obtained:

Mineral Content

Formation 1	7.6	11.1	6.8	9.8	4.9	6.1	15.1
Formation 2	4.7	6.4	4.1	3.7	3.9		

Do the data provide strong evidence that formation 1 has a higher mineral content than formation 2? Test with α near .05.

SOLUTION   To use the rank-sum test, first we rank the combined sample and determine the sum of ranks for the second sample, which has the smaller size. The observations from the second sample and their ranks are underlined here for quick identification:

Combined ordered values	3.7	3.9	4.1	4.7	4.9	6.1	6.4	6.8	7.6	9.8	11.1	15.1
Ranks	1	2	3	4	5	6	7	8	9	10	11	12

The observed value of the rank-sum statistic is

$$W_S = 1 + 2 + 3 + 4 + 7 = 17$$

We wish to test the null hypothesis that the two population distributions are identical versus the alternative hypothesis that the second population, corresponding to W_S, lies to the left of the first. The rejection region is therefore at the lower tail of W_S.

Reading Appendix B, Table 7 with smaller sample size $= 5$ and larger sample size $= 7$, we find $P[W_S \le 21] = .037$ and $P[W_S \le 22] = .053$. Hence, the rejection region with $\alpha = .053$ is established as $W_S \le 22$. Because the observed value falls in this region, the null hypothesis is rejected at $\alpha = .053$. In fact, it would be rejected if α were as low as $P[W_S \le 17]$, which is smaller than .009.

Example 4   Comparing Two Flame-Retardant Materials

Flame-retardant materials are tested by igniting a paper tab on the hem of a dress worn by a mannequin. One response is the vertical length of damage to the fabric measured in inches. The following data (courtesy of B. Joiner) for 5 samples, each taken from two fabrics, are obtained by researchers at the National Bureau of Standards as part of a larger cooperative study.

Fabric *A*	5.7	7.3	7.6	6.0	6.5
Fabric *B*	4.9	7.4	5.3	4.6	6.2

Do the data provide strong evidence that a difference in flammability exists between the two fabrics? Test with α near .05.

SOLUTION   The sample sizes are equal, so that we can take the rank sum of either sample as the test statistic. We compute the rank sum for the second sample.

Ordered values	4.6	4.9	5.3	5.7	6.0	6.2	6.5	7.3	7.4	7.6
Ranks	1	2	3	4	5	6	7	8	9	10

$$W_S = 1 + 2 + 3 + 6 + 9 = 21$$

Because the alternative hypothesis is two-sided, the rejection region includes both tails of W_S. From Appendix B, Table 7, we find that

$$P[W_S \geq 37] = .028 = P[W_S \leq 18]$$

Thus with $\alpha = .056$, the rejection region is $W_S \geq 37$ or $W_S \leq 18$. The observed value does not fall in the rejection region so the null hypothesis is not rejected at $\alpha = .056$.

LARGE SAMPLE APPROXIMATION

When the sample sizes are large, the null distribution of the rank-sum statistic is approximately normal and the test can therefore be performed using the normal table. Specifically, with W_A denoting the rank sum of the sample of size n_A, suppose that both n_A and n_B are large. Then W_A is approximately normally distributed. Under H_0, the distribution of W_A has

$$\text{Mean} = \frac{n_A(n_A + n_B + 1)}{2}$$

$$\text{Variance} = \frac{n_A n_B(n_A + n_B + 1)}{12}$$

Large Sample Approximation to the Rank-Sum Statistic

$$Z = \frac{W_A - n_A(n_A + n_B + 1)/2}{\sqrt{n_A n_B(n_A + n_B + 1)/12}}$$

is approximately $N(0, 1)$ when H_0 is true.

The rejection region for the Z statistic can be determined by using the standard normal table.

Example 5   The Error When Using the Large Sample Approximation

Investigate the amount of error involved in the large sample approximation to the distribution of the rank-sum statistic when $n_A = 9, n_B = 10$, and $\alpha = .05$.

SOLUTION   The approximate one-sided rejection region is

$$R: \frac{W_A - 9(20)/2}{\sqrt{9 \times 10 \times 20/12}} = \frac{W_A - 90}{12.247} \geq 1.645$$

which simplifies to $R: W_A \geq 110.1$. From Appendix B, Table 7, we find $P[W_S \geq 110] = .056$ and $P[W_S \geq 111] = .047$, which are quite close to $\alpha = .05$. The error decreases with increasing sample sizes.

HANDLING TIED OBSERVATIONS

In the preceding examples, observations in the combined sample are all distinct and therefore the ranks are determined without any ambiguity. Often, however, due to imprecision in the measuring scale or a basic discreteness of the scale, such as a five-point preference rating scale, observed values may be repeated in one or both samples. For example, consider the two samples

Sample 1	20	24	22	24	26
Sample 2	26	28	26	30	18

The ordered combined sample is

$$18 \quad 20 \quad 22 \quad \underbrace{24 \quad 24}_{\text{Tie}} \quad \underbrace{26 \quad 26 \quad 26}_{\text{Tie}} \quad 28 \quad 30$$

Here two ties are present; the first has 2 elements, and the second 3. The two positions occupied by 24 are eligible for the ranks 4 and 5, and we assign the average rank $(4 + 5)/2 = 4.5$ to each of these observations. Similarly, the three tied observations 26, eligible for the ranks 6, 7, and 8, are each assigned the average rank $(6 + 7 + 8)/3 = 7$. After assigning average ranks to the tied observations and usual ranks to the other observations, the rank-sum statistic can then be calculated. When ties are present in small samples, the distribution in Appendix B, Table 7 no longer holds exactly. It is best to calculate the null distribution of W_S under the tie structure or at least to modify the variance in the standardized statistic for use in large samples. See Lehmann [1] for details.

Exercises

15.1 Independent random samples of sizes $n_A = 4$ and $n_B = 2$ are taken from two continuous populations.

(a) Enumerate all possible collections of ranks associated with the smaller sample in the combined sample ranking. Attach probabilities to these rank collections under the null hypothesis that the populations are identical.

(b) Obtain the null distribution of W_S = sum of ranks of the smaller sample. Verify that the tail probabilities agree with the tabulated values.

15.2 Independent samples of sizes $n_A = 2$ and $n_B = 2$ are taken from two continuous populations.

(a) Enumerate all possible collections of ranks associated with population A. Also attach probabilities to these rank collections assuming that the populations are identical.

(b) Obtain the null distribution of W_A.

15.3 Using Appendix B, Table 7, find:

(a) $P[W_S \geq 39]$ when $n_A = 5, n_B = 6$.

(b) $P[W_S \leq 15]$ when $n_A = 6, n_B = 4$.

(c) The point c such that $P[W_S \geq c]$ is close to .05 when $n_A = 7, n_B = 7$.

15.4 Using Appendix B, Table 7, find:

(a) $P[W_S \geq 57]$ when $n_A = 6, n_B = 8$.

(b) $P[W_S \leq 31]$ when $n_A = 8, n_B = 6$.

(c) $P[W_S \geq 38$ or $W_S \leq 22]$ when $n_A = 5$ and $n_B = 6$.

(d) The point c such that $P[W_S \leq c]$ is close to .05 when $n_A = 4, n_B = 7$.

(e) The points c_1 and c_2 such that $P[W_S \leq c_1] = P[W_S \geq c_2]$ is about .025 when $n_A = 7, n_B = 9$.

15.5 Given the data

Population A	2.1	5.3	3.7
Population B	2.7	3.2	

(a) Evaluate W_A.

(b) Evaluate W_S.

15.6 The following data pertain to the serum calcium measurements in units of IU/L and the serum alkaline phosphate measurements in units of $\mu g/ml$ for two breeds of pigs, Chester White and Hampshire:

Chester White

Calcium	116	112	82	63	117	69	79	87
Phosphate	47	48	57	75	65	99	97	110

Hampshire

Calcium	62	59	80	105	60	71	103	100
Phosphate	230	182	162	78	220	172	79	58

Using the Wilcoxon rank-sum procedure, test if the serum calcium level is different for the two breeds.

15.7 Referring to the data in Exercise 15.6, is there strong evidence of a difference in the serum phosphate level between the two breeds?

15.8 A project (courtesy of Howard Garber) is constructed to prevent the decline of intellectual performance in children who have a high risk of the most common type of mental retardation, called cultural-familial. It is believed that this can be accomplished by a comprehensive family intervention program. Seventeen children in the high-risk category are chosen in early childhood and given special schooling until the age of $4\frac{1}{2}$. Another 17 children in the same high-risk category form the control group. Measurements of the psycholinguistic quotient (PLQ) are recorded for the control and the experimental groups at the age of $4\frac{1}{2}$ years.

Do the data at the bottom of the page strongly indicate improved PLQs for the children who received special schooling? Use the Wilcoxon rank-sum test with a large sample approximation: Use $\alpha = .05$.

15.9 The possible synergetic effect of insecticides and herbicides is a matter of concern to many environmentalists. It is feared that farmers who apply both herbicides and insecticides to a crop may enhance the toxicity of the insecticide beyond the desired level. An experiment is conducted with a particular insecticide and herbicide to determine the toxicity of the treatments.

Treatment 1: A concentration of .25 μg per gram of soil of insecticide with no herbicide.

Treatment 2: Same dosage of insecticide used in treatment 1 plus 100 μg of herbicide per gram of soil.

PLQ at Age $4\frac{1}{2}$ Years

Experimental group	105.4	118.1	127.2	110.9	109.3	121.8	112.7	120.3	
Control group	79.6	87.3	79.6	76.8	79.6	98.2	88.9	70.9	
Experimental group	110.9	120.0	100.0	122.8	121.8	112.9	107.0	113.7	103.6
Control group	87.0	77.0	96.4	100.0	103.7	61.2	91.1	87.0	76.4

Several batches of fruit flies are exposed to each treatment, and the mortality percent is recorded as a measure of toxicity. The following data are obtained:

Treatment 1	Treatment 2
40	36
28	49
31	56
38	25
43	37
46	30
29	41
18	

Determine if the data strongly indicate different toxicity levels among the treatments.

15.10 Morphologic measurements of a particular type of fossil excavated from two geological sites provided the following data:

Site A	Site B
1.49	1.31
1.32	1.46
2.01	1.86
1.59	1.58
1.76	1.64

Do the data strongly indicate that fossils at the sites differ with respect to the particular morphology measured?

15.11 If $n_A = 1$ and $n_B = 9$, find
   (a) The rank configuration that most strongly supports H_1: Population A is shifted to the right of population B.
   (b) The null probability of $W_A = 10$.
   (c) Is it possible to have $\alpha = .05$ with these sample sizes?

15.12 One aspect of a study of gender differences involves the play behavior of monkeys during the first year of life (courtesy of H. Harlow, U. W. Primate Laboratory). Six male and six female monkeys are observed in groups of four families during several ten-minute test sessions. The mean total number of times each monkey initiates play with another age mate is recorded.

Males	3.64	3.11	3.80	3.58	4.55	3.92
Females	1.91	2.06	1.78	2.00	1.30	2.32

   (a) Plot the observations.
   (b) Test for equality using the Wilcoxon rank-sum test with α approximately .05.
   (c) Determine the significance probability.

3. MATCHED PAIR COMPARISONS

In the presence of extensive dissimilarity in the experimental units, two treatments can be compared more efficiently if alike units are paired and the two treatments applied one to each member of the pair. In this section, we discuss two nonparametric tests, the **sign test** and the **Wilcoxon signed-rank test**, that can be safely applied to paired differences when the assumption of normality is suspect. The data structure of a matched pair experiment is given in Table 3, where the observations on the ith pair are denoted by (X_{1i}, X_{2i}). The null hypothesis of primary interest is that there is no difference, or

$$H_0: \text{No difference in the treatment effects}$$

TABLE 3 Data Structure of Matched Pair Sampling

Pair	Treatment A	Treatment B	Difference $A - B$
1	X_{11}	X_{21}	D_1
2	X_{12}	X_{22}	D_2
.	.	.	.
.	.	.	.
.	.	.	.
n	X_{1n}	X_{2n}	D_n

THE SIGN TEST

This nonparametric test is notable for its intuitive appeal and ease of application. As its name suggests, the sign test is based on the signs of the response differences D_i. The test statistic is

S = number of pairs in which treatment A has a higher response than treatment B

= number of positive signs among the differences $D_1, \ldots, D_n$

   When the two treatment effects are actually alike, the response difference D_i in each pair is as likely to be positive as it is to be negative. Moreover, if measurements are made on a continuous scale, the possibility of identical responses in a pair can be neglected. The null hypothesis is then formulated as

$$H_0: P[+] = .5 = P[-]$$

If we identify a plus sign as a success, the test statistic S is simply the number of successes in n trials and therefore has a binomial distribution with $p = .5$ under H_0. If the alternative hypothesis states that treatment A has higher responses than treatment B, which is translated $P[+] > .5$, then large values of S should be in the rejection region. For two-sided alternatives $H_1: P[+] \neq .5$, a two-tailed test should be employed.

Example 6 Applying the Sign Test to Compare Two Types of Spark Plugs

Mileage tests are conducted to compare a new versus a conventional spark plug. A sample of 12 cars ranging from subcompacts to full-sized sedans are included in the study. The gasoline mileage for each car is recorded, once with the conventional plug and once with the new plug. The results are given in Table 4. Test the null hypothesis of no difference versus the one-sided alternative that the new plug is better. Use the sign test and take $\alpha \leq .05$.

TABLE 4   Mileage Data

Car Number	New A	Conventional B	Difference $A - B$
1	26.4	24.3	+ 2.1
2	10.3	9.8	+ .5
3	15.8	16.9	− 1.1
4	16.5	17.2	− .7
5	32.5	30.5	+ 2.0
6	8.3	7.9	+ .4
7	22.1	22.4	− .3
8	30.1	28.6	+ 1.5
9	12.9	13.1	− .2
10	12.6	11.6	+ 1.0
11	27.3	25.5	+ 1.8
12	9.4	8.6	+ .8

SOLUTION   We are to test

$$H_0: \text{No difference between } A \text{ and } B, \text{ or } P[+] = .5$$

versus the one-sided alternative

$$H_1: \text{The new plug } A \text{ is better than the conventional plug } B, \text{ or } P[+] > .5$$

Looking at the differences $A - B$, we can see that there are 8 plus signs in the sample of size $n = 12$. Thus, the observed value of the sign test statistic is $S = 8$. We will reject H_0 for large values of S. Consulting the binomial table for $n = 12$ and $p = .5$, we find $P[S \geq 9] = .073$ and $P[S \geq 10] = .019$. If we wish to control α below .05, the rejection region should be established at $S \geq 10$. The observed value $S = 8$ is too low to be in the rejection region, so that at the level of significance $\alpha = .019$, the data do not sustain the claim of mileage improvement.

The significance probability of the observed value is $P[S \geq 8] = .194$.

An application of the sign test does not require the numerical values of the differences to be calculated. The number of positive signs can be obtained by glancing at the data. Even when a response cannot be measured on a well-defined numerical scale, we can often determine which of the two responses in a pair is better. This is the only information that is required to conduct a sign test.

For large samples, the sign test can be performed by using the normal approximation to the binomial distribution. With large n, the binomial distribution with $p = .5$ is close to the normal distribution with mean $n/2$ and standard deviation $\sqrt{n/4}$.

Large Sample Approximation to the Sign Test Statistic

Under H_0,

$$Z = \frac{S - n/2}{\sqrt{n/4}}$$

is approximately distributed as $N(0, 1)$.

Example 7   Applying the Sign Test to a Large Sample of Beer Preferences

In a TV commercial filmed live, 100 persons tasted two beers A and B and each selected their favorite. A total of $S = 57$ preferred beer A. Does this provide strong evidence that A is more popular?

SOLUTION   According to the large sample approximation,

$$Z = \frac{S - n/2}{\sqrt{n/4}} = \frac{57 - 50}{\sqrt{25}} = 1.4$$

The significance probability $P[Z > 1.4] = .0808$ is not small enough to provide strong support to the claim that beer A is more popular.

HANDLING TIES

When the two responses in a pair are exactly equal, we say that there is a tie. Because a tied pair has zero difference, it does not have a positive or a negative sign. In the presence of ties, the sign test is performed by discarding the tied pairs, thereby reducing the sample size. For instance, when a sample of $n = 20$ pairs has 10 plus signs, 6 minus signs, and 4 ties, the sign test is performed with the effective sample size $n = 20 - 4 = 16$ and $S = 10$.

THE WILCOXON SIGNED-RANK TEST

We have already noted that the sign test extends to ordinal data for which the responses in a pair can be compared without being measured on a numerical scale. However, when numerical measurements are available, the sign test may result in a considerable loss of information because it includes only the signs of the differences and disregards their magnitudes. Compare the two sets of paired differences plotted in the dot diagrams in Figure 3. In both cases, there are $n = 6$ data points with 4 positive signs, so that the sign test will lead to identical

(a)   *(b)*

Figure 3   Two plots of paired differences with the same number of + signs but with different locations for the distributions.

conclusions. However, the plot in Figure 3*b* exhibits more of a shift toward the positive side, because the positive differences are farther away from zero than the negative differences. Instead of attaching equal weights to all the positive signs, as is done in the sign test, we should attach larger weights to the plus signs that are farther away from zero. This is precisely the concept underlying the signed-rank test.

In the signed-rank test, the paired differences are ordered according to their numerical values without regard to signs, and then the ranks associated with the positive observations are added to form the test statistic. To illustrate, we refer to the mileage data given in Example 6 where the paired differences appear in the last column of Table 4. We attach ranks by arranging these differences in increasing order of their **absolute** values and record the corresponding signs.

Paired differences	2.1	.5	-1.1	$-.7$	2.0	.4	$-.3$	1.5	$-.2$	1.0	1.8	.8
Ordered absolute values	.2	.3	.4	.5	.7	.8	1.0	1.1	1.5	1.8	2.0	2.1
Ranks	1	2	3	4	5	6	7	8	9	10	11	12
Signs	$-$	$-$	$+$	$+$	$-$	$+$	$+$	$-$	$+$	$+$	$+$	$+$

The signed-rank statistic T^+ is then calculated as

$$T^+ = \text{sum of the ranks associated with positive observations}$$
$$= 3 + 4 + 6 + 7 + 9 + 10 + 11 + 12$$
$$= 62$$

If the null hypothesis of no difference in treatment effects is true, then the paired differences $D_1, D_2, \ldots, D_n$ constitute a random sample from a population that is symmetric about zero. On the other hand, the alternate hypothesis that treatment A is better asserts that the distribution is shifted from zero toward positive values. Under H_1, not only are more plus signs anticipated, but the positive signs are also likely to be associated with larger ranks. Consequently, T^+ is expected to be large under the one-sided alternative, and we select a rejection region in the upper tail of T^+.

Steps in the Signed-Rank Test

1. Calculate the differences $D_i = X_{1i} - X_{2i}$, $i = 1, \ldots, n$.
2. Assign ranks by arranging the absolute values of the D_i in increasing order; also record the corresponding signs.
3. Calculate the signed-rank statistic T^+ = sum of ranks of positive differences D_i.
4. Set the rejection region at the upper tail, lower tail, or at both tails of T^+, according to whether treatment A is stated to have a higher, lower, or different response than treatment B under the alternative hypothesis.

Selected tail probabilities of the null distribution of T^+ are given in Appendix B, Table 8 for $n = 3$ to $n = 15$.

USING APPENDIX B, TABLE 8

By symmetry of the distribution around $n(n + 1)/4$, we obtain

$$P[T^+ \geq x] = P[T^+ \leq x^*]$$

when $x^* = n(n + 1)/2 - x$. The x and x^* values in Appendix B, Table 8 satisfy this relation. To illustrate the use of this table, we refer once again to the mileage data given in Example 6. There, $n = 12$ and the observed value of T^+ is found to be 62. From the table, we find $P[T^+ \geq 61] = .046$. Thus, the null hypothesis is rejected at the level of significance $\alpha = .046$, and a significant mileage improvement using the new type of spark plug is indicated.

$$P = P[T^+ \geq x] = P[T^+ \leq x^*]$$
$$n = 12$$

	x	P	x^*
	56	·	22
	·	·	·
	·	·	·
	·	·	·
$\rightarrow$	61	.046	17
	62		16
	·	·	·
	·	·	·
	·	·	·
	68	·	10

With increasing sample size n, the null distribution of T^+ is closely approximated by a normal curve, with mean $n(n + 1)/4$ and variance $n(n + 1)(2n + 1)/24$.

Large Sample Approximation to Signed-Rank Statistic

$$Z = \frac{T^+ - n(n + 1)/4}{\sqrt{n(n + 1)(2n + 1)/24}}$$

is approximately distributed as $N(0, 1)$.

This result can be used to perform the signed-rank test with large samples.

Example 8   Applying the Signed-rank Test to Compare Spark Plugs

Refer to the mileage data in Example 6. Obtain the significance probability for the signed-rank test using (a) the exact distribution in Appendix B, Table 8 and (b) the large sample approximation.

SOLUTION   (a)   For the mileage data, $T^+ = 62$ and $n = 12$. From Appendix B, Table 8, the exact significance probability is $P[T^+ \geq 62] = .039$.

(b)   The normal approximation to this probability uses

$$z = \frac{62 - 12(13)/4}{\sqrt{12(13)(25)/24}} = \frac{23}{12.75} = 1.804$$

From the normal table, we approximate $P[T^+ \geq 62]$ by $P[Z \geq 1.804] = .036$.

The normal approximation improves with increasing sample size.

*HANDLING TIES

In computing the signed-rank statistic, ties may occur in two ways: Some of the differences D_i may be zero or some nonzero differences D_i may have the same absolute value. The first type of tie is handled by discarding the zero values after ranking. The second type of tie is handled by assigning the average rank to each observation in a group of tied observations with nonzero differences D_i.

See Lehmann [1] for instructions on how to modify the critical values to adjust for ties.

Exercises

15.13 In a taste test of two chocolate chip cookie recipes, 13 out of 18 subjects favored recipe A. Using the sign test, find the significance probability when H_1 states that recipe A is preferable.

15.14 Two critics rate the service at six award-winning restaurants on a continuous 0-to-10 scale. Is there a difference between the critics' ratings?

(a) Use the sign test with α below .05.

(b) Find the significance probability.

| | Service Rating | |
Restaurant	Critic 1	Critic 2
1	6.1	7.3
2	5.2	5.5
3	8.9	9.1
4	7.4	7.0
5	4.3	5.1
6	9.7	9.8

15.15 A social researcher interviews 25 newly married couples. Each husband and wife are independently asked the question: "How many children would you like to have?" The following data are obtained.

| | Answer of | |
Couple	Husband	Wife
1	3	2
2	1	1
3	2	1
4	2	3
5	5	1
6	0	1
7	0	2
8	1	3
9	2	2
10	3	1
11	4	2
12	1	2
13	3	3
14	2	1
15	3	2

| | Answer of | |
Couple	Husband	Wife
16	2	2
17	0	0
18	1	2
19	2	1
20	3	2
21	4	3
22	3	1
23	0	0
24	1	2
25	1	1

Do the data show a significant difference of opinion between husbands and wives regarding an ideal family? Use the sign test with α close to .05.

15.16 Two computer specialists estimated the amount of computer memory (in gigabytes) required by five different offices.

Office	Specialist A	Specialist B
1	5.2	7.1
2	6.4	6.8
3	3.2	3.1
4	2.0	2.9
5	8.1	12.3

Apply the sign test, with α under .10, to determine if specialist B estimates higher than specialist A.

15.17 Use Appendix B, Table 8, to find:

(a) $P[T^+ \geq 54]$ when $n = 11$.

(b) $P[T^+ \leq 32]$ when $n = 15$.

(c) The value of c so that $P[T^+ \geq c]$ is nearly .05 when $n = 14$.

15.18 Use Appendix B, Table 8, to find:

(a) $P[T^+ \geq 65]$ when $n = 12$.

(b) $P[T^+ \leq 10]$ when $n = 10$.

(c) The value c such that $P[T^+ \geq c] = .039$ when $n = 8$.

(d) The values c_1 and c_2 such that $P[T^+ \leq c_1] = P[T^+ \geq c_2] = .027$ when $n = 11$.

15.19 Referring to Exercise 15.14, apply the Wilcoxon signed-rank test with α near .05.

15.20 Referring to Exercise 15.16, apply the Wilcoxon signed-rank test with α under .10.

15.21 The null distribution of the Wilcoxon signed-rank statistic T^+ is determined from the fact that under the null hypothesis of a symmetric distribution about zero, each of the ranks 1, 2, . . . , n is equally likely to be associated with a positive sign or a negative sign. Moreover, the signs are independent of the ranks.

(a) Considering the case $n = 3$, identify all $2^3 = 8$ possible associations of signs with the ranks 1, 2, and 3, and determine the value of T^+ for each association.

(b) Assigning the equal probability of $\frac{1}{8}$ to each case, obtain the distribution of T^+ and verify that the tail probabilities agree with the tabulated values.

15.22 A sample size of $n = 30$ yielded the Wilcoxon signed-rank statistic $T^+ = 325$. What is the significance probability if the alternative is two-sided?

15.23 In Example 13 of Chapter 10, we presented data on the blood pressure of 15 persons before and after they took a pill.

Before	After	Difference
70	68	2
80	72	8
72	62	10
76	70	6
76	58	18
76	66	10
72	68	4
78	52	26
82	64	18

Before	After	Difference
64	72	− 8
74	74	0
92	60	32
74	74	0
68	72	− 4
84	74	10

(a) Perform a sign test, with α near .05, to determine if blood pressure has decreased after taking the pill.

(b) Perform a Wilcoxon signed-rank test to determine if blood pressure has decreased after taking the pill.

15.24 Charles Darwin performed an experiment to determine if self-fertilized and cross-fertilized plants have different growth rates. Pairs of *Zea mays* plants, one self- and the other cross-fertilized, were planted in pots, and their heights were measured after a specified period of time. The data Darwin obtained were

Pair	Plant height (in $\frac{1}{8}$ inches)	
	Cross-	Self-
1	188	139
2	96	163
3	168	160
4	176	160
5	153	147
6	172	149
7	177	149
8	163	122
9	146	132
10	173	144
11	186	130
12	168	144
13	177	102
14	184	124
15	96	144

Source: C. Darwin, *The Effects of Cross- and Self-Fertilization in the Vegetable Kingdom*, D. Appleton and Co., New York, 1902.

(a) Calculate the paired differences and plot a dot diagram for the data. Does the assumption of normality seem plausible?

(b) Perform the Wilcoxon signed-rank test to determine if cross-fertilized plants have a higher growth rate than self-fertilized plants.

4. MEASURE OF CORRELATION BASED ON RANKS

Ranks may also be employed to determine the degree of association between two random variables. These two variables could be mathematical ability and musical aptitude or the aggressiveness scores of first- and second-born sons on a psychological test. We encountered this same general problem in Chapter 3, where we introduced Pearson's product moment correlation coefficient

$$r = \frac{\sum_{i=1}^{n} (X_i - \overline{X})(Y_i - \overline{Y})}{\sqrt{\sum_{i=1}^{n} (X_i - \overline{X})^2} \sqrt{\sum_{i=1}^{n} (Y_i - \overline{Y})^2}}$$

as a measure of association between X and Y. Serving as a descriptive statistic, r provides a numerical value for the amount of linear dependence between X and Y.

Structure of the Observations

The n pairs (X_1, Y_1), (X_2, Y_2), . . . , (X_n, Y_n) are independent, and each pair has the same continuous bivariate distribution. The $X_1, \ldots, X_n$ are then ranked among *themselves*, and the $Y_1, \ldots, Y_n$ are ranked among *themselves*:

Pair no.	1	2	$\cdots$	n
Ranks of X_i	R_1	R_2	$\cdots$	R_n
Ranks of Y_i	S_1	S_2	$\cdots$	S_n

Before we present a measure of association, we note a few simplifying properties. Because each of the ranks, 1, 2, . . . , n must occur exactly once in the set $R_1, R_2, \ldots, R_n$, it can be shown that

$$\overline{R} = \frac{1 + 2 + \cdots + n}{n} = \frac{n + 1}{2}$$

$$\sum_{i=1}^{n} (R_i - \overline{R})^2 = \frac{n(n^2 - 1)}{12}$$

for all possible outcomes. Similarly,

$$\bar{S} = \frac{n+1}{2} \quad \text{and} \quad \sum_{i=1}^{n} (S_i - \bar{S})^2 = \frac{n(n^2-1)}{12}$$

A measure of correlation is defined by C. Spearman that is analogous to Pearson's correlation r, except that Spearman replaces the observations with their ranks. Spearman's rank correlation r_{Sp} is defined by

$$r_{Sp} = \frac{\sum_{i=1}^{n} (R_i - \bar{R})(S_i - \bar{S})}{\sqrt{\sum_{i=1}^{n} (R_i - \bar{R})^2} \sqrt{\sum_{i=1}^{n} (S_i - \bar{S})^2}} = \frac{\sum_{i=1}^{n} \left(R_i - \frac{n+1}{2}\right)\left(S_i - \frac{n+1}{2}\right)}{n(n^2-1)/12}$$

This rank correlation shares the properties of r that $-1 \leq r_{Sp} \leq 1$ and that values near $+1$ indicate a tendency for the larger values of X to be paired with the larger values of Y. However, the rank correlation is more meaningful, because its interpretation does not require the relationship to be linear.

Spearman's Rank Correlation

$$r_{Sp} = \frac{\sum_{i=1}^{n} \left(R_i - \frac{(n+1)}{2}\right)\left(S_i - \frac{(n+1)}{2}\right)}{n(n^2-1)/12}$$

1. $-1 \leq r_{Sp} \leq 1$.
2. r_{Sp} near $+1$ indicates a tendency for the larger values of X to be associated with the larger values of Y. Values near -1 indicate the opposite relationship.
3. The association need not be linear; only an increasing/decreasing relationship is required.

Example 9 Calculating Spearman's Rank Correlation

An interviewer in charge of hiring large numbers of data entry persons wishes to determine the strength of the relationship between ranks given on the basis of an interview and scores on an aptitude test. The data for six applicants are

Interview rank	5	2	3	1	6	4
Aptitude score	47	32	29	28	56	38

Calculate r_{Sp}.

SOLUTION There are 6 ranks, so that $\bar{R} = (n + 1)/2 = 7/2 = 3.5$ and $n(n^2 - 1)/12 = 35/2 = 17.5$. Ranking the aptitude scores, we obtain

Interview R_i	5	2	3	1	6	4
Aptitude S_i	5	3	2	1	6	4

Thus,

$$\sum_{i=1}^{n} \left(R_i - \frac{n + 1}{2} \right)\left(S_i - \frac{n + 1}{2} \right)$$
$$= (5 - 3.5)(5 - 3.5) + (2 - 3.5)(3 - 3.5) + \cdots + (4 - 3.5)(4 - 3.5)$$
$$= 1.5(1.5) + (-1.5)(-.5) + \cdots + (.5)(.5)$$
$$= 16.5$$

and

$$r_{Sp} = \frac{16.5}{17.5} = .943$$

The relationship between interview rank and aptitude score appears to be quite strong.

Figure 4 helps to stress the point that r_{Sp} is a measure of any monotone relationship, not merely a linear relation.

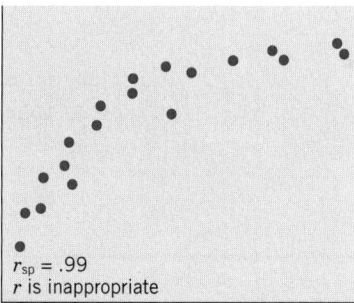

$r_{sp} = .99$
r is inappropriate

Figure 4 r_{Sp} is a measure of any monotone relationship.

A large sample approximation to the distribution of r_{Sp} is available.

If X and Y are independent,
$$\sqrt{n - 1}\, r_{Sp} \text{ is approximately distributed as } N(0, 1)$$
provided that the sample size is large.

This approximation leads to a convenient form of a test for independence. Reject

$$H_0: X \text{ and } Y \text{ are independent}$$

in favor of

$$H_1: \text{Large values of } X \text{ and } Y \text{ tend to occur together} \\ \text{and small values tend to occur together}$$

if

$$\sqrt{n - 1}\, r_{\mathrm{Sp}} \geq z_\alpha$$

Recall that z_α is the upper α point of a standard normal distribution. Two-tailed tests can also be conducted.

Example 10   Establishing Dependence When Large X and Y Tend to Occur Together

The grade point average (GPA) and Scholastic Achievement Test (SAT) scores for 40 applicants yielded $r_{\mathrm{Sp}} = .4$. Do large values of GPA and SAT tend to occur together? That is, test for lack of independence using $\alpha = .05$.

SOLUTION   For $\alpha = .05$, the rejection region is $\sqrt{n - 1}\, r_{\mathrm{Sp}} \geq z_{.05} = 1.96$. Since

$$\sqrt{n - 1}\, r_{\mathrm{Sp}} = \sqrt{39}(.4) = 2.498$$

we reject $H_0: X$ and Y are independent at level $\alpha = .05$. Large values of GPA and SAT tend to occur together.

Exercises

15.25   Calculate Spearman's rank correlation from the data

x	3.1	5.4	4.7
y	2.8	3.5	4.6

15.26   Determine Spearman's rank correlation from the data

x	13.2	18.7	19.4
y	6.3	9.2	8.5

15.27   The following scores are obtained on a test of dexterity and aggression administered to a random sample of 10 high-school seniors.

Student	1	2	3	4	5	6	7	8	9	10
Dexterity	23	29	45	36	49	41	30	15	42	38
Aggression	45	48	16	28	38	21	36	18	31	37

Evaluate Spearman's statistic.

15.28   Referring to Example 10, determine the significance probability of $r_{\mathrm{Sp}} = .4$, using the one-sided test, when $n = 40$.

5. CONCLUDING REMARKS

In contrast to nonparametric procedures, Student's t and the chi-square statistic $(n - 1)S^2/\sigma^2$ were developed to make inferences about the parameters μ and σ of a normal population. These *normal-theory parametric* procedures can be seriously affected when the sample size is small and the underlying distribution is not normal. Drastic departures from normality can occur in the forms of conspicuous asymmetry, sharp peaks, or heavy tails. For instance, a t test with an intended level of significance of $\alpha = .05$ may have an actual Type I error probability far in excess of this value. These effects are most pronounced for small or moderate samples sizes precisely when it is most difficult to assess the shape of the population. The selection of a parametric procedure leaves the data analyst with the question: Does my normality assumption make sense in the present situation? To avoid this risk, a nonparametric method could be used in which inferences rest on the safer ground of distribution-free properties.

When the data constitute measurements on a meaningful numerical scale and the assumption of normality holds, parametric procedures are certainly more efficient in the sense that tests have higher power than their nonparametric counterparts. This brings to mind the old adage, "You get what you pay for." A willingness to assume more about the population form leads to improved inference procedures. However, trying to get too much for your money by assuming more about the population than is reasonable can lead to the "purchase" of invalid conclusions. A choice between the parametric and nonparametric approach should be guided by a consideration of loss of efficiency and the degree of protection desired against possible violations of the assumptions.

Tests are judged by two criteria: control of the Type I error probability and the power to detect alternatives. Nonparametric tests guarantee the desired control of the Type I error probability α, whatever the form of the distribution. However, a parametric test established at $\alpha = .05$ for a normal distribution may suffer a much larger α when a departure from normality occurs. This is particularly true with small sample sizes. To achieve universal protection, nonparametric tests, quite expectedly, must forfeit some power to detect alternatives when normality actually prevails. As plausible as this argument sounds, it is rather surprising that the loss in power is often marginal with such simple procedures as the Wilcoxon rank-sum test and the signed-rank test.

Finally, the presence of dependence among the observations affects the usefulness of nonparametric and parametric methods in much the same manner. When either method is used, the level of significance of the test may seriously differ from the nominal value selected by the analyst.

> *Caution:* When successive observations are dependent, nonparametric test procedures lose their distribution-free property, and conclusions drawn from them can be seriously misleading.

Reference

1. E. L. Lehmann, *Nonparametrics: Statistical Methods Based on Ranks*, Holden-Day, San Francisco, 1975.

USING STATISTICS WISELY

1. A one-sample nonparametric test will provide valid inferences with a small sample size where it may not be possible to check the assumption of normality. Of course, the power of the rank test will generally be less than the normal theory paired t test when normality holds.

2. When the two sample sizes are small, it is a good idea to conduct a two-sample Wilcoxon rank-sum test. If software is available, also obtain the corresponding confidence interval for the difference in location. These provide a baseline comparison for the result based on the t distribution.

3. If the sample sizes are large enough so the dot diagrams reveal a difference in both location and spread, the Wilcoxon rank-sum test is not appropriate.

4. Remember that nonparametric tests can produce invalid inferences if there is time dependence between the observations.

KEY IDEAS AND FORMULAS

Nonparametric tests obtain their distribution-free character because rank orders of the observations do not depend on the shape of the population distribution.
The Wilcoxon rank-sum test, based on the test statistic

$$W_A = \text{sum of ranks of the } n_A \text{ observations}$$
$$\text{from population } A, \text{ among all}$$
$$n_A + n_B \text{ observations}$$

applies to the comparison of two populations. It uses the combined sample ranks.
In the paired-sample situation, equality of treatments can be tested using either the sign test based on the statistic

$$S = \text{No. of positive differences}$$

or the Wilcoxon signed-rank based on the statistic

$$T^+ = \text{sum, over positive differences, of the}$$
$$\text{ranks of their absolute values}$$

The level of a nonparametric test holds whatever the form of the (continuous) population distribution.
Any tie in the observations requires specific treatments.

TECHNOLOGY

Nonparametric tests and confidence intervals

MINITAB

One sample—inference about median

Start with the data in *C1*. To find a 95% confidence interval for the median using the sign test:

> **Stat > Nonparametrics > 1-Sample Sign.**
> Type *C1* in **Variables.** Click **Confidence interval** and type *0.95* in **Level.**
> Click **OK.**

To test a hypothesis concerning the median, instead of **Confidence interval:**

> Click **Test Median** and choose the form of the **Alternative** hypothesis. You cannot set the level.

To find a 95% confidence interval for the median using the Wilcoxon signed-rank test:

> **Stat > Nonparametrics > 1-Sample Wilcoxon.**
> Type *C1* in **Variables.** Click **Confidence interval** and type *0.95* in **Level.**
> Click **OK.**

To test a hypothesis concerning the median, instead of **Confidence interval:**

> Click **Test Median** and choose form of the **Alternative** hypothesis.

Two-sample Wilcoxon test for equality of populations

Start with the data from the first population in *C1* and the data from the second in *C2*. To test at level $\alpha = .05$:

> **Stat > Nonparametrics > Mann-Whitney.**
> Type *C1* and *C2* in **Variables.** Type *0.95* in **Confidence level** and select form of the **Alternative.** Click **OK.**

The output includes a confidence interval for the difference in locations.

6. REVIEW EXERCISES

15.29 Evaluate W_A for the data

Treatment *A*	90	32	81
Treatment *B*	67	99	43

15.30 Using Appendix B, Table 7, find:

(a) $P[W_S \geq 42]$ when $n_1 = 5, n_2 = 7$.

(b) $P[W_S \leq 25]$ when $n_1 = 6, n_2 = 6$.

(c) $P[W_S \geq 81$ or $W_S \leq 45]$ when $n_1 = 10, n_2 = 7$.

(d) The point c such that $P[W_S \geq c] = .036$ when $n_1 = 8, n_2 = 4$.

(e) The points c_1 and c_2 such that $P[W_S \geq c_2] = P[W_S \leq c_1] = .05$ when $n_1 = 3$, $n_2 = 9$.

15.31 (a) Evaluate all possible rank configurations associated with treatment A when $n_A = 3$ and $n_B = 2$.

(b) Determine the null distribution of W_A.

15.32 Five finalists in a figure-skating contest are rated by two judges on a 10-point scale as follows:

Contestants	A	B	C	D	E
Judge 1	6	9	2	8	5
Judge 2	8	10	4	7	3

Calculate the Spearman's rank correlation r_{Sp} between the two ratings.

15.33 Using Appendix B, Table 8, find:

(a) $P[T^+ \geq 28]$ when $n = 8$.

(b) $P[T^+ \leq 5]$ when $n = 9$.

(c) The point c such that $P[T^+ \leq c]$ is approximately .05 when $n = 13$.

15.34 Referring to Exercise 15.32, calculate:

(a) The sign test statistic.

(b) The significance probability when the alternative is that Judge 2 gives higher scores than Judge 1.

15.35 In a study of the cognitive capacities of non-human primates, 19 monkeys of the same age are randomly divided into two groups of 10 and 9. The groups are trained by two different teaching methods to recollect an acoustic stimulus. The monkeys' scores on a subsequent test are seen below.

Do the data strongly indicate a difference in the recollection abilities of monkeys trained by the two methods? Use the Wilcoxon rank-sum test with α close to .10.

15.36 A mixture of compounds called phenolics occurs in wood waste products. It has been found that when phenolics are present in large quantities, the waste becomes unsuitable for use as a livestock feed. To compare two species of wood, a dairy scientist measures the percentage content of phenolics from 6 batches of waste of species A and 7 batches of waste of species B. The following data are obtained.

Percentage of Phenolics

Species A	2.38	4.19	1.39	3.73	2.86	1.21	
Species B	4.67	5.38	3.89	4.67	3.58	4.96	3.98

Use the Wilcoxon rank-sum test to determine if the phenolics content of species B is significantly higher than that of species A. Use α close to .05.

15.37 (a) Calculate Spearman's rank correlation for the data on Chester Whites in Exercise 15.6.

(b) Test for independence of calcium and phosphate levels using the rejection region

$$\sqrt{n-1}\, r_{Sp} \geq 1.96 \quad \text{or} \quad \leq -1.96$$

(c) What is the approximate level of significance?

15.38 Given the following data on the pairs (x, y),

x	10	7	8
y	15	13	9

evaluate r_{Sp}.

Memory Scores

Method 1	167	149	137	178	179	155	164	104	151	150
Method 2	98	127	140	103	116	105	100	95	131	

15.39 Refer to Exercise 15.38. Evaluate:

(a) Sign test statistic.

(b) Signed-rank statistic.

*15.40 *Confidence interval for median using the sign test.* Let $X_1, \ldots, X_n$ be a random sample from a continuous population whose median is denoted by M. For testing $H_0: M = M_0$, we can use the sign test statistic $S = $ No. of $X_i > M_0$, $i = 1, \ldots, n$. H_0 is rejected at level α in favor of $H_1: M \neq M_0$ if $S \leq r$ or $S \geq n - r + 1$, where r is the largest integer satisfying

$$\sum_{x=0}^{r} b(x;\, n,\, .5) \leq \alpha/2$$

If we repeat this test procedure for all possible values of M_0, a $100(1 - \alpha)\%$ confidence interval for M is then the range of values M_0 so that S is in the acceptance region. Ordering the observations from smallest to largest, verify that this confidence interval becomes

$(r + 1)$st smallest to $(r + 1)$st largest observation

(a) Refer to Example 6. Using the sign test, construct a confidence interval for the population median of the differences $A - B$, with a level of confidence close to 95%.

(b) Repeat part (a) using Darwin's data given in Exercise 15.24.

A1

Summation Notation

A1.1 SUMMATION AND ITS PROPERTIES

The addition of numbers is basic to our study of statistics. To avoid a detailed and repeated writing of this operation, the symbol Σ (the Greek capital letter *sigma*) is used as mathematical shorthand for the operation of addition.

Summation Notation Σ

The notation $\displaystyle\sum_{i=1}^{n} x_i$ represents the sum of n numbers $x_1, x_2, \ldots, x_n$ and is read as **the sum of all x_i with i ranging from 1 to n**.

$$\sum_{i=1}^{n} x_i = x_1 + x_2 + \cdots + x_n$$

The term following the sign Σ indicates the quantities that are being summed, and the notations on the bottom and the top of the Σ specify the range of the terms being added. For instance,

$$\sum_{i=1}^{3} x_i = x_1 + x_2 + x_3$$

$$\sum_{i=1}^{4} (x_i - 3) = (x_1 - 3) + (x_2 - 3) + (x_3 - 3) + (x_4 - 3)$$

Example   Suppose that the four measurements in a data set are given as $x_1 = 2$, $x_2 = 5$, $x_3 = 3$, $x_4 = 4$. Compute the numerical values of

(a) $\displaystyle\sum_{i=1}^{4} x_i$ (b) $\displaystyle\sum_{i=1}^{4} 6$ (c) $\displaystyle\sum_{i=1}^{4} 2x_i$

(d) $\displaystyle\sum_{i=1}^{4} (x_i - 3)$ (e) $\displaystyle\sum_{i=1}^{4} x_i^2$ (f) $\displaystyle\sum_{i=1}^{4} (x_i - 3)^2$

SOLUTION

(a) $\displaystyle\sum_{i=1}^{4} x_i = x_1 + x_2 + x_3 + x_4 = 2 + 5 + 3 + 4 = 14$

(b) $\displaystyle\sum_{i=1}^{4} 6 = 6 + 6 + 6 + 6 = 4(6) = 24$

(c) $\displaystyle\sum_{i=1}^{4} 2x_i = 2x_1 + 2x_2 + 2x_3 + 2x_4 = 2\left(\displaystyle\sum_{i=1}^{4} x_i\right)$

$$= 2 \times 14 = 28$$

(d) $\displaystyle\sum_{i=1}^{4} (x_i - 3) = (x_1 - 3) + (x_2 - 3)$

$$+ (x_3 - 3) + (x_4 - 3)$$
$$= \sum_{i=1}^{4} x_i - 4(3) = 14 - 12 = 2$$

(e) $\displaystyle\sum_{i=1}^{4} x_i^2 = x_1^2 + x_2^2 + x_3^2 + x_4^2 = 2^2 + 5^2 + 3^2 + 4^2 = 54$

(f) $\displaystyle\sum_{i=1}^{4} (x_i - 3)^2 = (x_1 - 3)^2 + (x_2 - 3)^2$

$$+ (x_3 - 3)^2 + (x_4 - 3)^2$$
$$= (2 - 3)^2 + (5 - 3)^2$$
$$+ (3 - 3)^2 + (4 - 3)^2$$
$$= 1 + 4 + 0 + 1 = 6$$

Alternatively, noting that $(x_i - 3)^2 = x_i^2 - 6x_i + 9$, we can write

$$\sum_{i=1}^{4} (x_i - 3)^2 = \sum_{i=1}^{4} (x_i^2 - 6x_i + 9)$$
$$= (x_1^2 - 6x_1 + 9) + (x_2^2 - 6x_2 + 9)$$
$$+ (x_3^2 - 6x_3 + 9) + (x_4^2 - 6x_4 + 9)$$
$$= \sum_{i=1}^{4} x_i^2 - 6\left(\sum_{i=1}^{4} x_i\right) + 4(9)$$
$$= 54 - 6(14) + 36 = 6$$

A few basic properties of the summation operation are apparent from the numerical demonstration in the example.

Some Basic Properties of Summation

If a and b are fixed numbers,

$$\sum_{i=1}^{n} b x_i = b \sum_{i=1}^{n} x_i$$

$$\sum_{i=1}^{n} (b x_i + a) = b \sum_{i=1}^{n} x_i + na$$

$$\sum_{i=1}^{n} (x_i - a)^2 = \sum_{i=1}^{n} x_i^2 - 2a \sum_{i=1}^{n} x_i + na^2$$

Exercises

1. Demonstrate your familiarity with the summation notation by evaluating the following expressions when $x_1 = 4, x_2 = -2, x_3 = 1$.

   (a) $\sum_{i=1}^{3} x_i$　　　　(b) $\sum_{i=1}^{3} 7$　　　　(c) $\sum_{i=1}^{3} 5x_i$

   (d) $\sum_{i=1}^{3} (x_i - 2)$　　(e) $\sum_{i=1}^{3} (x_i - 3)$　　(f) $\sum_{i=1}^{3} (x_i - 2)^2$

   (g) $\sum_{i=1}^{3} x_i^2$　　　　(h) $\sum_{i=1}^{3} (x_i - 3)^2$　　(i) $\sum_{i=1}^{3} (x_i^2 - 6x_i + 9)$

2. Five measurements in a data set are $x_1 = 4, x_2 = 3, x_3 = 6, x_4 = 5, x_5 = 7$. Determine

   (a) $\sum_{i=1}^{5} x_i$　　　　(b) $\sum_{i=2}^{3} x_i$　　　　(c) $\sum_{i=1}^{5} 2$

   (d) $\sum_{i=1}^{5} (x_i - 6)$　　(e) $\sum_{i=1}^{5} (x_i - 6)^2$　　(f) $\sum_{i=1}^{4} (x_i - 5)^2$

A1.2 SOME BASIC USES OF Σ IN STATISTICS

Let us use the summation notation and its properties to verify some computational facts about the sample mean and variance.

$$\Sigma (x_i - \bar{x}) = 0$$

The total of the deviations about the sample mean is always zero. Since $\bar{x} = (x_1 + x_2 + \cdots + x_n)/n$, we can write

$$\sum_{i=1}^{n} x_i = x_1 + x_2 + \cdots + x_n = n\bar{x}$$

Consequently, whatever the observations,

$$\sum_{i=1}^{n} (x_i - \bar{x}) = (x_1 - \bar{x}) + (x_2 - \bar{x}) + \cdots + (x_n - \bar{x})$$
$$= x_1 + x_2 + \cdots + x_n - n\bar{x}$$
$$= n\bar{x} - n\bar{x} = 0$$

We could also verify this directly with the second property for summation in A1.1, when $b = 1$ and $a = -\bar{x}$.

ALTERNATIVE FORMULA FOR s^2

By the quadratic rule of algebra,

$$(x_i - \bar{x})^2 = x_i^2 - 2\bar{x}\, x_i + \bar{x}^2$$

Therefore,

$$\Sigma (x_i - \bar{x})^2 = \Sigma x_i^2 - \Sigma 2\bar{x}\, x_i + \Sigma \bar{x}^2$$
$$= \Sigma x_i^2 - 2\bar{x} \Sigma x_i + n\bar{x}^2$$

Using $(\Sigma x_i)/n$ in place of $\bar{x}$, we get

$$\Sigma (x_i - \bar{x})^2 = \Sigma x_i^2 - \frac{2(\Sigma x_i)^2}{n} + \frac{n(\Sigma x_i)^2}{n^2}$$
$$= \Sigma x_i^2 - \frac{2(\Sigma x_i)^2}{n} + \frac{(\Sigma x_i)^2}{n}$$
$$= \Sigma x_i^2 - \frac{(\Sigma x_i)^2}{n}$$

We could also verify this directly from the third property for summation, in A1.1, with $a = \bar{x}$.

This result establishes that

$$s^2 = \frac{\displaystyle\sum_{i=1}^{n} (x_i - \bar{x})^2}{n - 1} = \frac{\Sigma x_i^2 - (\Sigma x_i)^2/n}{n - 1}$$

so the two forms of s^2 are equivalent.

SAMPLE CORRELATION COEFFICIENT

The sample correlation coefficient and slope of the fitted regression line contain a term

$$S_{xy} = \sum_{i=1}^{n} (x_i - \bar{x})(y_i - \bar{y})$$

which is a sum of the products of the deviations. To obtain the alternative form, first note that

$$(x_i - \bar{x})(y_i - \bar{y}) = x_i y_i - x_i \bar{y} - \bar{x} y_i + \bar{x}\bar{y}$$

We treat $x_i y_i$ as a single number, with index i, and conclude that

$$\Sigma (x_i - \bar{x})(y_i - \bar{y}) = \Sigma x_i y_i - \Sigma x_i \bar{y} - \Sigma \bar{x} y_i + \Sigma \bar{x}\bar{y}$$
$$= \Sigma x_i y_i - \bar{y} \Sigma x_i - \bar{x} \Sigma y_i + n\bar{x}\bar{y}$$

Since $\bar{x} = (\Sigma x_i)/n$ and $\bar{y} = (\Sigma y_i)/n$,

$$\Sigma (x_i - \bar{x})(y_i - \bar{y}) = \Sigma x_i y_i - \frac{(\Sigma y_i)}{n} \Sigma x_i - \frac{(\Sigma x_i)}{n} \Sigma y_i$$
$$+ \frac{n(\Sigma x_i)}{n} \frac{(\Sigma y_i)}{n}$$
$$= \Sigma x_i y_i - \frac{(\Sigma x_i)(\Sigma y_i)}{n}$$

Consequently, either $\Sigma (x_i - \bar{x})(y_i - \bar{y})$ or $\Sigma x_i y_i - (\Sigma x_i)(\Sigma y_i)/n$ can be used for the calculation of S_{xy} with similar choices for S_{xx} and S_{yy}.

Rules for Counting

Some basic rules for counting can help us calculate probabilities. We begin with an example. Brazil (B), United States (U.S.), Mexico (M), Canada (C), and Japan (J) are in the final round of a volleyball tournament. How many ways can a winner and second place finisher be selected?

To systematically address this problem, we create a **tree diagram** where the possibilities for first place are represented by the initial 5 branches. For each choice of a first place finisher, there are four choices for second place. These are represented by the sets of 4 branches in the second stage. For instance, the left-most branch of the tree diagram depicts the case where Brazil is first and the United States second and so on.

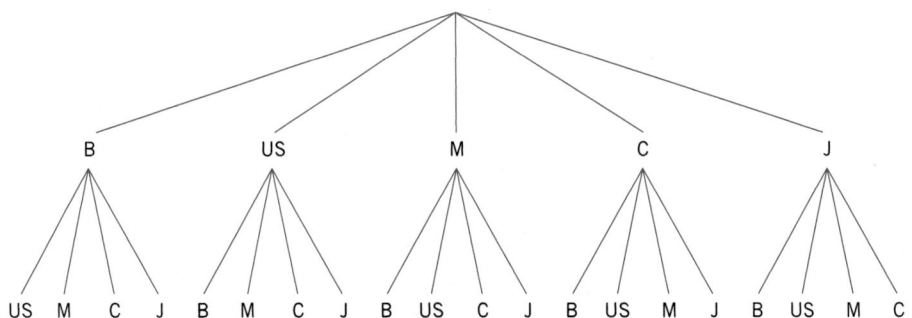

There are $5 \times 4 = 20$ terminal points or ways for these teams to finish first and second. Generalizing the tree diagram to k choices followed by m choices, we obtain the product rule for counting.

Product Rule

An experiment consists of two parts where the first part has k distinct possible outcomes and, for each outcome of the first part, there are m distinct possible outcomes for the second part. Then, there are $k \times m$ distinct possible outcomes to the experiment.

The product rule is readily extended to experiments with more than two parts. Suppose we are interested in the winner, second, and third place finishers in the volleyball tournament. As above, by the product rule, there are $5 \times 4 = 20$ distinct possibilities for selecting a first and second place volleyball team. Again, by the product rule, each of these 20 outcomes must be paired with one of the 3 outcomes for a third place finisher so there are $20 \times 3 = 5 \times 4 \times 3 = 60$ possibilities for the first, second, and third place finishers.

The term **permutation** means an ordering or arrangement of objects. The number of different orderings or arrangements that can be formed with r objects selected from a group of n distinct objects is called the **number of permutations of r out of n** and is denoted by P_r^n.

In the volleyball example, $P_2^5 = 5 \times 4 = 20$ for the first two places and $P_3^5 = 5 \times 4 \times 3 = 60$ for the first three places. When r objects are selected, the first has n possibilities and the last $n - r + 1$ possibilities.

Rule of Permutations or Arrangements

The number of different arrangements of r objects selected from a group of n distinct objects is denoted by P_r^n,

$$P_r^n = n(n - 1) \cdots (n - r + 1)$$

If all n distinct objects are arranged in order, the number of possible arrangements is $P_n^n = n(n - 1) \times \cdots \times 2 \times 1$, which is the product n times $n - 1$ on down through 2×1. This special product, of decreasing integers starting with n, is called **n factorial** and is denoted by $n!$.

When order is not important, we speak about collections of objects. The number of different collections of r objects selected from a group of n distinct objects is called the **number of combinations of r out of n.**

Rule of Combinations

The number of combinations, or different collections, of r objects selected from a group of n distinct objects is denoted by $\binom{n}{r}$,

$$\binom{n}{r} = \frac{P_r^n}{r!} = \frac{n(n-1)\cdots(n-r+1)}{r!} = \frac{n!}{r!(n-r)!}$$

The general formula for combinations is obtained by the following correspondence and an application of the product rule.

$$\begin{bmatrix} \text{Arrange } r \text{ objects} \\ \text{selected from } n \end{bmatrix} \quad \begin{array}{c} \text{is the} \\ \text{same as} \end{array} \quad \begin{bmatrix} \text{first select } r \\ \text{objects from } n \end{bmatrix} \quad \begin{array}{c} \text{and} \\ \text{then} \end{array} \quad \begin{bmatrix} \text{arrange the } r \\ \text{selected objects} \end{bmatrix}$$

$$P_r^n \qquad = \qquad \binom{n}{r} \qquad \times \qquad r!$$

According to the last form, $\binom{n}{r}$ is symmetric in r and $n - r$. Selecting r objects to be in the collection is the same as choosing which $n - r$ to exclude.

If 2 of the 5 volleyball teams must be selected to play another match, so the order of selection is not important, there are 10 ways to select 2 to play or the 3 teams which will not play.

$$\binom{5}{2} = \frac{5 \times 4}{2 \times 1} = 10 = \frac{5 \times 4 \times 3 \times 2 \times 1}{2 \times 1 \times (3 \times 2 \times 1)} = \frac{5 \times 4 \times 3}{3 \times 2 \times 1} \text{ which equals } \binom{5}{3}$$

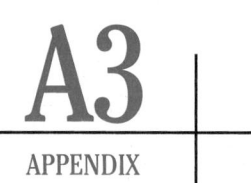

A3
APPENDIX

Expectation and
Standard Deviation—Properties

The expected value of a discrete random variable is a summation of the products (value × probability). The key properties of expectations are then all inherited from the properties of summation. In this appendix, we indicate this development for some of the most useful properties of expectation and variance. The interested reader can consult Bhattacharyya and Johnson[1] for more details.

A3.1   EXPECTED VALUE AND STANDARD DEVIATION OF $cX + b$

The units of the random variable X may be changed by multiplying by a constant, for example,

$$X = \text{height in feet}, \qquad 12X = \text{height in inches}$$

or by adding a constant, for example,

$$X = \text{temperature (°F)}$$
$$X - 32 = \text{degrees above freezing (°F)}$$

[1]G. K. Bhattacharyya and Richard A. Johnson, *Statistical Concepts and Methods*, John Wiley & Sons, New York, 1978.

The mean and standard deviation of the new random variables are related to $\mu = E(X)$ and $\sigma = sd(X)$.

If X is multiplied by a constant c,

Random Variable	Mean	sd
cX	$c\mu$	$|c|\sigma$

If a constant b is added to X,

Random Variable	Mean	sd
$X + b$	$\mu + b$	σ (unchanged)

Notice that adding a constant to a random variable leaves the standard deviation unchanged.

Example   Let X have mean $= 3 = \mu$ and standard deviation $= 5 = \sigma$. Find the mean and sd of (a) $X + 4$, (b) $2X$, (c) $-X$, and (d) $\frac{1}{5}(X - 3)$.

SOLUTION   By the foregoing properties,

Random Variable	Mean	sd
$X + 4$	$3 + 4 = 7$	5
$2X$	$2(3) = 6$	$2(5) = 10$
$-X = (-1)X$	$(-1)3 = -3$	$|-1|5 = 5$

Finally, $(X - 3)$ has mean $3 - 3 = 0$ and sd $= 5$, so $\frac{1}{5}(X - 3)$ has mean $= \frac{1}{5}(0) = 0$ and sd $= \frac{1}{5}(5) = 1$.

Any random variable having $E(X) = \mu$ and $Var(X) = \sigma^2$ can be converted to a

Standardized variable

$$Z = \frac{X - \mu}{\sigma}$$

The standardized variable Z has mean $= 0$ and variance $= 1$. This was checked for $Z = \dfrac{X - 3}{5}$ in the example above.

*VERIFICATION OF THE MEAN AND sd EXPRESSIONS FOR $cX + b$

Consider the random variable $cX + b$, which includes the two cases above. The choice $b = 0$ gives cX and the choice $c = 1$ gives $X + b$. We restrict our verification to discrete random variables, where probability $f(x_i)$ is attached to x_i. Because $cX + b$ takes value $cx_i + b$ with probability $f(x_i)$,

$$(\text{value} \times \text{probability}) = (cx_i + b)f(x_i) = cx_i f(x_i) + bf(x_i)$$

and

$$\begin{aligned}
\text{mean} &= \Sigma \,(\text{value} \times \text{probability}) \\
&= \Sigma \, cx_i f(x_i) + \Sigma \, bf(x_i) \\
&= c \,\Sigma \, x_i f(x_i) + b \,\Sigma \, f(x_i) = c\mu + b \cdot 1
\end{aligned}$$

Next,

$$\begin{aligned}
\text{deviation} &= (\text{random variable}) - (\text{mean}) \\
&= (cX + b) - (c\mu + b) = cX - c\mu = c(X - \mu)
\end{aligned}$$

so

$$\begin{aligned}
\text{variance} &= \Sigma \,(\text{deviation})^2 \times \text{probability} \\
&= \Sigma \, c^2 (x_i - \mu)^2 f(x_i) \\
&= c^2 \,\Sigma \,(x_i - \mu)^2 f(x_i) = c^2 \sigma^2
\end{aligned}$$

Taking the positive square root yields $\text{sd}(cX + b) = |c|\sigma$.
Finally, taking $c = 1/\sigma$ and $b = -\mu/\sigma$, we obtain

$$cX + b = \frac{1}{\sigma} X - \frac{\mu}{\sigma} = \frac{X - \mu}{\sigma} = Z$$

so the standardized variable Z has

$$\text{mean} \quad c\mu + b = \frac{1}{\sigma} \mu - \frac{\mu}{\sigma} = 0$$

$$\text{sd} \quad c\sigma = \frac{1}{\sigma} \sigma = 1$$

A3.2 ALTERNATIVE FORMULA FOR σ^2

An alternative formula for σ^2 often simplifies the numerical calculations. By definition,

$$\sigma^2 = \Sigma \,(\text{deviation})^2 (\text{probability}) = \Sigma \,(x_i - \mu)^2 f(x_i)$$

but σ^2 can also be expressed as

$$\sigma^2 = \Sigma x_i^2 f(x_i) - \mu^2$$

To deduce the second form, we first expand the square of the deviation:

$$(x_i - \mu)^2 = x_i^2 - 2\mu x_i + \mu^2$$

Then, multiply each term on the right-hand side by $f(x_i)$ and sum:

First term		$= \Sigma x_i^2 f(x_i)$	
Second term	$-2\mu \Sigma x_i f(x_i) =$	$-2\mu^2$	since $\Sigma x_i f(x_i) = \mu$
Third term	$+ \mu^2 \Sigma f(x_i) =$	μ^2	since $\Sigma f(x_i) = 1$
Result:		$\sigma^2 = \Sigma x_i^2 f(x_i) - \mu^2$	

Example Calculate σ^2 by both formulas.

SOLUTION

	Calculation $\Sigma (x - \mu)^2 f(x)$					Calculation $\Sigma x^2 f(x) - \mu^2$		
x	$f(x)$	$xf(x)$	$(x - 2)^2$	$(x - 2)^2 f(x)$	x	$f(x)$	$xf(x)$	$x^2 f(x)$
1	.4	.4	1	.4	1	.4	.4	.4
2	.3	.6	0	0	2	.3	.6	1.2
3	.2	.6	1	.2	3	.2	.6	1.8
4	.1	.4	4	.4	4	.1	.4	1.6
		$\overline{2.0} = \mu$		$\overline{1.0} = \sigma^2$			$\overline{2.0} = \mu$	5.0

$$\sigma^2 = 5 - 2^2 = 1$$

A3.3 PROPERTIES OF EXPECTED VALUE FOR TWO RANDOM VARIABLES

The concept of expectation extends to two or more variables. With two random variables:

1. $E(X + Y) = E(X) + E(Y)$ (additivity or sum law of expectation).
2. If X and Y are independent, then $E(XY) = E(X)E(Y)$.

Remark: Property 1 holds quite generally, independence is not required.

*DEMONSTRATION

We verify both (1) and (2) assuming independence. Independence implies that $P[X = x, Y = y] = P[X = x]P[Y = y]$ for all outcomes (x, y). That is, the distribution of probability over the pairs of possible values (x, y) is specified by the product $f_X(x)f_Y(y) = P[X = x, Y = y]$.

The expected value, $E(X + Y)$, is obtained by multiplying each possible value $(x + y)$ by the probability $f_X(x)f_Y(y)$ and summing

$$E(X + Y) = \sum_x \sum_y (x + y)f_X(x)f_Y(y)$$

$$= \sum_x \sum_y x f_X(x)f_Y(y) + \sum_x \sum_y y f_X(x)f_Y(y)$$

$$= \left(\sum_x x f_X(x)\right)\left(\underbrace{\sum_y f_Y(y)}_{= \ 1}\right) + \left(\underbrace{\sum_x f_X(x)}_{= \ 1}\right)\left(\sum_y y f_Y(y)\right)$$

$$= E(X) + E(Y)$$

Next,

$$E(XY) = \sum_x \sum_y xy\, f_X(x)f_Y(y)$$

$$= \sum_x x f_X(x)\left(\sum_y y f_Y(y)\right) = E(X)E(Y) \qquad \square$$

Under the proviso that the random variables are independent, variances also add.

3. If X and Y are independent,

$$\text{Var}(X + Y) = \text{Var}(X) + \text{Var}(Y)$$

*DEMONSTRATION

We set $\mu_1 = E(X)$ and $\mu_2 = E(Y)$, so by property 1,

$$E(X + Y) = \mu_1 + \mu_2$$

Then, since variance is the expected value of (variable − mean)^2,

$$\begin{aligned}
\text{Var}(X + Y) &= E(X + Y - \mu_1 - \mu_2)^2 \\
&= E[(X - \mu_1)^2 + (Y - \mu_2)^2 + 2(X - \mu_1)(Y - \mu_2)] \\
&= E(X - \mu_1)^2 + E(Y - \mu_2)^2 + 2E(X - \mu_1)(Y - \mu_2) \\
&\quad \text{(by the sum law of expectation, property 1)} \\
&= \text{Var}(X) + \text{Var}(Y)
\end{aligned}$$

This last step follows since

$$\begin{aligned}
E(X - \mu_1)(Y - \mu_2) &= E(XY - \mu_1 Y - X\mu_2 + \mu_1\mu_2) \\
&= E(XY) - \mu_1 E(Y) - \mu_2 E(X) + \mu_1\mu_2 \\
&= E(XY) - \mu_1\mu_2 \\
&= E(X)E(Y) - \mu_1\mu_2 = \mu_1\mu_2 - \mu_1\mu_2 = 0 \quad \text{by property 2} \quad \square
\end{aligned}$$

A4

APPENDIX

The Expected Value and Standard Deviation of $\overline{X}$

Some basic properties of the sampling distribution of $\overline{X}$ can be expressed in terms of the population mean and variance when the observations form a random sample. Let

$$\mu = \text{population mean}$$
$$\sigma^2 = \text{population variance}$$

In a random sample, the random variables $X_1, \ldots, X_n$ are independent, and each has the distribution of the population. Consequently, each observation has mean μ and variance σ^2, or

$$E(X_1) = \cdots = E(X_n) = \mu$$
$$\text{Var}(X_1) = \cdots = \text{Var}(X_n) = \sigma^2$$

Next,

$$\overline{X} = \frac{1}{n}(X_1 + \cdots + X_n)$$

and n is a constant. Using the additivity properties of expectation and variance discussed in Appendix A3, we obtain

$$E(\overline{X}) = \frac{1}{n} E(X_1 + \cdots + X_n)$$

$$= \frac{1}{n} [E(X_1) + \cdots + E(X_n)] \quad \text{(mean of sum = sum of means)}$$

$$= \frac{1}{n} [\mu + \cdots + \mu] = \frac{n\mu}{n} = \mu$$

$$\text{Var}(\overline{X}) = \frac{1}{n^2} \text{Var}(X_1 + \cdots + X_n)$$

$$= \frac{1}{n^2} [\text{Var}(X_1) + \cdots + \text{Var}(X_n)] \quad \text{(variances add due to independence)}$$

$$= \frac{1}{n^2} [\sigma^2 + \cdots + \sigma^2] = \frac{n\sigma^2}{n^2} = \frac{\sigma^2}{n}$$

Furthermore, taking the square root yields

$$\text{sd}(\overline{X}) = \sqrt{\text{Var}(\overline{X})} = \frac{\sigma}{\sqrt{n}}$$

Tables

TABLE 1 Random Digits

Row										
1	0695	7741	8254	4297	0000	5277	6563	9265	1023	5925
2	0437	5434	8503	3928	6979	9393	8936	9088	5744	4790
3	6242	2998	0205	5469	3365	7950	7256	3716	8385	0253
4	7090	4074	1257	7175	3310	0712	4748	4226	0604	3804
5	0683	6999	4828	7888	0087	9288	7855	2678	3315	6718
6	7013	4300	3768	2572	6473	2411	6285	0069	5422	6175
7	8808	2786	5369	9571	3412	2465	6419	3990	0294	0896
8	9876	3602	5812	0124	1997	6445	3176	2682	1259	1728
9	1873	1065	8976	1295	9434	3178	0602	0732	6616	7972
10	2581	3075	4622	2974	7069	5605	0420	2949	4387	7679
11	3785	6401	0540	5077	7132	4135	4646	3834	6753	1593
12	8626	4017	1544	4202	8986	1432	2810	2418	8052	2710
13	6253	0726	9483	6753	4732	2284	0421	3010	7885	8436
14	0113	4546	2212	9829	2351	1370	2707	3329	6574	7002
15	4646	6474	9983	8738	1603	8671	0489	9588	3309	5860
16	7873	7343	4432	2866	7973	3765	2888	5154	2250	4339
17	3756	9204	2590	6577	2409	8234	8656	2336	7948	7478
18	2673	7115	5526	0747	3952	6804	3671	7486	3024	9858
19	0187	7045	2711	0349	7734	4396	0988	4887	7682	8990
20	7976	3862	8323	5997	6904	4977	1056	6638	6398	4552

TABLE 1   *(Continued)*

Row										
21	5605	1819	8926	9557	2905	0802	7749	0845	1710	4125
22	2225	5556	2545	7480	8804	4161	0084	0787	2561	5113
23	2549	4166	1609	7570	4223	0032	4236	0169	4673	8034
24	6113	1312	5777	7058	2413	3932	5144	5998	7183	5210
25	2028	2537	9819	9215	9327	6640	5986	7935	2750	2981
26	7818	3655	5771	4026	5757	3171	6435	2990	1860	1796
27	9629	3383	1931	2631	5903	9372	1307	4061	5443	8663
28	6657	5967	3277	7141	3628	2588	9320	1972	7683	7544
29	4344	7388	2978	3945	0471	4882	1619	0093	2282	7024
30	3145	8720	2131	1614	1575	5239	0766	0404	4873	7986
31	1848	4094	9168	0903	6451	2823	7566	6644	1157	8889
32	0915	5578	0822	5887	5354	3632	4617	6016	8989	9482
33	1430	4755	7551	9019	8233	9625	6361	2589	2496	7268
34	3473	7966	7249	0555	6307	9524	4888	4939	1641	1573
35	3312	0773	6296	1348	5483	5824	3353	4587	1019	9677
36	6255	4204	5890	9273	0634	9992	3834	2283	1202	4849
37	0562	2546	8559	0480	9379	9282	8257	3054	4272	9311
38	1957	6783	4105	8976	8035	0883	8971	0017	6476	2895
39	7333	1083	0398	8841	0017	4135	4043	8157	4672	2424
40	4601	8908	1781	4287	2681	6223	0814	4477	3798	4437
41	2628	2233	0708	0900	1698	2818	3931	6930	9273	6749
42	5318	8865	6057	8422	6992	9697	0508	3370	5522	9250
43	6335	0852	8657	8374	0311	6012	9477	0112	8976	3312
44	0301	8333	0327	0467	6186	1770	4099	9588	5382	8958
45	1719	9775	1566	7020	4535	2850	0207	4792	6405	1472
46	8907	8226	4249	6340	9062	3572	7655	6707	3685	1282
47	6129	5927	3731	1125	0081	1241	2772	6458	9157	4543
48	7376	3150	8985	8318	8003	6106	4952	8492	2804	3867
49	9093	3407	4127	9258	3687	5631	5102	1546	2659	0831
50	1133	3086	9380	5431	8647	0910	6948	2257	0946	1245
51	4567	0910	8495	2410	1088	7067	8505	9083	4339	2440
52	6141	8380	2302	4608	7209	5738	9765	3435	9657	6061
53	1514	8309	8743	3096	0682	7902	8204	7508	8330	1681
54	7277	1634	7866	9883	0916	6363	5391	6184	8040	3135
55	4568	4758	0166	1509	2105	0976	0269	0278	7443	2431
56	9200	7599	7754	4534	4532	3102	6831	2387	4147	2455
57	3971	8149	4431	2345	6436	0627	0410	1348	6599	1296
58	2672	9661	2359	8477	3425	8150	6918	8883	1518	4708
59	1524	3268	3798	3360	2255	0371	7610	9114	9466	0901
60	6817	9007	5959	0767	1166	7317	7502	0274	6340	0427

TABLE 1   *(Continued)*

Row										
61	6762	3502	9559	4279	9271	9595	3053	4918	7503	5169
62	5264	0075	6655	4563	7112	7264	3240	2150	8180	1361
63	5070	8428	5149	2137	8728	9110	2334	9709	8134	3925
64	1664	3379	5273	9367	6950	6828	1711	7082	4783	0147
65	6962	7141	1904	6648	7328	2901	6396	9949	6274	1672
66	7541	4289	4970	2922	6670	8540	9053	3219	8881	1897
67	5244	4651	2934	6700	8869	0926	4191	1364	0926	2874
68	2939	3890	0745	2577	7931	3913	7877	2837	2500	8774
69	4266	6207	8083	6564	5336	5303	7503	6627	6055	3606
70	7848	5477	5588	3490	0294	3609	1632	5684	1719	6162
71	3009	1879	0440	7916	6643	9723	5933	0574	2480	6893
72	9865	7813	7468	8493	3293	1071	7183	9462	2363	6529
73	1196	1251	2368	1262	5769	9450	7485	4039	4985	6612
74	1067	3716	8897	1970	8799	5718	4792	7292	4589	4554
75	5160	5563	6527	7861	3477	6735	7748	4913	6370	2258
76	4560	0094	8284	7604	1667	9286	2228	9507	1838	4646
77	7697	2151	4860	0739	4370	3992	8121	2502	7670	4470
78	8675	2997	9783	7306	4116	6432	7233	4611	7121	9412
79	3597	3520	5995	0892	3470	4581	1068	8801	1254	8607
80	4281	8802	5880	6212	6818	8162	0052	1755	7107	5197
81	0101	0907	9057	2263	0059	8553	7855	7758	1020	1264
82	8179	0109	4412	6044	7167	4209	5250	4570	1984	8276
83	8980	9662	9333	6598	2990	8173	1753	1135	1409	2042
84	3050	2450	9252	6724	2697	7933	9540	3700	6561	2790
85	4465	1307	8782	6763	9202	5594	7166	7050	4462	0426
86	1925	5402	1379	3556	5109	4846	9827	2881	5574	9027
87	8753	4602	1838	4624	4632	2512	2652	4804	1624	5116
88	2645	9197	4541	4822	7883	3352	3202	0906	3676	8141
89	4287	5473	4493	7086	4271	9140	3315	7073	4533	0653
90	5280	5426	7240	2154	7952	3804	8097	9328	8069	6894
91	9553	3136	2112	1369	5562	7360	5530	8074	6488	3682
92	2975	7924	0253	3503	9383	9454	3320	3234	9255	3527
93	2596	7274	8967	8138	6868	0385	4467	3792	3844	8700
94	4192	7440	6410	6064	4561	0411	9187	9940	2866	3345
95	3980	8594	9935	8560	0229	8778	2386	7852	4031	0627
96	1822	1177	6846	3997	5822	9188	2479	7951	3051	0110
97	8415	2623	2358	8895	5125	0173	3182	4151	4419	9049
98	2123	5798	5444	3282	8022	3931	4429	6028	5385	6845
99	1754	4076	3507	3705	7459	7544	6127	4820	3760	6476
100	3967	9997	0695	3562	9997	2934	8469	9706	4763	7132

TABLE 1   *(Continued)*

Row										
101	7604	6645	6633	6288	5488	8355	9295	9637	5410	0452
102	6357	0216	1685	4308	0391	1517	1952	0108	1258	5498
103	5241	0554	6072	2412	1915	4451	0633	0449	9059	6873
104	9683	0618	2433	0154	0816	9885	3562	7392	4406	2994
105	8073	7718	9374	0965	8861	0018	2152	1736	5187	9347
106	3685	5901	6296	7748	6815	8033	5646	8691	3885	1550
107	9354	1854	1914	2592	9939	2468	0190	5882	3964	6938
108	2604	3040	9664	3962	4600	1314	8163	7869	2059	8203
109	9371	8390	6971	4931	1142	8588	2240	9256	7805	0153
110	5463	5569	1657	2797	9026	7754	8501	1953	1364	7787
111	5832	6510	1728	0531	9770	5790	8294	2702	4318	2494
112	6977	1478	4053	5836	5773	5706	8840	6575	6984	0196
113	6653	3177	7173	1053	8117	5818	2177	7524	3839	2438
114	2043	3329	3149	8591	8213	7941	0324	0275	2808	5787
115	1892	6495	7363	8840	6126	5749	5841	5564	3296	8176
116	4279	6686	2795	2572	6915	5770	0723	5003	6124	0041
117	9018	3226	1024	4455	4743	8634	7086	9462	5603	4961
118	6588	0445	5301	0442	7270	4287	9827	7666	4020	6061
119	3258	2829	5949	6280	9178	3614	8680	6705	1311	2408
120	9213	0161	4449	9084	8199	7330	4284	5061	1971	1008

TABLE 2   Cumulative Binomial Probabilities

$$P[X \leq c] = \sum_{x=0}^{c} \binom{n}{x} p^x (1 - p)^{n-x}$$

							p					
		.05	.10	.20	.30	.40	.50	.60	.70	.80	.90	.95
	c											
$n = 1$	0	.950	.900	.800	.700	.600	.500	.400	.300	.200	.100	.050
	1	1.000	1.000	1.000	1.000	1.000	1.000	1.000	1.000	1.000	1.000	1.000
$n = 2$	0	.902	.810	.640	.490	.360	.250	.160	.090	.040	.010	.002
	1	.997	.990	.960	.910	.840	.750	.640	.510	.360	.190	.097
	2	1.000	1.000	1.000	1.000	1.000	1.000	1.000	1.000	1.000	1.000	1.000
$n = 3$	0	.857	.729	.512	.343	.216	.125	.064	.027	.008	.001	.000
	1	.993	.972	.896	.784	.648	.500	.352	.216	.104	.028	.007
	2	1.000	.999	.992	.973	.936	.875	.784	.657	.488	.271	.143
	3	1.000	1.000	1.000	1.000	1.000	1.000	1.000	1.000	1.000	1.000	1.000

TABLE 2 *(Continued)*

		p										
	c	.05	.10	.20	.30	.40	.50	.60	.70	.80	.90	.95
$n = 4$	0	.815	.656	.410	.240	.130	.063	.026	.008	.002	.000	.000
	1	.986	.948	.819	.652	.475	.313	.179	.084	.027	.004	.000
	2	1.000	.996	.973	.916	.821	.688	.525	.348	.181	.052	.014
	3	1.000	1.000	.998	.992	.974	.938	.870	.760	.590	.344	.185
	4	1.000	1.000	1.000	1.000	1.000	1.000	1.000	1.000	1.000	1.000	1.000
$n = 5$	0	.774	.590	.328	.168	.078	.031	.010	.002	.000	.000	.000
	1	.977	.919	.737	.528	.337	.188	.087	.031	.007	.000	.000
	2	.999	.991	.942	.837	.683	.500	.317	.163	.058	.009	.001
	3	1.000	1.000	.993	.969	.913	.813	.663	.472	.263	.081	.023
	4	1.000	1.000	1.000	.998	.990	.969	.922	.832	.672	.410	.226
	5	1.000	1.000	1.000	1.000	1.000	1.000	1.000	1.000	1.000	1.000	1.000
$n = 6$	0	.735	.531	.262	.118	.047	.016	.004	.001	.000	.000	.000
	1	.967	.886	.655	.420	.233	.109	.041	.011	.002	.000	.000
	2	.998	.984	.901	.744	.544	.344	.179	.070	.017	.001	.000
	3	1.000	.999	.983	.930	.821	.656	.456	.256	.099	.016	.002
	4	1.000	1.000	.998	.989	.959	.891	.767	.580	.345	.114	.033
	5	1.000	1.000	1.000	.999	.996	.984	.953	.882	.738	.469	.265
	6	1.000	1.000	1.000	1.000	1.000	1.000	1.000	1.000	1.000	1.000	1.000
$n = 7$	0	.698	.478	.210	.082	.028	.008	.002	.000	.000	.000	.000
	1	.956	.850	.577	.329	.159	.063	.019	.004	.000	.000	.000
	2	.996	.974	.852	.647	.420	.227	.096	.029	.005	.000	.000
	3	1.000	.997	.967	.874	.710	.500	.290	.126	.033	.003	.000
	4	1.000	1.000	.995	.971	.904	.773	.580	.353	.148	.026	.004
	5	1.000	1.000	1.000	.996	.981	.938	.841	.671	.423	.150	.044
	6	1.000	1.000	1.000	1.000	.998	.992	.972	.918	.790	.522	.302
	7	1.000	1.000	1.000	1.000	1.000	1.000	1.000	1.000	1.000	1.000	1.000
$n = 8$	0	.663	.430	.168	.058	.017	.004	.001	.000	.000	.000	.000
	1	.943	.813	.503	.255	.106	.035	.009	.001	.000	.000	.000
	2	.994	.962	.797	.552	.315	.145	.050	.011	.001	.000	.000
	3	1.000	.995	.944	.806	.594	.363	.174	.058	.010	.000	.000
	4	1.000	1.000	.990	.942	.826	.637	.406	.194	.056	.005	.000
	5	1.000	1.000	.999	.989	.950	.855	.685	.448	.203	.038	.006
	6	1.000	1.000	1.000	.999	.991	.965	.894	.745	.497	.187	.057
	7	1.000	1.000	1.000	1.000	.999	.996	.983	.942	.832	.570	.337
	8	1.000	1.000	1.000	1.000	1.000	1.000	1.000	1.000	1.000	1.000	1.000
$n = 9$	0	.630	.387	.134	.040	.010	.002	.000	.000	.000	.000	.000
	1	.929	.775	.436	.196	.071	.020	.004	.000	.000	.000	.000
	2	.992	.947	.738	.463	.232	.090	.025	.004	.000	.000	.000
	3	.999	.992	.914	.730	.483	.254	.099	.025	.003	.000	.000

TABLE 2   *(Continued)*

		.05	.10	.20	.30	.40	.50	.60	.70	.80	.90	.95
	c											
	4	1.000	.999	.980	.901	.733	.500	.267	.099	.020	.001	.000
	5	1.000	1.000	.997	.975	.901	.746	.517	.270	.086	.008	.001
	6	1.000	1.000	1.000	.996	.975	.910	.768	.537	.262	.053	.008
	7	1.000	1.000	1.000	1.000	.996	.980	.929	.804	.564	.225	.071
	8	1.000	1.000	1.000	1.000	1.000	.998	.990	.960	.866	.613	.370
	9	1.000	1.000	1.000	1.000	1.000	1.000	1.000	1.000	1.000	1.000	1.000
$n = 10$	0	.599	.349	.107	.028	.006	.001	.000	.000	.000	.000	.000
	1	.914	.736	.376	.149	.046	.011	.002	.000	.000	.000	.000
	2	.988	.930	.678	.383	.167	.055	.012	.002	.000	.000	.000
	3	.999	.987	.879	.650	.382	.172	.055	.011	.001	.000	.000
	4	1.000	.998	.967	.850	.633	.377	.166	.047	.006	.000	.000
	5	1.000	1.000	.994	.953	.834	.623	.367	.150	.033	.002	.000
	6	1.000	1.000	.999	.989	.945	.828	.618	.350	.121	.013	.001
	7	1.000	1.000	1.000	.998	.988	.945	.833	.617	.322	.070	.012
	8	1.000	1.000	1.000	1.000	.998	.989	.954	.851	.624	.264	.086
	9	1.000	1.000	1.000	1.000	1.000	.999	.994	.972	.893	.651	.401
	10	1.000	1.000	1.000	1.000	1.000	1.000	1.000	1.000	1.000	1.000	1.000
$n = 11$	0	.569	.314	.086	.020	.004	.000	.000	.000	.000	.000	.000
	1	.898	.697	.322	.113	.030	.006	.001	.000	.000	.000	.000
	2	.985	.910	.617	.313	.119	.033	.006	.001	.000	.000	.000
	3	.998	.981	.839	.570	.296	.113	.029	.004	.000	.000	.000
	4	1.000	.997	.950	.790	.533	.274	.099	.022	.002	.000	.000
	5	1.000	1.000	.988	.922	.753	.500	.247	.078	.012	.000	.000
	6	1.000	1.000	.998	.978	.901	.726	.467	.210	.050	.003	.000
	7	1.000	1.000	1.000	.996	.971	.887	.704	.430	.161	.019	.002
	8	1.000	1.000	1.000	.999	.994	.967	.881	.687	.383	.090	.015
	9	1.000	1.000	1.000	1.000	.999	.994	.970	.887	.678	.303	.102
	10	1.000	1.000	1.000	1.000	1.000	1.000	.996	.980	.914	.686	.431
	11	1.000	1.000	1.000	1.000	1.000	1.000	1.000	1.000	1.000	1.000	1.000
$n = 12$	0	.540	.282	.069	.014	.002	.000	.000	.000	.000	.000	.000
	1	.882	.659	.275	.085	.020	.003	.000	.000	.000	.000	.000
	2	.980	.889	.558	.253	.083	.019	.003	.000	.000	.000	.000
	3	.998	.974	.795	.493	.225	.073	.015	.002	.000	.000	.000
	4	1.000	.996	.927	.724	.438	.194	.057	.009	.001	.000	.000
	5	1.000	.999	.981	.882	.665	.387	.158	.039	.004	.000	.000
	6	1.000	1.000	.996	.961	.842	.613	.335	.118	.019	.001	.000
	7	1.000	1.000	.999	.991	.943	.806	.562	.276	.073	.004	.000
	8	1.000	1.000	1.000	.998	.985	.927	.775	.507	.205	.026	.002
	9	1.000	1.000	1.000	1.000	.997	.981	.917	.747	.442	.111	.020
	10	1.000	1.000	1.000	1.000	1.000	.977	.980	.915	.725	.341	.118
	11	1.000	1.000	1.000	1.000	1.000	1.000	.998	.986	.931	.718	.460
	12	1.000	1.000	1.000	1.000	1.000	1.000	1.000	1.000	1.000	1.000	1.000

TABLE 2 *(Continued)*

	c	.05	.10	.20	.30	.40	.50	.60	.70	.80	.90	.95
							p					
$n = 13$	0	.513	.254	.055	.010	.001	.000	.000	.000	.000	.000	.000
	1	.865	.621	.234	.064	.013	.002	.000	.000	.000	.000	.000
	2	.975	.866	.502	.202	.058	.011	.001	.000	.000	.000	.000
	3	.997	.966	.747	.421	.169	.046	.008	.001	.000	.000	.000
	4	1.000	.994	.901	.654	.353	.133	.032	.004	.000	.000	.000
	5	1.000	.999	.970	.835	.574	.291	.098	.018	.001	.000	.000
	6	1.000	1.000	.993	.938	.771	.500	.229	.062	.007	.000	.000
	7	1.000	1.000	.999	.982	.902	.709	.426	.165	.030	.001	.000
	8	1.000	1.000	1.000	.996	.968	.867	.647	.346	.099	.006	.000
	9	1.000	1.000	1.000	.999	.992	.954	.831	.579	.253	.034	.003
	10	1.000	1.000	1.000	1.000	.999	.989	.942	.798	.498	.134	.025
	11	1.000	1.000	1.000	1.000	1.000	.998	.987	.936	.766	.379	.135
	12	1.000	1.000	1.000	1.000	1.000	1.000	.999	.990	.945	.746	.487
	13	1.000	1.000	1.000	1.000	1.000	1.000	1.000	1.000	1.000	1.000	1.000
$n = 14$	0	.488	.229	.044	.007	.001	.000	.000	.000	.000	.000	.000
	1	.847	.585	.198	.047	.008	.001	.000	.000	.000	.000	.000
	2	.970	.842	.448	.161	.040	.006	.001	.000	.000	.000	.000
	3	.996	.956	.698	.355	.124	.029	.004	.000	.000	.000	.000
	4	1.000	.991	.870	.584	.279	.090	.018	.002	.000	.000	.000
	5	1.000	.999	.956	.781	.486	.212	.058	.008	.000	.000	.000
	6	1.000	1.000	.988	.907	.692	.395	.150	.031	.002	.000	.000
	7	1.000	1.000	.998	.969	.850	.605	.308	.093	.012	.000	.000
	8	1.000	1.000	1.000	.992	.942	.788	.514	.219	.044	.001	.000
	9	1.000	1.000	1.000	.998	.982	.910	.721	.416	.130	.009	.000
	10	1.000	1.000	1.000	1.000	.996	.971	.876	.645	.302	.044	.004
	11	1.000	1.000	1.000	1.000	.999	.994	.960	.839	.552	.158	.030
	12	1.000	1.000	1.000	1.000	1.000	.999	.992	.953	.802	.415	.153
	13	1.000	1.000	1.000	1.000	1.000	1.000	.999	.993	.956	.771	.512
	14	1.000	1.000	1.000	1.000	1.000	1.000	1.000	1.000	1.000	1.000	1.000
$n = 15$	0	.463	.206	.035	.005	.000	.000	.000	.000	.000	.000	.000
	1	.829	.549	.167	.035	.005	.000	.000	.000	.000	.000	.000
	2	.964	.816	.398	.127	.027	.004	.000	.000	.000	.000	.000
	3	.995	.944	.648	.297	.091	.018	.002	.000	.000	.000	.000
	4	.999	.987	.836	.515	.217	.059	.009	.001	.000	.000	.000
	5	1.000	.998	.939	.722	.403	.151	.034	.004	.000	.000	.000
	6	1.000	1.000	.982	.869	.610	.304	.095	.015	.001	.000	.000
	7	1.000	1.000	.996	.950	.787	.500	.213	.050	.004	.000	.000
	8	1.000	1.000	.999	.985	.905	.696	.390	.131	.018	.000	.000
	9	1.000	1.000	1.000	.996	.966	.849	.597	.278	.061	.002	.000
	10	1.000	1.000	1.000	.999	.991	.941	.783	.485	.164	.013	.001
	11	1.000	1.000	1.000	1.000	.998	.982	.909	.703	.352	.056	.005
	12	1.000	1.000	1.000	1.000	1.000	.996	.973	.873	.602	.184	.036

TABLE 2 *(Continued)*

		.05	.10	.20	.30	.40	.50	.60	.70	.80	.90	.95
	c											
	13	1.000	1.000	1.000	1.000	1.000	1.000	.995	.965	.833	.451	.171
	14	1.000	1.000	1.000	1.000	1.000	1.000	1.000	.995	.965	.794	.537
	15	1.000	1.000	1.000	1.000	1.000	1.000	1.000	1.000	1.000	1.000	1.000
n = 16	0	.440	.185	.028	.003	.000	.000	.000	.000	.000	.000	.000
	1	.811	.515	.141	.026	.003	.000	.000	.000	.000	.000	.000
	2	.957	.789	.352	.099	.018	.002	.000	.000	.000	.000	.000
	3	.993	.932	.598	.246	.065	.011	.001	.000	.000	.000	.000
	4	.999	.983	.798	.450	.167	.038	.005	.000	.000	.000	.000
	5	1.000	.997	.918	.660	.329	.105	.019	.002	.000	.000	.000
	6	1.000	.999	.973	.825	.527	.227	.058	.007	.000	.000	.000
	7	1.000	1.000	.993	.926	.716	.402	.142	.026	.001	.000	.000
	8	1.000	1.000	.999	.974	.858	.598	.284	.074	.007	.000	.000
	9	1.000	1.000	1.000	.993	.942	.773	.473	.175	.027	.001	.000
	10	1.000	1.000	1.000	.998	.981	.895	.671	.340	.082	.003	.000
	11	1.000	1.000	1.000	1.000	.995	.962	.833	.550	.202	.017	.001
	12	1.000	1.000	1.000	1.000	.999	.989	.935	.754	.402	.068	.007
	13	1.000	1.000	1.000	1.000	1.000	.998	.982	.901	.648	.211	.043
	14	1.000	1.000	1.000	1.000	1.000	1.000	.997	.974	.859	.485	.189
	15	1.000	1.000	1.000	1.000	1.000	1.000	1.000	.997	.972	.815	.560
	16	1.000	1.000	1.000	1.000	1.000	1.000	1.000	1.000	1.000	1.000	1.000
n = 17	0	.418	.167	.023	.002	.000	.000	.000	.000	.000	.000	.000
	1	.792	.482	.118	.019	.002	.000	.000	.000	.000	.000	.000
	2	.950	.762	.310	.077	.012	.001	.000	.000	.000	.000	.000
	3	.991	.917	.549	.202	.046	.006	.000	.000	.000	.000	.000
	4	.999	.978	.758	.389	.126	.025	.003	.000	.000	.000	.000
	5	1.000	.995	.894	.597	.264	.072	.011	.001	.000	.000	.000
	6	1.000	.999	.962	.775	.448	.166	.035	.003	.000	.000	.000
	7	1.000	1.000	.989	.895	.641	.315	.092	.013	.000	.000	.000
	8	1.000	1.000	.997	.960	.801	.500	.199	.040	.003	.000	.000
	9	1.000	1.000	1.000	.987	.908	.685	.359	.105	.011	.000	.000
	10	1.000	1.000	1.000	.997	.965	.834	.552	.225	.038	.001	.000
	11	1.000	1.000	1.000	.999	.989	.928	.736	.403	.106	.005	.000
	12	1.000	1.000	1.000	1.000	.997	.975	.874	.611	.242	.022	.001
	13	1.000	1.000	1.000	1.000	1.000	.994	.954	.798	.451	.083	.009
	14	1.000	1.000	1.000	1.000	1.000	.999	.988	.923	.690	.238	.050
	15	1.000	1.000	1.000	1.000	1.000	1.000	.998	.981	.882	.518	.208
	16	1.000	1.000	1.000	1.000	1.000	1.000	1.000	.998	.977	.833	.582
	17	1.000	1.000	1.000	1.000	1.000	1.000	1.000	1.000	1.000	1.000	1.000
n = 18	0	.397	.150	.018	.002	.000	.000	.000	.000	.000	.000	.000
	1	.774	.450	.099	.014	.001	.000	.000	.000	.000	.000	.000
	2	.942	.734	.271	.060	.008	.001	.000	.000	.000	.000	.000

p

TABLE 2   *(Continued)*

		.05	.10	.20	.30	.40	.50	.60	.70	.80	.90	.95
	c						*p*					
	3	.989	.902	.501	.165	.033	.004	.000	.000	.000	.000	.000
	4	.998	.972	.716	.333	.094	.015	.001	.000	.000	.000	.000
	5	1.000	.994	.867	.534	.209	.048	.006	.000	.000	.000	.000
	6	1.000	.999	.949	.722	.374	.119	.020	.001	.000	.000	.000
	7	1.000	1.000	.984	.859	.563	.240	.058	.006	.000	.000	.000
	8	1.000	1.000	.996	.940	.737	.407	.135	.021	.001	.000	.000
	9	1.000	1.000	.999	.979	.865	.593	.263	.060	.004	.000	.000
	10	1.000	1.000	1.000	.994	.942	.760	.437	.141	.016	.000	.000
	11	1.000	1.000	1.000	.999	.980	.881	.626	.278	.051	.001	.000
	12	1.000	1.000	1.000	1.000	.994	.952	.791	.466	.133	.006	.000
	13	1.000	1.000	1.000	1.000	.999	.985	.906	.667	.284	.028	.002
	14	1.000	1.000	1.000	1.000	1.000	.996	.967	.835	.499	.098	.011
	15	1.000	1.000	1.000	1.000	1.000	.999	.992	.940	.729	.266	.058
	16	1.000	1.000	1.000	1.000	1.000	1.000	.999	.986	.901	.550	.226
	17	1.000	1.000	1.000	1.000	1.000	1.000	1.000	.998	.982	.850	.603
	18	1.000	1.000	1.000	1.000	1.000	1.000	1.000	1.000	1.000	1.000	1.000
n = 19	0	.377	.135	.014	.001	.000	.000	.000	.000	.000	.000	.000
	1	.755	.420	.083	.010	.001	.000	.000	.000	.000	.000	.000
	2	.933	.705	.237	.046	.005	.000	.000	.000	.000	.000	.000
	3	.987	.885	.455	.133	.023	.002	.000	.000	.000	.000	.000
	4	.998	.965	.673	.282	.070	.010	.001	.000	.000	.000	.000
	5	1.000	.991	.837	.474	.163	.032	.003	.000	.000	.000	.000
	6	1.000	.998	.932	.666	.308	.084	.012	.001	.000	.000	.000
	7	1.000	1.000	.977	.818	.488	.180	.035	.003	.000	.000	.000
	8	1.000	1.000	.993	.916	.667	.324	.088	.011	.000	.000	.000
	9	1.000	1.000	.998	.967	.814	.500	.186	.033	.002	.000	.000
	10	1.000	1.000	1.000	.989	.912	.676	.333	.084	.007	.000	.000
	11	1.000	1.000	1.000	.997	.965	.820	.512	.182	.023	.000	.000
	12	1.000	1.000	1.000	.999	.988	.916	.692	.334	.068	.002	.000
	13	1.000	1.000	1.000	1.000	.997	.968	.837	.526	.163	.009	.000
	14	1.000	1.000	1.000	1.000	.999	.990	.930	.718	.327	.035	.002
	15	1.000	1.000	1.000	1.000	1.000	.998	.977	.867	.545	.115	.013
	16	1.000	1.000	1.000	1.000	1.000	1.000	.995	.954	.763	.295	.067
	17	1.000	1.000	1.000	1.000	1.000	1.000	.999	.990	.917	.580	.245
	18	1.000	1.000	1.000	1.000	1.000	1.000	1.000	.999	.986	.865	.623
	19	1.000	1.000	1.000	1.000	1.000	1.000	1.000	1.000	1.000	1.000	1.000
n = 20	0	.358	.122	.012	.001	.000	.000	.000	.000	.000	.000	.000
	1	.736	.392	.069	.008	.001	.000	.000	.000	.000	.000	.000
	2	.925	.677	.206	.035	.004	.000	.000	.000	.000	.000	.000
	3	.984	.867	.411	.107	.016	.001	.000	.000	.000	.000	.000
	4	.997	.957	.630	.238	.051	.006	.000	.000	.000	.000	.000
	5	1.000	.989	.804	.416	.126	.021	.002	.000	.000	.000	.000

TABLE 2 *(Continued)*

		.05	.10	.20	.30	.40	.50	.60	.70	.80	.90	.95
							p					
	c											
	6	1.000	.998	.913	.608	.250	.058	.006	.000	.000	.000	.000
	7	1.000	1.000	.968	.772	.416	.132	.021	.001	.000	.000	.000
	8	1.000	1.000	.990	.887	.596	.252	.057	.005	.000	.000	.000
	9	1.000	1.000	.997	.952	.755	.412	.128	.017	.001	.000	.000
	10	1.000	1.000	.999	.983	.872	.588	.245	.048	.003	.000	.000
	11	1.000	1.000	1.000	.995	.943	.748	.404	.113	.010	.000	.000
	12	1.000	1.000	1.000	.999	.979	.868	.584	.228	.032	.000	.000
	13	1.000	1.000	1.000	1.000	.994	.942	.750	.392	.087	.002	.000
	14	1.000	1.000	1.000	1.000	.998	.979	.874	.584	.196	.011	.000
	15	1.000	1.000	1.000	1.000	1.000	.994	.949	.762	.370	.043	.003
	16	1.000	1.000	1.000	1.000	1.000	.999	.984	.893	.589	.133	.016
	17	1.000	1.000	1.000	1.000	1.000	1.000	.996	.965	.794	.323	.075
	18	1.000	1.000	1.000	1.000	1.000	1.000	.999	.992	.931	.608	.264
	19	1.000	1.000	1.000	1.000	1.000	1.000	1.000	.999	.988	.878	.642
	20	1.000	1.000	1.000	1.000	1.000	1.000	1.000	1.000	1.000	1.000	1.000
$n = 25$	0	.277	.072	.004	.000	.000	.000	.000	.000	.000	.000	.000
	1	.642	.271	.027	.002	.000	.000	.000	.000	.000	.000	.000
	2	.873	.537	.098	.009	.000	.000	.000	.000	.000	.000	.000
	3	.966	.764	.234	.033	.002	.000	.000	.000	.000	.000	.000
	4	.993	.902	.421	.090	.009	.000	.000	.000	.000	.000	.000
	5	.999	.967	.617	.193	.029	.002	.000	.000	.000	.000	.000
	6	1.000	.991	.780	.341	.074	.007	.000	.000	.000	.000	.000
	7	1.000	.998	.891	.512	.154	.022	.001	.000	.000	.000	.000
	8	1.000	1.000	.953	.677	.274	.054	.004	.000	.000	.000	.000
	9	1.000	1.000	.983	.811	.425	.115	.013	.000	.000	.000	.000
	10	1.000	1.000	.994	.902	.586	.212	.034	.002	.000	.000	.000
	11	1.000	1.000	.998	.956	.732	.345	.078	.006	.000	.000	.000
	12	1.000	1.000	1.000	.983	.846	.500	.154	.017	.000	.000	.000
	13	1.000	1.000	1.000	.994	.922	.655	.268	.044	.002	.000	.000
	14	1.000	1.000	1.000	.998	.966	.788	.414	.098	.006	.000	.000
	15	1.000	1.000	1.000	1.000	.987	.885	.575	.189	.017	.000	.000
	16	1.000	1.000	1.000	1.000	.996	.946	.726	.323	.047	.000	.000
	17	1.000	1.000	1.000	1.000	.999	.978	.846	.488	.109	.002	.000
	18	1.000	1.000	1.000	1.000	1.000	.993	.926	.659	.220	.009	.000
	19	1.000	1.000	1.000	1.000	1.000	.998	.971	.807	.383	.033	.001
	20	1.000	1.000	1.000	1.000	1.000	1.000	.991	.910	.579	.098	.007
	21	1.000	1.000	1.000	1.000	1.000	1.000	.998	.967	.766	.236	.034
	22	1.000	1.000	1.000	1.000	1.000	1.000	1.000	.991	.902	.463	.127
	23	1.000	1.000	1.000	1.000	1.000	1.000	1.000	.998	.973	.729	.358
	24	1.000	1.000	1.000	1.000	1.000	1.000	1.000	1.000	.996	.928	.723
	25	1.000	1.000	1.000	1.000	1.000	1.000	1.000	1.000	1.000	1.000	1.000

TABLE 3   Standard Normal Probabilities

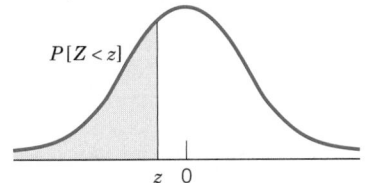

$P[Z < z]$

z	.00	.01	.02	.03	.04	.05	.06	.07	.08	.09
− 3.5	.0002	.0002	.0002	.0002	.0002	.0002	.0002	.0002	.0002	.0002
− 3.4	.0003	.0003	.0003	.0003	.0003	.0003	.0003	.0003	.0003	.0002
− 3.3	.0005	.0005	.0005	.0004	.0004	.0004	.0004	.0004	.0004	.0003
− 3.2	.0007	.0007	.0006	.0006	.0006	.0006	.0006	.0005	.0005	.0005
− 3.1	.0010	.0009	.0009	.0009	.0008	.0008	.0008	.0008	.0007	.0007
− 3.0	.0013	.0013	.0013	.0012	.0012	.0011	.0011	.0011	.0010	.0010
− 2.9	.0019	.0018	.0018	.0017	.0016	.0016	.0015	.0015	.0014	.0014
− 2.8	.0026	.0025	.0024	.0023	.0023	.0022	.0021	.0021	.0020	.0019
− 2.7	.0035	.0034	.0033	.0032	.0031	.0030	.0029	.0028	.0027	.0026
− 2.6	.0047	.0045	.0044	.0043	.0041	.0040	.0039	.0038	.0037	.0036
− 2.5	.0062	.0060	.0059	.0057	.0055	.0054	.0052	.0051	.0049	.0048
− 2.4	.0082	.0080	.0078	.0075	.0073	.0071	.0069	.0068	.0066	.0064
− 2.3	.0107	.0104	.0102	.0099	.0096	.0094	.0091	.0089	.0087	.0084
− 2.2	.0139	.0136	.0132	.0129	.0125	.0122	.0119	.0116	.0113	.0110
− 2.1	.0179	.0174	.0170	.0166	.0162	.0158	.0154	.0150	.0146	.0143
− 2.0	.0228	.0222	.0217	.0212	.0207	.0202	.0197	.0192	.0188	.0183
− 1.9	.0287	.0281	.0274	.0268	.0262	.0256	.0250	.0244	.0239	.0233
− 1.8	.0359	.0351	.0344	.0336	.0329	.0322	.0314	.0307	.0301	.0294
− 1.7	.0446	.0436	.0427	.0418	.0409	.0401	.0392	.0384	.0375	.0367
− 1.6	.0548	.0537	.0526	.0516	.0505	.0495	.0485	.0475	.0465	.0455
− 1.5	.0668	.0655	.0643	.0630	.0618	.0606	.0594	.0582	.0571	.0559
− 1.4	.0808	.0793	.0778	.0764	.0749	.0735	.0721	.0708	.0694	.0681
− 1.3	.0968	.0951	.0934	.0918	.0901	.0885	.0869	.0853	.0838	.0823
− 1.2	.1151	.1131	.1112	.1093	.1075	.1056	.1038	.1020	.1003	.0985
− 1.1	.1357	.1335	.1314	.1292	.1271	.1251	.1230	.1210	.1190	.1170
− 1.0	.1587	.1562	.1539	.1515	.1492	.1469	.1446	.1423	.1401	.1379
− .9	.1841	.1814	.1788	.1762	.1736	.1711	.1685	.1660	.1635	.1611
− .8	.2119	.2090	.2061	.2033	.2005	.1977	.1949	.1922	.1894	.1867
− .7	.2420	.2389	.2358	.2327	.2297	.2266	.2236	.2206	.2177	.2148
− .6	.2743	.2709	.2676	.2643	.2611	.2578	.2546	.2514	.2483	.2451
− .5	.3085	.3050	.3015	.2981	.2946	.2912	.2877	.2843	.2810	.2776
− .4	.3446	.3409	.3372	.3336	.3300	.3264	.3228	.3192	.3156	.3121
− .3	.3821	.3783	.3745	.3707	.3669	.3632	.3594	.3557	.3520	.3483
− .2	.4207	.4168	.4129	.4090	.4052	.4013	.3974	.3936	.3897	.3859
− .1	.4602	.4562	.4522	.4483	.4443	.4404	.4364	.4325	.4286	.4247
− .0	.5000	.4960	.4920	.4880	.4840	.4801	.4761	.4721	.4681	.4641

TABLE 3   *(Continued)*

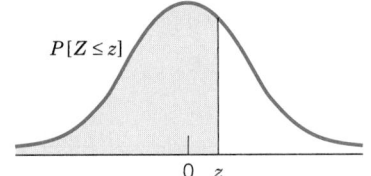

$P[Z \leq z]$

0   z

z	.00	.01	.02	.03	.04	.05	.06	.07	.08	.09
.0	.5000	.5040	.5080	.5120	.5160	.5199	.5239	.5279	.5319	.5359
.1	.5398	.5438	.5478	.5517	.5557	.5596	.5636	.5675	.5714	.5753
.2	.5793	.5832	.5871	.5910	.5948	.5987	.6026	.6064	.6103	.6141
.3	.6179	.6217	.6255	.6293	.6331	.6368	.6406	.6443	.6480	.6517
.4	.6554	.6591	.6628	.6664	.6700	.6736	.6772	.6808	.6844	.6879
.5	.6915	.6950	.6985	.7019	.7054	.7088	.7123	.7157	.7190	.7224
.6	.7257	.7291	.7324	.7357	.7389	.7422	.7454	.7486	.7517	.7549
.7	.7580	.7611	.7642	.7673	.7703	.7734	.7764	.7794	.7823	.7852
.8	.7881	.7910	.7939	.7967	.7995	.8023	.8051	.8078	.8106	.8133
.9	.8159	.8186	.8212	.8238	.8264	.8289	.8315	.8340	.8365	.8389
1.0	.8413	.8438	.8461	.8485	.8508	.8531	.8554	.8577	.8599	.8621
1.1	.8643	.8665	.8686	.8708	.8729	.8749	.8770	.8790	.8810	.8830
1.2	.8849	.8869	.8888	.8907	.8925	.8944	.8962	.8980	.8997	.9015
1.3	.9032	.9049	.9066	.9082	.9099	.9115	.9131	.9147	.9162	.9177
1.4	.9192	.9207	.9222	.9236	.9251	.9265	.9279	.9292	.9306	.9319
1.5	.9332	.9345	.9357	.9370	.9382	.9394	.9406	.9418	.9429	.9441
1.6	.9452	.9463	.9474	.9484	.9495	.9505	.9515	.9525	.9535	.9545
1.7	.9554	.9564	.9573	.9582	.9591	.9599	.9608	.9616	.9625	.9633
1.8	.9641	.9649	.9656	.9664	.9671	.9678	.9686	.9693	.9699	.9706
1.9	.9713	.9719	.9726	.9732	.9738	.9744	.9750	.9756	.9761	.9767
2.0	.9772	.9778	.9783	.9788	.9793	.9798	.9803	.9808	.9812	.9817
2.1	.9821	.9826	.9830	.9834	.9838	.9842	.9846	.9850	.9854	.9857
2.2	.9861	.9864	.9868	.9871	.9875	.9878	.9881	.9884	.9887	.9890
2.3	.9893	.9896	.9898	.9901	.9904	.9906	.9909	.9911	.9913	.9916
2.4	.9918	.9920	.9922	.9925	.9927	.9929	.9931	.9932	.9934	.9936
2.5	.9938	.9940	.9941	.9943	.9945	.9946	.9948	.9949	.9951	.9952
2.6	.9953	.9955	.9956	.9957	.9959	.9960	.9961	.9962	.9963	.9964
2.7	.9965	.9966	.9967	.9968	.9969	.9970	.9971	.9972	.9973	.9974
2.8	.9974	.9975	.9976	.9977	.9977	.9978	.9979	.9979	.9980	.9981
2.9	.9981	.9982	.9982	.9983	.9984	.9984	.9985	.9985	.9986	.9986
3.0	.9987	.9987	.9987	.9988	.9988	.9989	.9989	.9989	.9990	.9990
3.1	.9990	.9991	.9991	.9991	.9992	.9992	.9992	.9992	.9993	.9993
3.2	.9993	.9993	.9994	.9994	.9994	.9994	.9994	.9995	.9995	.9995
3.3	.9995	.9995	.9995	.9996	.9996	.9996	.9996	.9996	.9996	.9997
3.4	.9997	.9997	.9997	.9997	.9997	.9997	.9997	.9997	.9997	.9998
3.5	.9998	.9998	.9998	.9998	.9998	.9998	.9998	.9998	.9998	.9998

TABLE 4   Percentage Points of *t* Distributions

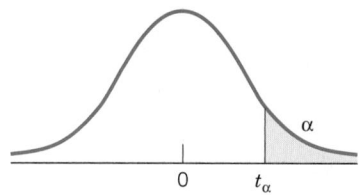

d.f.	.25	.10	.05	.025	.01	.00833	.00625	.005
1	1.000	3.078	6.314	12.706	31.821	38.204	50.923	63.657
2	.816	1.886	2.920	4.303	6.965	7.649	8.860	9.925
3	.765	1.638	2.353	3.182	4.541	4.857	5.392	5.841
4	.741	1.533	2.132	2.776	3.747	3.961	4.315	4.604
5	.727	1.476	2.015	2.571	3.365	3.534	3.810	4.032
6	.718	1.440	1.943	2.447	3.143	3.287	3.521	3.707
7	.711	1.415	1.895	2.365	2.998	3.128	3.335	3.499
8	.706	1.397	1.860	2.306	2.896	3.016	3.206	3.355
9	.703	1.383	1.833	2.262	2.821	2.933	3.111	3.250
10	.700	1.372	1.812	2.228	2.764	2.870	3.038	3.169
11	.697	1.363	1.796	2.201	2.718	2.820	2.981	3.106
12	.695	1.356	1.782	2.179	2.681	2.779	2.934	3.055
13	.694	1.350	1.771	2.160	2.650	2.746	2.896	3.012
14	.692	1.345	1.761	2.145	2.624	2.718	2.864	2.977
15	.691	1.341	1.753	2.131	2.602	2.694	2.837	2.947
16	.690	1.337	1.746	2.120	2.583	2.673	2.813	2.921
17	.689	1.333	1.740	2.110	2.567	2.655	2.793	2.898
18	.688	1.330	1.734	2.101	2.552	2.639	2.775	2.878
19	.688	1.328	1.729	2.093	2.539	2.625	2.759	2.861
20	.687	1.325	1.725	2.086	2.528	2.613	2.744	2.845
21	.686	1.323	1.721	2.080	2.518	2.601	2.732	2.831
22	.686	1.321	1.717	2.074	2.508	2.591	2.720	2.819
23	.685	1.319	1.714	2.069	2.500	2.582	2.710	2.807
24	.685	1.318	1.711	2.064	2.492	2.574	2.700	2.797
25	.684	1.316	1.708	2.060	2.485	2.566	2.692	2.787
26	.684	1.315	1.706	2.056	2.479	2.559	2.684	2.779
27	.684	1.314	1.703	2.052	2.473	2.552	2.676	2.771
28	.683	1.313	1.701	2.048	2.467	2.546	2.669	2.763
29	.683	1.311	1.699	2.045	2.462	2.541	2.663	2.756
30	.683	1.310	1.697	2.042	2.457	2.536	2.657	2.750
40	.681	1.303	1.684	2.021	2.423	2.499	2.616	2.704
60	.679	1.296	1.671	2.000	2.390	2.463	2.575	2.660
120	.677	1.289	1.658	1.980	2.358	2.428	2.536	2.617
∞	.674	1.282	1.645	1.960	2.326	2.394	2.498	2.576

TABLE 5   Percentage Points of χ^2 Distributions

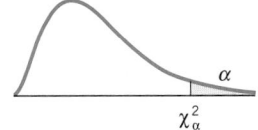

d.f. α	.99	.975	.95	.90	.50	.10	.05	.025	.01
1	.0002	.001	.004	.02	.45	2.71	3.84	5.02	6.63
2	.02	.05	.10	.21	1.39	4.61	5.99	7.38	9.21
3	.11	.22	.35	.58	2.37	6.25	7.81	9.35	11.34
4	.30	.48	.71	1.06	3.36	7.78	9.49	11.14	13.28
5	.55	.83	1.15	1.61	4.35	9.24	11.07	12.83	15.09
6	.87	1.24	1.64	2.20	5.35	10.64	12.59	14.45	16.81
7	1.24	1.69	2.17	2.83	6.35	12.02	14.07	16.01	18.48
8	1.65	2.18	2.73	3.49	7.34	13.36	15.51	17.53	20.09
9	2.09	2.70	3.33	4.17	8.34	14.68	16.92	19.02	21.67
10	2.56	3.24	3.94	4.87	9.34	15.99	18.31	20.48	23.21
11	3.05	3.81	4.57	5.58	10.34	17.28	19.68	21.92	24.72
12	3.57	4.40	5.23	6.30	11.34	18.55	21.03	23.34	26.22
13	4.11	5.01	5.89	7.04	12.34	19.81	22.36	24.74	27.69
14	4.66	5.62	6.57	7.79	13.34	21.06	23.68	26.12	29.14
15	5.23	6.26	7.26	8.55	14.34	22.31	25.00	27.49	30.58
16	5.81	6.90	7.96	9.31	15.34	23.54	26.30	28.85	32.00
17	6.41	7.56	8.67	10.09	16.34	24.77	27.59	30.19	33.41
18	7.01	8.23	9.39	10.86	17.34	25.99	28.87	31.53	34.81
19	7.63	8.90	10.12	11.65	18.34	27.20	30.14	32.85	36.19
20	8.26	9.59	10.85	12.44	19.34	28.41	31.41	34.17	37.57
21	8.90	10.28	11.59	13.24	20.34	29.62	32.67	35.48	38.93
22	9.54	10.98	12.34	14.04	21.34	30.81	33.92	36.78	40.29
23	10.20	11.69	13.09	14.85	22.34	32.01	35.17	38.08	41.64
24	10.86	12.40	13.85	15.66	23.34	33.20	36.42	39.36	42.98
25	11.52	13.11	14.61	16.47	24.34	34.38	37.65	40.65	44.31
26	12.20	13.84	15.38	17.29	25.34	35.56	38.89	41.92	45.64
27	12.88	14.57	16.15	18.11	26.34	36.74	40.11	43.19	46.96
28	13.56	15.30	16.93	18.94	27.34	37.92	41.34	44.46	48.28
29	14.26	16.04	17.71	19.77	28.34	39.09	42.56	45.72	49.59
30	14.95	16.78	18.49	20.60	29.34	40.26	43.77	46.98	50.89
40	22.16	24.42	26.51	29.05	39.34	51.81	55.76	59.34	63.69
50	29.71	32.35	34.76	37.69	49.33	63.17	67.50	71.42	76.15
60	37.48	40.47	43.19	46.46	59.33	74.40	79.08	83.30	88.38
70	45.44	48.75	51.74	55.33	69.33	85.53	90.53	95.02	100.43
80	53.54	57.15	60.39	64.28	79.33	96.58	101.88	106.63	112.33
90	61.75	65.64	69.13	73.29	89.33	107.57	113.15	118.14	124.12
100	70.06	74.22	77.93	82.36	99.33	118.50	124.34	129.56	135.81

TABLE 6 Percentage Points of $F(v_1, v_2)$ Distributions

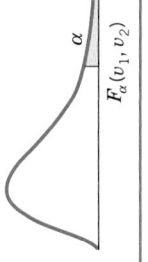

$\alpha = .10$

v_2 \ v_1	1	2	3	4	5	6	7	8	9	10	12	15	20	25	30	40	60
1	39.86	49.50	53.59	55.83	57.24	58.20	58.91	59.44	59.86	60.19	60.71	61.22	61.74	62.05	62.26	62.53	62.79
2	8.53	9.00	9.16	9.24	9.29	9.33	9.35	9.37	9.38	9.39	9.41	9.42	9.44	9.45	9.46	9.47	9.47
3	5.54	5.46	5.39	5.34	5.31	5.28	5.27	5.25	5.24	5.23	5.22	5.20	5.18	5.17	5.17	5.16	5.15
4	4.54	4.32	4.19	4.11	4.05	4.01	3.98	3.95	3.94	3.92	3.90	3.87	3.84	3.83	3.82	3.80	3.79
5	4.06	3.78	3.62	3.52	3.45	3.40	3.37	3.34	3.32	3.30	3.27	3.24	3.21	3.19	3.17	3.16	3.14
6	3.78	3.46	3.29	3.18	3.11	3.05	3.01	2.98	2.96	2.94	2.90	2.87	2.84	2.81	2.80	2.78	2.76
7	3.59	3.26	3.07	2.96	2.88	2.83	2.78	2.75	2.72	2.70	2.67	2.63	2.59	2.57	2.56	2.54	2.51
8	3.46	3.11	2.92	2.81	2.73	2.67	2.62	2.59	2.56	2.54	2.50	2.46	2.42	2.40	2.38	2.36	2.34
9	3.36	3.01	2.81	2.69	2.61	2.55	2.51	2.47	2.44	2.42	2.38	2.34	2.30	2.27	2.25	2.23	2.21
10	3.29	2.92	2.73	2.61	2.52	2.46	2.41	2.38	2.35	2.32	2.28	2.24	2.20	2.17	2.16	2.13	2.11
11	3.23	2.86	2.66	2.54	2.45	2.39	2.34	2.30	2.27	2.25	2.21	2.17	2.12	2.10	2.08	2.05	2.03
12	3.18	2.81	2.61	2.48	2.39	2.33	2.28	2.24	2.21	2.19	2.15	2.10	2.06	2.03	2.01	1.99	1.96
13	3.14	2.76	2.56	2.43	2.35	2.28	2.23	2.20	2.16	2.14	2.10	2.05	2.01	1.98	1.96	1.93	1.90
14	3.10	2.73	2.52	2.39	2.31	2.24	2.19	2.15	2.12	2.10	2.05	2.01	1.96	1.93	1.91	1.89	1.86
15	3.07	2.70	2.49	2.36	2.27	2.21	2.16	2.12	2.09	2.06	2.02	1.97	1.92	1.89	1.87	1.85	1.82
16	3.05	2.67	2.46	2.33	2.24	2.18	2.13	2.09	2.06	2.03	1.99	1.94	1.89	1.86	1.84	1.81	1.78
17	3.03	2.64	2.44	2.31	2.22	2.15	2.10	2.06	2.03	2.00	1.96	1.91	1.86	1.83	1.81	1.78	1.75
18	3.01	2.62	2.42	2.29	2.20	2.13	2.08	2.04	2.00	1.98	1.93	1.89	1.84	1.80	1.78	1.75	1.72
19	2.99	2.61	2.40	2.27	2.18	2.11	2.06	2.02	1.98	1.96	1.91	1.86	1.81	1.78	1.76	1.73	1.70
20	2.97	2.59	2.38	2.25	2.16	2.09	2.04	2.00	1.96	1.94	1.89	1.84	1.79	1.76	1.74	1.71	1.68
21	2.96	2.57	2.36	2.23	2.14	2.08	2.02	1.98	1.95	1.92	1.87	1.83	1.78	1.74	1.72	1.69	1.66
22	2.95	2.56	2.35	2.22	2.13	2.06	2.01	1.97	1.93	1.90	1.86	1.81	1.76	1.73	1.70	1.67	1.64
23	2.94	2.55	2.34	2.21	2.11	2.05	1.99	1.95	1.92	1.89	1.84	1.80	1.74	1.71	1.69	1.66	1.62
24	2.93	2.54	2.33	2.19	2.10	2.04	1.98	1.94	1.91	1.88	1.83	1.78	1.73	1.70	1.67	1.64	1.61
25	2.92	2.53	2.32	2.18	2.09	2.02	1.97	1.93	1.89	1.87	1.82	1.77	1.72	1.68	1.66	1.63	1.59
26	2.91	2.52	2.31	2.17	2.08	2.01	1.96	1.92	1.88	1.86	1.81	1.76	1.71	1.67	1.65	1.61	1.58
27	2.90	2.51	2.30	2.17	2.07	2.00	1.95	1.91	1.87	1.85	1.80	1.75	1.70	1.66	1.64	1.60	1.57
28	2.89	2.50	2.29	2.16	2.06	2.00	1.94	1.90	1.87	1.84	1.79	1.74	1.69	1.65	1.63	1.59	1.56
29	2.89	2.50	2.28	2.15	2.06	1.99	1.93	1.89	1.86	1.83	1.78	1.73	1.68	1.64	1.62	1.58	1.55
30	2.88	2.49	2.28	2.14	2.05	1.98	1.93	1.88	1.85	1.82	1.77	1.72	1.67	1.63	1.61	1.57	1.54
40	2.84	2.44	2.23	2.09	2.00	1.93	1.87	1.83	1.79	1.76	1.71	1.66	1.61	1.57	1.54	1.51	1.47
60	2.79	2.39	2.18	2.04	1.95	1.87	1.82	1.77	1.74	1.71	1.66	1.60	1.54	1.50	1.48	1.44	1.40
120	2.75	2.35	2.13	1.99	1.90	1.82	1.77	1.72	1.68	1.65	1.60	1.55	1.48	1.45	1.41	1.37	1.32
∞	2.71	2.30	2.08	1.94	1.85	1.77	1.72	1.67	1.63	1.60	1.55	1.49	1.42	1.38	1.34	1.30	1.24

$F_\alpha(v_1, v_2)$

TABLE 6 (Continued)

$\boxed{\alpha = .05}$

$F_\alpha(v_1, v_2)$

v_2 \\ v_1	1	2	3	4	5	6	7	8	9	10	12	15	20	25	30	40	60
1	161.5	199.5	215.7	224.6	230.2	234.0	236.8	238.9	240.5	241.9	243.9	246.0	248.0	249.3	250.1	251.1	252.2
2	18.51	19.00	19.16	19.25	19.30	19.33	19.35	19.37	19.38	19.40	19.41	19.43	19.45	19.46	19.46	19.47	19.48
3	10.13	9.55	9.28	9.12	9.01	8.94	8.89	8.85	8.81	8.79	8.74	8.70	8.66	8.63	8.62	8.59	8.57
4	7.71	6.94	6.59	6.39	6.26	6.16	6.09	6.04	6.00	5.96	5.91	5.86	5.80	5.77	5.75	5.72	5.69
5	6.61	5.79	5.41	5.19	5.05	4.95	4.88	4.82	4.77	4.74	4.68	4.62	4.56	4.52	4.50	4.46	4.43
6	5.99	5.14	4.76	4.53	4.39	4.28	4.21	4.15	4.10	4.06	4.00	3.94	3.87	3.83	3.81	3.77	3.74
7	5.59	4.74	4.35	4.12	3.97	3.87	3.79	3.73	3.68	3.64	3.57	3.51	3.44	3.40	3.38	3.34	3.30
8	5.32	4.46	4.07	3.84	3.69	3.58	3.50	3.44	3.39	3.35	3.28	3.22	3.15	3.11	3.08	3.04	3.01
9	5.12	4.26	3.86	3.63	3.48	3.37	3.29	3.23	3.18	3.14	3.07	3.01	2.94	2.89	2.86	2.83	2.79
10	4.96	4.10	3.71	3.48	3.33	3.22	3.14	3.07	3.02	2.98	2.91	2.85	2.77	2.73	2.70	2.66	2.62
11	4.84	3.98	3.59	3.36	3.20	3.09	3.01	2.95	2.90	2.85	2.79	2.72	2.65	2.60	2.57	2.53	2.49
12	4.75	3.89	3.49	3.26	3.11	3.00	2.91	2.85	2.80	2.75	2.69	2.62	2.54	2.50	2.47	2.43	2.38
13	4.67	3.81	3.41	3.18	3.03	2.92	2.83	2.77	2.71	2.67	2.60	2.53	2.46	2.41	2.38	2.34	2.30
14	4.60	3.74	3.34	3.11	2.96	2.85	2.76	2.70	2.65	2.60	2.53	2.46	2.39	2.34	2.31	2.27	2.22
15	4.54	3.68	3.29	3.06	2.90	2.79	2.71	2.64	2.59	2.54	2.48	2.40	2.33	2.28	2.25	2.20	2.16
16	4.49	3.63	3.24	3.01	2.85	2.74	2.66	2.59	2.54	2.49	2.42	2.35	2.28	2.23	2.19	2.15	2.11
17	4.45	3.59	3.20	2.96	2.81	2.70	2.61	2.55	2.49	2.45	2.38	2.31	2.23	2.18	2.15	2.10	2.06
18	4.41	3.55	3.16	2.93	2.77	2.66	2.58	2.51	2.46	2.41	2.34	2.27	2.19	2.14	2.11	2.06	2.02
19	4.38	3.52	3.13	2.90	2.74	2.63	2.54	2.48	2.42	2.38	2.31	2.23	2.16	2.11	2.07	2.03	1.98
20	4.35	3.49	3.10	2.87	2.71	2.60	2.51	2.45	2.39	2.35	2.28	2.20	2.12	2.07	2.04	1.99	1.95
21	4.32	3.47	3.07	2.84	2.68	2.57	2.49	2.42	2.37	2.32	2.25	2.18	2.10	2.05	2.01	1.96	1.92
22	4.30	3.44	3.05	2.82	2.66	2.55	2.46	2.40	2.34	2.30	2.23	2.15	2.07	2.02	1.98	1.94	1.89
23	4.28	3.42	3.03	2.80	2.64	2.53	2.44	2.37	2.32	2.27	2.20	2.13	2.05	2.00	1.96	1.91	1.86
24	4.26	3.40	3.01	2.78	2.62	2.51	2.42	2.36	2.30	2.25	2.18	2.11	2.03	1.97	1.94	1.89	1.84
25	4.24	3.39	2.99	2.76	2.60	2.49	2.40	2.34	2.28	2.24	2.16	2.09	2.01	1.96	1.92	1.87	1.82
26	4.23	3.37	2.98	2.74	2.59	2.47	2.39	2.32	2.27	2.22	2.15	2.07	1.99	1.94	1.90	1.85	1.80
27	4.21	3.35	2.96	2.73	2.57	2.46	2.37	2.31	2.25	2.20	2.13	2.06	1.97	1.92	1.88	1.84	1.79
28	4.20	3.34	2.95	2.71	2.56	2.45	2.36	2.29	2.24	2.19	2.12	2.04	1.96	1.91	1.87	1.82	1.77
29	4.18	3.33	2.93	2.70	2.55	2.43	2.35	2.28	2.22	2.18	2.10	2.03	1.94	1.89	1.85	1.81	1.75
30	4.17	3.32	2.92	2.69	2.53	2.42	2.33	2.27	2.21	2.16	2.09	2.01	1.93	1.88	1.84	1.79	1.74
40	4.08	3.23	2.84	2.61	2.45	2.34	2.25	2.18	2.12	2.08	2.00	1.92	1.84	1.78	1.74	1.69	1.64
60	4.00	3.15	2.76	2.53	2.37	2.25	2.17	2.10	2.04	1.99	1.92	1.84	1.75	1.69	1.65	1.59	1.53
120	3.92	3.07	2.68	2.45	2.29	2.18	2.09	2.02	1.96	1.91	1.83	1.75	1.66	1.60	1.55	1.50	1.43
∞	3.84	3.00	2.61	2.37	2.21	2.10	2.01	1.94	1.88	1.83	1.75	1.67	1.57	1.51	1.46	1.39	1.32

TABLE 7 Selected Tail Probabilities for the Null Distribution of Wilcoxon's Rank-Sum Statistic

$$P = P[W_s \geq x] = P[W_s \leq x^*]$$

Smaller Sample Size = 2

Larger Sample Size

	3			4			5			6	
x	P	x^*	x	P	x^*	x	P	x^*	x	P	x^*
8	.200	4	10	.133	4	11	.190	5	13	.143	5
9	.100	3	11	.067	3	12	.095	4	14	.071	4
10	0	2	12	0	2	13	.048	3	15	.036	3
						14	0	2	16	0	2

	7			8			9			10	
x	P	x^*	x	P	x^*	x	P	x^*	x	P	x^*
15	.111	5	16	.133	6	18	.109	6	19	.136	7
16	.056	4	17	.089	5	19	.073	5	20	.091	6
17	.028	3	18	.044	4	20	.036	4	21	.061	5
18	0	2	19	.022	3	21	.018	3	22	.030	4
			20	0	2	22	0	2	23	.015	3

Smaller Sample Size = 3

Larger Sample Size

	3			4			5			6	
x	P	x^*	x	P	x^*	x	P	x^*	x	P	x^*
13	.200	8	16	.114	8	18	.125	9	20	.131	10
14	.100	7	17	.057	7	19	.071	8	21	.083	9
15	.050	6	18	.029	6	20	.036	7	22	.048	8
16	0	5	19	0	5	21	.018	6	23	.024	7
						22	0	5	24	.012	6
									25	0	5

	7			8			9			10	
x	P	x^*	x	P	x^*	x	P	x^*	x	P	x^*
22	.133	11	24	.139	12	27	.105	12	29	.108	13
23	.092	10	25	.097	11	28	.073	11	30	.080	12
24	.058	9	26	.067	10	29	.050	10	31	.056	11
25	.033	8	27	.042	9	30	.032	9	32	.038	10
26	.017	7	28	.024	8	31	.018	8	33	.024	9
27	.008	6	29	.012	7	32	.009	7	34	.014	8
28	0	5	30	.006	6				35	.007	7
			31	0	5						

TABLE 7   *(Continued)*

Smaller Sample Size = 4

Larger Sample Size											
4			5			6			7		
x	P	x*	x	P	x*	x	P	x*	x	P	x*
22	.171	14	25	.143	15	28	.129	16	31	.115	17
23	.100	13	26	.095	14	29	.086	15	32	.082	16
24	.057	12	27	.056	13	30	.057	14	33	.055	15
25	.029	11	28	.032	12	31	.033	13	34	.036	14
26	.014	10	29	.016	11	32	.019	12	35	.021	13
27	0	9	30	.008	10	33	.010	11	36	.012	12
			31	0	9				37	.006	11

8			9			10		
x	P	x*	x	P	x*	x	P	x*
34	.107	18	36	.130	20	39	.120	21
35	.077	17	37	.099	19	40	.094	20
36	.055	16	38	.074	18	41	.071	19
37	.036	15	39	.053	17	42	.053	18
38	.024	14	40	.038	16	43	.038	17
39	.014	13	41	.025	15	44	.027	16
40	.008	12	42	.017	14	45	.018	15
			43	.010	13	46	.012	14
						47	.007	13

Smaller Sample Size = 5

Larger Sample Size											
5			6			7			8		
x	P	x*	x	P	x*	x	P	x*	x	P	x*
34	.111	21	37	.123	23	41	.101	24	44	.111	26
35	.075	20	38	.089	22	42	.074	23	45	.085	25
36	.048	19	39	.063	21	43	.053	22	46	.064	24
37	.028	18	40	.041	20	44	.037	21	47	.047	23
38	.016	17	41	.026	19	45	.024	20	48	.033	22
39	.008	16	42	.015	18	46	.015	19	49	.023	21
			43	.009	17	47	.009	18	50	.015	20
									51	.009	19

TABLE 7   *(Continued)*

	Larger Sample Size			
	9		10	
x	P	x*	x	P	x*
47	.120	28	51	.103	29
48	.095	27	52	.082	28
49	.073	26	53	.065	27
50	.056	25	54	.050	26
51	.041	24	55	.038	25
52	.030	23	56	.028	24
53	.021	22	57	.020	23
54	.014	21	58	.014	22
55	.009	20	59	.010	21

Smaller Sample Size = 6

	Larger Sample Size										
	6			7			8			9	
x	P	x*	x	P	x*	x	P	x*	x	P	x*
47	.120	31	51	.117	33	55	.114	35	59	.112	37
48	.090	30	52	.090	32	56	.091	34	60	.091	36
49	.066	29	53	.069	31	57	.071	33	61	.072	35
50	.047	28	54	.051	30	58	.054	32	62	.057	34
51	.032	27	55	.037	29	59	.041	31	63	.044	33
52	.021	26	56	.026	28	60	.030	30	64	.033	32
53	.013	25	57	.017	27	61	.021	29	65	.025	31
54	.008	24	58	.011	26	62	.015	28	66	.018	30
			59	.007	25	63	.010	27	67	.013	29
									68	.009	28

	10	
x	P	x*
63	.110	39
64	.090	38
65	.074	37
66	.059	36
67	.047	35
68	.036	34
69	.028	33
70	.021	32
71	.016	31
72	.011	30
73	.008	29

TABLE 7   *(Continued)*

Smaller Sample Size = 7

Larger Sample Size											
7			8			9			10		
x	P	x*	x	P	x*	x	P	x*	x	P	x*
63	.104	42	67	.116	45	72	.105	47	76	.115	50
64	.082	41	68	.095	44	73	.087	46	77	.097	49
65	.064	40	69	.076	43	74	.071	45	78	.081	48
66	.049	39	70	.060	42	75	.057	44	79	.067	47
67	.036	38	71	.047	41	76	.045	43	80	.054	46
68	.027	37	72	.036	40	77	.036	42	81	.044	45
69	.019	36	73	.027	39	78	.027	41	82	.035	44
70	.013	35	74	.020	38	79	.021	40	83	.028	43
71	.009	34	75	.014	37	80	.016	39	84	.022	42
			76	.010	36	81	.011	38	85	.017	41
						82	.008	37	86	.012	40
									87	.009	39

Smaller Sample Size = 8

Larger Sample Size								
8			9			10		
x	P	x*	x	P	x*	x	P	x*
80	.117	56	86	.100	58	91	.102	61
81	.097	55	87	.084	57	92	.086	60
82	.080	54	88	.069	56	93	.073	59
83	.065	53	89	.057	55	94	.061	58
84	.052	52	90	.046	54	95	.051	57
85	.041	51	91	.037	53	96	.042	56
86	.032	50	92	.030	52	97	.034	55
87	.025	49	93	.023	51	98	.027	54
88	.019	48	94	.018	50	99	.022	53
89	.014	47	95	.014	49	100	.017	52
90	.010	46	96	.010	48	101	.013	51
						102	.010	50

TABLE 7   *(Continued)*

Smaller Sample Size = 9						Smaller Sample Size = 10		
Larger Sample Size						Larger Sample Size		
9			10			10		
x	P	x^*	x	P	x^*	x	P	x^*
100	.111	71	106	.106	74	122	.109	88
101	.095	70	107	.091	73	123	.095	87
102	.081	69	108	.078	72	124	.083	86
103	.068	68	109	.067	71	125	.072	85
104	.057	67	110	.056	70	126	.062	84
105	.047	66	111	.047	69	127	.053	83
106	.039	65	112	.039	68	128	.045	82
107	.031	64	113	.033	67	129	.038	81
108	.025	63	114	.027	66	130	.032	80
109	.020	62	115	.022	65	131	.026	79
110	.016	61	116	.017	64	132	.022	78
111	.012	60	117	.014	63	133	.018	77
112	.009	59	118	.011	62	134	.014	76
			119	.009	61	135	.012	75
						136	.009	74

Source: Adapted from C. Kraft and C. van Eeden, *A Nonparametric Introduction to Statistics,* Macmillan, New York, 1968.

TABLE 8   Selected Tail Probabilities for the Null Distribution
of Wilcoxon's Signed-Rank Statistic

$$P = P[T^+ \geq x] = P[T^+ \leq x^*]$$

x	P	x^*	x	P	x^*	x	P	x^*	x	P	x^*
	$n = 3$			$n = 4$			$n = 5$			$n = 6$	
5	.250	1	8	.188	2	12	.156	3	17	.109	4
6	.125	0	9	.125	1	13	.094	2	18	.078	3
7	0		10	.062	0	14	.062	1	19	.047	2
			11	0		15	.031	0	20	.031	1
						16	0		21	.016	0
									22	0	

x	P	x^*	x	P	x^*	x	P	x^*	x	P	x^*
	$n = 7$			$n = 8$			$n = 9$			$n = 10$	
22	.109	6	27	.125	9	34	.102	11	40	.116	15
23	.078	5	28	.098	8	35	.082	10	41	.097	14
24	.055	4	29	.074	7	36	.064	9	42	.080	13
25	.039	3	30	.055	6	37	.049	8	43	.065	12
26	.023	2	31	.039	5	38	.037	7	44	.053	11
27	.016	1	32	.027	4	39	.027	6	45	.042	10
28	.008	0	33	.020	3	40	.020	5	46	.032	9
			34	.012	2	41	.014	4	47	.024	8
			35	.008	1	42	.010	3	48	.019	7
									49	.014	6
									50	.010	5

x	P	x^*	x	P	x^*	x	P	x^*	x	P	x^*
	$n = 11$			$n = 12$			$n = 13$			$n = 14$	
48	.103	18	56	.102	22	64	.108	27	73	.108	32
49	.087	17	57	.088	21	65	.095	26	74	.097	31
50	.074	16	58	.076	20	66	.084	25	75	.086	30
51	.062	15	59	.065	19	67	.073	24	76	.077	29
52	.051	14	60	.055	18	68	.064	23	77	.068	28
53	.042	13	61	.046	17	69	.055	22	78	.059	27
54	.034	12	62	.039	16	70	.047	21	79	.052	26
55	.027	11	63	.032	15	71	.040	20	80	.045	25
56	.021	10	64	.026	14	72	.034	19	81	.039	24
57	.016	9	65	.021	13	73	.029	18	82	.034	23
58	.012	8	66	.017	12	74	.024	17	83	.029	22
59	.009	7	67	.013	11	75	.020	16	84	.025	21
			68	.010	10	76	.016	15	85	.021	20
						77	.013	14	86	.018	19
						78	.011	13	87	.015	18
						79	.009	12	88	.012	17
									89	.010	16

TABLE 8 *(Continued)*

n	=	15
x	P	x*
83	.104	37
84	.094	36
85	.084	35
86	.076	34
87	.068	33
88	.060	32
89	.053	31
90	.047	30
91	.042	29
92	.036	28
93	.032	27
94	.028	26
95	.024	25
96	.021	24
97	.018	23
98	.015	22
99	.013	21
100	.011	20
101	.009	19

Source: Adapted from C. Kraft and C. van Eeden, *A Nonparametric Introduction to Statistics,* Macmillan, New York, 1968.

Data Bank

The Jump River Electric Company serves several counties in northern Wisconsin. Much of the area is forest and lakes. The data on power outages from a recent summer include date, time, duration of outage (hours), and cause.

TABLE D.1 Power Outages

Date	Time	Duration	Cause
6/11	4:00 PM	5.50	Trees and limbs
6/12	8:00 PM	1.50	Trees and limbs
6/16	8:30 AM	2.00	Trees and limbs
6/17	5:30 AM	2.00	Trees and limbs
6/17	5:00 PM	8.00	Windstorm
6/21	4:30 PM	2.00	Trees and limbs
6/26	3:00 AM	3.00	Trees and limbs
6/26	2:00 PM	1.75	Unknown
6/26	12:00 AM	2.00	Lightning blew up transformer
7/03	6:00 PM	1.50	Trees and limbs
7/04	9:00 AM	2.50	Unknown
7/04	5:00 PM	1.50	Trees and limbs
7/05	5:00 AM	1.50	Trees and limbs
7/08	7:00 AM	3.50	Lightning
7/20	12:00 AM	1.50	Unknown
7/21	9:00 AM	1.50	Animal
7/28	12:00 PM	1.00	Animal

TABLE D.1   *(Continued)*

Date	Time	Duration	Cause
7/30	7:00 PM	1.00	Squirrel on transformer
7/31	7:30 AM	0.50	Squirrel on cutout
8/04	6:00 AM	2.50	Trees and limbs
8/06	8:00 PM	2.00	Beaver
8/09	5:30 AM	1.50	Fuse-flying squirrel
8/11	12:00 AM	3.00	Beaver-cut trees
8/13	1:00 AM	1.00	Unknown
8/15	12:30 AM	1.50	Animal
8/18	8:00 AM	1.50	Trees and limbs
8/20	4:00 AM	2.00	Transformer fuse
8/25	9:00 AM	1.00	Animal
8/26	2:00 AM	10.00	Lightning
8/27	3:00 AM	1.00	Trees and limbs

Madison recruits for the fire department need to complete a timed test that simulates working conditions. It includes placing a ladder against a building, pulling out a section of fire hose, dragging a weighted object, and crawling in a simulated attic environment. The times, in seconds, for recruits to complete the test for a Madison firefighter are shown in Table D.2.

TABLE D.2   Time to Complete Firefighters Physical Test (seconds)

425	389	380	421	438	331	368	417	403	416	385	315
427	417	386	386	378	300	321	286	269	225	268	317
287	256	334	342	269	226	291	280	221	283	302	308
296	266	238	286	317	276	254	278	247	336	296	259
270	302	281	228	317	312	327	288	395	240	264	246
294	254	222	285	254	264	277	266	228	347	322	232
365	356	261	293	354	236	285	303	275	403	268	250
279	400	370	399	438	287	363	350	278	278	234	266
319	276	291	352	313	262	289	273	317	328	292	279
289	312	334	294	297	304	240	303	255	305	252	286
297	353	350	276	333	285	317	296	276	247	339	328
267	305	291	269	386	264	299	261	284	302	342	304
336	291	294	323	320	289	339	292	373	410	257	406
374	268										

Natural resource managers have attempted to use the Satellite Landsat Multispectral Scanner data for improved land-cover classification. The intensities of reflected light recorded on the near-infrared band of a thermatic mapper are given in Table D.3. Table D.3a gives readings from areas known to consist of forest and Table D.3b readings from urban areas.

TABLE D.3a   Near Infrared Light Reflected from Forest Areas

77	77	78	78	81	81	82	82	82	82	82	83	83	84	84	84	84	85	
86	86	86	86	86	87	87	87	87	87	87	87	89	89	89	89	89	89	89
90	90	90	91	91	91	91	91	91	91	91	91	91	93	93	93	93	93	93
94	94	94	94	94	94	94	94	94	94	94	94	95	95	95	95	95	96	96
96	96	96	96	97	97	97	97	97	97	97	97	97	98	99	100	100	100	100
100	100	100	100	100	101	101	101	101	101	101	102	102	102	102	102			
102	103	103	104	104	104	105	107											

TABLE D.3b   Near Infrared Light Reflected from Urban Areas

71	72	73	74	75	77	78	79	79	79	79	80	80	80	81	81	81	82	82	82	82
84	84	84	84	84	84	85	85	85	85	85	85	86	86	87	88	90	91	94		

Beginning accounting students need to learn to audit in a computerized environment. A sample of beginning accounting students took a test that is summarized by two scores shown in Table D.4: the Computer Attitude Scale (CAS), based on 20 questions, and the Computer Anxiety Rating Scale (CARS), based on 19 questions. (Courtesy of Douglas Stein.) Males are coded as 1 and females as 0.

TABLE D.4   Computer Attitude and Anxiety Scores

Gender	CAS	CARS	Gender	CAS	CARS
0	2.85	2.90	1	3.30	3.47
1	2.60	2.32	1	2.90	3.05
0	2.20	1.00	1	2.60	2.68
1	2.65	2.58	0	2.25	1.90
1	2.60	2.58	0	1.90	1.84
1	3.20	3.05	1	2.20	1.74
1	3.65	3.74	0	2.30	2.58
0	2.55	1.90	0	1.80	1.58
1	3.15	3.32	1	3.05	2.47
1	2.80	2.74	1	3.15	3.32
0	2.40	2.37	0	2.80	2.90
1	3.20	3.11	0	2.35	2.42
0	3.05	3.32	1	3.70	3.47
1	2.60	2.79	1	2.60	4.00
1	3.35	2.95	0	3.50	3.42
0	3.75	3.79	0	2.95	2.53
0	3.00	3.26	1	2.80	2.68
1	2.80	3.21			

Data were collected on students taking the course Conditioning 1, designed to introduce students to a variety of training techniques to improve cardio-respiratory fitness, muscular strength, and flexibility (Table D.5). (Courtesy of K. Baldridge.)

c1   Gender (1 = male, 2 = female)
c2   Pretest percent body fat
c3   Posttest percent body fat
c4   Pretest time to run 1.5 miles (seconds)
c5   Posttest time to run 1.5 miles (seconds)
c6   Pretest time to row 2.5 kilometers (seconds)
c7   Posttest time to row 2.5 kilometers (seconds)
c8   Pretest number of situps completed in 1 minute
c9   Posttest number of situps completed in 1 minute

TABLE D.5   Physical Fitness Improvement

Gender	Pretest % Fat	Posttest % Fat	Pretest Run	Posttest Run	Pretest Row	Posttest Row	Pretest Situps	Posttest Situps
1	15.1	12.4	575	480	621	559	60	67
1	17.1	18.5	766	672	698	595	42	45
2	25.5	15.0	900	750	840	725	32	36
2	19.5	17.0	715	610	855	753	28	38
2	21.7	20.6	705	585	846	738	46	54
1	17.7	18.5	820	670	630	648	18	41
2	22.7	17.2	880	745	860	788	22	33
2	26.6	22.4	840	725	785	745	29	39
2	36.4	32.5	1065	960	780	749	27	40
1	9.7	8.0	630	565	673	588	32	49
2	31.0	25.0	870	780	746	689	39	54
2	6.0	6.0	580	494	756	714	37	49
2	25.1	22.8	1080	806	852	838	43	38
1	15.1	13.4	720	596	674	576	48	62
2	23.0	21.1	780	718	846	783	30	38
2	9.7	9.7	945	700	890	823	38	44
1	7.0	6.1	706	657	652	521	52	63
1	16.6	15.1	650	567	740	615	38	47
2	21.7	19.5	686	662	762	732	29	47
2	23.7	21.7	758	718	830	719	35	42
2	24.7	21.7	870	705	754	734	24	37
1	11.6	8.9	480	460	640	587	55	60
2	19.5	16.0	715	655	703	731	36	44
1	4.8	4.2	545	530	625	571	40	45
2	31.1	21.6	840	790	745	728	30	36

TABLE D.5   *(Continued)*

Gender	Pretest % Fat	Posttest % Fat	Pretest Run	Posttest Run	Pretest Row	Posttest Row	Pretest Situps	Posttest Situps
1	23.2	19.9	617	622	637	620	35	32
1	12.5	12.2	635	600	805	736	37	52
2	29.3	21.7	790	715	821	704	41	42
2	19.5	16.0	750	702	1043	989	42	45
1	10.7	9.8	622	567	706	645	40	46
2	26.6	21.7	722	725	741	734	49	61
2	27.6	23.7	641	598	694	682	40	47
1	25.9	18.8	708	609	685	593	20	21
1	27.5	16.0	675	637	694	682	35	37
1	9.4	7.6	618	566	610	579	36	50
1	12.5	9.8	613	552	610	575	46	50
2	27.5	30.9	705	660	746	691	31	36
2	18.3	15.0	853	720	748	694	31	34
1	6.6	4.7	496	476	623	569	45	45
2	23.7	21.7	860	750	758	711	42	55
2	23.2	22.1	905	636	759	726	26	30
2	18.0	14.8	900	805	823	759	31	29
2	22.7	18.3	767	741	808	753	28	32
1	19.9	14.8	830	620	632	586	39	45
1	6.6	5.7	559	513	647	602	44	49
1	14.3	9.8	699	652	638	602	34	35
1	21.7	18.5	765	735	674	615	41	48
1	13.9	9.4	590	570	599	571	40	42
1	20.1	19.3	770	672	675	611	37	48
1	15.1	8.9	602	560	656	578	47	58
2	14.8	11.0	741	610	768	687	34	41
2	23.4	19.1	723	641	711	695	36	41
2	23.7	16.3	648	601	802	740	44	51
1	28.0	17.7	842	702	790	765	26	29
1	4.7	4.7	558	540	660	600	40	46
2	26.6	23.7	750	565	720	670	36	52
2	21.7	18.3	608	592	707	697	42	46
1	11.6	8.9	537	495	610	572	55	60
2	22.7	18.3	855	694	800	712	41	50
2	21.7	18.3	630	614	785	743	50	55
2	25.9	21.9	902	820	771	717	34	38
1	28.4	21.7	780	664	756	703	57	64
2	21.9	20.8	665	670	673	667	41	44
2	27.5	22.7	675	646	689	674	49	50
1	5.7	3.8	473	472	551	546	53	53
2	22.7	22.0	715	682	678	672	40	43
2	33.2	25.7	795	740	817	721	30	31
2	16.0	13.6	688	615	811	705	45	49
1	14.3	11.6	530	497	589	570	39	50

TABLE D.5   *(Continued)*

Gender	Pretest % Fat	Posttest % Fat	Pretest Run	Posttest Run	Pretest Row	Posttest Row	Pretest Situps	Posttest Situps
1	22.7	19.5	840	705	788	780	29	32
2	25.0	18.3	690	618	816	701	42	50
1	4.2	3.2	545	527	577	543	55	59
2	25.7	21.7	760	727	849	724	39	44
1	13.9	9.4	620	515	689	580	32	41
1	12.2	7.6	605	564	661	614	35	38
1	21.7	20.6	688	625	750	686	39	32
1	7.0	5.1	590	529	631	619	60	65
2	34.6	27.5	720	694	690	698	34	45
1	3.2	3.2	500	459	644	599	35	37
1	6.1	5.1	540	492	579	546	56	60
2	28.4	21.7	885	825	804	733	33	36

Grizzly bears are magnificent animals that weigh up to several hundred pounds and can run over 30 mph in short distances. Males range over several miles but the females tend to stay closer to home. Grizzly bears usually keep to themselves. However, several grizzly bear attacks on humans are reported each year. Because of their large size, they can menace hikers and campers in some circumstances. Even so, it is important that a healthy grizzly bear population be maintained for future generations.

Harry Reynolds, of the Alaska Game and Fish Department, has studied grizzly bears for over 20 years in an effort to protect them and learn what they need to survive. He and his colleagues typically spot bears from an airplane. Once a bear is located, they try to shoot it with a dart so that it falls asleep for a few minutes. During this brief period, they check its overall health and take measurements of the size of the bear. They also roll the bear onto a tarp, which is then lifted by a scale attached to a tripod. Age is given in years, weight in pounds, and the other size measurements in centimeters.

Occasionally, the researchers have to run to a vehicle for protection when the sedated bear awakes too quickly. These measurements are not easy to make!

TABLE D.6 Grizzly Bear Data

Bear	Sex	Age	Weight	Length	Neck	Girth	Head Length	Head Width
1	M	9	245	199	84	135	24.5	39.0
2	M	9	200	198	83	129	22.5	39.3
3	F	4	102	174	65	104	20.3	33.4
4	F	5	180	183	57	101	17.9	32.7
5	F	5	225	170	57	100	18.6	33.3
6	F	5	230	188	62	103	19.3	32.9
7	F	13	315	185	62	118	22.1	36.9
8	F	19	280	193	63	114	22.2	35.5
9	F	5	235	177	56	93	19.4	33.4
10	F	2	70	152	46	80	16.8	31.3
11	M	2	91	157	58	102	17.3	30.6
12	F	6	280	181	61	100	19.4	34.5
13	M	11	465	204	84	132	23.7	38.8
14	M	4	125	191	60	97	20.1	33.4
15	F	6	265	189	59	100	20.4	33.3
16	F	3	107	156	45	81	16.3	28.1
17	F	2	68	158	43	77	16.0	29.9
18	M	2	75	163	45	77	17.8	29.7
19	M	2	82	164	48	80	16.7	32.3
20	F	6	245	176	66	104	20.5	34.5
21	F	4	215	176	55	103	19.1	32.0
22	M	2	84	169	53	89	19.9	31.5
23	F	17	109	176	62	112	19.0	32.3
24	F	5	240	185	61	103	20.5	33.6
25	F	5	247	175	57	102	18.5	33.3

TABLE D.6   *(Continued)*

Bear	Sex	Age	Weight	Length	Neck	Girth	Head Length	Head Width
26	F	3	130	148	45	77	13.6	27.4
27	F	3	130	166	53	86	17.5	30.0
28	F	7	280	177	57	106	18.7	33.1
29	M	11	620	224	95	164	24.2	42.0
30	M	3	150	151	49	85	18.0	32.4
31	F	1	115	134	45	78	14.4	26.5
32	F	1	115	137	43	77	14.0	26.8
33	F	4	285	188	64	110	19.6	34.0
34	M	4	415	199	77	137	23.0	38.5
35	F	1	120	135	49	77	14.0	26.3
36	M	6	578	219	90	134	24.8	38.6
37	M	2	140	148	56	88	17.0	29.9
38	F	2	125	131	48	80	16.3	27.2
39	M	4	265	178	63	102	18.8	32.2
40	F	2	135	153	49	83	16.2	28.7
41	M	2	212	171	51	88	17.8	31.4
42	F	2	190	168	49	81	17.4	30.3
43	M	5	359	198	76	114	21.4	36.6
44	F	7	187	166	57	95	19.4	32.4
45	M	1	62	114	37	66	12.4	26.9
46	F	17	283	188	67	113	20.5	37.0
47	M	1	126	145	46	84	15.2	27.4
48	F	4	196	168	56	90	18.2	31.5
49	M	4	345	192	69	112	20.3	35.2
50	M	2	195	175	53	92	18.2	30.2
51	M	1	100	125	41	73	14.1	24.5
52	F	1	50	107	33	58	12.1	21.1
53	M	1	100	131	45	74	15.1	26.7
54	F	1	65	119	39	53	13.2	22.5
55	F	1	105	136	42	71	15.0	25.5
56	M	1	110	144	43	70	15.4	26.5
57	F	1	100	133	45	66	16.4	25.2
58	M	1	63	116	38	62	13.3	23.5
59	F	1	70	123	41	63	13.5	24.2
60	F	1	96	139	46	76	14.5	25.6
61	M	1	95	142	48	76	14.4	25.6

Salmon fisheries support a primary industry in Alaska and their management is of high priority. Salmon are born in freshwater rivers and streams, but then swim out into the ocean for a few years before returning to spawn and die. In order to identify the origins of mature fish and equably divide the catch of returning salmon between Alaska and the Canadian provinces, researchers have studied the growth of their scales. The growth the first year in freshwater is measured by the width of the growth rings for that period of life, and marine growth is measured by the width of the growth rings for the first year in the ocean environment. The scales are first magnified 100 times and then the measurements are made in hundredths of an inch. A set of these measurements, collected by the Alaska Department of Fish and Game, are given in Table D.7. (Courtesy of K. Jensen and B. Van Alen.)

TABLE D.7   Radius of Growth Zones for Freshwater and First Marine Year

Males		Females	
Freshwater Growth	First Year Marine Growth	Freshwater Growth	First Year Marine Growth
147	444	131	405
139	446	113	422
160	438	137	428
99	437	121	469
120	405	139	424
151	435	144	402
115	394	161	440
121	406	107	410
109	440	129	366
119	414	123	422
130	444	148	410
110	465	129	352
127	457	119	414
100	498	134	396
115	452	139	473
117	418	140	398
112	502	126	434
116	478	116	395
98	500	112	334
98	589	117	455
83	480	97	439
85	424	134	511
88	455	88	432
98	439	99	381
74	423	105	418
58	411	112	475
114	484	98	436
88	447	80	431

TABLE D.7   *(Continued)*

Males		Females	
Freshwater Growth	First Year Marine Growth	Freshwater Growth	First Year Marine Growth
77	448	139	515
86	450	97	508
86	493	103	429
65	495	93	420
127	470	85	424
91	454	60	456
76	430	115	491
44	448	113	474
42	512	91	421
50	417	109	451
57	466	122	442
42	496	68	363

The U. S. Department of Agriculture and State Agriculture Experiment Stations cooperate on the investigation of barleys for possible use in brewing processes. One year, the malt extract (%) was obtained for 40 different varieties from one experiment station.

TABLE D.8   Malt Extract

75.3	77.9	77.6	76.6	78.3	77.9	77.5	77.6	77.1	78.0
77.9	76.3	75.7	77.4	77.4	76.9	77.9	77.4	78.1	77.4
76.4	79.1	80.0	76.9	78.5	78.4	77.8	80.4	75.9	77.0
79.2	76.2	77.0	75.9	77.9	78.4	76.7	76.4	76.6	77.4

An ongoing study of wolves is being conducted at the Yukon-Charley Rivers National Preserve. Here are some of the physical characteristics of wolves that were observed. (Courtesy of John Burch National Park Service, Fairbanks, Alaska.)

TABLE D.9 Physical Characteristics of Wolves

Sex	Age	Weight (lb)	Body Length (cm)	Tail Length (cm)	Canine Length (mm)
M	4	71	134	44	28.7
F	2	57	123	46	27.0
F	4	84	129	49	27.2
M	4	93	143	46	30.5
M	4	101	148	48	32.3
M	1	84	127	42	25.8
M	2	88	136	47	26.6
M	3	117	146	46	29.1
F	2	90	143	43	27.1
M	4	86	142	51	29.2
F	6	71	124	42	28.2
F	8	71	125	42	27.8
M	0	86	139	54	24.0
M	2	93	140	45	29.0
M	2	86	133	44	29.3
F	3	77	122	45	27.4
F	2	68	125	51	27.3
M	3	106	123	53	29.0
F	0	73	122	43	24.5

Researchers studying sleep disorders needed to obtain data from the general population to serve as a reference set. They randomly selected state employees who worked in certain divisions, and many agreed to spend a night in the sleep lab and be measured. Female = 0 and male = 1. The body weight index (BMI) is a person's weight (kg) divided by the square of his or her height (m). The percent of rapid eye movement (PREM) sleep is the percent of total sleep time that is spent in rapid eye movement sleep. The number of breathing pauses per hour (BPH) is the total number of breathing pauses divided by the total hours of sleep. The snoring frequency (SNORF) is a response to the survey question, "According to what others have told you, please estimate how often you snore?" The possible responses were (1) rarely, (2) sometimes, (3) irregular pattern but at least once a week, (4) three to five nights a week, or (5) every or almost every night. (Courtesy of T. Young.)

TABLE D.10   Sleep Data

Gender	Age	BMI	PREM	BPH	SNORF	Gender	Age	BMI	PREM	BPH	SNORF
0	41	24.65	0.200	0.00	5	0	55	25.10	0.326	0.19	0
0	49	23.14	0.162	0.00	4	0	31	28.17	0.124	0.21	1
0	39	27.61	0.308	0.00	5	0	58	32.68	0.216	5.76	1
0	51	33.50	0.083	0.00	4	0	43	20.07	0.191	0.00	0
0	32	43.36	0.205	2.79	4	0	53	24.92	0.189	7.46	0
0	37	31.04	0.109	3.87	5	0	52	29.10	0.194	0.69	2
0	36	32.42	0.154	0.00	5	0	43	26.40	0.077	0.00	2
0	30	31.06	0.328	3.42	5	0	41	20.13	0.138	0.12	1
0	44	44.15	0.227	4.39	5	0	34	23.77	0.260	0.00	0
0	43	38.29	0.104	5.14	5	0	41	29.06	0.163	0.31	2
0	34	21.63	0.184	0.00	5	0	49	24.45	0.164	0.00	0
0	47	34.78	0.215	1.13	5	0	41	21.15	0.142	0.41	0
0	40	25.16	0.206	0.00	2	0	47	19.47	0.224	0.00	1
0	49	37.02	0.123	2.58	5	0	59	28.76	0.176	0.99	0
0	54	31.25	0.313	0.00	0	0	34	26.77	0.109	0.00	0
0	38	32.58	0.227	0.13	5	0	42	40.06	0.191	0.14	0
0	36	34.65	0.172	0.00	4	0	44	24.99	0.316	1.01	2
0	39	29.55	0.223	2.84	5	0	32	17.86	0.174	0.00	0
0	45	31.96	0.146	0.00	4	0	31	25.99	0.109	0.00	0
0	49	21.64	0.250	0.00	5	0	35	20.42	0.227	0.25	2
0	47	30.67	0.244	3.98	5	0	58	23.03	0.202	0.00	1
0	33	23.43	0.228	0.00	2	0	38	41.14	0.220	1.24	2
0	43	30.24	0.193	0.00	1	0	40	22.49	0.149	0.00	0
0	43	37.83	0.230	7.11	5	0	43	23.71	0.149	0.00	0
0	39	22.46	0.141	0.17	5	1	44	30.22	0.162	2.33	5
0	50	33.53	0.094	3.66	5	1	59	27.39	0.103	0.00	3
0	57	24.35	0.174	0.33	4	1	51	24.34	0.203	1.30	5
0	36	37.64	0.000	25.36	2	1	49	27.17	0.179	0.00	5
0	34	28.03	0.205	0.00	1	1	29	28.09	0.228	0.00	4
0	43	25.24	0.223	0.00	1	1	40	30.06	0.180	3.18	5
0	59	33.71	0.175	4.71	0	1	45	40.82	0.082	2.34	5

TABLE D.10   *(Continued)*

Gender	Age	BMI	PREM	BPH	SNORF	Gender	Age	BMI	PREM	BPH	SNORF
1	55	32.46	0.082	38.00	5	1	47	24.00	0.220	0.53	5
1	51	22.95	0.175	3.29	4	1	36	31.76	0.196	6.73	5
1	38	21.97	0.123	0.00	5	1	59	27.12	0.255	6.18	5
1	36	25.57	0.179	0.00	4	1	42	24.34	0.137	1.46	5
1	39	26.67	0.216	0.24	5	1	59	35.14	0.160	3.56	5
1	38	28.95	0.231	0.55	5	1	37	26.53	0.122	3.80	4
1	42	35.65	0.139	0.20	5	1	55	31.28	0.170	0.58	4
1	34	43.65	0.131	47.52	4	1	31	35.61	0.066	19.60	4
1	55	42.70	0.191	5.41	5	1	53	27.55	0.083	8.01	4
1	42	25.04	0.242	0.00	5	1	41	34.92	0.207	15.28	4
1	47	31.28	0.147	7.26	5	1	45	33.20	0.228	4.37	5
1	46	31.02	0.124	22.55	5	1	41	29.94	0.134	2.42	5
1	39	25.43	0.169	10.35	5	1	50	26.26	0.179	0.16	5
1	37	26.29	0.178	7.05	5	1	40	34.94	0.174	8.76	5
1	41	29.59	0.259	1.37	5	1	54	28.60	0.197	0.00	4
1	58	29.24	0.170	3.89	5	1	39	31.55	0.135	0.81	5
1	57	28.72	0.152	2.33	5	1	49	22.84	0.109	1.53	2
1	33	28.56	0.126	1.09	5	1	46	18.89	0.114	0.00	0
1	58	30.04	0.160	0.18	4	1	39	25.70	0.233	0.00	0
1	54	35.43	0.117	1.24	5	1	41	26.89	0.247	1.94	1
1	46	34.78	0.151	19.47	5	1	46	31.29	0.230	0.00	2
1	46	25.66	0.223	0.00	3	1	40	29.76	0.000	0.71	0
1	55	31.80	0.087	4.58	5	1	44	26.71	0.163	0.39	2
1	56	29.74	0.162	18.77	5	1	41	24.69	0.209	0.16	0
1	35	24.43	0.423	0.16	5	1	37	26.58	0.074	0.00	0
1	49	23.33	0.196	21.58	4	1	44	23.61	0.234	0.00	0
1	50	26.45	0.199	0.00	4	1	44	29.86	0.147	5.13	2
1	42	27.38	0.155	2.63	5	1	41	28.41	0.097	0.19	1
1	46	27.06	0.123	0.00	5	1	46	23.12	0.156	0.19	0
1	37	24.82	0.175	0.18	4	1	34	29.88	0.195	0.00	0
1	38	30.64	0.190	70.82	5	1	41	18.02	0.149	0.00	0
1	44	37.61	0.254	7.61	5	1	44	20.90	0.181	0.21	0
1	48	26.85	0.219	0.42	4	1	39	22.39	0.223	0.00	1
1	53	28.23	0.103	0.33	4	1	35	29.23	0.152	0.00	2
1	55	29.48	0.150	4.57	5	1	38	23.83	0.163	0.00	2
1	56	27.76	0.185	7.82	5	1	47	22.23	0.129	0.76	1
1	57	32.08	0.205	28.26	5	1	47	28.04	0.146	2.53	0
1	58	27.55	0.128	26.37	5	1	50	32.82	0.183	12.86	0
1	37	31.98	0.241	5.38	1	1	48	20.09	0.083	1.59	2
1	30	24.09	0.188	0.68	5	1	38	26.63	0.229	1.81	2
1	51	33.80	0.169	0.90	5	1	37	30.72	0.203	2.13	2
1	44	28.08	0.130	0.53	5	1	50	24.73	0.188	0.00	2
1	45	30.83	0.105	6.86	4	1	37	33.35	0.274	17.78	2
						1	44	24.70	0.170	0.42	2
						1	49	18.24	0.223	0.00	1
						1	41	24.86	0.216	2.31	0

Insecticides, including the long banned DDT, which imitate the human reproductive hormone estrogen, may cause serious health problems in humans and animals. Researchers examined the reproductive development of young alligators hatched from eggs taken from (1) Lake Apopka, which is adjacent to an EPA Superfund site, and (2) Lake Woodruff, which acted as a control. The contaminants at the first lake, including DDT, were thought to have caused reproductive disorders in animals living in and around the lake. The concentrations of the sex steroids estradiol and testosterone in the blood of alligators were determined by radioimmunoassay both at about six months of age and then again after the alligators were stimulated with LH, a pituitary hormone. (Courtesy of L. Guillette.)

The data are coded as (* indicates missing):

x_1 = 1, Lake Apopka, and = 0, Lake Woodruff

x_2 = 1 male and = 0 female

x_3 = E2 = estradiol concentration (pg/ml)

x_4 = T = testosterone concentration (pg/ml)

x_5 = LHE2 = estradiol concentration after receiving LH (pg/ml)

x_6 = LHT = testosterone concentration after receiving LH (pg/ml)

TABLE D.11   Alligator Data

Lake Apopka						Lake Woodruff					
Lake x_1	Sex x_2	E2 x_3	T x_4	LHE2 x_5	LHT x_6	Lake x_1	Sex x_2	E2 x_3	T x_4	LHE2 x_5	LHT x_6
1	1	38	22	134	15	0	1	29	47	46	10
1	1	23	24	109	28	0	1	64	20	82	76
1	1	53	8	16	12	0	1	19	60	*	*
1	1	37	6	220	13	0	1	36	75	19	72
1	1	30	7	114	11	0	1	27	12	118	95
1	0	60	19	184	7	0	1	16	54	33	64
1	0	73	23	143	13	0	1	15	33	99	19
1	0	64	16	228	13	0	1	72	53	29	20
1	0	101	8	163	10	0	1	85	100	72	0
1	0	137	9	83	7	0	0	139	20	82	2
1	0	88	7	200	12	0	0	74	4	170	75
1	0	73	19	220	21	0	0	83	18	125	45
1	0	257	8	194	37	0	0	35	43	19	76
1	0	138	10	221	3	0	0	106	9	142	5
1	0	220	15	101	5	0	0	47	52	24	62
1	0	88	10	141	7	0	0	38	8	68	20
						0	0	65	15	32	50
						0	0	21	7	140	4
						0	0	68	16	110	3
						0	0	70	16	58	18
						0	0	112	14	78	5

Answers to Selected Odd-Numbered Exercises

CHAPTER 1

1.1 Population: Entire set of responses from all U.S. teens 13 to 17. Sample: responses from the 1055 teens contacted.

1.3 (a) person living in Chicago, (b) whether a person is an illegal alien(yes) or not(no). (c) entire set of yes/no responses from all persons living in Chicago.

1.5 (a) An undergraduate at the school (b) Collection of ownership/non-ownership status for all undergraduates (c) Not representative since joggers more likely to own MP3 player.

1.7 Population: The entire collection of responses, from all residents, of their favorite establishment. Sample: the responses from the persons who filled out the form. This is a self selected sample, and not representative, as only those with strong opinions would fill out form.

1.9 (a) anecdotal, (b) sample based, (c) sample based

1.11 "Too long" is not well defined. Could, for instance, say half the time it takes longer than 5 minutes.

1.19 (b) Including votes only from those with strong enough opinions to call in, like big dogs jumping the fence, would produce unrepresentative results.

CHAPTER 2

2.1 (a) 20.9% (c) 38.6%, 51.4%, 79.1%

2.3 The frequency table for blood type:

Blood Type	Frequency	Relative Frequency
O	16	$0.40 = \frac{16}{40}$
A	18	$0.45 = \frac{18}{40}$
B	4	$0.10 = \frac{4}{40}$
AB	2	$0.05 = \frac{2}{40}$
Total	40	1.00

2.5 (a) The relative frequencies are Drive alone 0.625, Car pool 0.075, Ride bus 0.175, and Other 0.125.

2.7 3 belongs to two classes.

2.9 5 doesn't belong to any class.

2.11 (a) Yes (d) No (e) No

2.13 (c) $9/50 = .18$

2.17 (c) $1/15 = .067$

2.19 (a) The frequency table

Class	Frequency
0.45–0.90	2
0.90–1.35	6
1.35–1.80	11
1.80–2.25	5
2.25–2.70	6

2.23 The stem-and-leaf display:

```
0 | 6
1 | 2234455567777889
2 | 000000222445567799
3 | 022444566
4 | 1167
5 | 12
```

2.25 The double-stem display:

```
0 | 6          3 | 022444
1 | 22344      3 | 566
1 | 55567777889  4 | 11
2 | 00000022244  4 | 67
2 | 5567799    5 | 12
```

2.27 The five-stem display:

```
15 | 5
15 |
15 | 9
16 |
16 |
16 |
16 | 7
16 | 8
17 | 0
17 | 22333
17 | 4
17 | 677
17 | 88
18 | 11
18 | 2
18 |
18 | 67
18 |
19 | 011
```

2.29 (a) Median = 3 and $\bar{x}$ = 3
(c) Median = 1 and $\bar{x}$ = 1

2.31 (a) $\bar{x}$ = 254 (b) Median since one large observation heavily influences the mean.

2.33 Claim ignores variability. July with 105 is hot.

2.35 (a) $\bar{x}$ = 8.48 (b) Either

2.37 Mean = 118.05 is center where sum of positive deviations balance the sum of negative deviations. Median = 117.00 has at least half of weights the same or smaller and at least half the same or larger.

2.39 439.57

2.43 181.5

2.45 (b) $\bar{x}$ = 26.62; sample median = 24
(c) Q_1 = 22, Q_2 = 24, and Q_3 = 30

2.47 (a) Sample median = 153
(b) Q_1 = 135.5, Q_2 = 153, and Q_3 = 166.5

2.49 Sample median = 94, Q_1 = 73 and Q_3 = 105

2.51 (a) Median = 110; Q_1 = 60, Q_3 = 340
(b) 90th percentile = 400

2.53 (b) Mean = 24.33°C; median = 24.44°C

2.57 (b) s^2 = 7 and s = 2.646

2.59 (b) s^2 = 18.667 and s = 4.32

2.63 (a) $s^2 = 1.30$ (c) $s^2 = 0.286$
2.67 (a) $s^2 = 155,226$ (b) $s = 393.99$
2.71 (a) Median $= 68.4$
      (b) $\bar{x} = 68.343$ (c) $s = 2.419$
2.73 (a) $s^2 = 239.322$ (b) 25.00
      (c) 30 is nearly twice as variable.
2.75 Interquartile range $= 31.0$
2.77 No. Extremes more variable than middle.
2.81 (b) The interval (14.446, 35.874) has
           proportion $= 0.72$, guidelines $=$
           0.68.
2.83 (a) -1.037 (b) 1.677
2.87 (a) Mean $= 37.25$, standard devia-
           tion $= 6.48$ (b) $22/24 = .917$
2.89 (a) Mean $= -8.81$, standard devia-
           tion $= 20.24$ (b) By itself, -67
           makes the mean over 4 days smaller.
2.91 (b) maximum $= 55$ for Truman,
           minimum $= -5$ for Clinton
2.95 Only the value 215 from the second
      period is outside of limits.
2.97 Upward trend is predominant feature.
      Control limits not appropriate.
2.99 The relative frequencies of the occupa-
      tion groups are:

Occupation Group	Relative Frequency 1980	2001
Goods producing	0.284	0.190
Service (private)	0.537	0.652
Government	0.179	0.158
Total	1.000	1.000

2.103 (a) Yes   (b) Yes   (c) No
2.105 (a) Mean $= 227.4$ pounds is center
      where sum of positive deviations balance
      the sum of negative deviations. Median $=$
      232.5 has at least half of weights the same
      or smaller and at least half the same or
      larger.   (b) 82.7   (c) 1.120
2.107 (a) Median $= 9$ (b) $\bar{x} = 9.033$
      (c) $s^2 = 3.895$
2.109 (a) $\bar{x} = 7$, $s = 2$
      (c) Mean $= -21$, standard
           deviation $= 6$.

2.115 (a) (0, 175.6)   (0, 265.2)
           (0, 354.8) with -3.6 changed to 0,
           etc.
      (b) proportions .907   .944   .963,
           respectively
2.117 (a) Median $= 4.505$, $Q_1 = 4.30$,
           $Q_3 = 4.70$
      (b) 4.935
      (c) $\bar{x} = 4.5074$, $s = 0.368$
2.119 (a) Median $= 6.3$, $Q_1 = 5.9$, $Q_3 =$
           6.9
      (b) $\bar{x} = 6.42$ and $s = .721$
2.121 (b) Not reasonable with time trend
2.131 $\bar{x} = 109.1$, $s = 65.8$

CHAPTER 3

3.1   (c) The pill seems to reduce the propor-
           tions of severe and moderate nausea.
3.3   Proportion is highest for B7.
3.7   (b)

		Major				
		B	H	P	S	Total
Male		.245	.082	.102	.286	.715
Female		.122	0	.082	.082	.286
Total		.367	.082	.184	.368	1.001

3.11 (b) Research hospital best for either
           condition.
3.15 (b) Positive—better players are paid
           more.
3.19 (b) $r = -.415$
3.21 (a) Figure (c)   (b) Figure (b)
      (c) Figure (a)
3.23 $r = -.605$
3.25 $r = .891$
3.27 (b) $r = .279$
3.33 (b) $r = .995$
3.37 (b) 16 or more
3.39 (a) $\hat{\beta}_0 = .52$ $\hat{\beta}_1 = .70$
3.41 (a) $\hat{\beta}_0 = 2.143$ $\hat{\beta}_1 = 1.085$

3.47   (c)

2-Wheel	4-Wheel	Total
.364	.636	1.000
.643	.357	1.000

3.49   (a)   negative. Only so much time
        (c) no relation
3.51   (b) $r = .988$
3.53   (b) $r = -.158$
3.55   (c) $\hat{y} = 2.15 + .37x$
3.57   (a) x = road roughness and y = gas consumption
3.61   (b) $r = .437$

CHAPTER 4

4.1   (a)   (ii), (v)
        (d)   (vi)
        (f)   (iii), (v)
4.5   (a) $\{0, 1\}$
        (b) $\{0, 1, \ldots, 344\}$
        (c) $\{t : 90 < t < 425.4\}$
4.7   (a) $S = \{BJ, BL, JB, JL, LB, LJ, BS,$
                $JS, LS, SB, SJ, SL\}$
        (b) $A = \{LB, LJ, LS\}$
                $B = \{JL, LJ, JS, SJ, LS, SL\}$
4.11   (a)   Yes
        (b)   No, because the total is more than 1
        (c)   Yes
4.13   .2
4.15   .743
4.19   (a) $S = \{e_1, e_2, e_3\}$ where $e_i =$ [the ticket has number i]; $P(e_1) = \frac{2}{8}$, $P(e_2) = \frac{3}{8}$, $P(e_3) = \frac{3}{8}$
        (b) $\frac{5}{8}$
4.21   (c) $P(A) = \frac{5}{36}, P(B) = \frac{1}{6}$, $P(C) = \frac{1}{2}, P(D) = \frac{1}{6}$
4.23   (b) $\frac{4}{9}$
4.25   (a) $\frac{1}{15}$ (b) $\frac{11}{15}$ (c) $\frac{12}{15}$

4.27 $S = \{N, YN, YYN, YYYN, YYYYN,$
        $YYYYY\}$
4.29   (b) $\frac{1}{7}$
4.31   (b) $P(A) = \frac{1}{2}, P(B) = \frac{1}{3}$
4.35   (b)   (i) $AB = \{e_6, e_7\}$
                (iii) $A\overline{B} = \{e_4, e_5\}$
4.37   (b) $AB = \{e_2, e_6, e_7\}, P(AB) = .42$
        (d) $\overline{A}\,\overline{C} = \{e_3, e_4\}, P(\overline{A}\,\overline{C}) = .24$
4.39   (b) $A \cup B = \{e_1, e_3, e_4\}, AB = \{e_3\}$
4.43   (b) $P(A\overline{B}) = .32, P(\overline{A}B) = .16,$
                $P(\overline{A}\,\overline{B}) = .32$
4.45   (b) $P(AB) = .12$
        (c) $P(A \cup B) = .47$
        (d) $P(A\overline{B} \cup \overline{A}B) = .27$
4.47   (a) $P(\overline{A}) = .65$ (b) $P(A\overline{B}) = .15$
4.49 $\frac{7}{18}$
4.51   (a) $P(B) = .35$
4.55   Higher, they are dependent.
4.57   (a)   .515   (b)   .485   (c)   .625
4.59 $P(BA) = .0099$,   not independent
4.63   (a) $P(A) = .4, P(B) = .4$
        (b)   Not independent because
                $P(AB) = .1 \neq P(A)P(B)$
4.65   (b)   4.8%   (c)   .360
4.67   (a)   .688   (c)   .769
4.69   (a) $\frac{3}{15} = .2$ (c) $\frac{6}{210}$
4.71   (a) $\frac{2}{9}$ (b) $\frac{7}{12}$
4.73   (a) $BC, P(BC) = 0$
        (c) $\overline{B}, P(\overline{B}) = .8$
4.75   .99991
4.77   (b)   .896
4.79   (b)   .227
4.81   (a)   .55   (b)   Yes
4.83   (a)   20   (c)   231   (e)   4060
4.85   (a)   210   (b)   90
4.87   (a)   462   (b)   210
4.91   (a)   .491   (b)   .084
4.93   No
4.97   (a)   .018   (b)   .393
4.99   (a)   495   (b)   .141
4.105   (a) $A = \{23, 24\}$
        (d) $A = \{t : 0 \leq t < 500.5\}$
4.107 $A = \{p : .10 < p < 1\}, p$ is percent

4.109 (a) .5   (b) .75   (c) .167

4.111 .5

4.115 (b) 2/3

4.117 (a) Either a faulty transmission or faulty brakes or both

(b) Transmission, brakes, and exhaust system all faulty

4.119 $P(A) = .28, P(AB) = .10,$
$P(A \cup B) = .52$

4.123 (a) .25   (b) .73

4.125 (a) .25   (c) .34

4.129 (a) $P(A)P(C) = .15, P(AC) = .15,$ independent

(b) $P(A\bar{B})P(C) = .1, P(A\bar{B}C) = .15,$ not independent

4.131 (a) .44   (b) .50
(c) Not independent

4.135 (c) .64   (d) .49

4.137 $\frac{1}{165} = .0061$

4.139 (a) .108   (b) .515

4.141 (a) .008

CHAPTER 5

5.1 (a) Discrete   (b) Continuous
(c) Continuous

5.3 (a)

Possible choices	x
{1, 3}	2
{1, 5}	4
{1, 6}	5
{1, 7}	6
{3, 5}	2

Possible choices	x
{3, 6}	3
{3, 7}	4
{5, 6}	1
{5, 7}	2
{6, 7}	1

(b)

x	1	2	3	4	5	6
$f(x)$	.2	.3	.1	.2	.1	.1

5.9

x	0	2	4
$f(x)$	.64	.20	.16

5.13 (a)

x	2	3	4	5	Total
$f(x)$	$\frac{1}{10}$	$\frac{2}{10}$	$\frac{3}{10}$	$\frac{4}{10}$	1

Yes, a probability distribution

(b)

x	1	2	3	4	Total
$f(x)$	$-\frac{1}{3}$	0	$\frac{1}{3}$	$\frac{2}{3}$	1

Not a probability distribution; $f(1)$ is negative.

5.15

x	0	1	2
$f(x)$	$\frac{1}{4}$	$\frac{1}{2}$	$\frac{1}{4}$

5.17

x	0	1	9
$f(x)$	$\frac{1}{5}$	$\frac{2}{5}$	$\frac{2}{5}$

5.21 $f(2) = .2, f(4) = .2, f(6) = .2$

5.25 (a) .90   (b) .63   (c) .78

5.27 (a) .45   (b) .25
(c) The capacity must be increased by 2 to a total of 4.

5.29 (b) $E(X) = 2, \sigma = 1$

5.31 $955.20

5.33 $\mu = 1.2, \sigma = .917$

5.37 (a) $E(X) = 1.381$ (b) $\sigma = 1.318$

5.41 (b)

x	0	1	2	3	4	Total
$f(x)$	.2401	.4116	.2646	.0756	.0081	1.0000

(c) $E(X) = 1.2$

5.43 Median = 2

5.45 (a) Trials are dependent
(c) Bernoulli model plausible

5.47 Identify S = yellow, F = other colors
(a) Yes, $p = .4$ (b) No (c) No

5.49 (a) Yes, $P(S) = .5$
(b) Not independent

5.51 (a) Bernoulli trials model is not appropriate.
(b) Appropriate

5.53 (a) $(\frac{2}{3})^4 = .198$ (b) $\frac{1}{81} = .012$

5.57 (c)

x	0	1	2
$f(x)$	$\frac{81}{256}$	$\frac{126}{256}$	$\frac{49}{256}$

5.59 (a) Yes, $n = 10, p = \frac{1}{6}$
     (b) No, because the number of trials is not fixed.

5.61 (b) .132
5.63 (c) .3852
5.65 (a) .242   (b) .959   (c) .759
5.69 (a) .233   (b) .014   (c) .014
5.71 (a) .163   (b) .537
5.75 .411
5.77 (a) .164   (c) .163
5.81 (b) Mean $= 12.308$, sd $= 2.919$
5.85 (c) Third and fifth weeks
5.87 (b) $P[X \le 3 \text{ or } X \ge 17] = .002$
5.89 (a) $-3, -1, 1, 3$
     (b) $X = -3: TTT$
        $X = -1: HTT, THT, TTH$
        $X = 1: HHT, HTH, THH$
        $X = 3: HHH$

5.91 (b)

y	0	1	2
$f(y)$	.593	.363	.043

5.93 (b)

x	3	4	5
$f(x)$	.2800	.3744	.3456

5.95

x	0	1	2	3
$f(x)$	.255	.509	.218	.018

5.99 (a) .002   (b) $-\$0.60$
5.101 (a) .30   (b) $\sigma = 1.342$
5.103 (a)

x	0	1	3
$f(x)$	$\frac{2}{6}$	$\frac{3}{6}$	$\frac{1}{6}$

     (b) 1.0
5.105 (a)

x	-15	-5	5	15
$f(x)$	.1458	.3936	.3543	.1063

     (b) $-\$0.79$ (c) No
5.109 (a) $F(3) = .44, F(4) = .72,$
     $F(5) = .90, F(6) = 1.00$

5.111 $\mu = 3.667, \sigma = .943$
5.113 (a) Not plausible, because the assumption of independence is questionable.
     (b) Plausible
5.115 .125
5.117 (a) .49   (b) .343   (c) .441
5.121 (a) Binomial distribution with $n = 6,$
        $p = .4$
     (b) .821, .047, 2.4
5.123 (b) 1.759   (c) .2211
5.125 $P[X10] = .048.$ This small probability casts doubt on the claim that $p = .7$
5.127 (a) .942   (c) .818   (e) 1.833
*5.129 $n = 6$
5.133 $P[X \ge 10] = .048, P[X = 10] = .031$
5.135 (b) 32 dollars

CHAPTER 6

6.1 (c) Probability density function
6.3 The interval 1.5 to 2 has higher probability.
6.5 Median $= 1.414$
6.9 (c) $Z = \dfrac{X161}{5}$
6.11 (b) $Z = \dfrac{X350}{8}$
6.13 (a) .7967   (c) .1515
6.15 (c) .8907   (d) .0510
6.17 (b) .8164   (d) .0695
6.19 (a) $-.842$ (c) $z = .97$
6.21 (a) $z = -.93$ (d) $z = .753$
6.23 (a) .6293   (d) $z = .524$
6.25 (a) .1056   (c) .0668
     (e) .9477
6.27 (a) $b = 139.8$
6.31 .876
6.33 .2024
6.35 (a) .0413
6.37 (a) (iii) .3830   (b) 16.5 to 17.5
6.41 (a) Normal .0426
     (c) With continuity correction .8147

6.47 (a) .0923 (b) .0728

6.51 .7161 with continuity correction

6.55 (a) 0.5
(b) First quartile $= .25$, second quartile $= .5$, and third quartile $= .75$

6.59 (a) $z = -1.38$ (c) $z = .64$

6.61 (a) .2676 (c) .0026

6.63 (a) .8092 (c) .1056
(e) .5074

6.65 (a) .2000 (c) .2047

6.67 .4013

6.69 (a) .0228

6.71 (b) 305.3 ounces

6.75 (a) .0455 using continuity correction

6.77 (a) .0401 using continuity correction

6.79 (b) .0418 with continuity correction

CHAPTER 7

7.1 (a) Statistic
(c) Parameter

7.5 (b)

$\bar{x}$	Probability $f(\bar{x})$
3	$\frac{1}{9}$
4	$\frac{2}{9}$
5	$\frac{3}{9}$
6	$\frac{2}{9}$
7	$\frac{1}{9}$
Total	1

7.7 No, very best pictures become sample.

7.9 (a) $X = 4$ for 6 dots

7.11 (a) $E(\bar{X}) = \mu = 79$, s.d.$(\bar{X}) = 4.5$

7.13 (a) s.d.$(\bar{X}) = 4$ (c) s.d.$(\bar{X}) = 1.0$

7.15 $E(\bar{X}) = 5$, s.d.$(\bar{X}) = \sqrt{\frac{4}{3}}$

7.17 (b) $E(\bar{X}) = 2.0$ (c) mean $= 2.0$, variance $= \frac{4}{36} = .0111$

7.19 (a) $N(37, 2.449)$

7.21 (a) .1587 (b) $N(8.1, .0707)$

7.23 (a) $\bar{X}$ is approximately $N(31{,}000, 500)$. (b) .1587

7.27 .681

7.31 (b) $\sigma = 1.136$

7.35 (b) Sampling distribution of R:

Value of R	Probability
0	4/16
2	6/16
4	4/16
6	2/16
Total	1

7.37 (c) mean $= 1$ variance $= \frac{6}{81} = .0074$

7.39 (b) Exactly a normal distribution
(c) .7698

7.41 (a) $N(12.1, (3.2)/3)$ (c) About 26% smaller than 10

7.45 (a) Approximately .662
(b) $(52.95, 57.05)$

7.47 (b) $\sigma = .781$ (d) .195

7.49 (b) 305.29

CHAPTER 8

8.1 (a) S.E. $= 1.784$, 95% error margin $= 3.50$
(c) S.E. $= 3.260$, 92% error margin 5.71

8.3 (a) $\bar{x} = 10.74$, estimated S.E. $= .221$

8.7 (a) 2.143 (b) 3.525

8.9 (a) $n = 246$ (b) $n = 171$

8.13 $n = 68$

8.15 $(80.10, 82.50)$

8.19 $(28.94, 31.46)$ grams

8.21 $(17.1, 19.5)$ days

8.25 $(8.00, 9.20)$ miles

8.27 $(68.58, 82.28)$ dollars

8.29 $(-.00047, .00467)$

8.31 $(1.521, 1.909)$ centimeters

8.33 (b) In long run, covers with proportion .95

8.35 (b) Yes, in the middle (d) No

8.37 (a) $H_0: \mu = 14$, $H_1: \mu < 14$ days
(b) $H_0: \mu = 2.50$, $H_1: \mu 2.50$ dollars

8.39 (a) Correct decision if $\mu = 14$. Wrong decision if $\mu < 14$, Type II error.
(b) Correct decision if $\mu = 2.50$ dollars. Wrong if $\mu > 2.50$, Type II error.

8.43 (a) $\alpha = .025$ (b) $c = 31.10$

8.45 $R: Z1.96.$ Observed $z = -1.88$; H_0 is not rejected at $\alpha = .05$.

8.47 (a) Reject H_0 (c) .01   (e) .005

8.51 (a) .0089, strong rejection of H_0
(c) .1032, support for H_1 is not strong.

8.53 $H_0: \mu = 3.0, H_1: \mu3.0,$
$R: Z \geq 1.645$
Observed $z = 1.44$; H_0 is not rejected.

8.55 $H_0: \mu = 3000, H_1: \mu \neq 3000,$
$R: |Z| \geq 1.96$
Observed $z = 2.13$; H_0 is rejected at $\alpha = .05$.

8.57 (a) $\hat{p} = .56, 95\%$ error margin $= .138$
(b) $\hat{p} = .183, 95\%$ error margin $= .037$

8.59 (a) $n = 632$ (c) $n = 752$

8.61 (a) $\hat{p} = .628$ (b) estimated S. E. $= .055$

8.63 (b) (.49, .53).

8.65 (a) $H_0: p = .26$ and $H_1: p > .26$

8.67 (c) (i) $H_0: p = .60$ versus $H_1: p \neq .60$
(ii) $\dfrac{\hat{p} - .6}{.0558}$ (iii) $R: |Z| \geq 2.33$

8.69 (a) $Z = \dfrac{\hat{p} - .4}{\sqrt{.4 \times .6/n}}$ $R: Z \leq -1.28$

8.71 (b) P–value $= .0244$

8.73 P–value $= .112$, not higher

8.75 (a) (.47, .55)   (b) (3889, 4490)

8.77 (.58, .85)

8.79 (a) $\bar{x} = 9.40$ S.E. $= .234$

8.81 (b) A factor of 16

8.83 (a) 2.8   (b) (124.6, 129.2)

8.85 (a) Correct

8.87 The alternative hypothesis H_1 is the assertion that is to be established; its opposite is the null hypothesis H_0.
(a) $H_0: \mu = 50, H_1: \mu < 50$
(d) $H_0: \mu = 16, H_1: \mu \neq 16$

8.91 (b) P–value $.014/2 = .007$

8.93 (b) $\sigma = .781$
(c) $H_0: \mu = .7$ versus $H_1: \mu > .7,$
$Z = \dfrac{\bar{X} - .7}{S/\sqrt{64}}, R: Z \geq 1.645.$
Observed $z = 2.80$; H_0 is rejected at $\alpha = .05$.

8.95 (a) $\hat{p} = .0875$ (b) .0124

8.97 (a) $R: Z \geq z_{.05} = 1.645.$
(b) Observed $z = 2.93$; H_0 is rejected at $\alpha = .05$.

8.99 (a) $\alpha = .0384$ (b) $c = .048$

8.101 (b) (.36, .48)

8.103 (a) $H_0: p = .8, H_1: p.8,$
$Z = \dfrac{\hat{p} - .8}{\sqrt{(.8 \times .2)/200}}, R: |Z| \geq 1.96,$
Observed $z = -4.24, H_0$ is rejected at $\alpha = .05$.
P–value $P[|Z| \geq 4.24] < .0001$ is extremely small. The genetic model is strongly contradicted.
(b) (.62, .74)

*8.105 (a) .5160   (b) .8925

8.107 (a) Larger, must cover more often.
(b) P–value .0041 or output .0045

8.109 (a) (1.55, 1.88)   (b) Observed $z = -2.464$, reject $H_0: \mu = 1.9$.

8.111 (90.54, 106.16)

CHAPTER 9

9.1 (a) 1.943   (b) -2.110

9.3 (b) 3.747   (c) -1.711

9.5 (a) 1.895   (b) -2.120 and 2.120

9.7 (b) Between .025 and .05
(c) Between .05 and .10

9.9 (a) (135.5, 144.5)
(b) Center $= 140,$ Length $= 9.0$
(c) Usually different since S is random

9.11 (128.17, 147.03)

9.13 (3.2, 4.0) kg

9.15 (a) $\bar{x} = 22.4, s = 5.98$
(b) (17.7, 27.1)

9.17 (5.372, 5.594)

9.19 $H_0: \mu = 620, H_1: \mu > 620, \quad T = \dfrac{\bar{X} - 620}{S/\sqrt{10}},$ d.f. $= 9, R: T \geq 1.833$
Observed $t = 1.90$; H_0 is rejected at $\alpha = .05$.

9.21 (a) Cannot tell, mean unknown (b) In the long run, 95% of the intervals obtained using this procedure will cover.

9.23 (b) (23.38, 29.86) feet

9.25 $H_0: \mu = 128, H_1: \mu > 128$
$T = \dfrac{\overline{X} - 128}{S/\sqrt{20}}$, d.f. $= 19, t_{.05} = 1.729$,
$R: T \geq 1.729$
Observed $t = 2.13, H_0$ is rejected at $\alpha = .05$.

9.27 $H_0: \mu = 83, H_1: \mu \neq 83$ pounds.
$T = \dfrac{\overline{X} - 83}{S/\sqrt{8}}$, d.f. $= 7,$
$t_{.025} = 2.365, R: T \leq -2.365$ or $T \geq 2.365$
Observed $t = -2.566; H_0$ is rejected at $\alpha = .05$.

9.29 (b) (83, 89)

9.31 (b) Reject H_0 when it is true (d) When H_0 prevails, H_0 will be rejected in about proportion .05 of the times.

9.33 (a) H_0 is rejected at $\alpha = .05$.

9.35 (a) (1.28, 12.28)
(c) Observed $t = -3.53; H_0$ is rejected at $\alpha = .05$.

9.39 (a) $\chi^2_{.10} = 23.54$
(b) $\chi^2_{.95} = 12.34$
(c) $\chi^2_{.10} = 40.26$
(d) $\chi^2_{.95} = 3.33$

9.41 (a) 3.3615   (b) (2.01, 9.70)

9.43 $H_0: \sigma = .6, H_1: \sigma < .6$
$\chi^2 = \dfrac{(n-1)S^2}{(.6)^2}$, d.f. $= 39,$
$R: \chi^2 \leq \chi^2_{.95} = 25.7$
Observed $\chi^2 = 24.4; H_0$ is rejected at $\alpha = .05$.

9.45 (1.56, 4.15)

9.47 (a) (15.99, 27.60)

9.49 (4.70, 10.83)

9.51 (a) 2.353   (c) -2.353

9.55 (39.4, 54.6)

9.57 (a) $\bar{x} = 74.7$, 95% error margin $= 8.58$
(b) (67.7, 81.7)

9.59 $H_0: \mu = 42, H_1: \mu < 42$
$T = \dfrac{\overline{X} - 42}{S/\sqrt{21}}$, d.f. $= 20,$
$R: T \leq -t_{.01} = -2.528$
Observed $t = -3.23, H_0$ is rejected at $\alpha = .01$.

9.63 (a) $H_0: \mu = 65, H_1: \mu > 65$
(b) $T = \dfrac{\overline{X} - 65}{S/\sqrt{9}}$, d.f. $= 8,$
$t_{.05} = 1.860, R: T \geq 1.860.$
(c) Observed $t = .18$; fail to reject H_0. The claim is not demonstrated.

9.65 $H_0: \mu = 1500, H_1: \mu > 1500$
$T = \dfrac{\overline{X} - 1500}{S/\sqrt{5}}$, d.f. $= 4, t_{.05} = 2.132, R: T \geq 2.132.$ Observed
$t = \dfrac{1620 - 1500}{90/\sqrt{5}} = 2.98;$
H_0 is rejected at $\alpha = .05$.

9.67 (a) (423.09, 456.05)
(b) $t = -1.99$, reject H_0

9.69 (a) $\chi^2_{.05} = 12.59$ (c) $\chi^2_{.95} = 1.64$

9.71 (a) $\chi^2 = \dfrac{(n-1)S^2}{1.0}$, d.f. $= 24,$
$\chi^2_{.05} = 36.42, R: \chi^2 \geq 36.42$
Observed $\chi^2 = 40.16; H_0$ is rejected.

9.75 (a) Cannot tell, mean unknown
(b) In the long run, 95% of the itervals obtained using this procedure will cover.

9.77 $t = 2.25$, reject H_0

CHAPTER 10

10.1 First group, using first letter of subject,
$\{B, C\}, \{B, E\}, \{B, H\}, \{B, P\}$
$\{C, E\}, \{C, H\}, \{C, P\}$
$\{E, H\}, \{E, P\}$
$\{H, P\}$

10.3 (a) Using first letter of name,
$\{(S, G), (T, E)\} \{(S, G), (T, R)\}$
$\{(S, G), (E, R)\} \{(S, J), (T, E)\}$

$\{(S, J), (T, R)\} \{(S, J), (E, R)\}$
$\{(J, G), (T, E)\} \{(J, G), (T, R)\}$
$\{(J, G), (E, R)\}$

(b) There are three sets each consisting of three pairs.

10.5 (a) $\bar{x} - \bar{y} = 7$, S.E. $= 2.48$
(b) $(2.1, 11.9)$

10.7 $H_0: \mu_1 - \mu_2 = 0$,
$H_1: \mu_1 - \mu_2 \neq 0$, $R: |Z| \geq 1.96$
Observed $z = -4.17$; null hypothesis is rejected at $\alpha = .05$.

10.9 (a) $H_0: \mu_1 - \mu_2 = 0$, $H_1: \mu_1 - \mu_2 < 0$
(b) $Z = \dfrac{\bar{X} - \bar{Y}}{\sqrt{\dfrac{S_1^2}{n_1} + \dfrac{S_2^2}{n_2}}}$, $R: Z \leq -1.645$
(c) Observed $z = -3.04$, H_0 is rejected, P–value $= .0012$

10.11 H_0 is rejected at $\alpha = .05$.

10.13 P–value is smaller than .001.

10.15 (a) 3.5 (b) 1.870 (c) $t = 1.31$

10.17 (a) $T = -3.37$
(b) $(-29.3, -6.7)$ pounds

10.19 $H_0: \mu_1 - \mu_2 = 0$, $H_1: \mu_1 - \mu_2 > 0$,
$Z = \dfrac{\bar{X} - \bar{Y}}{\sqrt{\dfrac{S_1^2}{n_1} + \dfrac{S_2^2}{n_2}}}$ Observed $z = 2.91$,
P–value $= P[Z \geq 2.91] = .0018$.
Strong evidence in support of H_1.

10.23 (a) $H_0: \mu_1 - \mu_2 = 0$,
$H_1: \mu_1 - \mu_2 < 0$, d.f. $= 18$,
$t_{.05} = 1.734$ so $R: T \leq -1.734$.
Observed $t = -1.98$; H_0 is rejected at $\alpha = .05$. The mean job time is significantly less.
(b) Normal populations with equal σ's
(c) $(-1.65, .05)$

10.27 (a) Reject H_0. Evidence is strong since P–value $= .004$.
(b) $t = 2.43$, reject H_0

10.29 We drew slips with α, β, τ so group 1 is $\{$alpha, beta, tau$\}$.

10.33 (a) $t = 1.035$ (b) d.f. $= 5$

10.37 (a) Observed $t = 2.87$; H_0 is rejected at $\alpha = .05$.
(b) $(1.52, 7.90)$.

10.39 $H_0: \delta = 0, H_1: \delta \neq 0, T = \dfrac{\bar{D}}{S_D/\sqrt{n}}$,
d.f. $= 8$, $R: |T| \geq 2.306$
Observed $t = 1.48$; H_0 is not rejected. The difference is not significant.

10.43 (b) $(.43, 6.77)$

10.45 (a) $z = 4.26$, reject H_0
(b) $(.17, .43)$

10.49 $H_0: p_1 = p_2, H_1: p_1 > p_2$ (suffix 1 refers to abused group), $R: Z \geq 2.33$.
Observed $z = 3.01$, H_0 is rejected at $\alpha = .01$. There is strong evidence in support of the conjecture.

10.51 $H_0: p_1 = p_2, H_1: p_1 \neq p_2$ (suffix 1 refers to "with carbolic acid"),
$R: |Z| \geq 1.96$.
Observed $z = 2.91$; H_0 is rejected at $\alpha = .05$. P–value $= .0036$.

10.55 (a) $H_0: p_1 = p_2, H_1: p_1 > p_2$ (suffix 1 refers to "smokers").
Observed $z = 4.29$, P–value less than .00001. The conjecture is strongly supported.

10.57 (b) $(-.14, -.08)$

10.59 (a) $(.17, .38)$ (b) $(-.045, .103)$

10.61 (b) Observed $z = 3.00$, difference is significant.

10.63 (a) $H_0: \mu_2 - \mu_1 = 10$,
$H_1: \mu_2 - \mu_1 > 10$
(c) Observed $z = 2.11$, H_0 is rejected. P–value $= .0174$.

10.65 (a) $(.13, 1.09)$
(b) $(4.31, 4.97)$

10.67 (a) 4.857 (b) $t = 2.03$

10.69 $(-12.3, 44.3)$

10.71 (a) $H_0: \mu_1 - \mu_2 = 0$,
$H_1: \mu_1 - \mu_2 \neq 0$ (suffix 1 refers to "city A"), $R: |Z| \geq 2.33$.
Observed $z = -4.97$; reject H_0.
(b) $(-7.93, -2.87)$

10.73 (a) Do not reject H_0 (b) $(-.022, .627)$

10.75 (a) $z = 2.00$, fail to reject H_0
(b) $(.68, 18.65)$

10.77 $(-3.55, -.45)$

10.79 (a) $t = -5.38$, reject H_0.

10.81 $H_0: p_1 = p_2, H_1: p_1 < p_2$ (suffix 1 refers to "seeded"), $R: Z \le c$.

$$Z = \frac{\hat{p}_1 - \hat{p}_2}{\sqrt{\hat{p}\hat{q}}\sqrt{\dfrac{1}{n_1} + \dfrac{1}{n_2}}}$$

Observed $z = -1.77, P\text{-value} = P[Z \le -1.77] = .0384$. Fairly strong evidence in support of the conjecture.

10.83 (a) $H_0: p_1 = p_2, H_1: p_1 > p_2$ (suffix 1 refers to "uremic"), with $\alpha = .01$, the rejection region is $R: Z \ge 2.33$. Observed $z = 2.64$, reject H_0 at $\alpha = .01$. Strong evidence of higher incidence.
(b) $(.05, .29)$.

10.85 (a) Fail to reject H_0

10.87 (a) Do not pool   (b) $(-58.0, -16.0)$

10.89 $z = 2.961$, reject H_0

10.91 $z = 13.13$, reject H_0

CHAPTER 11

11.1 Intercept $= 2$, slope $= 3$

11.3 (a) $x = $ duration of training, $y = $ measure of performance
(c) $x = $ level of humidity, $y = $ growth rate of fungus

11.5 $\beta_0 = 7, \beta_1 = -5$ and $\sigma = 3$

11.7 (a) $E(Y) = -1$, s.d.$(Y) = 4$

11.11 (b) No, only the mean is higher.

11.13 (d) The fitted line is $\hat{y} = 8.9 - 1.8x$

11.15 (b) SSE $= 7.10$
(c) $s^2 = 1.775$

11.17 (b) $\hat{\beta}_1 = .75, \hat{\beta}_0 = 3.00$

11.19 (a) $\hat{\beta}_1 = .1152, \hat{\beta}_0 = 5.04$
(c) $s^2 = .091$

11.21 (a) $\hat{\beta}_1 = .5033, \hat{\beta}_0 = 89.61$
(c) $s^2 = 31.37$

11.25 (a) $\hat{\beta}_1 = -.615, \hat{\beta}_0 = 4.845, s^2 = .0513$
(b) $H_0: \beta_1 = 0, H_1: \beta_1 \ne 0,$ $R: |T| \ge t_{.025} = 3.182,$

d.f. $= 3$ Observed $t = -13.8$, reject H_0 at the 5% level.
(c) $(3.06, 3.55)$.

11.27 $(-.756, -.474)$

11.29 (b) $\hat{y} = 28.52 + .869x$
(c) $(.729, 1.009)$

11.31 (a) $\hat{\beta}_1 = 1.595, \hat{\beta}_0 = 2.846$
(b) $s^2 = .116$
(c) $H_0: \beta_1 = 1.3, H_1: \beta_1 > 1.3,$ $R: T \ge t_{.05} = 1.771$, d.f. $= 13$ Observed $t = 1.775$, reject H_0.

11.33 (a) $Y = \beta_0 + \beta_1 x + e$ and fit $\hat{y} = 994 + .10373x$, with $\widehat{\text{s.d.}}(e) = \hat{\sigma} = 299.4$.
(b) Observed t-ratio $= 3.48$ with P-value .002. Reject $H_0: \beta_1 = 0$ at the 5% level.

11.35 (a) $Y = \beta_0 + \beta_1 x + e$ and fit $\hat{y} = .3381 + .83099x$ with $\widehat{\text{s.d.}}(e) = \hat{\sigma} = .1208$.
(b) Observed t-ratio $= 9.55$ with P-value approximately zero. Reject $H_0: \beta_1 = 0$ at 5% level.

11.37 (a) $\hat{y} = 24.78 + 1.413x$
(c) $(29.4, 31.4)$

11.39 .305

11.41 $r^2 = .568$

11.43 (a) .968   (b) $r = .984$

11.45 Proportion explained $= r^2 = .799$.

11.49 $x = $ (leaf length) $\times$ (leaf width) is the area of a rectangle containing leaf. Slope should be less than one.

11.51 (b) SSE $= 16.989$
(c) $s^2 = 2.427$

11.53 (a) $\hat{y} = 73.18 - 2.214x$
(b) $H_0: \beta_1 = -2, H_1: \beta < -2,$ $R: T \le t_{.05} = -1.771$, d.f. $= 13$. Observed $t = -.54$, fail to reject H_0.

11.55 (b) $H_0: \beta_1 = 0, H_1: \beta_1 > 0,$ $R: T \ge t_{.05} = 1.943$, d.f. $= 6$. Observed $t = 7.42$, reject H_0.
(d) $(622.4, 785.8)$ dollars

11.57 (b) $\hat{y} = 17.114 - 1.558x$
    (c) $(-1.09, -2.03)$
11.59 $r = -.945, .893$ explained
11.61 (a) $\hat{\beta}_0 = -1.071, \hat{\beta}_1 = 2.7408$
    (c) Estimated $S.E. (\hat{\beta}_0) = 2.751$
    (e) $r^2 = .828$
11.63 (a) $\hat{y} = -87.17 + 1.2765x$
    (b) Observed $t = 5.92$, d.f. $= 16$;
    reject $H_0: \beta_1 = 0.$
11.65 (a) $\hat{y} = 24.96 + 3.306x$
    (c) 74.55 degrees
11.67 (b) MARINGRW $= 478 - 0.236$
    FRESHGRW

CHAPTER 12

12.1 (b) $\hat{y} = 3.92 - .53x$
    (c) $r^2 = .842$, but relation is not a straight line.
12.3 (a) $y' = \dfrac{1}{y^{1/3}}, x' = x$
12.5 (b) No. beetles $= 6.14 - 0.899$
    $\log_e$ (distance)
    (c) $(-1.52, -.28)$
12.7 $E(Y) = -11$
12.9 (a) $\beta_2: (-5.06, -3.34)$
    (b) $t = 2.09 \geq t_{.05} = 1.740$,
    reject $H_0: \beta_1 = 6.$
12.11 (a) $\hat{\beta}_0 = -.0810, \hat{\beta}_1 = .64588$, and
    $\hat{\beta}_2 = .8046$
    (c) $R^2 = .865$ (d) $s^2 = .01023$
12.13 (a) $t = 7.02 \geq t_{.025} = 2.074$, reject H_0.
    (c) 1.951
12.15 (b) $\hat{y} = 46.55 - 11.77 \log_{10}(x)$,
    $r^2 = .749$
    (c) $(-17.66, -5.88)$
12.17 (c) $\widehat{\log_{10} y} = -1.16 + .0305x$,
    $r^2 = .988$
12.19 (a) $(1.82, 2.04)$
    (b) $T = -1.975 \leq t_{.05} = -1.717$,
    reject H_0
12.21 (b) The residuals tend to increase over time. Errors dependent.
12.25 (a) $\hat{y} = 117.9 - 16.05x + 3.702x^2$
    (b) 99.4%

12.27 (a) C3 $= -.167 + .237$C1
    $r^2 = .925$ (c) $(9.67, 11.32)$
12.29 (a) $\hat{y} = 50.4 + .1907x_2$ and $r^2 = .03$
    (c) Even three variables do not predict well.
12.33 (b) $\hat{y} = 9.999 + .155$(gender) $+ .9015$(initial no.) but P–value for gender large

CHAPTER 13

13.1 Observed $\chi^2 = 13.90$, d.f. $= 5$, $\chi^2_{.05} = 11.07$. The model of a fair die is contradicted.
13.3 Observed $\chi^2 = 2.00$, d.f. $= 3$, $\chi^2_{.05} = 7.81$. The model is not contradicted.
13.5 Observed $\chi^2 = 17.336$, d.f. $= 3$, $\chi^2_{.025} = 9.35$. Strong evidence of mislabeling.
13.7 Observed $\chi^2 = 156.18$, d.f. $= 5$, $\chi^2_{.01} = 15.09$. Reject H_0 at $\alpha = .01$.
13.9 (a)

x	0	1	2	3
Probability	.216	.432	.288	.064

    (b) Observed $\chi^2 = 1.098$, d.f. $= 2$, $\chi^2_{.05} = 5.99$. The binomial model is not contradicted.
13.11 (c) With $\alpha = .05$ and d.f. $= 1$, $\chi^2_{.05} = 3.84$
    Observed $\chi^2 = 9.298$, reject H_0.
13.13 Observed $\chi^2 = 5.580$, d.f. $= 3$, $\chi^2_{.05} = 7.81$. The two brands are not shown to be significantly different in quality.
13.15 (b) Observed $\chi^2 = 9.780$, d.f. $= 3$, $\chi^2_{.05} = 7.81$. Reject H_0
    (c) $p_1: (.53, .79)$
13.17 (a) Observed $z = 3.00$, reject H_0.
    P–value $= .0013$
13.21 Observed $\chi^2 = 15.734$, d.f. $= 2$, $\chi^2_{.05} = 5.99$. Reject H_0. The appeals decision and type of representation are dependent.
13.23 Observed $\chi^2 = 27.847$, d.f. $= 2$, $\chi^2_{.01} = 9.21$. Reject the null hypothesis of independence at $\alpha = .01$.

13.25 Observed $\chi^2 = 4.134$, d.f. $= 2$, $\chi^2_{.05} = 5.99$. Fail to reject the null hypothesis of independence.

13.27 (b) Observed $\chi^2 = 27.167$ d.f. $= 2$, $\chi^2_{.01} = 9.21$. Reject H_0.

13.29 Observed $\chi^2 = 16.44$, d.f. $= 9$, $\chi^2_{.05} = 16.92$. Fail to reject H_0.

13.31 Observed $\chi^2 = 4.977$, d.f. $= 3$, $\chi^2_{.10} = 6.25$. Fail to reject H_0. The stated conjecture is not contradicted.

13.33 Observed $\chi^2 = 48.242$, d.f. $= 1$, $\chi^2_{.01} = 6.63$. The null hypothesis of homogeneity is rejected even at $\alpha = .01$.

13.35 (a) Observed $\chi^2 = 12.053$ statistic, d.f. $= 3$, $\chi^2_{.05} = 7.81$. Reject H_0. Drugs are different.
　　　(b) Drug 1: (.32, .55)
　　　　　 Drug 3: (.26, .48)

13.37 (a) Observed $z = 2.78$ and P–value $= .0027$. Strong evidence for H_1.
　　　(b) (.09, .48)

13.39 Observed $\chi^2 = 0.153$, d.f. $= 2$, $\chi^2_{.05} = 5.99$. Not significant.

13.41 Observed $\chi^2 = 2.764$, d.f. $= 1$, $\chi^2_{.05} = 3.84$. Fail to reject H_0.

13.43 Observed $\chi^2 = 18.338$, d.f. $= 1$, $\chi^2_{.01} = 6.63$. Reject H_0.

CHAPTER 14

14.1 (b)–(d)

ANOVA Table

Source	Sum of Squares	d.f.
Treatment	28	3
Error	20	4
Total	48	7

14.3 (d)

ANOVA Table

Source	Sum of Squares	d.f.
Treatment	32	2
Error	22	8
Total	54	10

14.5

ANOVA Table

Source	Sum of Squares	d.f.
Treatment	30	3
Error	26	8
Total	56	11

14.7

ANOVA Table

Source	Sum of Squares	d.f.
Magazines	4414.80	2
Error	148196.39	57
Total	152611.19	

14.9 (a) $F_{.10}(3, 5) = 3.62$
　　　(b) $F_{.10}(3, 10) = 2.73$

14.11 $F_{.05}(5, 30) = 2.53$ and $F = 2.46$, fail to reject H_0.

14.13 $F_{.05}(2, 9) = 4.26$ and $F = 8.26$, reject H_0.

14.15 $F_{.05}(2, 8) = 4.46$ and $F = 5.82$, reject H_0.

14.17 (a) With 26 d.f., $t_{.0083} = 2.559$
　　　(b) With df $= 26$, $t_{.005} = 2.779$

14.19 $t_{.00833} = 2.46$ with 66 d.f.
　　　For $\mu_1 - \mu_2$: t interval 4.0 ± 2.68, multiple-t interval 4.0 ± 3.29

14.21 $\dfrac{t_{.05}}{t_{.005}} = .595$ for 15 d.f.

14.23 (b) Treatment SS $= 56$, Block SS $= 78$, Residual SS $= 64$
　　　(c) Treatment d.f. $= 2$, Block d.f. $= 3$, and Residual d.f. $= 2 \cdot 3 = 6$

14.25

Source of Variation	Degrees of Freedom	Sum of Squares	Mean Square	F
Treatment	2	56	28	2.62
Blocks	3	78	26	2.44
Error	6	64	10.667	
Total	11	198		

Fail to reject equal treatment means. Blocks not significant.

14.27 (a) Randomize the position of the loaves in the oven.
(b) $F_{.05}(2, 8) = 4.46 > 3.92$, no significant treatment difference
$F_{.05}(4, 8) = 3.84 < 5.31$, block effects are significant.

14.29 $F_{.05}(5, 25) = 2.603 < 106.79$, treatments are highly significant.
$F_{.05}(5, 25) < 6.11$, block effects are significant.

14.33 (a) $F_{.05}(7, 13) = 2.83$
(c) $F_{.10}(7, 12) = 2.28$

14.35 $F = 20.85$ with P–value .000

14.37 (b) Treatment SS $= 56$, Block SS $= 138$, Residual SS $= 32$
(c) Treatment d.f. $= 2$, Block d.f. $= 3$, Residual d.f. $= 2 \cdot 3 = 6$

CHAPTER 15

15.1 (a) Probability $= \frac{1}{15}$ for each rank collection.
(b) $P[W_B = 5] = \frac{2}{15}$

15.3 (a) $P[W_S \geq 39] = .063$
(c) $c = 66$

15.5 (a) $W_A = 1 + 4 + 5 = 10$
(b) $W_S = 2 + 3 = 5$

15.7 $W_S = 47$, reject H_0.

15.9 $W_S = 62$, fail to reject H_0.

15.11 (a) BBBBBBBBBA
(b) $P[W_A = 10] = .1$

15.13 .048

15.15 $S = 11$; with 7 ties, $n = 18$; do not reject H_0.

15.17 (a) .034   (c) $c = 79$

15.19 (a) $T^+ = 17$, fail to reject H_0 at $\alpha = .062$.

15.21 (b) $P[T^+ = t] = .125$ for $t = 0, 1, 2$ and $4, 5, 6$

15.23 (a) $S = 11$; with $n = 13$, reject H_0.
(b) $T^+ = 83$, reject H_0.

15.25 $r_{Sp} = 0.5$

15.27 $z = -.60$, do not reject H_0 ($n = 10$ is not large).

15.29 $W_A = 10$

15.31 (a) Each triple has probability $\frac{1}{10}$
(b) $P[W_A = 8] = .2$

15.33 (a) .098   (c) $c = 21$

15.35 $W_S = 51 \leq 69$, reject H_0.

15.37 (a) $r_{Sp} = -0.524$
(b) Do not reject independence.
(c) $\alpha = .05$ ($n = 8$ may be too small)

15.39 (a) $S = 3$ (b) $T^+ = 6$

Index

TABLE 4 Percentage Points of t Distributions

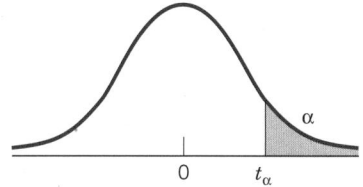

d.f.	α .25	.10	.05	.025	.01	.00833	.00625	.005
1	1.000	3.078	6.314	12.706	31.821	38.204	50.923	63.657
2	.816	1.886	2.920	4.303	6.965	7.649	8.860	9.925
3	.765	1.638	2.353	3.182	4.541	4.857	5.392	5.841
4	.741	1.533	2.132	2.776	3.747	3.961	4.315	4.604
5	.727	1.476	2.015	2.571	3.365	3.534	3.810	4.032
6	.718	1.440	1.943	2.447	3.143	3.287	3.521	3.707
7	.711	1.415	1.895	2.365	2.998	3.128	3.335	3.499
8	.706	1.397	1.860	2.306	2.896	3.016	3.206	3.355
9	.703	1.383	1.833	2.262	2.821	2.933	3.111	3.250
10	.700	1.372	1.812	2.228	2.764	2.870	3.038	3.169
11	.697	1.363	1.796	2.201	2.718	2.820	2.981	3.106
12	.695	1.356	1.782	2.179	2.681	2.779	2.934	3.055
13	.694	1.350	1.771	2.160	2.650	2.746	2.896	3.012
14	.692	1.345	1.761	2.145	2.624	2.718	2.864	2.977
15	.691	1.341	1.753	2.131	2.602	2.694	2.837	2.947
16	.690	1.337	1.746	2.120	2.583	2.673	2.813	2.921
17	.689	1.333	1.740	2.110	2.567	2.655	2.793	2.898
18	.688	1.330	1.734	2.101	2.552	2.639	2.775	2.878
19	.688	1.328	1.729	2.093	2.539	2.625	2.759	2.861
20	.687	1.325	1.725	2.086	2.528	2.613	2.744	2.845
21	.686	1.323	1.721	2.080	2.518	2.601	2.732	2.831
22	.686	1.321	1.717	2.074	2.508	2.591	2.720	2.819
23	.685	1.319	1.714	2.069	2.500	2.582	2.710	2.807
24	.685	1.318	1.711	2.064	2.492	2.574	2.700	2.797
25	.684	1.316	1.708	2.060	2.485	2.566	2.692	2.787
26	.684	1.315	1.706	2.056	2.479	2.559	2.684	2.779
27	.684	1.314	1.703	2.052	2.473	2.552	2.676	2.771
28	.683	1.313	1.701	2.048	2.467	2.546	2.669	2.763
29	.683	1.311	1.699	2.045	2.462	2.541	2.663	2.756
30	.683	1.310	1.697	2.042	2.457	2.536	2.657	2.750
40	.681	1.303	1.684	2.021	2.423	2.499	2.616	2.704
60	.679	1.296	1.671	2.000	2.390	2.463	2.575	2.660
120	.677	1.289	1.658	1.980	2.358	2.428	2.536	2.617
∞	.674	1.282	1.645	1.960	2.326	2.394	2.498	2.576